Library of Congress Cataloging-in-Publication Data

Consolidation of soils.

(ASTM special technical publication; 892)
Papers presented at the ASTM Symposium on the
Consolidation Behavior of Soils.
Includes bibliographies and index.
"ASTM publication code number (PCN) 04-892000-38."
1. Soil consolidation test—Congresses. I. Yong,
R. N. (Raymond Nen) II. Townsend, Frank C. III. ASTM
Committee D-18 on Soil and Rock. IV. ASTM Symposium on
the Consolidation Behavior of Soils (1985: Ft. Lauderdale,
Fla.) V. Series.
TA710.5.C68   1986   624.1′51362   86-1126
ISBN 0-8031-0446-4

NOTE

The Society is not responsible, as a body,
for the statements and opinions
advanced in this publication.

Printed in Baltimore, Md.
May 1986

CONSOLIDATION OF SOILS: TESTING AND EVALUATION

A symposium
sponsored by ASTM
Committee D-18
on Soil and Rock
Ft. Lauderdale, Fla., 24 Jan. 1985

ASTM SPECIAL TECHNICAL PUBLICATIO
R. N. Yong, McGill University, and
F. C. Townsend, University of Florida,
editors

ASTM Publication Code Number (PCN)
04-892000-38

1916 Race Street, Philadelph

Foreword

The ASTM Symposium on Consolidation Behavior of Soils was held in Ft. Lauderdale, Florida, on 24 January 1985. ASTM Committee D-18 on Soil and Rock served as sponsor. Symposium co-chairmen were R. N. Yong, McGill University, and F. C. Townsend, University of Florida. This volume, *Consolidation of Soils: Testing and Evaluation,* has been edited by Messrs. Yong and Townsend.

Related
ASTM Publications

Strength Testing of Marine Sediments: Laboratory and *In Situ* Measurements, STP 883 (1985), 04-883000-38

Hydraulic Barriers in Soil and Rock, STP 874 (1985), 04-874000-38

Measurement of Rock Properties at Elevated Pressures and Temperatures, STP 869 (1985), 04-869000-38

Laterally Loaded Deep Foundations: Analysis and Performance, STP 835 (1984), 04-835000-38

Testing of Peats and Organic Soils, STP 820 (1983), 04-820000-38

A Note of Appreciation
to Reviewers

The quality of the papers that appear in this publication reflects not only the obvious efforts of the authors but also the unheralded, though essential, work of the reviewers. On behalf of ASTM we acknowledge with appreciation their dedication to high professional standards and their sacrifice of time and effort.

ASTM Committee on Publications

ASTM Editorial Staff

Allan S. Kleinberg
Janet R. Schroeder
Kathleen A. Greene
Bill Benzing

Contents

SUMMARY AND EVALUATION

Introduction

It is noteworthy that the last full documentation of consolidation performance of soils in the laboratory and field occurred over 20 years ago (*Design of Foundations for Control of Settlements*, ASCE Soil Mechanics and Foundation Division, New York, 1964) and that no ASTM Special Technical Publication has specifically addressed this complex subject in the intervening period. During these past two decades, considerable progress in developing a better appreciation of soil performance has led to the design of new and innovative techniques in assessing soil properties and characteristics pertinent to compression/consolidation behavior.

In addition to the well documented historic concerns dealing with the less-than-accurate correlations between predicted and measured consolidation behavior of field soils, recent encounters with significant problem situations have highlighted the need for a comprehensive documentation of consolidation testing and evaluation. The advent of computers for solving previously restricted numerical analytical schemes, and transducers coupled with data acquisition systems to provide continuous monitoring, have become commonplace. Problem soils, such as soft clays, gaseous soils, organic soils, and slimes, require not only specialized test equipment and techniques but also the development of pertinent theories and field validation tools.

In recognition of the above, the ASTM Symposium on the Consolidation Behavior of Soils, held in Ft. Lauderdale, Florida, 24 January 1985, was planned as a companion to the ASCE Symposium on Sedimentation/Consolidation Models; Predictions and Validation, held in San Francisco in October 1984. It was hoped that through these symposia the advances made in addressing and solving the many issues of consolidation testing, modelling, prediction, and validation would be fully communicated to the profession at large.

To handle properly the difficult task of bringing forward recent studies pertinent to the overall problem of consolidation testing, modelling, prediction, and validation, it was decided that studies which address theory, prediction, and validation would be better accommodated by the ASCE symposium format. The studies and concerns that relate to testing and methods of evaluation and interpretation of test data, being of direct importance to ASTM, should, it was felt, be developed into a format which would provide a record of recent developments in the area of consolidation testing.

The objectives of the ASTM symposium (and of this volume) were as follows: (1) to review the state of the art of consolidation testing, with particular emphasis on developments made in the last two decades, (2) to establish and assess requirements for consolidation testing of problem soils not previously considered in detail via actual studies with new techniques and procedures, (3) to compare and evaluate the various new methods and test equipment used for determination of consolidation behavior, (4) to study the viability of the various data reduction models and methods of application of data and measurements for assessment consolidation behavior, (5) to identify the shortfalls and areas of needed study for development of consolidation testing and evaluation, and (6) to provide a focal point for the development and improvement of ASTM consolidation testing standards.

There have been many advances made in (1) equipment and instrumentation capabilities, (2) test methodologies and data gathering and control systems, and (3) methods for data handling and analysis. All of these combine to provide a better capability of securing a more reliable analysis of the problem of consolidation determination. The ASTM Symposium on Consolidation Behavior of Soils was developed as a two-session presentation which included State-of-the-Art (SOA) presentations, General Reports dealing with papers submitted to the symposium, and General Discussions by the authors, dealing with specific points and issues identified by the moderators. The program details were as follows:

Session I—Theory and Laboratory Testing Requirements

Chairman: R. T. Donaghe, U.S. Army Engineer Waterways Experiment
                  Station, Vicksburg, Mississippi
Moderator: R. N. Yong, McGill University, Montreal
State-of-the-Art R. E. Olson, University of Texas, Austin, "Consolidation
  Speaker: Testing"
Reporters: V. P. Drnevich, University of Kentucky, Lexington
                  V. Silvestri, École Polytechnique, Montreal
General Discussion by authors, reporters, and SOA speaker

Session II—Evaluation and Special Tests

Chairman: E. T. Selig, University of Massachusetts, Amherst
Moderator: F. C. Townsend, University of Florida, Gainesville
State-of-the-Art C. B. Crawford, DBR/NRC, Ottawa, "Evaluation and Inter-
  Speaker: pretation of Soil Consolidation Tests"
Reporters: D. Bloomquist, University of Florida, Gainesville
                  J. F. Peters, U.S. Army Engineer Waterways Experiment Sta-
                  tion, Vicksburg, Mississippi
General Discussion by authors, reporters, and SOA speaker

The two State-of-the-Art papers open this volume. Next appear four General

Reports; these summarize the 31 Technical Papers which follow. Finally, the volume closes with two retrospective evaluations. Significant issues and problems identified during the course of the symposium, and those problems of concern to the active researcher and practitioner, are reviewed by the editors. A brief general discussion on consolidation theory and testing is provided by S. Leroueil and M. Kabbaj.

The editors wish to record their appreciation to all those who participated in the symposium and who contributed to this volume, to the reviewers of the submitted papers, to ASTM Committee D-18 on Soil and Rock for sponsoring the event through its Subcommittee D18.05 on Structural Properties of Soils, and to the editorial staff of ASTM. Without their combined efforts, this STP would not have reached the high quality level associated with ASTM publications. Acknowledgments are also made to Professors H. Y. Ko, A. S. Saada, E. T. Selig, and R. L. Schiffman, for their input and assistance in the planning of the symposium.

R. N. Yong
Geotechnical Research Centre, McGill University, Montreal, Canada, H3A 2K6; symposium co-chairman and co-editor

F. C. Townsend
Department of Civil Engineering, University of Florida, Gainesville, Fla. 32611; symposium co-chairman and co-editor

State-of-the-Art Papers

Roy E. Olson[1]

State of the Art: Consolidation Testing

REFERENCE: Olson, R. E., "State of the Art: Consolidation Testing," *Consolidation of Soils: Testing and Evaluation, ASTM STP 892,* R. N. Yong and F. C. Townsend, Eds., American Society for Testing and Materials, Philadelphia, 1986, pp. 7–70.

ABSTRACT: The state of the art of consolidation testing is addressed except that conditions involving generalized stress states were excluded and large-strain problems were assigned to a companion paper. Terzaghi's classical theory was found to fit primary data rather well and the root time fitting method was preferred. Errors associated with boundary impedance and ring friction can be major but are generally understood and can be minimized. Existence of nonlinear stress-strain relationships can perturb the shape of S-t curves significantly and lead to problems when pore pressures are used to determine c_v. Radial flow tests should be used for field problems involving horizontal flow; theories are presented to use in reducing laboratory data. Secondary effects can be addressed using linear theory and trial solutions, but such effects may be much more apparent in the laboratory than in the field. Effects of partial saturation are understood qualitatively and can influence laboratory data greatly. Continuous loading tests lead to overprediction of effective stresses but apparently to reasonable values of c_v. Soil properties change significantly as a result of sampling disturbance and storage time; techniques for correcting for these effects are presented. The usefulness of laboratory data is examined using a case history.

KEY WORDS: case history, clay, consolidation, continuous loading, impedance, laboratory, partial saturation, radial flow, ring friction, sampling disturbance, secondary consolidation, soil, storage, temperature effects, testing, theory

The term "consolidation testing" is here interpreted to refer to efforts to measure generalized stress-strain-time relationships for soils under conditions of partial or complete dissipation of excess pore water pressures. Unfortunately, even the bibliography for such a general topic would exceed the acceptable length of a paper and thus restrictions in the range of the topic become necessary.

Firstly, consideration is restricted to soils in which there are significant delays in the dissipation of excess pore water pressures (clays and possibly silts).

Secondly, any attempt to consider behavior under generalized stress states leads to an intolerable expansion in length. Attention is therefore directed towards one-dimensional behavior.

Thirdly, consolidation testing serves to measure soil properties for use in a theory of some type, and thus the testing must be coupled with a theory without devoting significant space to the theory.

[1] L. P. Gilvin Professor of Civil Engineering, University of Texas, Austin, TX 78712.

Fourthly, papers submitted for publication in this ASTM Special Technical Publication were excluded from this review.

Fifthly, organizers of the program assigned the topic of finite-strain consolidation testing to a companion state-of-the-art paper.[2] It may be noted that various aspects of one-dimensional consolidation testing, including finite strains, have recently been addressed by Znidarcic et al [1].

Finally, it is important to relate the properties of small samples in the laboratory to masses of earth in the field. One is confronted with two problems. The first relates to sampling disturbances and the second to the variations in properties within soil deposits. Both topics will be considered but in a restricted fashion. Field case histories must be examined to demonstrate practical usefulness of laboratory data.

Brief Early History

Consolidation tests are usually performed to determine stress-strain relationships in confined compression and to obtain parameters for use in Terzaghi's [2] theory or one of its progeny. It may be of interest to note that the term "consolidation," as a technical term, dates back at least to 1809 [3] and was widely used in the late 1800s [4].

Early experiments to determine the consolidation characteristics of remolded clays were reported by Spring [5] and Frontard [6], and in 1910 Moran, referenced by Enkeboll [7], performed one-dimensional consolidation tests using undisturbed samples.

Variables Used in One-Dimensional Analyses

The required soil properties must relate to the stress-strain-time response of the soil under appropriate boundary conditions. It is usually convenient to use effective stress as the independent variable, and seek stress-strain and stress-permeability relationships. Equations are presented here for later reference.

Properties are conveniently defined in terms of the vertical effective stress, σ' [8]. Strain, ϵ, is conveniently defined in the usual engineering sense as $\Delta L/L_0$ where L_0 is the original length (height) and ΔL is positive for compression. The one-dimensional stress-strain curve may be plotted on a natural scale, and have a local slope of m_v where

$$m_v = d\epsilon/d\sigma' \qquad (1)$$

(m_v is the reciprocal of the confined compression modulus), or plotted on a semilogarithmic scale of ϵ versus $\log_{10}\sigma'$, in which case the local slope is

$$R = d\epsilon/d \, (\log_{10}\sigma') \qquad (2)$$

[2] C. B. Crawford, this publication, pp. 71–103.

The stress-strain curves may be corrected in ways to be discussed and the resulting curves used to calculate compressions, ΔS, of strata in the field:

$$\Delta S = L_0 \Delta \epsilon = L_0 \int m_v d\sigma' = L_0 \int R \, d (\log \sigma') \tag{3}$$

For ranges in stress where m_v or R can be considered constant, they come out from under the integration sign; for example,

$$\Delta S = R \, H_0 \log \frac{\sigma'_2}{\sigma'_1} \tag{4}$$

where H_0 is commonly used in place of L_0 to denote initial thickness of a layer, and σ'_1 and σ'_2 are the effective stresses at the initial and final condition on a linear part of the curve of ϵ versus $\log \sigma'$. Where the stress-strain relationship is nonlinear, the curve may be approximated as a series of straight lines and

$$\Delta S = H_0 \Sigma m_v \Delta \sigma' = H_0 \Sigma R \log \frac{\sigma'_2}{\sigma'_1} \tag{5}$$

where m_v or R are constant only over the segments from σ'_1 to σ'_2 where σ'_1 and σ'_2 are different for each segment. In many cases it is sufficient to assume a bilinear relationship and use special symbols R_r and R_c for the slopes of the recompression and virgin curves (Fig. 1); thus

$$\Delta S = R_r H_0 \log \frac{\sigma'_c}{\sigma'_0} + R_c H_0 \log \frac{\sigma'_f}{\sigma'_c} \tag{6}$$

where σ'_c is the stress at the intersection of the two straight lines, and R_r and R_c are termed the recompression ratio and compression ratio, respectively. Note that σ'_c is not necessarily the maximum previous consolidation pressure, σ'_{max}, and that the value of σ'_{max} does not enter into the analysis unless the engineer arbitrarily chooses to make σ'_c equal to σ'_{max}.

The definition of strain as $\Delta L/L$ has apparently concerned some engineers because of the use of the total length in the denominator. A different strain scale may be defined using

$$\Delta e = \Delta L/L_s \tag{7}$$

where Δe is strain and L_s is the length that the solid phase of the soil would have (V_s/A where V_s is the volume of solids in a specimen of area A and height L_0). In Eq 7, ΔL and Δe are both negative for compression. The revised stress-strain curves can be plotted to natural (Δe–σ') or semi-logarithmic scales (Δe–$\log \sigma'$). Linear sections have slopes a_v and C for the natural (arithmetic) and semi-

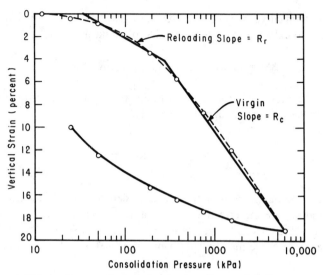

FIG. 1—*Stress-strain curve for clay from Hammond, Indiana.*

logarithmic plots, respectively, and previous equations can be regenerated using Δe; for example, Eq 6 becomes

$$\Delta S = C_r H_s \log \frac{\sigma'_c}{\sigma'_0} + C_c H_s \log \frac{\sigma'_f}{\sigma'_c} \qquad (8)$$

where C_r and C_c are the recompression index and compression index, respectively, and H_s is traditionally used in place of L_s. If Δe exists, then clearly there must be a variable e. From Eq 7 it can be developed that e is the ratio of the volume of voids to the volume of solids. The term *void ratio* is thus used for e. Once e is defined, it can be shown that

$$H_s = H/(1 + e_0) \qquad (9)$$

Equation 9 is generally substituted into Eq 8.

It should be noted that Eqs 6 and 8 (and their equivalents for natural plots) give identical values of ΔS; thus the choice of whether to plot stress-strain curves using strains (ϵ) or void ratios (e) is a matter of personal choice. For practical work, ϵ is to be preferred because it eliminates any need to determine specific gravities or even to weigh the specimen and it is easier to prepare standard forms [9]. Use of e does not provide for a large strain correction.

Estimation of time rates of consolidation in the field is generally performed using some form of the theory originally developed by Terzaghi [2]. For one-dimensional vertical compression of a saturated soil, if water flow is governed by Darcy's law, soil properties are constant, strains are small, and the soil is linearly elastic:

$$\Delta S = \Delta S_u U = \Delta S_u \left[1 - \sum_{m=0,1,\ldots}^{\infty} \frac{2}{M^2} \exp\left(-M^2 T\right) \right] \tag{10}$$

where ΔS is the compression at any time, ΔS_u is the ultimate compression, U is the average degree of consolidation, and

$$M = (2m + 1) \tag{11}$$

and T is the time factor given by

$$T = c_v t / H \tag{12}$$

where c_v is the coefficient of consolidation, t is time, and H is the drainage distance (maximum distance that any molecule of water must flow to escape the system). In turn,

$$c_v = k/m_v \gamma_w = k (1 + e)/a_v \gamma_w \tag{13}$$

where k is the hydraulic conductivity, γ_w is the unit weight of water, m_v and a_v are slopes of natural stress-strain curves using linear strain and void ratio, respectively, and e is void ratio.

Standard Technique for Performing One-Dimensional Vertical-Flow Consolidation Tests

The parameters used to describe the one-dimensional consolidation characteristics of clays are usually measured using small samples in the laboratory. The standard technique is covered in ASTM Test for One-Dimensional Consolidation Properties of Soils (D 2435) and in most textbooks on geotechnical engineering and soil testing. Several aspects of the standard methods are reviewed here as a basis for expanded discussion of selected aspects of the procedures later.

The test is performed using a disk-shaped sample of clay with a diameter of at least 50 mm (2 in.) and an initial thickness of at least 13 mm (0.5 in.). The specimen is laterally confined by a relatively rigid ring with a smooth inside surface. The soil is loaded on its flat surfaces and drains through one (single drainage) or both (double drainage) faces. Drainage is provided through porous "stones" which are generally either ceramic or metal. A sheet of filter paper is often placed between the clay and the stone to prevent particles of the soil from being flushed into the stone and clogging its pores.

The soil is loaded in a series of increments, where each load is twice the previous one. The compression of the sample is measured as a function of time under each load, and successive loads are applied either after primary consolidation has been completed or after some preselected period of time, typically 24 h. After consolidation under the peak pressure, the sample is unloaded, usually in a series of decrements in which each pressure is one fourth of the previous pressure. Expansion of the sample may or may not be measured as a function of time during swelling. On occasion the load may be reduced to the final swelling pressure without waiting for swelling to occur under intermediate loads. The sample should be allowed to swell to equilibrium under the final pressure, which should be small, say 20 kPa (100 psf).

If the sample was under the water table in the field, or will be inundated later in the field, the sample is also inundated in the laboratory test. The most common procedure is to apply a nominal seating load, the magnitude of which depends on the strength of the soil, typically between about 1 kPa (20 psf) for very soft clays and peats to perhaps 10 kPa (200 psf) for stiff clays, set the deformation measuring device, typically a dial indicator and/or a linear variable differential transformer (LVDT), in place and take an initial reading, and then inundate the sample. The load is then adjusted to maintain a constant thickness, after correcting for apparatus deflection. When equilibrium is established, the loading stage begins. For stiff, fissured clays it is preferable to start with a seating load equal to the effective overburden pressure (total pressure for a sample above the water table) so some of the fissures that opened up because of stress relief will be closed again before the deformation-measuring equipment is zeroed.

When the sample has swollen to equilibrium under the final pressure, the apparatus is dismantled as quickly as possible, the sample is separated from the stones and filter paper, any free water on the sample is removed, and a wet weight of the sample (and ring typically) is obtained. The sample (and ring typically) is dried in an oven at 110°C.

The initial void ratio is calculated using the known volume of the ring and the volume of solids, and the void ratio under subsequent effective stresses is calculated using Eq 7 where ΔL is the measured compression corrected for apparatus compression.

To save costs, some laboratories only reduce data (calculate c_v) for one load increment. Such a procedure may be justified in specific cases, but it should be remembered that c_v varies with effective stress (see later discussion).

Curve Fitting To Obtain c_v

Standard Methods

The laboratory time-settlement curves are used to evaluate c_v. If laboratory time-settlement curves are developed in accord with theory (Eq 10), the engineer could enter the measured S-t curve at a point where, say, 50% of the settlement had occurred and read off the time, t_{50}. From Eq 12:

$$c_v = T_{50} H_{50}^2 / t_{50} \qquad (14)$$

From Eq 10, T_{50} = 0.197. For a small strain case (one of the assumptions), H is essentially a constant but the average value, H_{50}, is typically used. It may be noted that if the current thickness is used in reducing data from laboratory tests, then the design engineer should use a numerical method that utilizes current thicknesses. If the engineer uses the original layer thicknesses in all calculations of time rates of consolidation, then laboratory data should also be reduced using the original thickness (H_0).

Unfortunately, experience shows that laboratory curves rarely match the theory at either the beginning or the end, and perhaps not in the middle either. Lack of fit at low times was first recognized by Casagrande in 1928 [10, p. 479]. The corrected starting point for the fitted theoretical curve is found using Fox's [11] approximation to Eq 10 for $U \leq 60\%$:

$$T = \frac{\pi}{4} U^2 \tag{15}$$

Equations 10 and 12 are inserted into Eq 15 to obtain:

$$\Delta S = [(c_v/H^2)(4/\pi)\Delta S_u^2]^{1/2}t^{1/2} \tag{16}$$

The easiest way to find the corrected zero point is to plot ΔS versus $\sqrt{t}$ (this approach was originated by Taylor [12,13] and publicized by Gilboy [14]) and draw a straight line through the linear part of the curve and extend it back to t = 0 to obtain ΔS_0, the corrected initial settlement. A less accurate and more cumbersome procedure is to plot ΔS versus log time [10] and pick two points, 1 and 2, on the part of the ΔS–log t curve satisfying Eq 15 (pick point 2 near t_{50}), with $t_1 = \frac{1}{4} t_2$, and note that:

$$\Delta S = \Delta S_1 - (\Delta S_2 - \Delta S_1) \tag{17}$$

Equation 17 is valid because $\Delta S_2 = 2\Delta S_1$ (from Eq 16).

The settlement at U = 100%, ΔS_{100}, can also be defined using either the $\sqrt{t}$ or log t plots. For $\sqrt{t}$ plots, ΔS_{100} = (10/9) ΔS_{90}, and ΔS_{90} is found by extending the early linear curve and drawing a second line with abscissa 15% greater (ratio of slopes is termed the curve fitting factor, CFF, and is $5\sqrt{T_{90}}/9\sqrt{T_{50}}$) to intersect the laboratory curve at U = 90% [13, p. 239, or most modern textbooks on geotechnical engineering or soil testing]. For log t plots, the 100% primary consolidation settlement is estimated to be at the intersection of straight lines drawn tangent to the laboratory curve at the point of inflection and out on the secondary curve [13, p. 241].

Once ΔS_{90} is found in the $\sqrt{t}$ plot, t_{45} may be read from the laboratory curve at $\Delta S_{45} = \frac{1}{2}(\Delta S_0 + \Delta S_{90})$, and Eq 14 is applied (but at t_{45} and with T_{45} = 0.159) to obtain c_v. When the log t method is used, it is easier to use $\Delta S_{50} = \frac{1}{2}(\Delta S_0 + \Delta S_{100})$ and Eq 14. It may be noted that if c_v is constant for this loading increment, as assumed, then it makes no difference whether c_v is fitted at t_{45}, t_{50} or any other point between ΔS_0 and ΔS_{100}. The correlation between the theoretical ΔS-t curve and the measured primary curve is best if the fitting is accomplished in the middle of the curve rather than at, say, t_{90}.

Experience shows that c_v from $\sqrt{t}$ plots almost always exceeds c_v from log t plots (Fig. 2), leading to questions regarding their relative advantages and disadvantages, as well as absolute accuracy. The following points appear valid:

1. For finding ΔS_0, the $\sqrt{t}$ method is to be preferred. In the log t method it is more difficult to find the portion of the ΔS-t curve that satisfies Eq 15.

2. For finding ΔS_{100}, the two methods are both theoretically valid if no secondary effects occur. The $\sqrt{t}$ method is valid if secondary effects are unimportant up to t_{90}. The log t method leads to an ambiguous result in the common case that the secondary curve is nonlinear on the ΔS-log t plot.

3. Semi-logarithmic paper is more readily available but square root paper is easily prepared. Through use of computers and plotters, paper is easily prepared to any desired scale and in any range.

A check on the validity of the theory and fitting method can be obtained. Equation 13 is solved for k (use the symbol k_c for calculated k), which is then plotted versus ϵ_{50} or e_{50}. Values of k are then measured [15] when the strain rate is small out on the secondary curve, and the measured k (k_m) is plotted versus the ϵ or e at the time of measurement. If the theory, and its application, are correct, then the two curves should be identical. Experience (Fig. 3) shows that k_c is almost always less than k_m and that the k_c from $\sqrt{t}$ is closer than the k_c from log t. The main explanation for the differences between k_c and k_m seems to be

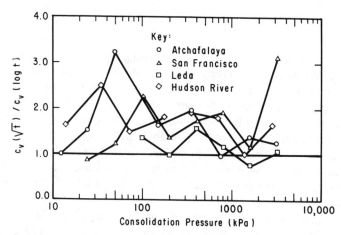

FIG. 2—*Comparison of coefficients of consolidation computed using the square root and logarithmic methods for clays from several sites. The symbol for the first pressure on the virgin consolidation curve is solid.*

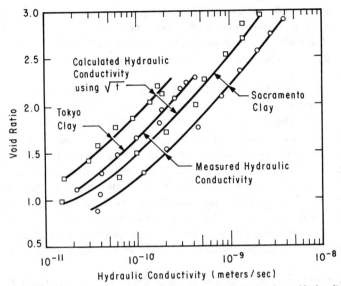

FIG. 3—*Comparison of measured and computed (Terzaghi's theory) values of hydraulic conductivity.*

that secondary effects lead to a delay in settlement in addition to that caused by the real k (presumably k_m). If all causes of delayed compression are lumped into k_c, then it is apparent that k_c must be less than the real value.

It may be noted that when Eq 13 is used for calculating k, the compressibility terms (m_v and a_v) must be defined in the range from ΔS_0 to ΔS_{100}, not from ΔS_{100} of the previous load to ΔS_{100} of the present load, because the second compressibility includes secondary compression.

Alternative Curve Fitting Methods

A few alternative techniques of curve fitting will be reviewed.

Naylor and Doran [*16*] proposed fitting in the range of 60% < U < 80% where only the first term in Eq 10 need be used. Thus Eq 10 can be written as

$$\ell n\,(1 - U) = \ell n\left(\frac{8}{\pi^2}\right) - \left(\frac{1}{4}\pi^2\,\frac{c_v}{H^2}\right)t \qquad (18)$$

and

$$1 - U = (\Delta S_f - \Delta S)/(\Delta S_f - \Delta S_i) \qquad (19)$$

Naylor and Doran suggest several trial methods for finding ΔS_i and ΔS_f. Once they are found, any point on the ΔS-t curve can be used to find c_v. The method is time consuming to apply.

Matlock and Dawson [*17*] note that the slope of the U-log t curve at its inflection point is 68% per log cycle of T. Thus a tangent line is drawn to the S-log t plot, ΔS for one cycle of time is measured, this value is divided by 0.68 to obtain the amount of primary compression, and this value is added to ΔS_0 to obtain ΔS_{100}. Any point can then be used to obtain c_v but Matlock and Dawson used $U = 68\%$ where $T = 0.38$.

Scott [18] plotted the ratio $U(T)/U(NT)$ versus T. The user selects an arbitrary value of N and then picks two points, say 1 and 2, on the S-t curve, such that $\Delta S_2/\Delta S_1 = N$. The value of T_1 is read from the $U(T)/U(NT)$ versus T curve and t_1 from the S-t curve, and Eq 12 is applied to obtain c_v. The method cannot be used with low values of t because the $U(T)/U(NT)$ curve is relatively flat, thus making it difficult to define T_1. Further, the only method of finding ΔS_0 is a trial solution, the correct value of ΔS_0 presumably being found when c_v is not dependent on t_1. The method is inconvenient because it requires trial solutions for ΔS_0, may break down if c_v changes naturally during consolidation, does not give a plot that can be examined visually to detect non-Terzaghian effects, is more difficult to apply than the standard methods, and does not work for small values of t.

Hansen [19] developed a method in which S is plotted versus $\sqrt{t}$ up to $U = 89\%$ and then against log t, with the settlement curve smooth and continuous at the intersection of the two scales. The actual time for 89% consolidation varies with c_v and H so an awkward procedure is developed to set up the scales. Although it is neater in a report to use Hansen's plot, it would appear less time consuming to just make two time-settlement curves, one with $\sqrt{t}$ and one with log t.

Wilson, Radforth, MacFarlane, and Lo [20], working with peat, recommended plotting log (de/dt) versus log t. The curve is initially rather flat but then turns downward. They estimated that $U = 100\%$ at the point of sharpest curvature even though substantial excess pore water pressures still existed. This method seems to have been superseded by Parkin's (1981) method.

Parkin [21] chose to plot log $\dot{U}$ versus T where

$$\dot{U} = dU/dT = -\Sigma\, 2 \exp\,(-M^2 T) \tag{20}$$

where M and T are given by Eqs 11 and 12, respectively, and the summation is carried out for $m = 0, 1, 2, 3, \ldots$. For values of T less than 0.2 the log $\dot{U}$ versus log T plot is linear and has a slope of 2 horizontal to 1 vertical, provided the log-log paper has the same scale factor in both directions.

From Eq 12:

$$\log t = \log\,(H/c_v) + \log T \tag{21}$$

so a plot using log t instead of log T is simply displaced along the horizontal (x) axis by a constant amount log (H/c_v). From Eq 10, and using:

$$\dot{S} = dS/dt \tag{22}$$

then:

$$\log \dot{S} = \log U - \log (H^2/s_u c_v) \tag{23}$$

and a plot using $\dot{S}$ must be parallel to the one using $\dot{U}$ but displaced by log $(H/S_u c_v)$.

Consequently, a plot of log $\dot{S}$ versus log t has the same slope as a plot of log $\dot{U}$ versus log T. Laboratory data can be plotted on log-log paper and overlaid with a log U – log T plot drawn on transparent log-log paper of the same scale factor; the two curves should match in regions where Terzaghi's theory is applicable. A common point is selected and the associated values of T, U, t, and S are recorded. Then Eq 12 is applied to calculate c_v and:

$$S_u = (H^2/c_v)(\dot{S}/\dot{U}) \tag{24}$$

Parkin's method has an advantage in that the corrected zero point and the 100% primary consolidation point need not be located. However, it requires calculation of $\dot{S}$ and use of an overlay procedure, and is thus more time consuming than usual procedures. The effects of secondary consolidation after primary are shown clearly but not in a fashion such that a parameter like C_α can be evaluated easily.

Sridharan and Rao [22] plotted T/U versus T and note the presence of a more-or-less linear region for $60\% < U < 95\%$. They find that lines drawn from the origin to the points where $U = 60\%$ and 90% have slopes of 2.02 and 1.35, respectively, times the slope of the "linear" T/U–T curve. Thus they plot $t/\Delta S$ versus t, define a slope (B) in the "linear" range, draw two lines from the origin with slopes of 2.02B and 1.35B to intersect the laboratory curve at $U = 60\%$ and 90%, respectively, and then determine ΔS_0, ΔS_{100}, and c_v in obvious manners. The method has the disadvantages that the user must plot $t/\Delta S$ versus t, that there is no truly linear part of a T/U–T curve (let alone a laboratory curve), and that ΔS_0 must be known to plot $t/\Delta S$ so the method, in principle, is iterative.

Rao and Kodandaramoswamy [23] note that beyond a time factor of 0.3 the T-U, thus t-ΔS, curve can be modeled as

$$\Delta S = t (a + bt) \tag{25}$$

Thus a plot of $t/\Delta S$ versus t should be more-or-less linear in this range [22]. At large times, $bt >> a$, thus leading to $\Delta S_{100} = 1/b$.

Discussion

Based on a survey of the literature it appears that these alternative methods have rarely been used. The main reasons are probably that many of the methods require trial solutions (a process difficult to justify economically) and are more complex than the standard solutions. Most of the methods are purported to work when the soil response deviates so greatly from that predicted by Terzaghi's theory that the standard methods do not work well. In such cases one wonders how the user is to justify use of a time consuming method to evaluate parameters in a theory that does not work.

Special Aspects of One-Dimensional Consolidation Testing

Several special aspects of one-dimensional consolidation testing will be considered in this section.

Rapid Loading

If the engineer uses the incremental loading procedure and is in need of useful data as quickly as possible, ϵ should be used in place of e, and succeeding loads can be applied as soon as consolidation is complete under the existing load.

For the case that Terzaghi's theory is valid, an assumption that applies to the entire discussion up to this point, the effect of early loading can be analyzed using a composite "stress surface" (my translation of the term *lastfläche* used by Terzaghi and Frohlich [4] to describe the distribution of hydrostatic excess pore water pressure) as suggested by Taylor [13, p. 236]. For the case at hand, consolidation is assumed to result from a residual sinusoidal stress surface and a new rectangular stress surface. The composite T-U relationship requires that:

$$U = \Sigma\, U_j W_j \qquad (26)$$

where U_j is the degree of consolidation for the jth stress surface and W_j is the ratio of the area under the jth stress surface to the area of the composite stress surface, all at the same T. The T-U relationship for a rectangular stress surface is given by Eq 10. For a sinusoidal stress surface [4]:

$$U_s = 1 - \exp\left(-1/4\,\pi\,T\right) \qquad (27)$$

The analyses show that if the load increment ratio is 1.0, reloading at $U = 90\%$ yields a subsequent T-U curve that is almost indistinguishable from the classical value (e.g., at $U = 50\%$ the classical T is 0.197), whereas T for loading at $U = 90\%$ is $T = 0.211$, a difference of only 7%. Data should be reduced using the proper value of T but a 7% error in c_v is not very critical in a practical problem. For load increment ratios of 0.5, $T = 0.225$ at $U = 50\%$ for the next load. Thus the theory indicates that loading at a degree of consolidation of 90% or more should yield acceptable results.

In a rapid loading test, settlement would be plotted against $\sqrt{t}$ and the standard construction used to locate ΔS_{90}. The next load would be applied at any time after t_{90}. Leonards and Ramiah [24] cite Lewis [25] and Northey [26] as using variable loading times and finding no effect on the e-σ' curve and either no effect on c_v or a slight decrease (about 15%) as the loading time increased. Newland

and Allely [27] appear to have been among the first to recommend application of successive loads as soon as 100% primary consolidation is achieved on any load. Leonards and Ramiah [24] tested remolded clay using loading times of 4 h, one day, and one week per load and found essentially the same e-σ' relationships but a c_v that was generally largest for one day/load and least for one week/load. We have used rapid loading (load at S_{100}) on a number of occasions and have found that the differences in measured soil properties are inconsistent and apparently represent variations in actual properties. See, for example, the stress-strain curves for duplicate standard (24 h loading) and rapid loading tests on Atchafalaya clay shown in Fig. 4 (strains were defined at S_{100}).

A somewhat similar, but less convenient, method was suggested by Su [28] based on a plot of ΔS versus log t and reloading at t_{90}.

Boundary Impedance

Newland and Allely [27] used relatively coarse porous stones and no filter paper and found that (1) c_v of the Whangamarino clay increased by a factor of about four when the drainage distance increased from about 3 mm (1/8 in.) to 64 mm (2.5 in.), and (2) the early part of the S-$\sqrt{t}$ curve was distinctly concave towards the S axis, both observations suggesting impeded drainage. When they used a finer stone and filter paper, c_v was independent of H and the early parts of the S-$\sqrt{t}$ curves were linear.

When stones are used in direct content with clays, the stones often become visibly clogged with soil. Baracos [29] presented scanning electron microscope photographs of such clogging.

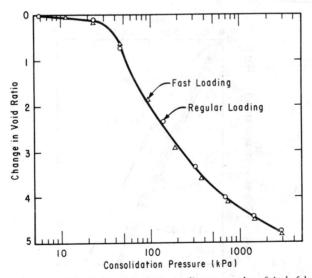

FIG. 4—*Comparative stress-strain curves for two adjacent samples of Atchafalaya clay, one loaded in the regular manner of one load per day, and one loaded as soon as* S_{100} *was attained.*

A method of analysis for cases with impeded drainage was presented by Gray [30]. For a double drained sample with a thickness H and hydraulic conductivity k, with boundary porous stones of thickness H_d and hydraulic conductivity k_d, the average degree of consolidation, U, is given by

$$U = 1 - \Sigma D_n E_n^2 \exp(-4r_n^2 T) \tag{28}$$

where r_n represents successive roots of the equation:

$$r_n = \tan^{-1}[2r_n I/(I^2 r_n^2 - 1)] \tag{29}$$

$$I = k H_d/k_d H \tag{30}$$

$$D_n = 2I^2/(r_n^2 I^2 + 2I + 1) \tag{31}$$

$$E_n = (1/r_n)(Ir_n \sin r_n - \cos r_n + 1) \tag{32}$$

and T is defined as in Eq 12. The factor I is here called the *boundary impedance*. For freely draining boundaries, $I = 0$. Theoretical curves of average degree of consolidation versus $\sqrt{t}$ are presented in Fig. 5. The effect of impedance is to induce an initial curvature in the U-$\sqrt{t}$ plot with the eventual formation of a displaced linear portion that is parallel to the curve for unimpeded drainage. An impedance of less than 0.01 should be sought.

Boundary impedance is minimized by using clean, reasonably fine grained,

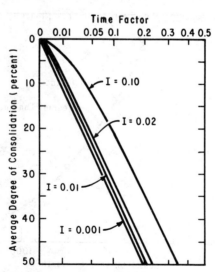

FIG. 5—*Effect of boundary impedance on the theoretical curve of average degree of consolidation versus square root of time factor.*

porous stones (similar to the Norton grade P2120). Use of filter paper between the soil and stones tends to keep the stones clean and also facilitates drainage. Whatman grade 54 is preferred.

Ring Friction

Friction between the soil and the consolidation ring reduces the mean stress in the soil, compared with the applied stress, during loading, and increases the mean stress during swelling. Taylor [12] analyzed friction in a simple but useful fashion. Let p be the applied stress (P is force) and q the average effective stress (Q is average force) at any depth z in the sample. If the local side shearing stress is τ, then:

$$\tau = \sigma'_h \mu = q \, K_0 \mu \qquad (33)$$

and over a height dz:

$$dQ = - (2/R) \, Q \, K_0 \, \mu \, dz \qquad (34)$$

At a depth z:

$$Q = P \exp \left(- \frac{2}{R} K_0 \, \mu \, z \right) \qquad (35)$$

The average stress, q, over a depth H is then:

$$\bar{q} = p \, \lambda \qquad (36)$$

where

$$\lambda = \frac{R/H}{2K_0\mu} \left[1 - \exp \left(- \frac{2K_0\mu}{R/H} \right) \right] \qquad (37)$$

As an example of the application of Eqs 36 and 37, consider a test using a fixed ring of height 20 mm (0.8 in.) and diameter 80 mm (3.2 in.) for a plastic clay with $\phi = 18$ deg. Then, for a normally consolidated sample [31]:

$$K_0 = 1 - \sin \phi = 0.69 \qquad (38)$$

The wall friction angle, δ, is assumed to be 90% of ϕ so μ is about 0.29. From Eq 36, $\lambda = 0.91$; that is, about 9% of the applied stress is lost in friction. For a floating ring, the average value of q would be for an H of 10 mm (0.4 in.) and $\lambda = 0.95$, a reduction in frictional loss of almost two times. If the soil had been rebounding in a fixed ring, the upper limit on K_0 would be the Rankine passive value:

$$K_0 = \tan^2 (45 + \phi/2) = 1.89 \qquad (39)$$

and $\lambda = 0.77$.

The equations are quite approximate in that the actual vertical stresses near the ring may differ significantly from the mean value, the deformation may not be enough to mobilize the full soil/ring friction, no consideration is given to creep, and the coefficients are approximate. Nevertheless, the equations indicate that frictional effects can be minimized by using wide thin specimens and by reducing the soil/ring friction. They also suggest that friction may be much larger during rebound than during initial compression, and will be less during initial reconsolidation than if the sample is loaded onto the virgin curve, then rebounded, and then reloaded in the same stress range [9,32].

Taylor [12] reported values of λ in the range of 0.90 to 0.95 for undisturbed and remolded Boston Blue clay. Leonards and Girault [33] measured the force needed to support the consolidation ring and obtained values of λ for normally consolidated clay on the order of 0.82 for steel rings, 0.89 for rings lined with tetrafluoroethylene, and 0.93 for tetrafluoroethylene coated with molybdenum disulfide grease. They found a tendency for λ to increase slightly with time. Burland and Roscoe [34] placed a load cell, with a 25.4-mm (1-in.)-diameter diaphragm, at the base of some 38.1-mm (1.5-in.)-diameter samples in consolidation rings and measured the base stress directly. They reported values of λ approaching zero for height/diameter ratios near 2½ and for ungreased brass tubes. Use of silicone grease reduced friction greatly initially but after some period of time (presumably on the order of a week) the friction began to increase. However, values of λ of less than, or equal to, 0.88 were reported even for apparently tall samples when greased rings were used. They found much larger amounts of friction during unloading.

Several unpublished theses at the Massachusetts Institute of Technology (MIT) between 1936 and 1943 indicated values of λ on the order of 0.90 to 0.95 for greased and ungreased metal rings. A large number of unpublished tests at the University of Illinois in 1959, with soils ranging from sand to plastic clays, indicated that friction decreased as clay plasticity increased. For soils ranging from sand to clay of low plasticity, rings of plastic (acrylic resin) had on the order of half the friction of cadmium-coated steel and use of silicone grease reduced friction by almost half again. For clay of high plasticity there was only a small reduction in friction (15%) when plastic rings were used in place of steel. A typical set of data is shown in Table 1.

It appears that for normally consolidated clays and rings of typical geometry, λ is on the order of 0.90 to 0.98 for greased plastic, 0.80 to 0.95 for ungreased plastic, and 0.70 to 0.90 for ungreased smooth steel. Large amounts of friction are likely during swelling of a sample previously consolidated in the same ring.

TABLE 1—*Results of ring friction tests with the Minus No. 4 sieve fraction of an Illinois glacial till. The soil is a clay of low plasticity. The rings were fixed and were either acrylic resin or cadmium-coated steel, and were 63.5 mm (2.5 in.) in diameter. Samples were about 15 mm (0.6 in.) thick.*

Consolidation Pressure kPa	Values of λ for		
	Cadmium-Coated Steel	Ungreased Plastic	Greased Plastic
56	0.60	0.85	0.98
140	0.70	0.90	0.98
281	0.74	0.89	0.97
562	0.72	0.90	0.97
1125	0.81	0.88	0.96
2250	0.82	0.88	0.95
4500	0.85	0.90	0.94
2250	. . .	0.83	0.75
1125	. . .	0.72	. . .
562	. . .	0.55	. . .
281	. . .	0.35	. . .

The flattening of the swelling curve at low stresses is probably caused by ring friction.

Effects of Temperature

Most observers have found that an increase in temperature causes a small reduction in void ratio at the time of the temperature change but a negligible change in compressibility [30,35–38], but Finn [39] reported an increase in void ratio when temperature increased. Habibagahi [40] found no effect of temperature on void ratio for inorganic Paulding soil and a slight decrease ($\Delta e = -0.06$) for the organic soil for an increase in temperature from 25 to 50°C.

Taken collectively, the data indicate relatively inconsequential changes in void ratio even for relatively large changes in temperature. However, increases in temperature during a period of constant stress, after primary consolidation is completed, always (in my experience) result in a clear reduction in void ratio. Habibagahi [40] explained this observation by pointing out that increases in temperature cause expansion of the consolidation ring and probably a sudden, temporary reduction in ring friction.

Increases in temperature reduce the viscosity of water, and thus increase k and c_v. Finn [39] observed that the increase in c_v for his soils was more than would be predicted by viscosity changes (c_v essentially doubled when temperature rose from 4.4 to 21.1°C (40 to 70°F)).

Rates of Dissipation of Pore Water Pressures

One-dimensional consolidation soil properties can be evaluated from pore water pressure observations as well as in terms of strains. The usual test procedure is to use a fixed ring, drain the pore water out through the top, and measure pore water pressure at the impervious base. The excess pore pressure at the base, $\bar{u}_b$, according to Terzaghi's [2] theory, is

$$\bar{u}_b/\bar{u}_i = \sum_{m=0}^{\infty} \frac{2}{M} (-1)^m \exp(-M^2 T) \tag{39}$$

where $\bar{u}_i$ is the initial (uniform) excess pore water pressure, and M and T are as defined in Eqs 11 and 12. From Eq 39, the time factor at which 50% of the initial excess pore water pressure has been dissipated is 0.379. Thus the time at which $\bar{u}_b$ is half of $\bar{u}_i$, say t_{50}, is determined and is used in Eq 13 to evaluate c_v. Other values of $\bar{u}_b/\bar{u}_i$ may also be used. Values of $\bar{u}_b/\bar{u}_i$ and associated time factors are given in Table 2.

Taylor [12] was perhaps the first investigator to measure $\bar{u}_b$. He found generally good agreement between measured and predicted values of $\bar{u}_b$. Leonards and Girault [33] found excellent agreement between measured values of $\bar{u}_b$ and the values predicted using values of c_v evaluated using the log t method if the load increment ratio (LIR) was 2 or 3, but at LIR = 0.15 the pore pressures dissipated much faster than predicted, the measured t_{50} being about two orders of magnitude smaller than the theoretical value.

Christie [41] developed a Terzaghi-type theory for base pore pressure, which included the effects of compliance of the measuring system. He also found faster-than-expected dissipation of pore pressure.

Raymond [42] showed that rapid dissipation of midplane pore water pressures can be caused by variation in soil properties (e.g., m_v and k) during consolidation.

Barden and Berry [43] and Barden [44] attribute differences between measured and computed values of $\bar{u}_b$ to a combination of nonlinear stress-strain curves, variations in k, and structural viscosity. Burland and Roscoe [34] and Mesri and Godlewski [45] came to rather similar conclusions.

Clearly, values of c_v evaluated from measured pore water pressures may differ significantly from values obtained from settlement observations.

TABLE 2—*Values of time factor associated with various values of midplane excess pore pressure ratio.*

$\bar{u}_b/\bar{u}_i$	T
0.1	1.031
0.2	0.750
0.3	0.586
0.4	0.469
0.5	0.379
0.6	0.305
0.7	0.241
0.8	0.185
0.9	0.130

Effects of Nonlinear Stress-Strain Properties on S-t *and* ū-t *Relationships*

Consideration is here restricted to primary effects alone. In the classical solution, the stress-strain curve is assumed to be linear and the hydraulic conductivity is assumed to be constant. Solutions have been obtained for a variety of cases in which specific relations between σ' and ϵ, and k and e, were assumed [43,46–48].

It seems easier and more general to simply model the process using a piecewise continuous approach [49]. A finite difference program of that type was written to analyze laboratory consolidation tests. To show the effects of nonlinear stress-strain relationships, analyses were performed for a 12-mm (0.5-in.)-thick, singly drained sample of clay. The clay was loaded from a stress of 69 kPa (10 psi) to 138 kPa (20 psi) in one step. The void ratio reduced from 1.00 to 0.98 as a result (small strain). The hydraulic conductivity was fixed at 2.2 × 10^{-8} cm/s (2 × 10^{-7} in./min). The three stress-strain curves shown in Fig. 6 were used. The calculated curves of U versus $\sqrt{t}$ and $\bar{u}/\bar{u}_b$ versus log t are given in Figs. 7 and 8, respectively.

The root plots have a significant linear range in all cases. However, for Curve 2 (normally consolidated soil, concave upwards e-σ' curve) the shape is such that 20% of the settlement appears to be secondary when in fact it is primary. Conversely, for Curve 3 (overconsolidated soil, concave downwards e-σ' curve), the end of primary consolidation appears to be 8% beyond the end of actual primary consolidation.

The coefficients of consolidation calculated using the $\sqrt{t}$, log t, and $\bar{u}_b$ methods

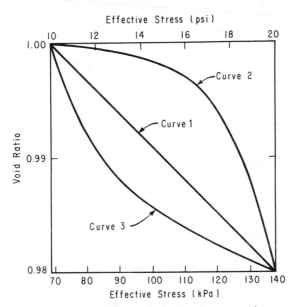

FIG. 6—*Stress-strain curves for use in subsequent analyses.*

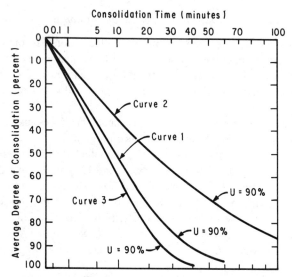

FIG. 7—*Predicted curves of average degree of consolidation (settlement) versus time (square root scale) for three samples with constant coefficient of consolidation but different stress-strain curves.*

are presented in Table 3. Use of the $\sqrt{t}$ and log t methods led to relatively constant values of c_v, but use of base pore pressures led to significant variations depending on the fitting point. It may be noted that these analyses involved constant hydraulic conductivity because of the essentially constant void ratio. The differing shapes of the e-σ' curves thus led to differing values of a_v and thus c_v (Eq 13). Local

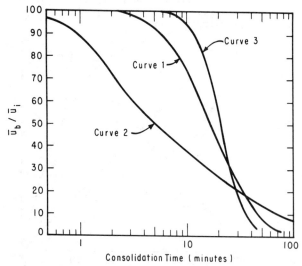

FIG. 8—*Predicted curves of base pore pressure versus logarithm of time for soil with a constant coefficient of consolidation but different shapes of stress-strain curves.*

TABLE 3—*Calculated values of* c_v *influenced by the shape of* σ'-e *curve and method of data reduction (see Fig. 8).*

Curve No.	Root Method	Log Method	c_v (m²/s) Use $\bar{u}_b$ at		
			10%	50%	90%
1	6.0×10^{-8}	6.0×10^{-8}	6.0×10^{-8}	6.0×10^{-8}	6.0×10^{-8}
2	4.3×10^{-8}	2.5×10^{-8}	3.7×10^{-7}	2.0×10^{-7}	4.0×10^{-8}
3	6.6×10^{-8}	7.7×10^{-8}	2.9×10^{-8}	4.8×10^{-8}	7.8×10^{-8}

variations in c_v within the sample ranged much more widely (e.g., from 5.3×10^{-3} cm²/s down to 3.0×10^{-4} cm²/s for the overconsolidated clay). The initial values of c_v for Curves 1, 2, and 3 (at $\sigma' = 68.9$ kPa [10 psi]) were 6.0×10^{-4}, 5.3×10^{-3}, and 1.5×10^{-4} cm²/s, respectively. These values are closest to the values calculated using $\bar{u}_b$ at 10% settlement and are closer to the $\sqrt{t}$ value than the log t value.

It may be noted that the $\sqrt{t}$ fitting method leads to a higher value of c_v (as is observed in most laboratory tests) for the overconsolidated sample but not for the case of the concave upwards e-σ' curve.

In the classical theory, the average degree of consolidation, U, is defined in terms of excess pore water pressures immediately after loading, $\bar{u}_i$, and at the time of interest, $\bar{u}$, using:

$$U = 1 - \frac{\int \bar{u}dz}{\int \bar{u}_i dz} \tag{40}$$

Use of Eq 40 to define U led to values of c_v that were considerably different from values obtained from settlements. These values are not included here because they are irrelevant in laboratory testing. The results indicate that values of c_v backed out of field values of $\bar{u}$ at several depths in a compressible layer may differ widely from values obtained from settlement observations.

Radial Flow Consolidation Testing

Many field problems involve at least partial flow of pore water in the horizontal directions. In the case of sand drains and drainage wicks the main flow may be horizontal. Perhaps the simplest testing procedure for such problems is to perform standard one-dimensional tests on samples rotated 90 deg [50–53]. However, such tests match only the flow direction; the deformation conditions are incorrect.

A preferable procedure is to test flat, disk-shaped specimens, with pressure applied to the horizontal flat faces and flow either into a central sand drain or outwards to a porous drainage layer. McKinlay [52] used standard cells but with impervious loading platens and a porous stainless steel ring. Aboshi and Monden [53] developed both inflow and outflow equipment. They used porous bronze rings for outflow but recommended against such tests because of large amounts of ring friction. For inflow tests they used central drains of sand and micaceous sand, and also a porous bronze rod, as did Wu, Chang, and Ali [54].

McKinlay [52] and Aboshi and Monden [53] used rigid loading plates. Rowe and Barden [55] used a flexible membrane and water pressure for loading, thus eliminating the need for a cumbersome load frame and making it possible to use larger samples; they used sample diameters of 76 mm (3 in.), 152 mm (6 in.), 254 mm (10 in.), and 508 mm (20 in.). Their device will be discussed in more detail later. Because of the availability of equipment using both rigid and flexible loading platens, data reduction schemes should be available using both equal strain and free strain approaches [56].

Da Silviera [57] published a curve of U_r versus log T_r where U_r is the average degree of consolidation and T_r is the radial time factor. Presumably a log t fitting method could be used with $T_r = 0.25$ at $U_r = 50\%$ but no method was recommended. McKinlay [52] recommended a root time method with time plotted to the 0.465 power and a curve fitting factor of 1.22. McKinlay implies that fitting should be done at $U_r = 90\%$ where $T_r = 0.3345$. Berry and Wilkinson [58] use U_r-$T_r^{0.5}$ plots for free strain radial inflow and find linear relationships for $20\% \leq U_r \leq 60\%$, and find a curve fitting factor of 1.17 at $U_r = 90\%$. No numerical values were given for T_r as a function of N ($N = R_E/R_W$ where R_E and R_W are the radii of the external boundary and the drainage well, respectively) nor was a method given to find the corrected zero point.

To present a consistent set of methods, we first need a set of T_r-U_r relationships for inflow and outflow, and for free strain and equal strain. For free strain outflow [52,57]:

$$U = 1 - 4 \sum_{n=1,2}^{\infty} (1/\alpha_n^2)\exp(\alpha_n^2 \, T_r) \tag{41}$$

where α_n represents successive roots of Bessel's function of the first kind and order zero:

$$J_0(\alpha_n) = 0 \tag{42}$$

Values of α_n have been tabulated by Abramowitz and Stegun [59] and T_r is the radial time factor:

$$T_r = c_r t/R_E^2 \tag{43}$$

where c_r is the radial coefficient of consolidation, t is time, and R_E is the radius of the external drainage boundary. In turn:

$$c_r = k_r(1 + e)/a_v\gamma_w \tag{44}$$

where k_r is the radial hydraulic conductivity and other terms are as defined previously (Eq 13).

For equal strain outflow, the method used by Barron [56] for inflow can be rederived as:

$$U_r = 1 - \exp(-8\,T_r) \tag{45}$$

For free strain inflow [56]:

$$U_r = 1 - \Sigma \frac{4A_n}{\alpha_n^2\,(N^2 - 1)(B_n - C_n)} \exp(-\alpha_n^2\,N^2 T_r) \tag{46}$$

where

$$A_n = [J_0(\alpha_n)Y_1(\alpha_n) - Y_0(\alpha_n)J_1(\alpha_n)]^2 \tag{47}$$

$$B_n = N^2[J_0(\alpha_n N)Y_0(\alpha_0) - Y_0(\alpha_n N)J_0(\alpha_n)]^2 \tag{48}$$

$$C_n = [J_0(\alpha_n)Y_1(\alpha_n) - Y_0(\alpha_n)J_1(\alpha_n)]^2 \tag{49}$$

and α_n represents successive roots of:

$$J_1(\alpha_n N)Y_0(\alpha_n) - Y_1\,(\alpha_n N)J_0(\alpha_n) = 0 \tag{50}$$

For equal strain inflow [56]:

$$U_r = 1 - \exp(-2T_r/F_N) \tag{51}$$

where

$$F_N = \frac{N^2}{N^2 - 1} \ell n\, N - \frac{3N^2 - 1}{4N^2} \tag{52}$$

and

$$N = R_E/R_w \tag{53}$$

where R_E is the external radius of the sample and R_W is the radius of the central drain well.

A set of log t methods could be developed but the awkward method of finding the corrected zero point and the arbitrary nature of the method used to find the point where primary consolidation is completed suggest that a root method should be sought. These methods involve a plot of volume change (or average compression) versus t^n, with an early linear relationship that passes through (or very close to) the origin of the theoretical curve, and the $U_r = 90\%$ point is found by drawing a line through the origin with a slope of CFF (curve fitting factor) times the slope of the early linear part of the experimental curve. For inflow tests, these factors depend on N (Eq 53). To avoid excessive tabulation of results, a standard N of 5 will be used. Values of the relevant factors are presented in Table 4 [60].

Barron [56] presented both free strain and equal strain solutions for inflow when the central drain is surrounded by an incompressible annular zone of incompressible soil. The thickness of this zone is defined using the dimensionless factor S ($= R_s/R_w$ where R_s and R_w are the outer radii of the zone of disturbed soil, the so-called "smear zone", and drainage well, respectively). The hydraulic conductivity of the smear zone is defined using the dimensionless factor K ($= k_s/k_r$ where k_s is the hydraulic conductivity in the smear zone and k_r is the radial hydraulic conductivity of unsmeared soil. Unfortunately, we now find that U_r depends on four variables (T_r, N, S, K) and it is difficult even to present results [50,58]. Berry and Wilkinson [58] also presented a solution for outflow with an external smear zone. In this case a single factor, S_p, can be used to represent both the thickness and hydraulic conductivity of the smear zone:

$$S_p = (k_s/k_h)/\log(R_E/R_s) \qquad (54)$$

Although it does not appear useful to attempt curve fitting procedures that account for smear zones, it is possible to assign reasonable thicknesses and properties of smear zones and calculate the ratio of the correct value of c_r to the value calculated when (erroneously) the smear zones are ignored (c_{ra}/c_{rm}). Berry and Wilkinson [58] presented such plots for inflow ($N = 10$ and 20) and outflow and showed c_{ra}/c_{rm} to be less than 2 for most likely conditions.

Trautwein et al [60] found that use of outflow tests led to spontaneous formation of a smear zone next to the rough external boundary as the soil deforms, and thus to c_r being consistently lower for outflow tests than for inflow tests. They performed outflow tests with a freely draining sand/mica outer zone and obtained

TABLE 4—*Factors recommended for use with radial flow consolidation tests (for inflow, N = 5).*

Factor	Outflow		Inflow	
	Equal Strain	Free Strain	Equal Strain	Free Strain
Root Power (n)	0.83	0.47	0.83	0.67
Curve Fitting Factor (CFF)	1.61	1.22	1.61	1.31
Time Factor (T_r) at $U_r = 45\%$	0.075	0.049	0.280	0.264

the same values of c_r for inflow and outflow. They generally recommend use of inflow tests and free strain theory.

Radial flow tests can be analyzed using measured pore water pressures but the method will not be considered further here because it is unlikely to find any practical use and suffers from the same problems discussed earlier for vertical flow tests.

Secondary Effects

It is convenient to use the term *primary* when referring to time rates of consolidation governed by the hydraulic conductivity of the soil. Effects of nonlinear stress-strain curves, time-dependent loading, impeded drainage, stratified soils, multidimensional flow, large strains, effective-stress dependent (void ratio dependent) hydraulic conductivities, anisotropy, and other such effects are included, and can lead to response that deviates widely from that predicted using Terzaghi's original theory. Remaining effects may loosely be termed *initial* and *secondary*.

In the laboratory, real initial compression may result from compression of the apparatus (porous stones, filter paper), compression of gas bubbles (not instantaneous), extrusion of soil, expansion of the ring, compression of water and solids (negligible), and other such effects. It should not simply be the difference between the actual thickness and the corrected initial thickness (the usual definition) because that may be influenced greatly by primary effects (previous paragraph) as well as secondary effects. Initial compression, after correction for apparatus deformation, should thus be negligible for a properly performed test with a saturated sample, and will not be considered further.

Secondary effects are then defined as all effects except those classified as initial and primary. Secondary effects may result from a variety of causes and it may be impractical to develop a theory based directly on real physical phenomena, as in the case of primary theory, but instead resort to rheologic models, the properties of which have little, if any, real world equivalents. In such cases, the properties of the rheologic elements cannot be measured independently, as can hydraulic conductivity in primary theory, and thus extrapolation from laboratory scale to field scale is more uncertain.

Data available at present make it clear that the elements in the rheologic models should be nonlinear. There has been negligible success in obtaining analytic solutions for nonlinear consolidation problems, even for primary consolidation, and numerical methods are used. The result is that curve fitting is very much more difficult than in the usual case, and users must have appropriate computer programs available to recalculate the theoretical curves. Such programs have been developed but they are not readily available to the profession at large.

Finally, measurement of the appropriate properties in the laboratory implies that design engineers will use these same models for design. Field problems typically involve time-dependent loading, layered systems, two-or-three-dimensional flow, and other such effects not included in the programs used to model laboratory tests. Only one such program is known to exist [47] and it does not appear to use a rheologic model of the type generally used to fit laboratory data.

Under these circumstances, and considering reasonable space limitations, it seems unreasonable to attempt a true state-of-the-art summary of methods used to evaluate parameters in theories of secondary consolidation. Instead, it seems useful to make a few observations about secondary effects in general, to review a relatively simple model, and to reference other work, all in the context of laboratory testing.

Secondary Compression Following Primary Consolidation

In its simplest form, secondary "compression" is sometimes considered to be the compression that follows primary compression, particularly when settlement-versus-log time plots are used. A typical curve of this type is shown in Fig. 9, where secondary compression represents about 60% of the total compression. Examination of many sets of test data indicates a range in the ratio of secondary compression to total compression of from essentially 0% for some tests using remolded clays to essentially 100% for some highly organic soils.

When the secondary slope is essentially linear, the amount of secondary compression in the laboratory can be represented using the slope; for example, if

$$\epsilon_\alpha = d\epsilon/d \log t \tag{55}$$

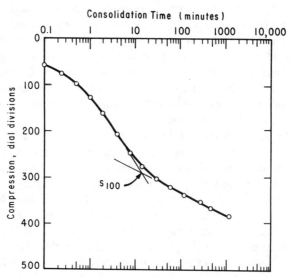

FIG. 9—*Typical curve of settlement versus logarithm of time (San Francisco Bay mud).*

where ϵ_α is the slope on a strain (ϵ)-log time (log t) diagram, then:

$$S_s = \epsilon_\alpha H_0 \log(t_2/t_1) \tag{56}$$

where H_0 is whatever reference height is used in defining strain, t_2 is the time at which secondary settlement (S_s) is calculated, and t_1 is the time at the tangent point of the backextended secondary curve and the primary consolidation curve. Mesri [61] has recommended appropriate symbolism to use in defining these slopes.

More-or-less linear secondary slopes in the laboratory [62] have been reported for up to about 3 years by Haefeli and Schaad [63] and 5.5 years by Cox [64] for clay and for up to a week in volcanic sand [65]. Thompson and Palmer [66] found the secondary slope to increase after about 1000 min, whereas Lo [67] found that secondary consolidation would stop in most clays after about three weeks and Dhowian and Edil [68] found it to stop even in peat after about 200 days. Experience indicates a wide range in shapes of secondary curves [33]. Experience also indicates that long-term laboratory tests are not easy to interprete because of effects such as temperature changes, ring friction, vibration, and chemical changes (particularly oxidation).

There seems to be little, if any, well documented evidence that secondary slopes in the field are the same as in the laboratory. Some existing field evidence involves soft, highly compressible upper layers over less compressible deeper layers. The long-term settlements that have been inferred to be secondary compression of the soft layer are more likely delayed primary compression of deeper layers.

In other cases where there is clear "secondary type" settlement in the field [69] there are no comparative laboratory slopes. In other cases [70] the presumed secondary settlement may actually have resulted from mass movement rather than volume change.

The existence of secondary compression following primary compression brings into question some laboratory testing issues that need immediate attention, particularly the uniqueness of the void ratio at the end of primary consolidation and the effect of secondary compression on the time-settlement response on the next loading.

Influence of Secondary Compression on Subsequent Void Ratios

A question arises as to the effect of secondary compression under one load on subsequent void ratios. Leonards [71] has suggested that the e-log σ' curve be plotted using:

$$e = e_i + \Sigma(e_{100} - e_0) \tag{57}$$

where e_i is the void ratio at the beginning of the test, and $e_{100} - e_0$ is the primary change in void ratio under successive loads. Such a procedure carries with it the assumption that secondary compression under any stress causes a lowering of the entire remaining e-log σ' relationship.

In the earliest study of the effect of loading at the time S reaches S_{100}, Newland and Allely [27] found that e_{100} appeared not to depend on whether loads were applied at e_{100} or at 24 h. The same conclusion was reached by Hansen and Inan [38], Walker [72], Andersland and Mathew [73], and is implicit in Bjerrum's models for long-term settlement [69]. We have used rapid loading procedures coupled with standard tests and find e_{100} to be independent of earlier secondary compression, provided that secondary compression under one pressure does not exceed primary compression under the next load. The void ratio should thus be calculated using the total soil compression to that point, with no correction for secondary compression under previous loads (Eq 57 should not be used).

Influence of Secondary Compression and Load Increment Ratio on Subsequent Time-Settlement Response

The effect of secondary compression under one pressure on the shape of time-settlement curve for the next pressure has been documented (e.g., Leonards and Girault [33], Leonards and Altschaeffl [74], and Raymond [42]). It seems clear that the ideas were known, if not published, by researchers at Harvard in the late 1930s.

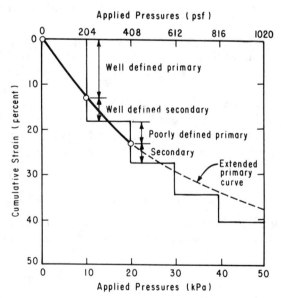

FIG. 10—*Consolidation curve for remolded peat, with loads applied in arithmetic progression, indicating the effect of secondary compression on subsequent behavior* [157].

If, for the present section, we ignore secondary effects during primary consolidation, it is clear that the reduction in void ratio, under one pressure, lowers the e-σ' point below the virgin curve (Fig. 10) so that the primary stress path involves a nearly horizontal reloading portion with a correspondingly high c_v (Eq 13). Thus, for a small load increment ratio, primary compression may conclude almost at once and nearly all the settlement may appear to be due to secondary effects. For a larger load increment ratio, where the soil strains out onto the virgin curve, one obtains an S-t response such as discussed earlier in the section on Effects of Nonlinear Stress-Strain Properties on the S-t and $\bar{u}$-t Relationships. During the late 1950s we observed that when we allowed considerable secondary compression to occur under one load, the S-$\sqrt{t}$ curve for the next load was often concave downwards, whereas when we loaded quickly the S-$\sqrt{t}$ curve was always linear and fitted Terzaghi's theory well. The earliest observations of such behavior appear to date back at least to the late 1930s. It is clear, therefore, that secondary settlement under one pressure exerts an influence on the primary curve for the next load.

Slopes of Secondary Curves

Because the S-log t curves are generally nonlinear it has not been possible to assign a simple analytic function to describe them properly. If the first cycle of log time is used, then a linear S-log t relationship can generally be assumed and Mesri's [61] symbolism followed. For insensitive, normally consolidated, clays, the slopes C_α or ϵ_α do not vary greatly with pressure. Secondary slopes are much lower for overconsolidated clay. For sensitive clays the slope is highest at the apparent maximum previous consolidation pressure and seems to be related to the slope of the primary e-log σ' curve [45,72]. Thus, for field application, the secondary slope should be measured at the correct stress.

Secondary Consolidation Theories

Space limitations prevent a review of the many theories of secondary consolidation, but these theories cannot be ignored because there is no *a priori* reason to believe that secondary effects exert the same influence in the field as in the laboratory. A theory would provide the scaling laws to be used in interpreting the laboratory data.

All secondary theories suffer a serious flaw in that they contain parameters that are not subject to independent determination. Primary theory suffers from no such defect in that the hydraulic conductivity can be measured directly [2].

The simplest method of taking into account secondary effects is to assume a linear S-log t curve in the secondary range and simply extend the calculated primary S-log t curve at that slope, starting at the time when dS/d log t of the primary curve equals the slope of the secondary curve. Such an approach dates back at least into the 1930s [62] and has been supported with limited field evidence (e.g., Moran et al [50]). Bjerrum [69] quantified the method using field data for buildings in Drammen, Norway, and Garlanger [75] set up formal analytical procedures which seemed to predict time-settlement behavior of thick laboratory specimens, based on data for thin samples, rather well [76]. However, there is no field data to indicate that secondary slopes measured during perhaps a day in the laboratory can be used to predict field behavior over a period of decades, and evidence has already been cited to indicate that secondary consolidation may cease in the laboratory after a few weeks to a few months.

To be most useful in extrapolating laboratory data to the field, the secondary model ought to be one that applies to a differential element so it can be used in integrations in space and time. The simplest such model, and the most widely known, was proposed by Gibson and Lo [77] and Lo [67]. It has the added advantage of having been applied to a field case history [78]. The rheologic model for the effective stress is a linear spring in series with a Kelvin body (Fig. 11). In this model, as pore water pressures dissipate, load is transferred to effective stress, leading to immediate compression in Element 1. The load also goes onto Element 2, the Kelvin element, where it is equal to the sum of the loads in the spring (p_{2s}) and dashpot (p_{2d}). As fluid escapes from the dashpot, load is transferred from the dashpot to the spring and secondary compression occurs. The additional properties to be measured in the laboratory are the dashpot constant λ and the secondary compressibility b. The average degree of consolidation, U, can perhaps be most easily written as [67,77]:

$$U = S_t/S_u = 1 + (8/\pi^2) \sum_{\text{odd } n} 1/n^2 \left\{ \left[\left(\frac{n^2\pi^2}{M} - x_1 \right)/(x_1 - x_2) \right] \exp(-x_2 t) \right.$$

$$\left. - \left[\left(\frac{n^2\pi^2}{M} - x_2 \right)/(x_1 - x_2) \right] \exp(-x_1 t) \right\} \quad (58)$$

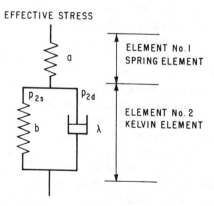

FIG. 11—*Rheologic model for response of soil structure.*

where

$$\frac{x_1}{x_2} = 1/2[(4MN + n^2\pi^2) \pm \sqrt{(4MN + n^2\pi^2)^2 - 16N\,n^2\pi^2}] \qquad (59)$$

where M is the ratio of ultimate settlement to primary settlement:

$$M = (a + b)/a \qquad (60)$$

N is:

$$N = (\lambda/b)(H^2/c_v) \qquad (61)$$

and S_t and S_u are the settlement at time t and ultimately, respectively. If a load is left in place long enough (Lo [67] reported that three weeks is usually adequate), then S_u is measured directly. Equation 58 can be rewritten to apply when all excess pore water pressure, $\bar{u}$, has dissipated.

$$\log\left(\frac{\epsilon_u - \epsilon_t}{\bar{\sigma}}\right) = \log b - 0.434\,\frac{\lambda}{b}\,t \qquad (62)$$

Equation 62 applies to a straight line in a plot of $\log\,[(\epsilon_u - \epsilon_t)/\bar{\sigma}]$ versus t, with an intercept $\log b$ and slope $= 0.434$, thus yielding b and λ. Then a is calculated from S_u and b, and c_v from the small time solution:

$$S_t = (2/\sqrt{\pi})\sqrt{c_v t}\;aq(1 + \lambda t/3a) \qquad (63)$$

Lo, Bozozuk, and Law [78] suggest an alternative method of finding λ and b in which Eq 62 is written as

$$\log\left(\frac{de}{dt}\right) = \log(\lambda\,q) - 0.434\frac{\lambda}{b}\,t \qquad (64)$$

and $\log\,(de/dt)$ is plotted versus t, the intercept being $\log\,(\lambda q)$ and the slope $= 0.434\lambda/b$, where q is the increment of stress.

In their original report, Gibson and Lo [77] found constant values of λ for any one stress. However, Lo, Bozozuk, and Law [78] found λ to change continuously during secondary consolidation (in agreement with our observations), thus complicating the application of the theory.

Berre and Iversen [76] performed laboratory consolidation tests on undisturbed samples of Drammen clay with drainage distances ranging from 19 mm (0.74 in.) to 450 mm (17.7 in.) and attempted to use the methods of Gibson and Lo [77], Barden [44], and Garlanger [75] in explaining their results. They concluded that Gibson and Lo's method did not fit their results well. The one curve they

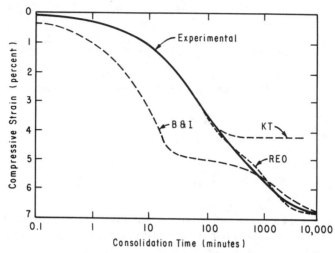

FIG. 12—*Experimental time-settlement curve for Drammen clay with three fitted theoretical curves.*

published is presented in Fig. 12, where the measured S-log t curve is labelled "experimental" and their fitted Gibson and Lo solution is labelled "B&I". They fitted Terzaghi's theory to their pore pressure data and generally obtained poor correlations between theory and measurement. In Fig. 12, the curve marked "KT" was fit by the author using the square root method [13]; this fits the primary curve quite well. The log t method was not used because there was no obvious way of finding S_{100}. The curve marked "REO" was calculated using Gibson and Lo's theory, using c_v and a from the square root fitting method, calculating b on the assumption that all consolidation would cease at 10 000 min, and finding the value of λ by trial. The "REO" curve could have been made to fit the secondary curve as well as it did the primary range by increasing b and altering λ, but that change would require that the measured curve turn downwards beyond 10 000 min where there were no substantiating experimental data.

We have already seen that Terzaghi's simple linear theory is an approximation of a more general theory involving nonlinear stress-strain curves, effective stress-dependent soil properties, etc. It has been widely used because of its simplicity and in spite of a lack of good corroborating field evidence. Gibson and Lo's theory is a linear extension to include secondary effects and can be expected to fit data, provided actual nonlinearities are not great. The main source of uncertainty in its use is the lack of an independent means of measuring λ.

Two analyses were performed using Gibson and Lo's [77] theory with $a = 8.3 \times 10^{-4}$ m^2/kN (5.7×10^{-3} in.^{2}/lb), $b = 6.82 \times 10^{-8}$ m^2/kN (4.7×10^{-7}in.^{2}/lb), $\lambda = 8.7 \times 10^{-7}$m-s/kN($1.0 \times 10^{-7}$min/psi), and $c_v = 3.4$ m^2/year (0.1 ft^2/day), one for a laboratory test with $H = 13$ mm (0.5 in.) and one for a field case with $H = 3$ m (10 ft). The calculated U-t curves are shown

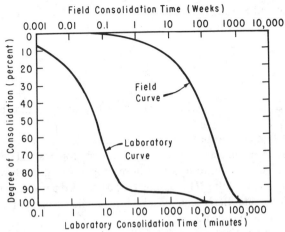

FIG. 13—*Theoretical curves of degree of consolidation versus time for the same soil in the laboratory and in the field. Based on the linear theory of Gibson and Lo [77].*

in Fig. 13. Apparently, consolidation was so retarded in the field, due to the large drainage distance, that all secondary compression occurred simultaneously with the primary consolidation; thus primary effects appeared to dominate. The inference that primary effects may dominate in the field was drawn by Gibson and Lo [77] and by others since [44,79–81].

Linear theories of the Gibson and Lo [77] type can be generalized by making the springs and dashpots nonlinear. The effects of nonlinearity in the spring of Element one (Fig. 11) have already been considered. Nonlinearities apparently exist in the spring and dashpot of Element 2, and they need to be evaluated if the nonlinear theories are to be useful.

In the case of primary consolidation, the m_v-σ' relationship (Element 1, Fig. 11) can be approximated directly from the e_{100}-σ' curve for thin samples and can be as nonlinear as desired (assume linear e-σ', linear e-log σ', bilinear e-log σ', . . . or use the actual curve in numerical analyses). Similarly, k can be measured and any desired relationship selected (constant k, linear e-log k, . . . general). No such option exists in the secondary part of the model, although the spring compressibility can be determined if each load is left in place long enough (weeks to years) for movement to stop. More practically, an analytic relationship is assumed for each element and only the constants in the equations then need to be known. For example, a linear spring was used by Barden [44,79], whereas Berry and Poskitt [81] assume a linear Δe-log σ' relationship. For the dashpot, Barden [44] assumed a form:

$$p_{2d} = b \left(-\frac{\partial e}{\partial t} \right) \tag{65}$$

whereas Berry and Poskitt [*81*] assumed:

$$p_{2d} = \frac{d\epsilon/dt}{\beta \sinh \alpha} \tag{66}$$

where b, n, β, and α are dashpot constants (they may, in turn be written as functions of σ', e, . . .).

Other models may also prove useful. For example, secondary consolidation may be viewed as being caused by drainage from micropores that develops when pore water pressures begin to dissipate in macropores (e.g., Adams [*82*] referenced by Berry and Poskitt [*81*]). DeJong [*83*] developed a mathematical solution for the problem, and Berry and Poskitt [*81*] applied a simplified version to consolidation of fibrous peat.

In principle, two variables might be used to define k, e.g., $k = 10^{**}(e - e_o)/A$, two variables each for the primary and secondary springs, e.g., $\epsilon = \epsilon_o + S \log \sigma'$, and two variables for the dashpot (Eqs 65 and 66); it should then be possible to define the nine constants by simultaneous solution of the governing equations at nine judiciously selected points on the S-t curve. No such solution appears to be available. Statistical techniques might be used [*84*], but it is not apparent that solutions have been developed which are unique.

It appears that further efforts to develop systematic methods for determination of nonlinear consolidation properties of soils from laboratory tests are warranted.

For most mechanisms proposed to explain secondary effects, one would expect a more noticeable secondary effect in the laboratory than in the field. The logical inference is that total settlements in the field are likely to be predicted most accurately if the laboratory stress-strain curve (e-σ') is defined using as long a consolidation time, per loading increment, as is practical, say 24 h, not at the end of laboratory primary consolidation.

Laboratory values of c_v (and k) are likely to be too low because retarding secondary effects are likely to be much more important in the laboratory than in the field due to the higher strain rate in the laboratory [*12,44,78,80*]. Thus it would be preferable (but perhaps uneconomical) to use values of hydraulic conductivity measured directly in the laboratory and to calculate c_v using Eq 13. It would be better yet to use the hydraulic conductivity measured in the field [*50, 85–87*].

It may be noted that laboratory experimental evidence exists to show that the S-log t curves for thick samples have shapes more nearly approximating those predicted using Terzaghi's theory than do curves from identical soil tested in thin samples [*43,88*].

Based on the foregoing discussion, a conflict develops when incremental loading laboratory tests are used. Use of rapid loading (apply next load as soon as S_{100} is reached under the present load) leads to a much reduced testing time and to a σ'-ϵ relationship that more nearly fits Terzaghi's theory. Use of long times per load may lead to more accurate σ'-ϵ relationships and probably to more correct values of c_v if k is measured and Eq 13 used.

Effects of Partial Saturation

Almost all interest in consolidation testing has involved essentially saturated soils and only a small amount of work has been published with partially saturated soils. Thus only a limited discussion of partially saturated soils is apropos.

It is convenient to divide partially saturated soils into soils with interconnected air voids and ones with occluded air bubbles.

Soils with Interconnected Air Voids

When the air voids are interconnected, the soil has a measureable air pressure, u_a, and a water potential that is measured as the water pressure in the measuring system [89], u_w. In such a system it is not correct to use Terzaghi's [8] effective stress equation [90] and, indeed, Terzaghi [8] was careful to restrict his equation to a two-phase system. Scalar substitutes have been suggested [91–93], but they fail to describe even one-dimensional compressibility adequately [94] and one is forced to define the compressibility not as a simple stress-strain curve (e-log σ') but rather as a surface in space where the three axes could be void ratio, capillary stress ($u_a - u_w$), and normal stress ($\sigma - u_a$), as suggested by Bishop and Blight [95] and Matyas and Radhakrishna [96]. The volumetric strain, v, must then be obtained from an equation of the form [97]:

$$v = C_1 \Delta(\sigma - u_a) + C_2 \Delta (u_a - u_w) \qquad (67)$$

where C_1 and C_2 are compressibilities which are each functions of both ($\sigma' - u_a$) and ($u_a - u_w$). Performance of these tests requires independent control of applied total stresses, and the pore air and pore water pressures. Such apparatus exists [96,98], but a discussion of apparatus and testing procedures is deemed beyond the scope of this paper.

For immediate use in design, the one-dimensional compression characteristics of partially saturated soils with interconnected air voids can be expressed in terms of the total applied stress, with the pore air pressure zero (atmospheric) and the pore water pressure negative but unknown. The soil sample is sealed sufficiently to eliminate evaporation but it is not inundated. The upper porous stone is dry to allow pore air to escape. The void ratio is then simply a function of applied total stress.

Application of an increment of pressure to such a system leads to very rapid initial compression [99], provided the air voids remain interconnected, followed by a creep that resembles secondary consolidation. A rheologic model has been proposed for this phase [100].

For such cases, consolidation in the field is likely to be so rapid that consolidation times need not be predicted, and the sole function of the laboratory test is to predict compressibilities. Compressibilities are often low because of the high densities of these soils and the stabilizing effects of the negative water pressures.

Soaking of such soils after they have been loaded can lead to substantial and sudden settlements which may be quite localized and thus cause damage beyond what would be expected for such a movement. If soaking under load is anticipated in the field, the soil sample should be loaded to the appropriate stress in the laboratory and soaked, with the resulting strains measured. If a range of stresses, at soaking time, is likely to exist in the field, the soil should be soaked at the lowest pressure, and then loaded under successive pressures. The ultimate settlement under successive loads is essentially the same as if the soil were loaded to the appropriate pressure and soaked [94]. Once the sample is soaked, the air voids become occluded.

Soils with Occluded Air Bubbles

Soils with occluded air bubbles are inherently in a state of disequilibrium, thus making it more difficult to develop rational solutions for problems. For spherical bubbles, the air pressure exceeds the adjacent water pressure by an amount Δu:

$$\Delta u = 2T_s/R \tag{68}$$

where T_s is water/air surface tension (units of energy per unit surface area) and R is the radius of curvature. The solubility of air in water is given, approximately, by Henry's law [101]:

$$C = hu_a \tag{69}$$

where C is the gas concentration (weight of gas per unit volume of water), h is one form of Henry's constant, and u_a is the air pressure. If there are two bubbles of different radii, the air pressure will be higher in the smaller bubble (Eq 68), leading to a higher concentration of dissolved gas near the small bubble (Eq 69) and thus to steady diffusion of gas from the small bubble to the larger one. The biggest bubble in the vicinity is the atmosphere so all occluded bubbles should diffuse into the atmosphere, being replaced by pore water if the soil is inundated. If not inundated, continuous air voids should form. As a result, soils with occluded bubbles are not in equilibrium. Of course, the rate of gas diffusion in pure water is quite low [102] and is apparently even lower in soils [103]. Thus a pseudo-equilibrium may exist for the time involved in laboratory tests, even with air bubbles of different diameter and thus different pressures.

If a sample with occluded bubbles is loaded and inundated in the laboratory, the pore water pressure is the hydrostatic value and the gas pressures are above atmospheric. If the gas bubbles are spherical, the effects of surface tension around each bubble precisely counterbalance the effects of the increased gas pressures and Terzaghi's [8] effective stress equation is again valid, but u becomes the pore water pressure, u_w. The e-log σ' curve is then determined in the same manner as for saturated soils.

However, time rates of compression will not follow Terzaghi's theory. Application of an increment of total stress leads to an immediate compression of gas bubbles, to further time-dependent creep of the soil mass into air void space (time dependent because of a viscous resistance of the soil to mass movement), a further time-dependent compression due to gradual dissolution of gas bubbles, to time-dependent dissipation of excess pore water pressures (with consequent re-expansion of gas bubbles), and to secondary consolidation.

In developing a "solution" for this case, it is necessary to make a number of assumptions, some of which are clearly quite approximate. We assume that the bubbles of air are uniformly distributed through the soil mass so we can consider differential elements. Because an inundated partially saturated soil is inherently in a state of disequilibrium, it is impossible to define bubble sizes except arbitrarily. To simplify the analysis we assume that the bubble air pressures are equal to each other and to the pore water pressure and that air bubbles dissolve or precipitate instantly when the air pressure changes.

The analysis involves two stages, the first being the (assumed) instantaneous compression when the stress is applied, the second being the time-dependent dissipation of pore pressure. Space permits only an abbreviated derivation and presentation of results.

It is convenient to put Henry's law in the form [101]:

$$V_{ad} = hV_w \qquad (70)$$

where V_{ad} is the volume of air dissolved in a volume of water V_w, and h is a modified Henry's constant with a numerical value that varies with temperature and type of gas [50,101]. The volume of air is the volume that the dissolved air would occupy if it were a free gas at a pressure equal to that in the air bubbles. Compression of a sample, that has a volume of water, V_w, and initial volume of free air, V_{afi}, through a volume change ΔV (positive for compression), then increases the initial pore air (and water) pressure from p to $p + u$ where [101, p. 43]:

$$u = p \, \Delta V / (hV_w + V_{afi} - \Delta V) \qquad (71)$$

This equation can be rewritten in more convenient terms using Δv for volumetric strain ($\Delta V / V_o$), S_{ri} for initial degree of saturation (ratio), w for water content (ratio), and G_s for specific gravity of solids:

$$u = p \, \Delta v / (D - \Delta v) \qquad (72)$$

where

$$D = \left(h + \frac{1 - S_{ri}}{S_{ri}} \right) \left[\frac{1}{1 + \dfrac{1 - S_{ri}}{S_{ri}} + \dfrac{1}{wG_s}} \right] \tag{73}$$

For the case of occluded air bubbles we assume the validity of Eq 72 and find:

$$m_v \Delta\sigma'^2 + (D - m_v\Delta\sigma + p\, m_v)\, \Delta\sigma' + D\Delta\sigma = 0 \tag{74}$$

Equation 74 is solved for $\Delta\sigma'$ and the initial compressive strain, ϵ_0, is then:

$$\epsilon_0 = m_v\Delta\sigma' \tag{75}$$

Solution of Eq 75 indicates that fairly large initial compressions are expected. For example, consider the case of a clay with a recompression ratio, R_r ($= d\epsilon/d \log \sigma'$) of 0.03, initial void ratio of 1.0, and G_s of 2.7, loaded from 69 to 138 kPa (10 to 20 psi). The initial compression, expressed as a percentage of the ultimate compression, varies with the initial degree of saturation as follows:

$$S_{ri}(\%) \quad 50 \ 60 \ 70 \ 80 \ 90 \ 95 \ 99$$

$$\epsilon_0/\epsilon_u(\%) \quad 92 \ 90 \ 87 \ 81 \ 70 \ 60 \ 43$$

The reason for the large initial compressions is the small amount of pore pressure generated (Eq 71). If the compressibility of the soil increases, ϵ_0/ϵ_u decreases. For example, if $R = 0.3$ and $S_r = 95\%$, $\epsilon_0/\epsilon_u = 27\%$.

Gould [50, pp. 70–71] derived a differential equation of one-dimensional consolidation assuming that the volume of dissolved gas remains constant during consolidation and that no free gas escapes. If, instead, we assume that dissolved gas escapes with the water and that a proportionate fraction of free gas also is squeezed out, then the flow rate, q (volume/time), is:

$$q = -kiA \left(1 + h\frac{\gamma_a}{\gamma_w} + \frac{1 - S_{ri}}{S_{ri}} \right) \tag{76}$$

In deriving Eq 76, dissolved air was assumed to have the same density as water. The usual derivation for Terzaghi's differential equation of one-dimensional consolidation can then be followed to yield:

$$\frac{\partial \bar{u}}{\partial t} = \frac{1}{m_v\gamma_w} \frac{\partial}{\partial z} \left[k \left(1 + h\frac{\gamma_a}{\gamma_w} + \frac{1 - S_r}{S_r} \right) \frac{\partial \bar{u}}{\partial z} \right] \tag{77}$$

In Eq 77, $h = 0.02$ for air, γ_a is about 1.2 g/L, and $\gamma_w = 1$ g/cm^3, so $h(\gamma_a/\gamma_w) << 1$ and can be ignored. If one makes the simplifying assumption that k and S_r are independent of z, Eq 77 reduces to Terzaghi's differential equation:

$$\frac{\partial \overline{u}}{\partial t} = c_v \frac{\partial^2 \overline{u}}{\partial z^2} \tag{78}$$

where

$$c_v = \frac{k(1 - S_r)}{m_v \gamma_w S_r} \tag{79}$$

and Terzaghi's usual solution and standard fitting methods may be applied. Of course, k of a partially saturated soil is less than k of a saturated soil, and $(1 - S_r)/S_r$ is less than one so the c_v value is expected to be lower than for a saturated equivalent soil.

It may be noted that Eq 78 can easily be solved using finite difference methods if k and S_r are expressed as functions of $\overline{u}$.

More sophisticated approaches to this problem were used by Barden [100,104], Fredlund and Hasan [105], and Fredlund [106].

Use of Backpressure to Saturate Specimens

Lowe et al [107] pointed out that stress relief may cause formation of gas bubbles in samples that were saturated in the field but contained a high percentage of dissolved gas. To return the samples to a saturated condition they advocated testing samples using elevated pore water pressures (i.e., with "backpressuring"). Backpressuring is also needed when excess pore water pressures are to be measured [108].

Saturating a sample may be accomplished by compressing a sealed sample until it is saturated or by pumping in water at constant volume. To saturate a sealed sample, we assume the validity of Terzaghi's [8] effective stress equation, calculate the needed $\Delta \overline{u}$ with Eq 71 ($\Delta V = V_{afl}$), and the associated $\Delta \sigma$ using:

$$\Delta \sigma = \frac{\Delta V}{V_0} / m_v \tag{80}$$

where V_0 is the original total volume, and find [101,109,110]:

$$\Delta \sigma = \frac{p}{h} \frac{1 - S_r}{S_r} + \frac{(1 - S_r)e_0}{a_v} \tag{81}$$

where e_0 is the original void ratio and a_v the compressibility ($-de/d\sigma'$).

To saturate a sample at constant volume, with water that was originally sat-

TABLE 5—*Pressures (kPa) needed to saturate samples with an initial void ratio of one.*

Degree of Saturation, %	Sealed Sample Values of $\Delta\sigma$ for a_v (kPa^{-1})			Values of $\Delta u = \Delta\sigma$ Constant Volume
	1.5×10^{-5}	1.5×10^{-4}	1.5×10^{-3}	
100	0	0	0	0
99	15.5	2.5	1.5	1.0
98	31.2	5.1	2.9	2.2
95	78.3	12.6	7.2	5.4
90	156.7	26.1	14.5	10.7
80	. . .	55.1	29.0	21.3
70	. . .	88.5	43.5	31.9
60	. . .	129.1	58.0	42.6

urated with air at one atmosphere of pressure, we need only calculate the pressure needed to dissolve all the original and added air in a quantity of water equal to the original void volume of the sample. The required pressure is:

$$u = \frac{p}{h}(1 - S_r) \tag{82}$$

Values of u from Eq 82 are also given in Table 5 ($h = 0.02$).

There are a number of questionable assumptions in the derivation of these equations, and the numerical values should be considered approximate. Nevertheless, they tell us that it is more efficient (less pressure required) to saturate by pumping in water at constant volume rather than by compressing a sealed sample, and that backpressures in the range of 276 to 690 kPa (40 to 100 psi) should be contemplated when the samples are initially almost saturated.

In the foregoing analyses, h was assigned a value of 0.02, which is appropriate for air at 20°C. Gould [50, p. 64] points out that h is about 20% lower for methane (more soluble) and 167 times lower for hydrogen sulfide so slightly lower backpressures may be required in organic soils.

Finally, Lee and Black [102] point out that dissolving air bubbles requires time. For example, for air bubbles in plastic tubes, it took 1000 min to dissolve a bubble 1.3 mm (0.05 in.) in diameter and about a month to dissolve a bubble 3.3 mm (0.13 in.) in diameter. Bubbles within a clay are presumably smaller, but diffusion rates are lowered by the presence of solid particles [103] and time is required to pump water into the sample.

Continuous Loading Tests

The incremental loading procedure generally leads to tests of one to two week duration, a period that is excessive in many practical cases. Further, points on the stress-strain curve are widely spaced and the point of sharp curvature (at σ'_c) for sensitive clays may not be defined with the desired accuracy. Use of a load-increment ratio of one half [111,112] reduces uncertainty in σ'_c by two but multiplies the testing time by two also. Finally, the incremental loading test does not lend itself to automatic data reduction. All these disadvantages can be reduced or eliminated with continuous loading tests.

Constant Rate of Deformation Tests

Continuous loading tests appear to have begun as an effort to gain resolution in defining the stress-strain curves of sensitive clays [113,114]. Tests were performed at constant rate of deformation (usually abbreviated to CRS for "constant rate of strain") because of the availability of loading presses used for triaxial shear testing. The rates of deformation used were low enough to yield peak excess pore water pressures less than about 8% of the applied total stress. The average effective stress, σ', was calculated using:

$$\sigma' = \sigma - \alpha \bar{u}_b \tag{83}$$

where $\bar{u}_b$ was the pore water pressure measured at the base and α was taken as 0.5.

Wahls and de Godoy [115] reported tests in which much higher rates of strain were used; thus higher values of $\bar{u}_b$ were measured and there was greater need for an accurate value of α. They note that in an incremental loading test α is 1.0 at the beginning but beyond about a time factor of 0.2, α has reached a steady state value of about 0.63. For the case of a stress increasing linearly with time they estimated that α began at about ⅔ and gradually dropped to a value near 0.64.

Wissa et al [108] developed both a sophisticated consolidation cell with provisions for backpressuring and a comprehensive theory. For a soil with a linear $\epsilon - \sigma'$ curve, they found that the strain (ϵ) at any dimensionless depth ($X = z/H$) and time ($T_v = c_v t/H$) was:

$$\epsilon(X,T_v) = rt[1 + F(X,T_v)] \tag{84}$$

where r is the deformation rate:

$$r = d\delta/dt \tag{85}$$

and

$$F(X_1,T_v) = (1/6T_v)(3X^2 - 6X + 2)$$

$$- (2/T_v) \sum_{n=1}^{\infty} (1 - n^2\pi^2)\cos(n\pi X)\exp(-n^2\pi^2 T_v) \tag{86}$$

The transient term disappears for $T_v \geq 0.5$. Thereafter, if the stress-strain curve is linear, the mean effective stress (σ') is:

$$\sigma' = \sigma - 2/3u_b \qquad (87)$$

where σ is the applied total stress and u_b is the pore water pressure at the impervious base. Further:

$$c_v = \frac{H^2}{2\bar{u}_b}\left(\frac{d\sigma}{dt}\right) \qquad (88)$$

where H is the current thickness of the sample, $\bar{u}_b$ is the excess pore water pressure at the impervious base, and $d\sigma/dt$ is the time rate of increase of total stress, and

$$k = 1/2\dot{\delta}H^2\gamma_w/\bar{u}_b \qquad (89)$$

where $\dot{\delta}$ is the deformation rate ($d\delta/dt$). The strain at any time is:

$$\epsilon = \dot{\delta}t \qquad (90)$$

or, the void ratio, e, is:

$$e = e_0 - \dot{\delta}t/(1 + e_0) \qquad (91)$$

where e_0 is the initial void ratio. Revised forms of Eqs 87 to 89 were provided for a soil where the ϵ-log σ' curve is assumed to be linear. The ratio of c_v for linear e-σ' to c_v for linear e-log σ' increased as u_b/σ'_0 increased (where σ'_0 is here used to denote the effective stress at the top of the sample). They found that:

$$\frac{c_v(\epsilon - \sigma')}{c_v(\epsilon - \log \sigma')} = -\frac{\log[1 - (\bar{u}_b/\sigma'_0)]}{0.434(\bar{u}_b/\sigma'_0)} \qquad (92)$$

For $\bar{u}_b/\sigma'_0$ less than 0.3, the ratio of c_v values was less than 1.2.

For the early stages of loading, where the transient terms in Eq 86 are significant, if the ϵ-σ' curve is linear, calculate F_3 from:

$$F_3 = \frac{\sigma'(H,t) - \sigma'(H,0)}{\sigma'(0,t) - \sigma'(0,0)} \qquad (93)$$

and find T_v from:

$$F_3 = \frac{6T_V + 1 - 12\Sigma(1/n^2\pi^2)\cos(n\pi)\exp(-n^2\pi^2T_v)}{3T_v + 1 - 6\Sigma(1/n^2\pi^2)\exp(-n^2\pi^2T_v)} \qquad (94)$$

A curve of F_3 versus T_v is prepared so that T can be read off and used in Eq 15 to obtain c_v. During the transient period, ϵ and e are still given by Eqs 90 and 91, respectively, and α in Eq 83 remains near ⅔ [115].

A finite strain solution, developed by Lee [116], indicates a need to restrict strain rates to obtain reasonable values of c_v.

Constant Gradient Tests

An alternative test method is to apply an initial incremental load to generate a pore pressure $\bar{u}_o$, allow dissipation until $\bar{u}$ at the base drops to a preselected value $\bar{u}_b$, and then load at a rate such that $\bar{u}_b$ is maintained constant. For a homogeneous soil, $\bar{u}$ at any depth is then independent of time, and Terzaghi's partial differential equation reduces to an ordinary differential equation. Lowe et al [117] developed apparatus for the test and solved the equations. Once the transient phase had passed (time factors beyond 0.08 if $\bar{u}_b = \bar{u}_o$), they derived equations identical to Eqs 87 (for σ'), 88 (for c_v), 90 (for ϵ), and 91 (for e).

Constant Rate of Loading Test

Viggiani [118] derived a solution for the case of a test in which the applied stress increases linearly with time. Again, a transient term disappears after a certain time factor, which is about two in this case, and Eqs 87 to 91 are again valid. Aboshi et al [119] derived a somewhat different solution.

General Case

Janbu et al [120] sought a general solution for any continuous loading test. Their solutions were:

$$M = \alpha_m \dot{q} H / \dot{\delta} \tag{95}$$

$$m_v = 1/M \tag{96}$$

$$k = \alpha_k \gamma_w H \, \dot{\delta} / 2\bar{u}_b \tag{97}$$

$$c_v = \alpha_c \dot{q} H^2 / 2\bar{u}_b \tag{98}$$

where H was the original thickness of the sample (single drainage), q is applied stress, δ is sample compression, γ_w is the unit weight of water, $\bar{u}_b$ is the excess pore water pressure at the impervious base, the super-dot denotes a time derivative (e.g., $\dot{q} = dq/dt$), and α_m, α_k, and α_c are dimensionless coefficients. "Exact" solutions for these coefficients are:

$$\alpha_m = (1/a)\tanh(a) \tag{99}$$

$$\alpha_k = 2[\cosh(a) - 1]/a\sinh(a) \tag{100}$$

$$\alpha_c = 2[\cosh(a) - 1]/a^2\cosh(a) \tag{101}$$

where:

$$a = \cosh^{-1}[1/(1 - \lambda)] \tag{102}$$

and:

$$\lambda = \ddot{u}_b/\dot{q} \tag{103}$$

Approximate solutions were also provided. Janbu et al [119] do not discuss transient stages but their derivation contains an assumption of constant λ (p. 653) so their equations must not apply under transient conditions. Presumably, ϵ, e, and σ' are calculated using Eqs 90, 91, and 87, respectively.

General Comments on Continuous Loading Tests

Continuous loading tests offer the advantages of more rapid testing, more-or-less continuous definition of properties (e, ϵ, k, c_v, m_v, . . .) as functions of effective stress, and automated collection and reduction of data. In the case of laboratories occupying expensive floor space, the more rapid testing may also lead to a reduction in required floor space. In addition, the strain rates during most of the primary phase of consolidation are less than for incremental loading tests and closer to, but greater than, loading rates encountered in the field. The prices paid for these advantages include much increased capital equipment costs, increased maintenance costs, and possibly a need for better trained technicians. Further, because excess pore water pressures are usually measured, the samples need to be backpressured (the time needed to backpressure samples and ensure saturation of the measuring system do not appear to be included in the low testing times usually cited by advocates of this type of testing). Further, secondary consolidation aspects are masked unless loads are periodically held constant, in which case the testing time is increased.

A question exists as to how properties measured in continuous loading tests differ from properties measured in incremental tests. The results are mixed. Janbu et al [120] reported that the maximum previous consolidation pressure, σ'_c, for most of their tests with Norwegian clays was only slightly increased when high loading rates were used. For clays of medium-to-low plasticity they advocated strain rates of 5 to 10%/h and loading times as low as 1 h. For their highly plastic Eberg clay they found that σ'_c increased from 55 kPa (1100 psf) to 90 kPa (1900 psf) when λ was increased from 0 to 0.6.

Other authors have also found e-log σ' curves to be similar for standard and continuous loading tests but with a tendency for σ' to be somewhat higher for the fastest tests [108,115,117,121].

Gorman et al [122] found a general tendency for σ'_c to be larger for continuous loading tests than for incremental tests. For eleven sets of samples from three sites in Kentucky, the ratio of σ'_c for CRS and CG tests, to σ'_c for incremental loading tests, was 1.35 but the scatter was large with the above ratio ranging from 0.4 to 3.1. Only one CRS or CG test was performed on soil from each tube, but duplicate standard tests were performed and the values of σ'_c differed by up to about two times between standard tests. In some cases the break point was too indistinct to define a useful σ'_c.

On the other hand, Crawford [114] found large effects of loading rate on Leda clay and Leroueil, Samson, and Bozozuk [111], and Leroueil, Tavenas, Samson, and Morin [112] found that σ'_c could be correlated directly with strain rate and values of σ'_c were typically about 30% higher in continuous loading tests compared with standard tests.

It seems clear that σ' at any given void ratio increases when strain rates increase, and the fractional increase may increase as plasticity increases.

The available data on the coefficient of consolidation are less clear. Comparisons with incremental tests are often uncertain because of lack of information on whether the log t or $\sqrt{t}$ method was used for data reduction. The existing comparisons are mixed. Thus, Smith and Wahls [121] found $c_v(\text{CRS})/c_v(\text{STD})$ to be about 1.0 to 2.0 for remolded Ca-montmorillonite and 1.0 to 1.5 for undisturbed Massena (NY) clay; Gorman et al [122] found the same ratio to be about ⅓ for their Site 1, about ½ for the normally consolidated range of Site 2. Lowe et al [117] found the constant gradient c_v to be between the incremental values for the log t and $\sqrt{t}$ methods. Gorman et al [122] found $c_v(\text{CRS})$ and $c_v(\text{CG})$ to be more or less equal for one site, $c_v(\text{CRS})/c_v(\text{CG}) = 1.5$ to 2.0 for another, and the c_v-log σ curves to cross over each other in two other cases (the CRS-c_v was lower in the normally consolidated range in both cases). Wissa et al [108] found c_v to increase as strain rate increased for remolded Boston Blue clay, but Smith and Wahl's [121] data indicate that c_v increased with increased strain rate for remolded calcium-montmorillonite, decreased with increased ϵ for remolded kaolinite, and seemed not significantly affected for undisturbed Massena clay.

Part of the problem is that reproducible values of c_v cannot be obtained in the reloading range for undisturbed samples, especially when the log t fitting method is used. Some of the CRS data may be defective because the transient phase was still in progress. Some of the incremental data may have been perturbed by effects discussed earlier (e.g., formation of less permeable soil zone near a drainage boundary). Gorman et al's [122] data indicate that differences between data from duplicate incremental tests on undisturbed samples were at least as large as the differences between them and data from CRS and CG tests. The c_v values are, of course, perturbed by the differences in the e-log σ' relationships (alters a_v and e in the c_v equation; different e also alters k).

Taken collectively, the available data suggest that c_v values from continuous loading tests are as reliable as those from incremental tests.

Effects of Sample Size

There are at least two aspects to the question of the optimum sample size. The first relates to size in terms of nonhomogeneities in the natural soil (for example, stratification and fissuring). The second relates to effects of sampling disturbance.

Experience shows that natural soils are generally stratified, and properties may vary substantially in short distances vertically. Many soils contain fissures and the fissures may be filled with a different soil. Soil properties vary across sites as well. If soils are quite nonhomogeneous, it may be pointless to do any consolidation testing and design may have to accommodate substantial uncertainties in soil behavior. If the soil is relatively homogeneous on a "large scale" but heterogeneous on a "small scale", it may be desirable to perform laboratory tests using "large scale" samples. The size of the samples then depends on the scale of the inhomogeneities.

For the case of purely vertical flow in an idealized soil composed of layers of sand and clay, with the flow perpendicular to the stratification, the flow rate is governed by the clay layers because there would be a negligible gradient across the sand layers. If the compressibility of the clay greatly exceeds that of the sand, as would be expected, then only the clay layers are relevant. Laboratory tests should be performed on the clay layers alone, and field analyses should use a thickness equal to the sum of the thickness of the clay layers. It is unlikely that such a case will be encountered in engineering practice.

If drainage is horizontal and the soil is composed of alternating layers of more pervious and less pervious soil, and the horizontal distance to a drainage layer (perhaps a drain well) is large compared with the thickness of layers, it seems apparent that water will drain vertically into the more pervious layers and then radially to the drain. The time rate of consolidation will then be controlled by the radial drainage characteristics of the more pervious layers. This problem was analyzed by Horne [123].

Rowe [124] applied Horne's equations to the problem of laboratory testing and found that the measured value of c_h in the laboratory, calculated assuming a homogeneous sample, increases as the sample size increases and finally levels out at a value that would be more applicable in the field than would be the value obtained with a small sample. Equations were developed for uniform interlayering of individually homogeneous layers, but even in this case it was difficult to develop any simple rules for deciding on sample size because of the number of variables involved and the difficulty in evaluating them for real samples. Rowe [124, p. 338] recommended performing tests on samples of increasingly large diameter and using the largest values of c_h in design.

Rowe [125] has published laboratory data showing values for the ratio of c_v for 250-mm-diameter by 90-mm-thick samples to the c_v for "conventional 76-mm-diameter" samples of about 2 to 100, the ratio tending to decrease as the consolidation pressure increased. Burghignoli and Calabresi [126] measured values of c_v of 0.5 to 5 × 10^{-4} cm/s for 2-cm-thick samples and 0.8 to 1.2 × 10^{-4} cm/s for 25-cm-thick samples.

It may be noted that Rowe developed a special hydraulic consolidation cell for such samples [55,127], with sample diameters of 8 to 50 cm (3 to 20 in.). The samples can be drained vertically with single or double drainage, can be drained radially to a central drain or an outer drainage boundary, can be loaded through a rigid cap or a flexible membrane, and can be backpressured. Although Rowe and Barden [55] indicate average frictional losses in both the sample and the convoluted loading diaphragm, Shields [128] found that loads applied by the convoluted diaphragm were only 60 to 80% of the expected values for the 7.6-cm (3-in.) cells and 80 to 90% for the 15.2-cm (6-in.) cell, errors he attributed largely to a diaphragm that was so stiff that it did not cover the whole surface. It appears that the errors must actually be due to membrane/wall friction. We have used 7.6 and 15.2 cm (3 and 6 in.) Rowe cells and found that volume change measurements were seriously perturbed because of the need to expel water from between the membrane and the wall and that the procedure for draining this area recommended by Rowe and Barden [55] does not completely solve the problem. However, the alternative of a mechanical loading system is not favored because of the large forces involved and other systems are more expensive.

At the other extreme, Karol [129] recommended speeding up tests by reducing sample size and used remolded samples with thicknesses of 12.7 to 25.4 mm (0.5 to 1.0 in.) to support his view. However, van Zelst [130] showed that standard techniques of sample preparation led to thicknesses of remolded soil on each face of 1.8 to 2.8 mm (0.07 to 0.11 in.), thus indicating that use of thin specimens could lead to large degrees of disturbance [131]. Evidence of disturbance during face trimming was also provided by Chan and Kenney [132] based on permeability testing, but they were probably more careful than usual and found an equivalent thickness of disturbed soil of about 0.64 mm (0.025 in.) per face.

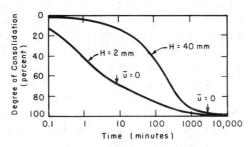

FIG. 14—*Influence of the drainage distance on the shape of the time-settlement curve and the apparent amount of secondary compression* [133].

Finally, it should be noted that the relative effects of secondary consolidation depend on the drainage distance so secondary effects are more noticeable for thin specimens than for thick ones. Curves in Fig. 14 [*133*] demonstrate this point well.

Disturbance

If laboratory consolidation tests are to be useful they must be performed using undisturbed samples, or it must be possible to reconstruct properties of undisturbed soil, or there must be no effects of disturbance. It seems clear that the properties of almost all natural soils are altered due to disturbance, that testing truly undisturbed samples is more a goal than a reality, and that methods of reconstructing properties leave much room for improvement. The topic is thus one of importance. The literature on the topic is vast, however, and the topic itself could be considered beyond the scope of a paper on laboratory testing, so the goal here will be only to emphasize certain important points.

The fact that disturbance alters soil properties has undoubtedly been recognized for thousands of years, but its effects on soil testing were largely ignored until this century. Terzaghi [*134*, p. 1196] stated that ". . . the Swedish Commission initiated the development of the technique for securing undisturbed samples", but Enkeboll [7] noted that Daniel E. Moran was performing one-dimensional consolidation tests in 1910 and used the results in design. Samples were reportedly taken in thin-walled brass tubes, thus minimizing disturbance, and tested in the same tubes, which were used as floating rings during testing.

The effect of sampling disturbance on the e-σ' curve may have first been shown by Casagrande [*135*] and efforts to reconstruct undisturbed e-σ' relationships were discussed by Casagrande [*136*], Rutledge [*137*], and Terzaghi [*138*] as well as many others during that same period. More recently, Schmertmann [*139*] presented a widely used method which is similar to the one proposed by Terzaghi [*138*]. The effect of disturbance is well known to be to round off the e-log σ' curve and reduce void ratios at any given effective stress. Representative data are shown in Fig. 15 for a soil from Green Bay, Wisconsin. The samples were taken using pushed, thin-walled tubes with fixed pistons.

The effects of disturbance on the e-log σ' relationship are often corrected by using some sort of construction (e.g., Schmertmann's [*139*] as in Peck et al [*140*, Fig. 3.6]). Such methods tend to work adequately for the cases for which they were developed, typically simple overconsolidation by erosion. They break down in various cases (e.g., for soils that are overconsolidated by cementation), in which case the reloading curve is quite flat and the reconstructed field curve is below the laboratory curve. The author's opinion is that the reconstructed field curve should simply be drawn in by eye with the obvious requirements that the field curve must pass through the point representing the field void ratio (e_o) and effective stress (σ'_o), lie above the laboratory curve, and tend to be parallel to, or merge with, the laboratory curve at high pressures. In the case of some

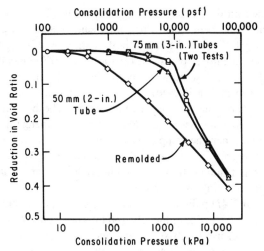

FIG. 15—*Consolidation (stress-strain) curves for samples taken using thin-walled, fixed piston, samplers from a site in Green Bay, Wisconsin.*

cemented soils, the cement keeps the soil from swelling in the field. Sampling disturbance ruptures the cement and causes the laboratory e-σ' curve to be above the field curve. In cases where accurate prediction of settlements is critical, the best possible samples should be used.

For some cemented soils even perfect sampling may cause yielding and rupture of the cement because of the change in state of total stress from a K_0-state to one with an isotropic state of stress [141]. However, Noorany and Poormand [142] consolidated a sample of San Francisco Bay mud out onto the virgin curve under K_0 conditions, removed all stresses to simulate perfect sampling, trimmed a sample into a consolidation ring, allowed it to swell under a low pressure, and performed a standard one-dimensional consolidation test. The one-dimensional consolidation test gave a virgin curve that was a precise extension of the previous K_0 virgin curve, and thus provided evidence that accurate e-log σ' curves can be obtained if samples of high enough quality are obtained (and no problems of gas bubble formation and expansion occur during unloading).

Most of the available data on sampling disturbance are in terms of undrained shearing strengths (none of that literature will be reviewed here). Effects of disturbance on the e-log σ' curve are difficult to express easily. For sensitive soils, numerical data may be the pressure at which the e-log σ' curve suddenly steepens (σ'_c) and an average slope of the reloading curve; e.g., R_r ($d\epsilon/d \log \sigma'$) or m_v ($d\epsilon/d\sigma'$). For insensitive soils it is really necessary to compare e-log σ' curves directly because the rounded nature of the curves [143] prevents useful definition of either σ_c or R_r.

In North America, more research has been published on sampling disturbance of Laurentian clays in Canada than for any other clay [110,141,144–148]. Taken collectively, the data indicate an improvement in sample quality as the sampler diameter increased in the range from 5.1 to 12.7 cm (2 to 5 in.). Fixed piston samplers appear to be better than open tube samplers (no direct data). Hand carved block samples appeared to be of best quality. Within a given sample, in the sampling tube, the upper part (say, the top 30 to 40% typically) is disturbed due to stress relief at the bottom of the borehole, due to shearing strains beneath the fixed piston (when one is used), due to a greater distance of movement in the tube, and perhaps due to moisture imbibation. The bottom of the sample (say, to a height above the base equal to the sample diameter) may be disturbed because of suction when the sampler is withdrawn. The outer radial edge is of course disturbed by shear along the inside of the sampler. The best consolidation specimens are thus trimmed from the middle or lower part of the tube sample and require edge trimming as well.

LaRochelle et al [141] point out that sensitive soils develop only minimal drag between the soil and the inside of the tube and thus do not need an inside clearance ratio. What is worse, use of a fixed piston and an inside clearance ratio causes development of a lateral suction and the resulting shearing strains probably increase disturbance. The same authors discuss a 200-mm (8-in.)-diameter sampler with no inside clearance ratio and an external coring tool to minimize suction during sampler withdrawal. They obtained e-log σ' curves identical to those from hand carved block samples.

Kenney [149] points out that the outer radial zone of disturbed soil is more compressible than the rest of the sample. When a rigid loading cap is used in a one-dimensional consolidation test, the vertical deformation is everywhere equal. Thus, for the case of a more compressible outer zone, the middle of the sample is forced to undergo radial mass movement, with a consequent increase in dis-

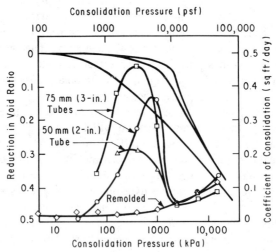

FIG. 16—*Effect of disturbance on the coefficients of consolidation for samples from Green Bay, Wisconsin.*

turbance. Kenney tested sensitive Norwegian clays by first applying an isotropic stress to compress the disturbed zone and then consolidating under one-dimensional (K_0) conditions and found an increase in σ'_c and reduction in R_r in support of his argument.

The effects of disturbance on time rates of consolidation are less well documented. However, from the definition of c_v (Eq 13) it is apparent that the disturbance-induced increase in compressibility (m_v or a_v) must cause a reduction in c_v, and a further reduction is likely to occur because of a reduction in k, due to reduction in e and possible smearing of drainage channels. As an example of such effects, the coefficients of consolidation for the samples from Green Bay, Wisconsin, are shown in Fig. 16.

A method to use in reconstructing the undisturbed values of c_v was suggested by Pelletier et al [150]. The field e-σ' curve is first reconstructed (as discussed previously) and a plot of a_v (from the field curve) versus σ' is prepared. Then, values of k are calculated from Eq 13, using the original laboratory data, and a plot of e-log k is prepared. The field values of k probably involve less secondary effects than do the laboratory values (see earlier discussion) so the field curve of e-log k is prepared with values of k greater than the laboratory values by perhaps 30 to 50% in the normally consolidated range, and with the k curve of the normally consolidated range extended smoothly into the overconsolidated range. For each value of σ', a value is selected for e (and thus k) and a_v and a value of c_v is calculated (Eq 13).

Disturbance also impacts secondary effects. Secondary shapes for the samples from Green Bay, Wisconsin, are shown in Fig. 17 as examples.

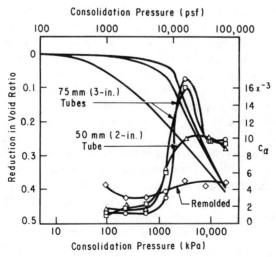

FIG. 17—*Effect of disturbance on the coefficient of secondary compression*, C_α ($=de/d(\log t)$), *for samples from Green Bay, Wisconsin.*

Effects of Storage

Storage of samples in the sampling tube often results in what appears to be a time-dependent increase in the degree of disturbance. For example, Arman and McManis [151] found that the apparent maximum previous consolidation pressure, σ'_c, of one soil decreased from 4.05 tsf just after sampling to 2.8 tsf after a year of storage, and the e-log σ' curve became more rounded. Bozozuk [144] reported about a 5% reduction in σ'_c for a block sample during a 16-month storage period, but Arman and McManis [151] and La Rochelle et al [148] report no change in σ'_c for block samples stored for periods of 1 and 3 years, respectively.

Part of the reason for time-dependent disturbance of the middle of the tube sample is probably moisture movement inwards from the disturbed outer zone but such movements should occur quickly, not extend over a year or more. In any case, La Rochelle et al [148] report a tendency for moisture to move inwards a small amount and then stop (scatter was large, however).

It seems clear that disturbance will be reduced if it is possible to extrude the samples in the field, trim away perhaps the outer 6 mm (¼ in.), wrap the rest in foil and plastic film, and store in suitably tight containers.

Comparison of Laboratory and Field Properties

Laboratory consolidation tests are generally performed to measure properties to be used in an analysis of a practical field problem. The ultimate check on the usefulness of the laboratory data is certainly a comparison between predictions and field measurements. Unfortunately, field conditions are characterized by special problems including uncertain stratigraphy, time-dependent loading, complex stress-states that involve mass movement, soil-structure interaction problems (an embankment is here considered to be "structure"), two- and three-dimensional drainage, and a host of others. Even the best documented case histories contain serious deficiencies. The comments of Leroueil and Tavenas [152] regarding the pitfalls of back analyses are in line with our own experiences. Specifically, it is possible to perform careful analyses and interpret the results honestly, and predict some field response, such as surface settlement, quite accurately, only to obtain additional data (e.g., pore pressures, lateral displacements, or changes in thicknesses of individual layers) at a later date and discover that the original good correlation resulted from counterbalancing errors.

Most published case histories are nearly useless. For example, the authors usually report a single value of c_v for a heterogeneous clay layer when it is obvious that the value of c_v varies from one part of a layer to another, and in each layer varies with effective stress. In many cases the field c_v is estimated from time rates of dissipation of pore water pressures. Information now available indicates that values of u are greatly effected by the shape of the stress strain curve. In addition, we have re-analyzed published case histories, using the published soil properties, and found wide discrepancies between our predicted field response and the response predicted by the author. Efforts to trace the sources for the discrepancies were not met with cooperation by the original authors. Comparisons of published case histories often show great conflicts which cannot be explained based on published data. Finally, space limitations make it impossible to present a general discussion of the topic. Rather than abandon the topic altogether, the results of some of our own observations will be used.

The main source of information will be an embankment over the Fore River near Portland, Maine [150,153]. The problem was a fairly typical one in that the loads were time dependent, the soil stratified, soil properties depended on effective stress, drain wells were installed after a period of one-dimensional consolidation, and there were a number of other complicating factors. Analyses were performed using a finite difference method which treated each layer as instantaneously homogeneous but allowed properties to vary with time. No account of secondary effects was possible, and no soil-structure interaction effects were included. The analysis was of the pseudo-three-dimensional type [154]. The analyses did account for time-dependent loading, stratified soil, two-dimensional flow, large strains, effective-stress-dependent soil properties, nonlinear e-log σ' curves, anisotropy, drains installed after consolidation began, partially penetrating drains, and an initial artesian condition.

Soil samples were taken using a 89-mm (3.5-in.)-diameter thin-walled fixed piston sampler. Consolidation tests were performed on samples 64 mm (2.5 in.) in diameter by 22 mm (0.87 in.) high, using floating rings, incremental loading with LIR = 1, and loads maintained for 24 h each. Curve fitting used the $\sqrt{t}$ method, and e-log σ' curves were plotted using e at the end of primary consolidation.

Soil properties were backed out of field data. It was relatively easy to pick reasonable soil properties to predict the correct surface settlements. It was considerably more difficult to match the time-dependent compressions of each layer separately (deep settlement markers were used). Pore water pressures were matched reasonably well, but the uncertainty in radial location of the piezometers made those field data suspect.

The theoretical curves of surface settlement and layer compression versus time are compared with field data in Fig. 18. The fitted field e-log σ' curves are compared with laboratory curves for four sublayers within the main compressible layer in Fig. 19. The field reloading slope is apparently flatter than the laboratory value, a finding in agreement with general experience [155]. The field values of σ'_c were slightly larger than the laboratory values. The field virgin e-log σ' curve generally became parallel to the laboratory curve. The difference between the laboratory and field e-log σ' curves are of a type and magnitude that could be explained by effects of disturbance of the laboratory samples.

The backfitted field values of c_v are compared with the laboratory values in Fig. 20. In the reloading range the field values of c_v generally exceed the laboratory values, as would be expected for partially disturbed samples (Fig. 16).

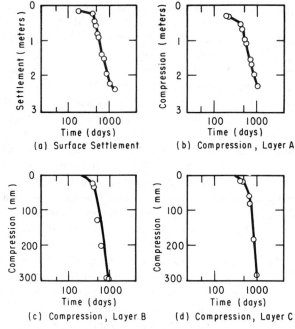

FIG. 18—*Field settlements and compressions* (solid lines) *compared with fitted theoretical points* (hollow symbols).

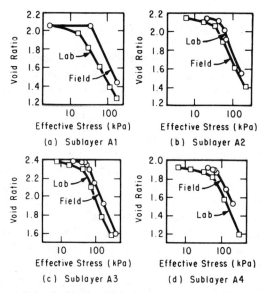

FIG. 19—*Field and laboratory stress-strain curves.*

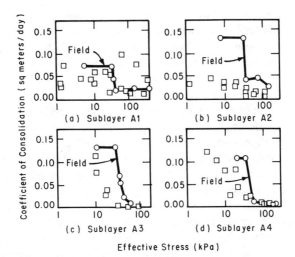

FIG. 20—*Field* (solid line) *and laboratory* (hollow boxes) *coefficients of consolidation.*

In the normally consolidated range the field and laboratory values of c_v are similar, as expected (Fig. 16).

The foregoing observations are similar to those we have made at other sites, including one involving peat [*156*]. The observations are similar to many in the literature, although some published field cases seem to indicate a predominant influence of secondary effects [*157–160*].

Summary and Conclusions

The following comments are the author's opinions on several aspects of laboratory consolidation testing. They are based on the foregoing review of the state of the art and, more particularly, on the author's experience in the laboratory and field.

Laboratory consolidation tests can provide a reliable basis for prediction of field response for saturated clays under conditions of essentially one-dimensional deformation. The major problems for most projects are more in the areas of predicting stratigraphy, construction schedules, and the consequences of mass movement than in either analysis or measurement of soil properties, although ignorance of the existing state of the art can lead to problems in these last two areas as well.

For projects involving large settlements, typically embankments on soft clay, sufficiently accurate field curves of (e or ϵ) versus σ', and c_v-σ' can be determined using samples of moderate quality (e.g., 76 mm [3-in.], open, pushed, thin-walled samplers). In the case of settlement-sensitive structures where loads must be less than σ'_c but conditions dictate that loads approach σ'_c, soil samples of high quality may be required to justify the use of shallow foundations.

The standard sample size for laboratory testing should remain a diameter on the order of 63 mm (2.5 in.) and a thickness of 19 mm (0.75 in.). Smaller samples may be used for uniform clays but should be discouraged. The increased effects of trimming disturbance and secondary effects should be considered for thin samples. Use of samples smaller than about 60 to 75 mm (2.5 to 3 in.) in diameter should be discouraged when data of good quality are desired. For soils that are fissured, contain thin stratification (radial flow), or other similar inhomogeneities, larger samples should be considered; samples with diameters in the range of 150 to 250 mm (6 to 10 in.) should be the norm for many such clays.

Disturbance seems to be minimized if an outer disturbed zone of soil is trimmed away as quickly as possible. Long-term storage in the sampling tubes has an effect similar to sampling disturbance.

The present incremental loading procedure with LIR = 1 seems adequate for almost all work. When the break at σ'_c is sharp and must be well defined, the test can be performed using a smaller LIR or a slow constant loading method.

The testing time can be reduced greatly by using a rapid loading procedure in which successive loads are applied as soon as the average degree of consolidation on the previous load passes, say, the 90% value. If small values of LIR, say 0.2, are used, it may be preferable to use rapid loading because otherwise it will be difficult to locate the primary part of the settlement-time curve.

It is faster, and just as accurate, to use strain, ϵ, in place of void ratio, e.

The author has personally developed a preference for the $\sqrt{t}$ fitting method because it gives both the zero and 100% primary points in an unambiguous fashion (a trial procedure must be used with the log t method to find the part of the curve, if any, where $t = KS^2$, and the 100% point is ambiguous in the common case that the secondary curve is nonlinear) and the fitted values of hydraulic conductivity are closer to the measured (real) values. However, either the $\sqrt{t}$ or log t method may be considered satisfactory. None of the other fitting methods can be recommended. They suffer from the disadvantages of requiring more data reduction effort and a better trained user and offer no apparent advantages. If the measured S-t response differs so far from Terzaghi's predicted response that the $\sqrt{t}$ and log t methods cannot be used, then no amount of data massaging will rectify the situation.

Porous stones should be clean and provide free drainage. For soft clays a sheet of filter paper should be used between the clay and the stone.

The error in mean effective stress, caused by ring friction, for a greased smooth steel ring and samples of typical geometry, averages about 10% in the normally consolidated range but is much larger for the rebound curves. Use of a thin tetrafluoroethylene lining produces a small reduction in friction.

The effect of nonlinear stress-strain response appears to a minor extent in the S-t curves when LIR = 1, but has a major effect on pore pressures. For the case of a σ'-ϵ curve concave towards the ϵ axis, pore pressures dissipate more rapidly than expected for linear σ'-ϵ curves.

Radial flow tests, with drainage to a central sand drain, should be used when drainage wicks are to be used in the field. The size of the sample should be selected after a preliminary examination of soil samples for inhomogeneities.

Secondary effects appear to exert much more influence on the S-t response in the laboratory than in the field. Accordingly, the field σ'-ϵ response probably corresponds to the long-term laboratory values of ϵ, say for times of a week per load. For routine loadings, at one day per load, the final recorded ϵ should thus be used in the σ'-ϵ curve (note, however, the counterbalancing, and usually more important, effect of sample disturbance). In the laboratory, if the post-primary secondary compression under one load is of the same magnitude as, or larger than, the primary compression of the next load, then settlement under the next load will appear as all secondary. Use of a series of loads, each held for a week or so, for soils such as peat, have led to the mistaken view that such soils experience no primary consolidation.

Secondary theories involving a representation of the effective stress by a Kelvin body, or perhaps a spring in series with a Kelvin body, may prove useful, but the spring(s) and dashpots should be nonlinear and useful curve fitting methods will probably involve use of a computer and statistics-based curve-fitting procedures. It remains to be shown that such theories can generally be used to improve on settlement forecasts.

Partially saturated soils with interconnected air voids can probably be tested adequately by simply compressing a sample, with provisions to minimize evaporation, with inundation only if it is to occur in the field. Consolidation prior to inundation is generally too fast to be of interest in the field. Presence of occluded air bubbles causes increased initial compression and delayed primary. Substantial amounts of backpressure and time are needed to saturate samples of partially saturated soil.

Continuous loading tests seem to be the wave of the future for busy laboratories because of the reduced testing time and the opportunity to fully automate the loading stage and the data reduction process. Evidence is accumulating that they lead to excessive values of σ' for any e, but this problem can be corrected using an appropriate soil rheologic model. The error helps to compensate for effects of sample disturbance in many cases.

References

[1] Znidarcic, D., Croce, P., Pane, V., Ko, H.-Y., Olsen, H. W., and Schiffman, R. L., *Geotechnical Testing Journal*, Vol. 7, No. 3, Sept. 1984, pp. 123–134.

[2] Terzaghi, K., *Mathematisch-naturwissenschaftliche Klasse*, Akademie der Wissenschaften in Wien, Sitzungsberichte, Part IIa, Vol. 132, No. 3/4, 1923, pp. 125–138.

[3] Telford, T., *Edinburgh Encyclopedia*, Vol. 15, 1830, p. 306.

[4] Terzaghi, K. and Frohlich, O. K., *Theorie der setzung von tonschichten*, Franz Deuticke, Leipzig, 1936.

[5] Spring, W., *Societe Geologique de Belgique*, Annales, Tome 28, 1901, pp. 117–127.

[6] Frontard, J., *Annales des Ponts et Chaussees*, 9th series, Vol. 22, 1914, pp. 173–280.

[7] Enkeboll, W., "Investigation of Consolidation and Structural Plasticity of Clay," D.Sc. thesis, Massachusetts Institute of Technology, Cambridge, Mass., 1946.

[8] Terzaghi, K. in *Proceedings,* International Conference on Soil Mechanics and Foundation Engineering, Cambridge, Vol. 1, 1936, pp. 54–56.

[9] Brumund, W. F., Jonas, E., and Ladd, C. C., "Estimation of Consolidation Settlement," Special Report 163, Transportation Research Board, 1976, pp. 4–12.

[10] Casagrande, A. and Fadum, R. E., *Transactions of ASCE,* Vol. 109, 1944, pp. 383–416.

[11] Fox, E. N. in *Proceedings,* Second International Conference on Soil Mechanics and Foundation Engineering, Vol. 1, 1948, pp. 41–42.

[12] Taylor, D. W., Serial 82, Massachusetts Institute of Technology, Cambridge, Mass., 1942.

[13] Taylor, D. W., *Fundamentals of Soil Mechanics,* Wiley, New York, 1948.

[14] Gilboy, G., *Engineering News Record,* 21, May 1936, pp. 732–734.

[15] Olson, R. E. and Daniel, D. E. in *Permeability and Groundwater Contaminant Transport, ASTM STP 746,* American Society for Testing and Materials, Philadelphia, 1981, pp. 18–64.

[16] Naylor, A. H. and Doran, I. G. in *Proceedings,* Second International Conference on Soil Mechanics and Foundation Engineering, Rotterdam, Vol. 1, 1948, pp. 34–40.

[17] Matlock, H. and Dawson, R. F. in *Symposium on Consolidation Testing of Soils, ASTM STP 126,* American Society for Testing and Materials, Philadelphia, 1951, pp. 43–52.

[18] Scott, R. F., *Journal of the Soil Mechanics and Foundations Division, Proceedings of ASCE,* Vol. 87, No. 1, 1961, pp. 29–39.

[19] Hansen, J. B. in *Proceedings,* Fifth International Conference on Soil Mechanics and Foundation Engineering, Vol. 1, 1961, pp. 133–136.

[20] Wilson, N. E., Radforth, N. W., MacFarlane, I. C., and Lo, M. B. in *Proceedings,* Sixth International Conference on Soil Mechanics and Foundation Engineering, Vol. 1, 1965, pp. 407–411.

[21] Parkin, A. K. in *Proceedings,* Tenth International Conference on Soil Mechanics and Foundation Engineering, Vol. 1, 1981, pp. 723–726.

[22] Sridharan, A. and Rao, A. S., *Geotechnical Testing Journal,* Vol. 4, No. 4, 1981, pp. 161–168.

[23] Rao, S. N. and Kodandaramoswamy, K., *Journal of the Geotechnical Engineering Division, Proceedings of ASCE,* Vol. 108, No. GT2, 1982, pp. 310–314.

[24] Leonards, G. A. and Ramiah, B. K. in *Papers on Soils—1959 Meetings, ASTM STP 254,* American Society for Testing and Materials, Philadelphia, 1960, pp. 116–130.

[25] Lewis, W. A., Department of Scientific and Industrial Research, Road Research Laboratory, Note RN/1349/WAL, 1950.

[26] Northey, R. D. in *Proceedings,* Second Australia-New Zealand Conference on Soil Mechanics, Christchurch, New Zealand, 1956, pp. 20–24.

[27] Newland, P. L. and Allely, B. H., *Geotechnique,* Vol. 10, 1960, pp. 62–74.

[28] Su, H. L., *Journal of the Soil Mechanics and Foundations Division, Proceedings of ASCE,* Vol. 84, No. SM3, Paper 1729, 1958.

[29] Baracos, A., *Geotechnique,* Vol. 26, No. 4, 1976, pp. 634–636.

[30] Gray, H. in *Proceedings,* International Conference on Soil Mechanics and Foundation Engineering, Vol. 2, 1936, pp. 138–141.

[31] Jaky, J. in *Proceedings,* Second International Conference on Soil Mechanics and Foundation Engineering, Rotterdam, Vol. 1, 1944, pp. 103–107.

[32] Burmister, D. M. in *Papers on Soils—1959 Meetings, ASTM STP 254,* American Society for Testing and Materials, Philadelphia, 1960, pp. 88–105.

[33] Leonards, G. A. and Girault, O. in *Proceedings,* Fifth International Conference on Soil Mechanics and Foundation Engineering, Vol. 1, 1961, pp. 213–218.

[34] Burland, J. B. and Roscoe, K. H., *Geotechnique,* Vol. 19, No. 3, 1969, pp. 335–356.

[35] Laguros, J. G. in *Effects of Temperature and Heat on Engineering Behavior of Soils,* Highway Research Board, Special Report 103, 1969, pp. 186–193.

[36] Mitchell, J. K. in *Effects of Temperature and Heat on Engineering Behavior of Soils,* Highway Research Board, Special Report 103, 1969, pp. 9–28.

[37] Plum, R. L. and Esrig, M. I. in *Effects of Temperature and Heat on Engineering Behavior of Soils,* Highway Research Board, Special Report 103, 1969, pp. 231–242.

[38] Hansen, J. B. and Inan, S. in *Proceedings,* Seventh International Conference on Soil Mechanics and Foundation Engineering, Vol. 1, 1969, pp. 45–53.

[39] Finn, F. N. in *Symposium on Consolidation Testing of Soils, ASTM STP 126,* American Society for Testing and Materials, Philadelphia, 1951, pp. 65–71.

[40] Habibagahi, K. in *Proceedings*, Eighth International Conference on Soil Mechanics and Foundation Engineering, Vol. 1.1, 1973, pp. 159–162.

[41] Christie, I. F. in *Proceedings*, Sixth International Conference on Soil Mechanics and Foundation Engineering, Vol. 1, 1965, pp. 198–202.

[42] Raymond, G. P., *Canadian Geotechnical Journal*, Vol. 3, No. 4, 1966b, pp. 217–234.

[43] Barden, L. and Berry, P. L., *Journal of the Soil Mechanics and Foundations Division, Proceedings of ASCE*, Vol. 91, No. SM5, 1965, pp. 15–35.

[44] Barden, L., *Geotechnique*, Vol. 15, No. 4, 1965, pp. 345–362.

[45] Mesri, G. and Godlewski, P. M., *Journal of the Geotechnical Engineering Division, Proceedings of ASCE*, Vol. 103, No. GT5, 1977, pp. 417–430.

[46] Raymond, G. P., *Journal of the Soil Mechanics and Foundations Division, Proceedings of ASCE*, Vol. 92, No. SM5, 1966a, pp. 1-20.

[47] Mesri, G. and Rokhsar, A., *Journal of the Geotechnical Engineering Division, Proceedings of ASCE*, Vol. 100, No. GT8, 1974, pp. 889–904.

[48] Samarasinghe, A. M., Huang, Y. H., and Drenevich, V. P., *Journal of the Geotechnical Engineering Division, Proceedings of ASCE*, Vol. 108, No. GT6, 1982, pp. 835–850.

[49] Olson, R. E. and Ladd, C. C., *Journal of the Geotechnical Engineering Divison, Proceedings of ASCE*, Vol. 105, No. 1, 1979, pp. 11–30.

[50] Moran, Proctor, Meuser, and Rutledge, Publ. PB 151692, U.S. Dept. of Commerce, Washington, DC, 1958.

[51] Ward, W. H., Samuels, S. G., and Butler, M. E., *Geotechnique*, Vol. 9, 1959, pp. 33–58.

[52] McKinlay, D. G. in *Proceedings*, Fifth International Conference on Soil Mechanics and Foundation Engineering, Vol. 1, 1961, pp. 225–228.

[53] Aboshi, H. and Monden, H. in *Proceedings*, Fifth International Conference on Soil Mechanics and Foundation Engineering, Vol. 1, 1961, pp. 559–562.

[54] Wu, T. H., Chang, N. Y., and Ali, E. M., *Journal of the Geotechnical Engineering Division, Proceedings of ASCE*, Vol. 104, No. GT7, 1978, pp. 889–905.

[55] Rowe, P. W. and Barden, L., *Geotechnique*, Vol. 16, No. 2, 1966, pp. 162–170.

[56] Barron, R. A., *Transactions of ASCE*, Vol. 113, 1948, pp. 718–754.

[57] da Silviera, I. in *Proceedings*, Third International Conference on Soil Mechanics and Foundation Engineering, Vol. 1, 1953, pp. 55–56.

[58] Berry, P. L. and Wilkinson, W. B., *Geotechnique*, Vol. 19, No. 2, 1969, pp. 253–284.

[59] Abromowitz, M. and Stegun, I. A., *Handbook of Mathematical Functions*, Applied Mathematics Series 55, National Bureau of Standards, Washington, DC, 1964.

[60] Trautwein, S. J., Olson, R. E., and Thomas, R. L. in *Proceedings*, Tenth International Conference on Soil Mechanics and Foundation Engineering, Stockholm, Vol. 1, 1981, pp. 811–814.

[61] Mesri, G., *Journal of the Soil Mechanics and Foundations Division, Proceedings of ASCE*, Vol. 99, No. SM1, 1973, pp. 123–137.

[62] Buisman, A. S. K. in *Proceedings*, International Conference on Soil Mechanics and Foundation Engineering, Cambridge, Vol. 1, 1936, pp. 103–105.

[63] Haefeli, R. and Schaad, W. in *Proceedings*, Second International Conference on Soil Mechanics and Foundation Engineering, Rotterdam, Vol. 3, 1948, pp. 23–29.

[64] Cox, J. B. in *Proceedings*, International Conference on Soil Mechanics and Foundation Engineering, Vol. 2, 1936, pp. 296–297.

[65] Croce, A. in *Proceedings*, Second International Conference on Soil Mechanics and Foundation Engineering, Vol. 1, 1948, pp. 166–169.

[66] Thompson, J. B. and Palmer, L. A. in *Symposium on Consolidation Testing of Soils, ASTM STP 126*, American Society for Testing and Materials, Philadelphia, 1951, pp. 4–7.

[67] Lo, K. Y., *Journal of the Soil Mechanics and Foundations Division, Proceedings of ASCE*, Vol. 87, No. SM4, 1961, pp. 61–68.

[68] Dhowian, A. W. and Edil, T. B., *Geotechnical Testing Journal*, Vol. 3, No. 3, 1980, pp. 105–114.

[69] Bjerrum, L., *Geotechnique*, Vol. 18, No. 2, 1967, pp. 83–118.

[70] Bjerrum, L., Jonson, W., and Ostenfeld, C. in *Proceedings*, Fourth International Conference on Soil Mechanics and Foundation Engineering, Vol. 2, 1957, pp. 14–18.

[71] Leonards, G. A., Special Report 163, Transportation Research Board, 1976, pp. 13–16.

[72] Walker, L. K., *Journal of the Soil Mechanics and Foundations Division, Proceedings of ASCE,* Vol. 95, No. SM1, 1969, pp. 167–188.

[73] Andersland, O. B. and Mathew, P. J., *Journal of the Soil Mechanics and Foundations Division, Proceedings of ASCE,* Vol. 99, No. SM5, 1973, pp. 365–374.

[74] Leonards, G. A. and Altschaeffl, A. G., *Journal of the Soil Mechanics and Foundations Division, Proceedings of ASCE,* Vol. 90, No. SM5, 1964, pp. 133–155.

[75] Garlanger, J. E., *Geotechnique,* Vol. 22, No. 1, 1972, pp. 71–78.

[76] Berre, T. and Iverson, K., *Geotechnique,* Vol. 22, No. 1, 1972, pp. 53–70.

[77] Gibson, R. E. and Lo, K. Y., Publication No. 41, Norwegian Geotechnical Institute, Oslo, 1961; see also *Acta Polytechnica Scandinavica,* No. 296, Ci 10, 1961.

[78] Lo, K. Y., Bozozuk, M., and Law, K. T., *Canadian Geotechnical Journal,* Vol. 13, No. 4, 1976, pp. 339–354.

[79] Barden, L., *Geotechnique,* Vol. 18, No. 1, 1968, pp. 1–24.

[80] Poskitt, T. J., *Geotechnique,* Vol. 18, No. 3, 1967, pp. 284–289.

[81] Berry, P. L. and Poskitt, T. J., *Geotechnique,* Vol. 22, No. 1, 1972, pp. 27–52.

[82] Adams, J. I. in *Proceedings,* Ninth Muskeg Research Conference and Ontario Hydro Research Quorum, Vol. 15, 1963, pp. 1–7.

[83] DeJong, G. D. J., *Geotechnique,* Vol. 18, No. 2, 1968, pp. 195–228.

[84] Juszkiewicz-Bednarczyk, B. and Werno, M. in *Proceedings,* Tenth International Conference on Soil Mechanics and Foundation Engineering, Vol. 1, 1981, pp. 179–183.

[85] Cedergren, H. R. and Weber, W. G., Jr., in *Field Testing of Soils, ASTM STP 322,* American Society for Testing and Materials, Philadelphia, 1962, pp. 248–264.

[86] Weber, W. J., Jr., Highway Research Record No. 243, 1968, pp. 49–61.

[87] Johnson, S. J., *Journal of the Soil Mechanics and Foundations Division, Proceedings of ASCE,* Vol. 96, No. SM1, 1970, pp. 145–173.

[88] Barden, L., discussion, *Journal of the Soil Mechanics and Foundations Division, Proceedings of ASCE,* Vol. 91, No. SM3, 1965, pp. 144–145.

[89] Olson, R. E. and Langfelder, L. J., *Journal of the Soil Mechanics and Foundations Division, Proceedings of ASCE,* Vol. 91, No. SM4, 1965, pp. 127–150.

[90] Donald, I. B. in *Proceedings,* Second Australia-New Zealand Conference on Soil Mechanics and Foundation Engineering, 1956, pp. 200–205.

[91] Bishop, A. W., *Teknisk Ukeblad,* Vol. 39, 1959, pp. 859–863.

[92] Bishop, A. W., Alpan, I., Blight, G. E., and Donald, I. B. in *Proceedings,* ASCE Research Conference on the Shear Strength of Cohesive Soils, 1960, pp. 503–532.

[93] Bishop, A. W. and Donald, I. B. in *Proceedings,* Fifth International Conference on Soil Mechanics and Foundation Engineering, Vol. 1, 1961, pp. 13–22.

[94] Jennings, J. E. and Burland, J. S., *Geotechnique,* Vol. 12, 1962, pp. 125–144.

[95] Bishop, A. W. and Blight, G. E., *Geotechnique,* Vol. 13, No. 3, 1963, pp. 177–197.

[96] Matyas, E. L. and Radhakrishna, H. S., *Geotechnique,* Vol. 18, 1968, pp. 432–448.

[97] Coleman, J. D., discussion, *Geotechnique,* Vol. 12, No. 4, 1962, pp. 348–350.

[98] Barden, L., Madedor, A. O., and Sides, G. R., *Journal of the Soil Mechanics and Foundations Division, Proceedings of ASCE,* Vol. 95, No. SM1, 1969, pp. 33–51.

[99] Yoshimi, Y. and Osterberg, J. O., *Journal of the Soil Mechanics and Foundations Division, Proceedings of ASCE,* Vol. 89, No. SM4, 1963, pp. 1–24.

[100] Barden, L., *Geotechnique,* Vol. 24, No. 4, 1974, pp. 605–625.

[101] Hilf, J. W., Technical Memorandum No. 654, Bureau of Reclamation, Denver, CO, 1956.

[102] Lee, K. L. and Black, D. K., *Journal of the Soil Mechanics and Foundations Division, Proceedings of ASCE,* Vol. 98, No. SM2, 1972, pp. 181–194.

[103] Barden, L. and Sides, G. R. in *Proceedings,* Third Asian Regional Conference on Soil Mechanics and Foundation Engineering, Vol. 1, 1967.

[104] Barden, L., *Geotechnique,* Vol. 15, No. 3, 1965, pp. 267–286.

[105] Fredlund, D. G. and Hasan, J. U., *Canadian Geotechnical Journal,* Vol. 16, No. 3, 1979, pp. 521–531.

[106] Fredlund, D. G. in *Proceedings,* NATO Advanced Study Institute, Mechanics of Fluids in Porous Media, Univ. of Delaware, Newark, July 1982, p. 49.

[107] Lowe, J., III, Zaccheo, P., and Feldman, H. S., *Journal of the Soil Mechanics and Foundations Division, Proceedings of ASCE,* Vol. 90, No. SM5, 1964, pp. 69–86.

[108] Wissa, A. E. Z., Christian, J. T., Davis, E. H., and Heiberg, S., *Journal of the Soil Mechanics and Foundations Division, Proceedings of ASCE*, Vol. 97, No. SM10, 1971, pp. 1393–1414.

[109] Bishop, A. W. and Eldin, G., *Geotechnique*, Vol. 2, 1950, pp. 13–32.

[110] Lowe, J., III, and Johnson, T. C. in *Proceedings*, Research Conference on the Shear Strength of Cohesive Soils, Boulder, ASCE, 1960, pp. 819–836.

[111] Leroueil, S., Samson, L., and Bozozuk, M., *Canadian Geotechnical Journal*, Vol. 20, No. 3, 1983, pp. 477–490.

[112] Leroueil, S., Tavenas, F., Samson, L., and Morin, P., *Canadian Geotechnical Journal*, Vol. 20, No. 4, 1983, pp. 803–816.

[113] Hamilton, J. J. and Crawford, C. B. in *Papers on Soils—1959 Meetings, ASTM STP 254*, American Society for Testing and Materials, Philadelphia, 1959, pp. 254–271.

[114] Crawford, C. B., *Journal of the Soil Mechanics and Foundations Division, Proceedings of ASCE*, Vol. 90, No. SM5, 1964, pp. 87–102.

[115] Wahls, H. E. and deGodoy, N. S., discussion, *Journal of the Soil Mechanics and Foundations Division, Proceedings of ASCE*, Vol. 91, No. SM3, 1965, pp. 147–152.

[116] Lee, K., *Geotechnique*, Vol. 31, No. 2, 1981, pp. 215–229.

[117] Lowe, J., III, Jonas, E., and Obrician, V., *Journal of the Soil Mechanics and Foundations Division, Proceedings of ASCE*, Vol. 95, No. SM1, 1969, pp. 77–97.

[118] Viggiani, C., discussion, *Journal of the Soil Mechanics and Foundations Division, Proceedings of ASCE*, Vol. 95, No. SM6, 1969, pp. 1574–1578.

[119] Aboshi, H., Yoshikuni, H., and Maruyama, S., *Soils and Foundations*, Vol. 10, No. 1, 1970, pp. 43–56.

[120] Janbu, N., Tokheim, O., and Senneset, K. in *Proceedings, Tenth International Conference on Soil Mechanics and Foundation Engineering*, Vol. 1, 1981, pp. 645–654.

[121] Smith, R. E. and Wahls, H. E., *Journal of the Soil Mechanics and Foundations Division, Proceedings of ASCE*, Vol. 95, No. SM2, 1969, pp. 519–539.

[122] Gorman, C. T., Hopkins, T. C., Dean, R. C., and Drenevich, V. P., *Geotechnical Testing Journal*, Vol. 1, No. 1, 1978, pp. 3–15.

[123] Horne, M. R., *International Journal of Mechanical Sciences*, Vol. 6, 1964, pp. 187–197.

[124] Rowe, P. W., *Geotechnique*, Vol. 14, No. 4, 1964, pp. 321–340.

[125] Rowe, P. W. in *Sampling of Soil and Rock, ASTM STP 483*, American Society for Testing and Materials, Philadelphia, 1971, pp. 77–108.

[126] Burghignoli, A. and Calabresi, G. in *Proceedings, Ninth International Conference on Soil Mechanics and Foundations Engineering*, Vol. 1, 1977, pp. 443–450.

[127] Shields, D. H. and Rowe, P. W., *Journal of the Soil Mechanics and Foundations Division, Proceedings of ASCE*, Vol. 91, No. SM1, 1965, pp. 15–23.

[128] Shields, D. H., *Geotechnique*, Vol. 26, No. 1, 1976, pp. 209–212.

[129] Karol, R. H. in *Symposium on Consolidation Testing of Soils, ASTM STP 126*, American Society for Testing and Materials, Philadelphia, 1951, pp. 53–62.

[130] Van Zelst, T. W. in *Proceedings*, Second International Conference on Soil Mechanics and Foundation Engineering, Vol. 7, 1949, pp. 52–61.

[131] Osterberg, J. O., discussion, in *Symposium on Consolidation Testing of Soils, ASTM STP 126*, American Society for Testing and Materials, Philadelphia, 1951, pp. 63–64.

[132] Chan, H. T. and Kenney, T. C., *Canadian Geotechnical Journal*, Vol. 10, No. 3, 1973, pp. 453–472.

[133] Barden, L., discussion, *Journal of the Soil Mechanics and Foundations Division, Proceedings of ASCE*, Vol. 91, No. SM3, 1965, pp. 144–145.

[134] Terzaghi, K., discussion, *Transactions of ASCE*, Vol. 109, 1944, pp. 1196–1201.

[135] Casagrande, A., *Contributions to Soil Mechanics: 1925–1940*, Boston Society of Civil Engineers, 1932, pp. 72–113.

[136] Casagrande, A. in *Proceedings*, International Conference on Soil Mechanics and Foundation Engineering, Vol. 3, 1936, pp. 60–64.

[137] Rutledge, P. C., *Transactions of ASCE*, Vol. 109, 1944, pp. 1155–1183.

[138] Terzaghi, K., *Transactions of ASCE*, Vol. 109, 1944, pp. 1196–1201.

[139] Schmertmann, J. H., *Transactions of ASCE*, Vol. 120, 1955, pp. 1201–1233.

[140] Peck, R. B., Hansen, W. E., and Thornburn, T. H., *Foundation Engineering*, Wiley, New York, 1973.

[141] La Rochelle, P., Sarrailh, J., Tavenas, F., Roy, M., and Leroueil, S., *Canadian Geotechnical Journal*, Vol. 18, No. 1, 1981, pp. 52–66.
[142] Noorany, I. and Poormand, I., *Journal of the Soil Mechanics and Foundations Division, Proceedings of ASCE*, Vol. 99, No. SM12, 1973, pp. 1184–1188.
[143] Morgenstern, N. R. and Thomson, S. in *Sampling of Soil and Rock, ASTM STP 483*, American Society for Testing and Materials, Philadelphia, 1971, pp. 180–191.
[144] Bozozuk, M. in *Sampling of Soil and Rock, ASTM STP 483*, American Society for Testing and Materials, Philadelphia, 1971, pp. 121–131.
[145] Eden, W. J. in *Sampling of Soil and Rock, ASTM STP 483*, American Society for Testing and Materials, Philadelphia, 1971, pp. 132–142.
[146] La Rochelle, P. and Lefebvre, G. in *Sampling of Soil and Rock, ASTM STP 483*, American Society for Testing and Materials, Philadelphia, 1971, pp. 143–163.
[147] Milovic, D. M. in *Sampling of Soil and Rock, ASTM STP 483*, American Society for Testing and Materials, Philadelphia, 1971, pp. 164–179.
[148] La Rochelle, P., Sarrailh, J., Roy, M., and Tavenas, F. A. in *Soil Specimen Preparation for Laboratory Testing, ASTM STP 599*, American Society for Testing and Materials, Philadelphia, 1976, pp. 126–146.
[149] Kenney, T. C., *Canadian Geotechnical Journal*, Vol. 5, No. 2, 1968, p. 97.
[150] Pelletier, J. H., Olson, R. E., and Rixner, J. J., *Geotechnical Testing Journal*, Vol. 2, No. 1, 1979, pp. 34–43.
[151] Arman, A. and McManis, K. L. in *Soil Specimen Preparation for Laboratory Testing, ASTM STP 599*, American Society for Testing and Materials, Philadelphia, 1976, pp. 66–87.
[152] Leroueil, S. and Tavenas, F. in *Proceedings*, Tenth International Conference on Soil Mechanics and Foundation Engineering, Vol. 1, 1981, pp. 185–190.
[153] Olson, R. E., Daniel, D. E., and Liu, T. K. in *Proceedings*, Geotechnical Engineering Division Specialty Conference, ASCE, Vol. 1, 1974, pp. 85–110.
[154] Schiffman, R. L., Chen, A. T-F., and Jordan, J. C., *Journal of the Soil Mechanics and Foundations Division, Proceedings of ASCE*, Vol. 95, No. SM1, 1969, pp. 285–312.
[155] Burn, K. N., *Canadian Geotechnical Journal*, Vol. 6, No. 1, 1969, pp. 33–45.
[156] Olson, R. E. in *Proceedings*, International Conference on Case Histories in Geotechnical Engineering, Univ. of Missouri-Rolla, S. Prakash, Ed., Vol. 2, 1984, pp. 625–630.
[157] Lake, J. R. in *Proceedings*, Conference on Pore Pressure and Suction in Soils, London, 1961, pp. 103–107.
[158] Lake, J. R. in *Proceedings*, European Conference on Soil Mechanics and Foundation Engineering, Vol. 1, 1963, pp. 351–357.
[159] Lewis, W. A., *Geotechnique*, Vol. 6, 1956, pp. 106–114.
[160] Lewis, W. A. in *Proceedings*, European Conference on Soil Mechanics and Foundation Engineering, Wiesbaden, Vol. 1, 1963, pp. 359–366.
[161] Cedergren, H. R. and Weber, W. G., Jr., in *Field Testing of Soils, ASTM STP 322*, American Society for Testing and Materials, Philadelphia, 1962, pp. 248–264.

DISCUSSION

G. Sällfors[1] and R. Larsson[2] (written discussion)—Professor Olson is to be congratulated for his success in summing up the most important findings in the tremendously large field of laboratory consolidation testing. His paper will no

[1] Associate Professor, Chalmers University of Technology.
[2] Research Engineer, Swedish Geotechnical Institute.

doubt be a key paper for many future researchers in the field of consolidation testing.

Professor Olson covers in detail the determination of the coefficient of consolidation (c_v), especially different plotting procedures for incremental tests. He also illustrates that different methods of interpretation consistently give different results. Many soil engineers have also found that the scatter in the c_v-value is often large compared with other geotechnical parameters. It is important to remember the Terzaghi consolidation equation and what c_v actually represents. The differential equation can be written as

$$\frac{\partial u}{\partial t} = M \frac{\partial}{\partial z}\left(\frac{k}{g\,\rho_w}\frac{\partial u}{\partial z}\right)$$

Now, if the permeability (k) is independent of depth (z), the equation can be written as

$$\frac{\partial u}{\partial t} = \frac{M \cdot k}{g\,\rho_w}\frac{\partial^2 u}{\partial z^2} \quad \text{or} \quad \frac{\partial u}{\partial t} = c_v \frac{\partial^2 u}{\partial z^2}$$

It is obvious that the c_v-value is nothing but the product of the modulus (M) and the permeability, divided by a constant. The only rational way then to determine the c_v-value is to determine the modulus and the permeability.

These parameters can be determined in incremental oedometer tests with measurement of the permeability after some load increments, or more easily and continuously by CRS tests. With these tests there is no need for elaborate curve plotting and fitting methods. In problems where the c_v-value is of no interest other tests may be more appropriate.

R. E. Olson (author's closure)—I agree with Messrs. Sallfors and Larsson that the coefficient of consolidation is a derived soil property and that substantial scatter in values can result from variations in the fundamental variables, k, e, and a_v (M and k in the formulation of Sallfors and Larsson). It may indeed be more rational to measure k directly and obtain a_v (or m_v, M, . . .) from the laboratory test. If structural viscosity has a much smaller effect on field behavior than it does in the laboratory, the usefulness of the data could be improved significantly by using such an approach. However, several problems develop. One involves economics and time. To be generally useful, it would be necessary to measure k at several void ratios to obtain an e-k relationship. Testing time and cost would be increased significantly. In any case, the accuracy of prediction of field behavior seems more controlled by uncertainties in stratigraphy; in loading rates (where the loading time represents a significant part of consolidation time); by use of nonrepresentative samples; by use of simplified theories that ignore mass flow, secondary effects, multidimensional consolidation, layering, effective-stress dependent soil properties, and so on; and there are numerous other problems. We should, of course, strive to perfect each aspect of the design process in principle, but on a practical project we should invest our funds in such a way as to maximize our overall benefit. It may be noted that useful results may result from measurement of k in the field [161] and using the laboratory data mainly to correct k for changing e. Measurement of k in the laboratory carries with it the possibility of introducing significant new errors [15].

Carl B. Crawford[1]

State of the Art: Evaluation and Interpretation of Soil Consolidation Tests

REFERENCE: Crawford, C. B., "**State of the Art: Evaluation and Interpretation of Soil Consolidation Tests,**" *Consolidation of Soils: Testing and Evaluation, ASTM STP 892,* R. N. Yong and F. C. Townsend, Eds., American Society for Testing and Materials, Philadelphia, 1986, pp. 71–103.

ABSTRACT: The consolidation characteristics of natural soils vary widely depending on their stress history, void ratio, and structure. The standard method of measuring consolidation properties involves the incremental loading of specimens of soil, but other methods using controlled loading techniques have been useful in improving the understanding of soil compressibility. The various test methods and their influence on the evaluation of consolidation properties are reviewed in the paper. The interpretation of preconsolidation pressure for some soils is greatly influenced by the rate of loading. Total consolidation settlements can be estimated reasonably well from good oedometer tests, but predictions of the rate of settlement are usually quite unreliable in the primary consolidation stage. The ability to evaluate and interpret consolidation tests has been improved where the tests have been carried out in conjunction with field observations. More research of this kind is needed to improve the usefulness of consolidation theories and testing.

KEY WORDS: consolidation, overconsolidation, preconsolidation pressure, soil mechanics, soil properties, soil structure, soil tests

The consolidation test is a model test in which a specimen of soil is subjected to pressure in order to predict the deformation that would occur to a stratum of soil under similar pressures in the ground. The success of the test depends on how well the model test represents the situation in nature. The loading and deformation in the conventional oedometer test, for example, is unidirectional, although this may not truly represent the situation in the field. In addition, it is not uncommon for a specimen 40 cm^3 in volume to be used to represent the properties of a soil mass of 400 m^3, a mass ten million times as great. Extrapolation, together with the disturbance caused to the specimen during sampling and trimming, might be expected to lead to gross errors in prediction, but this is not normally the case, except when settlements are small. Bjerrum [1] concluded from a comparison of predictions and observed settlements in soft clays

---

[1] National Research Council of Canada, Division of Building Research, Ottawa, Ontario, Canada.

that when settlements are greater than 20 cm, the agreement is good. For settlements less than 20 cm, the error increases and predictions are usually too high.

This volume deals with laboratory testing for consolidation properties and the present paper with the evaluation and interpretation of such test results. The important values obtained from the test are the preconsolidation pressure, the compression indices, and the rate of consolidation. The main factors influencing the evaluation and interpretation of consolidation test results are the quality of the laboratory specimen (such as the degree of disturbance, size and shape) and the method of testing, including the influence of apparatus, rate of loading, and the environment of the test.

The testing of soils is an important part of soil engineering, but only a part; it is essential, therefore, that the test method and the test interpretation be related to real soils and to real engineering problems. In a previously unpublished statement revealed by Peck [2] in his Rankine Lecture, Terzaghi described this as the "experimental method" and illustrated it as follows:

Base the design on whatever information can be secured. Make a detailed inventory of all the possible differences between reality and the assumptions. Then compute, on the basis of the original assumptions, various quantities that can be measured in the field. For instance, if assumptions have been made regarding pressure in the water beneath a structure, compute the pressure at various easily accessible points, measure it, and compare the results with the forecast. Or, if assumptions have been made regarding stress-deformation properties, compute displacements, measure them, and make a similar comparison. On the basis of the results of such measurements, gradually close the gaps in knowledge and, if necessary, modify the design during construction. Soil mechanics provides us with the knowledge required for practical application of this "learn-as-you-go" method.

Historical Developments

There is an enormous literature on consolidation, beginning with Terzaghi's initial work. In a series of eight weekly articles published by *Engineering News-Record* in 1925, Terzaghi described his experiments with soil, including one article on the "Settlement and Consolidation of Clay" [3]. Figure 1 shows his test curves and computed pressure changes with time. He recognized the similarity of moisture flow in soils to heat flow through materials and thus evolved the theory of consolidation.

The understanding of the consolidation phenomenon in soils was greatly increased by the six years of concentrated effort of Taylor at the Massachusetts Institute of Technology (MIT) [4]. His study was directed primarily at the differences between the rates of compression observed in laboratory samples and the rates given by theory. A modification to the original Terzaghi theory was developed, taking into account the time-dependent plastic resistance of the soil structure, independent of the hydrodynamic time lag. This is illustrated in Fig. 2, in which the upper line approximates the end of primary consolidation and the lower lines correspond to the pressure/void ratio after increasing amounts of secondary consolidation.

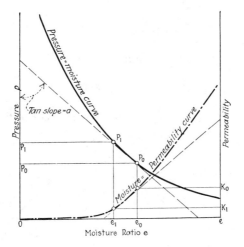

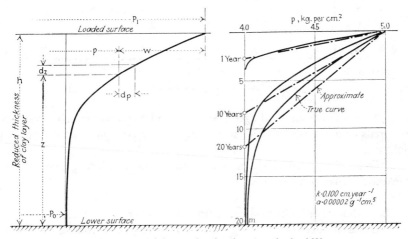

FIG. 1—*Consolidation of a clay deposit under load* [3].

Twenty-five years later Bjerrum [5], in the Seventh Rankine Lecture, provided a useful interpretation to the observations of Taylor. He introduced the term "delayed consolidation" to describe the secondary consolidation that occurs in a natural deposit under constant effective stress.

In his state-of-the-art report to the 8th International Conference on Soil Mechanics and Foundation Engineering, prepared just before his untimely death, Bjerrum [1] elaborated on the views expressed earlier in the Rankine Lecture. His discussion was limited to homogeneous, flocculated soft clays, classified as:

1. Normally consolidated (NC) young clays, recently deposited and at equilibrium under their own weight but not yet undergoing significant delayed or secondary consolidation.

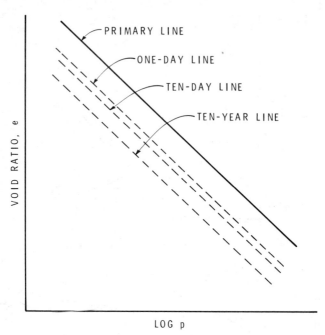

FIG. 2—*Semilogarithmic plot of pressure versus void ratio* [4].

2. Normally consolidated aged clays, left under constant effective stress for hundreds or thousands of years and having undergone significant delayed consolidation.

3. Overconsolidated clays, previously subjected to a load greater than the present effective overburden stress.

The performance under load of the first two categories is shown in Fig. 3. An undisturbed specimen of a "young" NC clay would follow the upper bold curve, with a preconsolidation pressure[2] (p_c) equal to the present effective overburden pressure (p_o). If the "young" NC clay is left intact for hundreds or thousands of years, it will continue to consolidate with time under constant effective stress, reaching a more stable arrangement of particles with greater strength or a "reserve resistance" against further compression. If an undisturbed specimen of this "aged" NC clay is subjected to a consolidation test, it will follow the lower bold curve with a measured $p_c \geq p_o$. This reasoning provides an explanation for measured p_c-values in excess of p_o, when there is no satisfactory geological explanation.

Bjerrum observed that the p_c/p_o ratio of clay deposits of the same age will increase with the amount of "delayed consolidation" and, because secondary consolidation increases with the plasticity of the clay, the p_c/p_o ratio will increase with the plasticity index. The correlation of the p_c/p_o ratio with the plasticity

---

[2] Later in the text effective stresses are indicated by σ'_c and σ'_o, etc.

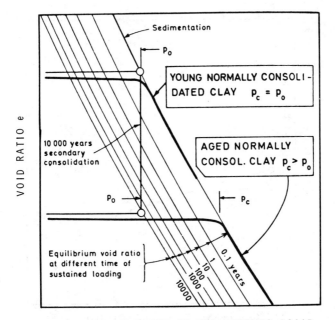

FIG. 3—*Geological history and compressibility of a "young" and an "aged" normally consolidated clay* [1].

index observed in some normally consolidated late glacial and post-glacial clays is shown in Fig. 4.

For those clay deposits which are known to be overconsolidated because of their geologic history, the pressure/void ratio is shown in Fig. 5. In this case, the previous maximum effective stress was p_1, the delayed consolidation occurred during a period of 10 000 years, and the present effective *in situ* stress is p_o. When such a specimen is loaded in the oedometer, the measured p_c exceeds even

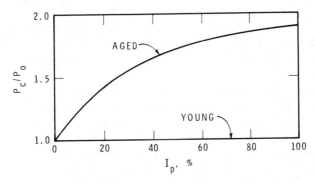

FIG. 4—*Typical values of p_c/p_o observed in normally consolidated late glacial and post-glacial clays* [1].

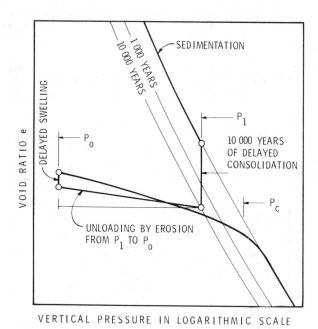

VERTICAL PRESSURE IN LOGARITHMIC SCALE

FIG. 5—*Geological history and compressibility of an overconsolidated clay* [1].

the previous maximum effective stress, due to the "reserve resistance" built up before the geologic unloading

The rationale behind Bjerrum's interpretation is impressive, but the field evidence to support the hypothesis has been limited and comes largely from Norway. For example, Berre and Iversen [6] measured the rates of compression and pore pressure dissipation on specimens (ranging from 1.75 to 45 cm in height) of a soft normally consolidated clay from the city of Drammen in Norway and found that their test results could be explained qualitatively by Bjerrum's delayed compression concept. Garlinger [7] developed a new theory based on Bjerrum's concept and applied it successfully to the prediction of settlements of three buildings in Drammen. Earlier, Foss [8] had shown that the settlement of the three buildings in Drammen was governed by the "reserve resistance" in the clay. Engesgaar [9] used the concept successfully in the design of a 15-story building in Drammen.

A comprehensive review of consolidation testing and interpretation was provided by the conference on "Settlement of Structures" organized by the British Geotechnical Society in 1974. Review papers by Simons [10] on normally consolidated and lightly overconsolidated soils, and by Butler [11] on heavily overconsolidated clays, draw attention to well over 200 references on the subject.

Simons [10] concluded that major uncertainties still exist in predicting settlements and that one of the most significant factors is the determination of preconsolidation pressure. For the prediction of settlement rates, he recommended tests for permeability on large samples or *in situ* because laboratory tests on small specimens generally give misleading results. This recommendation was based on the work of Rowe [12], who found that soil fabric (the arrangement of soil constituents) has an enormous influence on permeability that can only be truly represented by tests on very large specimens. Simons considered settlements due to secondary consolidation to be important in many cases, but methods of rational analysis are not available. Quoted values for the coefficient of secondary consolidation (C_α) generally ranged up to 0.02 in a relationship with water content in which $C_\alpha = 0.00018 \, (W_n\%)$ on average. He identified sample disturbance as a major influence on test results. Butler [11] emphasized the use of stress-path methods for estimating settlements of heavily overconsolidated clays.

Variations in Consolidation Characteristics of Natural Soils

It is difficult, for a number of reasons, to generalize on the consolidation characteristics of natural soils. A primary reason is the variation in the soils themselves and in their environment [12]. Consider, for example, the Gault Clay described by Samuels [13] and Mexico City Clay [14]. The compressibility characteristics of almost all natural soils are encompassed by these two extreme types. The Gault Clay has a natural void ratio of about 0.75 and an estimated preconsolidation pressure of about 7000 kN/m². The Mexico City Clay has a natural void ratio of about 14 and a preconsolidation pressure of less than 100 kN/m². In Mexico City, the Gault Clay would be considered virtually incompressible.

A second hazard in attempting to generalize from published results is that, unlike the examples just mentioned, there is often very little information on the basic properties of the soils being discussed. Furthermore, sampling and testing techniques have improved so much over the years that relating modern results to those of 30 years ago may be misleading. In the early days of soil testing, remolded soils were often used for research because undisturbed samples were not easy to obtain. In the author's view, test results on remolded soils are often deceiving unless they are considered in conjunction with tests on undisturbed samples of the same deposit. Also, the degree of soil disturbance and the influence of disturbance have to be considered in analyzing test results.

Influence of Test Method

For some kinds of soils the method of testing appears not to have a great influence on the result. Samuels [13] stated that the manner in which undisturbed specimens of Gault Clay were tested for strength had no noticeable effect on the results, but he did not even mention how the consolidation tests were performed; it probably did not matter very much. He reported that care was taken to prevent initial swelling and to minimize apparatus and bedding errors and he was concerned about sample disturbances. In the test result shown in the paper, the total compression of the specimen was less than 7%. In a typical test of Mexico City clay, however, the total compression of the specimen was about 80%. Mesri et al [14] were probably not so concerned about bedding and apparatus errors, but they investigated other aspects of the testing method—for example, the influence of sustained loading.

Test Methods

Incremental Loading

The conventional and most common method of testing is to apply daily increments of vertical load to a submerged specimen contained in a rigid ring, with drainage permitted through porous stones at the top and bottom. The ratio of load increment to existing load ($\Delta\sigma/\sigma$) is usually 1; that is, the load is doubled each day. Such a test normally lasts for one or two weeks. From observations of the rate of deformation under each load increment, the coefficient of consolidation may be computed for use in estimating the rate of settlement in the field. The time-deformation curve is also used to distinguish between primary and secondary consolidation and to determine the coefficient of secondary consolidation (i.e., the compression per log cycle of time after pore pressures have essentially dissipated). The stress-deformation curve, usually plotted from values of void ratio versus the log of pressure for each increment, is used to estimate the preconsolidation pressure and the compression index (i.e., the slope of the virgin compression curve on a semi-log plot).

Variations to the conventional method include lower load increment ratios (say $\Delta\sigma/\sigma = \frac{1}{2}$), especially in the vicinity of the preconsolidation pressure, in order to obtain better definition of the e–log σ' curve. Sometimes load increments are applied at the end of primary consolidation rather than on a daily basis. This is often the case for low values of $\Delta\sigma/\sigma$, in order to reduce the total testing time. It is also common, especially in research studies, to permit drainage at one face only, while measuring pore pressures at the other. Another variation is to reload the specimen in a single increment to the vertical effective stress in the ground before sampling and then to apply another increment to bring it to the vertical effective stress expected from the field loading. And, finally, the single-stage loading test has been used in which a single increment load, considerably greater than the σ'_c-value, is applied to a specimen. A discontinuity in the effective-stress–time plot is interpreted as the preconsolidation pressure, but the test has little additional value.

These variations in the test procedure often have a great effect on the estimated preconsolidation pressure, especially for soft sensitive clays. The compression index appears to be affected little by test variations. Sällfors [15] reported a higher recompression modulus (i.e., less compression) at higher rates of strain, but Leroueil et al [16] were unable to detect any effect.

Constant Rate of Strain (CRS)

The constant-rate-of-strain test was first used about 25 years ago [17]. Since that time many CRS tests have been carried out, primarily for research purposes. The great advantage of the CRS test is that it gives a continuous definition of the pressure/void ratio curve and it permits an evaluation of the influence of strain rate on test results. Specimens are prepared and mounted in the same way as for incremental loading, but the loading is usually applied in a testing frame at a predetermined rate with pore pressure measurements at the base.

Smith and Wahls [18] developed a simple theory, based on several simplifying assumptions, to determine the coefficient of consolidation (c_v) from CRS tests. Test results on specimens of one undisturbed clay and two remolded clays gave good agreement between CRS and conventional incremental tests, as long as the base pore pressures did not exceed 50% of the applied pressure. In general, the calculated c_v-values for the CRS tests were slightly higher than those for incremental tests. The authors recognized that the method does not address the strain rate effect, but for slow rates of strain this did not appear to be important for the soils tested. Wissa et al [19] reached similar conclusions on CRS tests on artificially sedimented Boston Blue Clay, but they found it necessary to keep the base pore pressure at less than 5% of applied pressure.

Controlled Gradient Test (CGT)

The controlled gradient test requires a special apparatus in which a back pressure can be applied to the pore water while maintaining a constant pore pressure differential across the specimen by varying the rate of loading. Under these controlled conditions, the e–log σ' curve and the c_v–log σ' curve can be drawn continuously. Consolidation parameters determined in this way on an undisturbed marine clay agreed reasonably well with those from conventional tests, provided the duration of the two types of test (rates of strain) was about the same [20].

Gorman et al [21] carried out conventional, CRS, and CG tests on undisturbed soils from three sites in Kentucky and found reasonable agreement between the stress-strain curves for the three types of test. Below the preconsolidation pressure, the calculated c_v-values from the CRS and CG tests were considered to be unreliable because steady-state conditions do not exist. The authors recognized that the pore pressure generated depends on the strain rate during the test and that this influences the applicability of the theory. For their rather low water content soils (average 26%), they found that when the base pore pressures were kept to less than 50% of applied load, the influence of strain rate was low.

Continuous Loading (CL)

The continuous loading test has been automated by Janbu et al [22] in order to maintain a constant ratio of pore pressure at the base to total applied pressure.

Data acquisition and the interpretation and plotting of results are computerized to give continuous plots of strain, modulus, coefficient of consolidation, and deformation rate.

Recompression Index

Consolidation test results in the recompression range (i.e., below the preconsolidation pressure) are generally unreliable, due to a number of factors such as the small movement, apparatus errors, sample disturbance, and swelling. On the basis of an extensive investigation of London Clay, Simons and Som [23] concluded that satisfactory results can be obtained only when initial swelling is positively prevented and apparatus and bedding errors are eliminated. They suggested that the soil should be tested by subjecting the specimen to the stress system initially prevailing in the ground and then applying as closely as possible the same stress changes as those to which the soil will be subjected in the field. This could be achieved by applying the concept of stress path testing proposed by Lambe [24].

Janbu et al [22] measured the modulus of recompression in a continuous loading consolidation test, but they did not give any comparisons with field settlements. The measured settlement of a mat foundation resting on a silty till soil in Toronto was found to be only about one tenth the value estimated from recompression in the oedometer [25]. In this case, a load-cycled triaxial compression test, where the specimen was maintained under equivalent overburden pressure, gave a more reasonable prediction of settlement.

Preconsolidation Pressure

A primary objective of the consolidation test is to estimate the maximum pressure that may be applied to the soil without causing large settlements. This pressure, called the preconsolidation pressure (σ'_c), is generally considered to indicate the maximum load on a natural soil during its geologic history, although the effect may also be caused by aging. For soils that have been heavily overconsolidated, the preconsolidation pressure is poorly defined on the e–$\log \sigma'$ curve, but for normally consolidated soils there is usually a rather dramatic change in compressibility at σ'_c. As mentioned earlier, the measurement of recompression of overconsolidated or stiff clays in the oedometer is not usually satisfactory and attention will therefore be limited to highly compressible normally consolidated or slightly overconsolidated clays.

The first consolidation tests were done on remolded soils, which had little structure, but when natural soils were tested there was a threshold stress level beyond which compressibility increased sharply. Based on laboratory observations of compression, rebound, and recompression of soil specimens, Casagrande [26] developed a graphical method for estimating maximum past pressure—a method that is still commonly used to interpret test results. An alternative method of merit was proposed by Schmertmann [27].

Accurate determination of preconsolidation pressure is particularly important for the highly structured sensitive clays commonly found in Eastern Canada and Scandinavia, because they are relatively incompressible at loads up to the σ'_c and very compressible at higher loads. Consequently, much research has been done on these soils. The term "structure" in these soils is considered to include the size, shape, and orientation of particles and the bonds between them. The Canadian sensitive clays appear to have more cementation between particles than the Norwegian soils.

The author first encountered the problem in 1952 while trying unsuccessfully to explain large settlements of a building by use of soil mechanics techniques. This led to laboratory studies of consolidation procedures [17] in which load increment ratios were varied from 1 to 1/10 and constant-rate-of-loading and constant-rate-of-straining techniques were tried. Small load increment ratios gave better definition to the e–log σ' curve. The tests at various rates of strain, however, showed that the rate of testing has a significant influence on the e–log σ' curve and that the preconsolidation pressure (σ'_c) is considerably lower for slow tests than for fast ones.

A subsequent series of tests on specimens carefully trimmed from block samples obtained in a tunnel at a depth of 10 m showed that the e–log σ' curves (Fig. 6) were displaced by a factor of two, depending on whether incremental loads were applied at the end of primary consolidation or after one week [28]. A second series of tests on specimens trimmed from block samples at another site in the

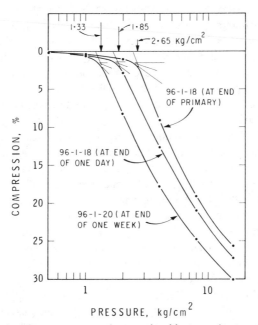

FIG. 6—*Compression-log pressure curves for normal and long-term incremental loading* [28].

same region was carried out at controlled rates of strain. At the fastest rate of strain, 50 times faster than the slowest rate, the pressure/void ratio curve was displaced sufficiently that the estimated preconsolidation pressure was 50% greater than that at slow rates [29]. These tests were on a much more overconsolidated soil than the first set. Further tests on samples from a higher elevation at this site included constant-rate-of-loading tests [30]. Three zones of influence were identified: the reloading zone, where time effects are minimal; the zone near the preconsolidation pressure, where time effects are at a maximum; and the high loading zone, where time effects are decreasing. The time effects during controlled rates of loading appear to be less than those under controlled rate of strain.

Leonards and Ramiah [31] also investigated time effects using remolded clays with incremental loading. For these clays they concluded that rate of loading had little effect on the e–log σ' curve, as long as primary consolidation was completed under each load increment. Interpretation could be improved by using small increments near σ'_c. They also observed that after a period of rest at a particular pressure, if a small pressure increment is added, a quasi-preconsolidation pressure is observed that is considerably larger than the maximum previous consolidation pressure.

Sällfors [15] concluded from oedometer tests on three different highly plastic clays that the measured preconsolidation pressure is time-dependent, decreasing with decreasing rates of strain, and that a constant-rate-of-strain test carried out in 24 h will give results equivalent to normal incremental tests. He also presented a new graphical method for determining σ'_c using an arithmetic e-p curve. Sällfors checked his laboratory tests with two full-scale constant-rate-of-loading tests, using a 9.6-m-diameter water tank placed on the ground surface. Settlements and pore water pressures under the tank were monitored during loading. Excess pore pressures were small during the early stages of loading and then increased rapidly at a load considered to be at the preconsolidation pressure. No further increase in effective stress occurred beyond the σ'_c, the additional load being carried by the pore water pressure. Good agreement was found between the laboratory tests and the field tests.

Janbu et al [22] favor continuous loading oedometer tests in which the ratio of the pore water pressure at the undrained base to the applied load is kept constant by varying the rate of loading. The idealized curve of modulus in relation to effective stress, shown in Fig. 7, illustrates the region of structure breakdown and its relation to the preconsolidation pressure (σ'_c). The authors also present a theory for deriving other consolidation parameters, such as the coefficient of consolidation (c_v). The test results show little influence of rate of loading, leading the authors to conclude that a satisfactory test of these clays can be carried out in less than one hour. Such rapid tests of the sensitive clays mentioned earlier would be quite unsatisfactory.

Leroueil et al [32] reported a series of conventional and special consolidation tests on samples from eleven widely scattered sites in the glacial Champlain Sea area of Eastern Canada. At every site but one, the natural water content was

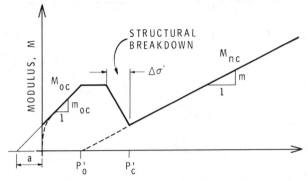

FIG. 7—*Idealized modulus-effective stress curve for overconsolidated clay* [22].

higher than the liquid limit; the liquidity index ranged from 0.9 to 2.7. More than 150 specimens were tested and the measured preconsolidation pressure (by incremental loading) ranged from 47 to 270 kPa. The testing methods were:

1. Conventional oedometer with drainage both ends, $\Delta\sigma/\sigma = 0.5$ at 24-h intervals.
2. Multiple-stage loading (MSL), similar to the conventional test but with pore pressures measured at the bottom.
3. Constant rate of strain (CRS).
4. Controlled gradient test (CGT).
5. Single-stage loading (SSL).

Leroueil et al found that the rate of strain had a substantial effect on the interpretation of preconsolidation pressure, as shown in the normalized curves of Fig. 8. Conventional and multistage loading at 24-h increments corresponds to the slowest rates of strain and consequently gave the lowest values of σ'_c. They concluded that there is no unique value for the preconsolidation pressure; the measured value depends on the method of testing. On the other hand, they concluded that for each clay at a given depth there is a unique preconsolidation-pressure/strain-rate relationship and that, from a practical point of view, the *in situ* σ'_c can be estimated from conventional or MSL (24-h loading) oedometer tests or preferably from CRS tests.

In a companion paper Morin et al [33] compared the *in situ* preconsolidation pressure at five sites with values determined in the laboratory using conventional oedometer tests. *In situ* σ'_c values were estimated in several ways: by observing the stress-strain changes in individual clay layers, by measuring changes in water content or pore pressures under applied loads, and by observing settlement under different heights of embankment. The field evidence indicated that the σ'_c mobilized *in situ* can be estimated satisfactorily using conventional oedometer tests on good-quality samples. Although the full-scale results are limited, Morin et al

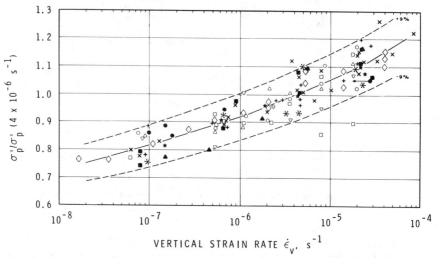

FIG. 8—*Normalized preconsolidation-pressure/strain-rate relationship* [32]. *(Illustration courtesy of Canadian Geotechnical Journal.)*

proposed as follows a small correction to laboratory values depending on the overconsolidation ratio (OCR):

$$\sigma'_c \ in \ situ \ = \ \alpha_1 \sigma'_c \qquad by \ conventional \ test$$

where

α_1 = 1.1 for OCR < 1.2,
α_1 = 1.0 for OCR = 1.2 to 2.5, and
α_1 = 0.9 for OCR = 2.5 to 4.5.

One of the five sites is discussed in detail in the paper by Leroueil et al [16]. This is the test fill at Gloucester, which was extensively instrumented for vertical settlement and pore pressure changes by the Division of Building Research of the National Research Council Canada [34]. From the observations *in situ* the threshold pressure under the full-scale load (Fig. 9) gave a σ'_c-value between 55 and 60 kPa. These values are compared with laboratory tests in Fig. 10 to show the influence of strain rate. The best agreement is with conventional tests using incremental loading at 24-h intervals, resulting in strain rates only about two orders of magnitude greater than the field rates.

In a recent paper, Mesri and Choi [35] conclude that there is a unique "end of primary" e–log σ'_c curve for any soft clay and the rate of compression is therefore of no consequence. This conclusion is based on tests lasting as long as one year, in which four 127-mm-high specimens were connected in series so that

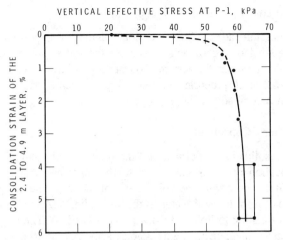

FIG. 9—*Consolidation strain of clay layer from 2.4 to 4.9 m depth* [16]. *(Illustration courtesy of Canadian Geotechnical Journal.)*

the maximum drainage path was 508 mm. Other tests on 25, 51, 76, and 152-mm-high specimens were carried out. Most of the tests were loaded isotropically in a pressure cell. By testing specimens in series, it was possible to sketch isochrones of excess pore water pressures from measurements at the quarter points of the simulated 508-mm-high specimens. The unique shape of the virgin compression curve is not surprising, but the conclusion also implies a unique preconsolidation pressure, which is at variance with other evidence given above.

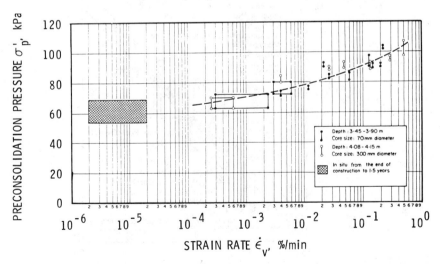

FIG. 10—*Preconsolidation-pressure/strain-rate relationship in laboratory and* in situ [16]. *(Illustration courtesy of Canadian Geotechnical Journal.)*

Compression Index

The compression index (C_c) is the slope of the linear portion of the virgin compression e–log σ'_c curve. Disturbance of the soil sample tends to decrease the C_c, but variations in the test method appear to have little influence on it. Mesri and Rokhsar [36] reported a relationship between C_c and water content for a selection of natural soils (Fig. 11).

Rate of Primary Consolidation

Primary consolidation is defined as "the reduction in volume of a soil mass by the application of a sustained load to the mass and due principally to a squeezing out of water from the void spaces of the mass and accompanied by a transfer of load from the soil water to the soil solids" (ASTM Standard Terms and Symbols Relating to Soil and Rock [ASTM D 653]). The original Terzaghi theory provided a mathematical expression for the rate of primary consolidation on the assumption that this is the total consolidation that would occur under a certain load (i.e., it neglected secondary consolidation). Two other important

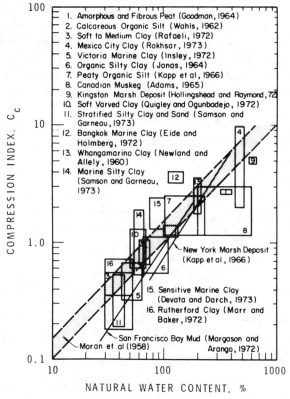

FIG. 11—*Compression index for natural soil deposits* [36].

assumptions are that the permeability and the compressibility of the soil remain constant during consolidation under a particular load increment. Experience has shown that these assumptions are not correct and in most cases lead to substantial discrepancies between computed and actual rates of settlement.

The rate of volume changes of a soil is expressed by the coefficient of consolidation:

$$c_v = k(1 + e)/\alpha_v\gamma_w$$

where

k = coefficient of permeability,

e = void ratio,

α_v = coefficient of compressibility, and

γ_w = unit weight of water.

ASTM Test for One-Dimensional Consolidation Properties of Soils (ASTM D 2435) describes two procedures for computing the coefficient of consolidation. Both procedures involve graphical interpretations of the magnitude of total primary consolidation using time-deformation plots for each load increment. For example

$$c_v = \frac{0.05\,H^2}{t_{50}}$$

where H is the sample height in metres for a doubly drained sample at 50% consolidation, and t_{50} is the time for 50% consolidation in years.

Much effort has been devoted to the extension of the classical consolidation theory to account for variations in permeability and compressibility of the soil as it consolidates and, more recently, to account for large strains [37]. Unfortunately, less attention has been paid to the influence of secondary consolidation.

Leonards and Altschaeffl [38] illustrated the influence of test procedure on the time-deformation curves obtained for undisturbed Mexico City clay (Fig. 12). The large load increment ratio ($\Delta\sigma/\sigma = 1$) gives a classical S-shaped Type 1 curve with most of the consolidation in the primary phase, but under a low load increment ratio, a Type III curve is obtained and most of the consolidation would be considered to be secondary. For the sensitive Leda clay even a load increment ratio of 1 results in a substantial amount of secondary consolidation (Fig. 13) [28].

Secondary Consolidation

It is now well understood that when a soil consolidates under load the rate of compression is slowed by two effects: a delay due to the time taken for the pore water to escape, and a delay due to the plastic resistance of the soil structure.

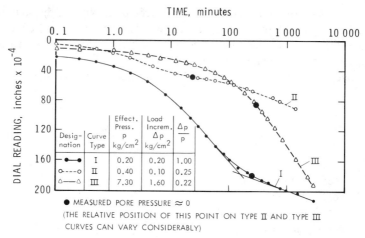

FIG. 12—*Effect of load-increment ratio on shape of dial reading-time curves—undisturbed Mexico City clay* [38].

During the so-called primary phase, the permeability of the soil is the predominant factor controlling the rate of volume change, and during the secondary phase it is the resistance of the structure that determines the rate. Secondary consolidation is defined as "the reduction in volume of a soil mass caused by the application of a sustained load to the mass and due principally to the adjustment of the internal structure of the soil mass after most of the load has been transferred from the soil water to the soil solids" (ASTM D 653-83). The rate of secondary consolidation is expressed by the coefficient

$$C_\alpha = \frac{\Delta e}{\Delta \log t}$$

This represents the change in void ratio over one log-cycle of time after the primary consolidation under a load increment is completed.

The separation of total consolidation into primary and secondary stages is rather arbitrary, but the concept has been useful for settlement analyses. In reality the "adjustment of the internal structure" is a creep mechanism that is time dependent and it must therefore have much influence during the primary stage. Taylor [4] attempted to develop a theory of consolidation that would account for the secondary effects observed in laboratory tests. It was his view that most secondary compression occurred after the primary compression was finished. Wahls [39] concluded from an extensive investigation that secondary consolidation occurs during the primary phase but reaches its maximum rate only after the primary is completed. Walker and Raymond [40] found that the secondary consolidation rates in laboratory tests on sensitive Leda clay appeared to be linearly dependent on the compression index over the entire pressure range, with

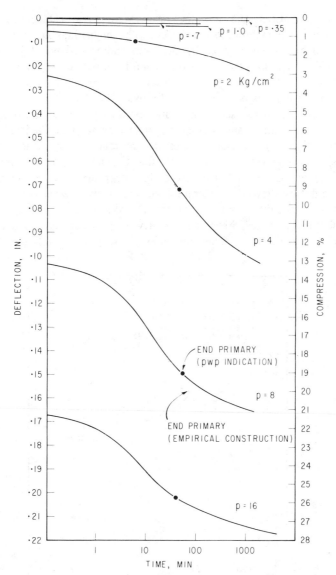

FIG. 13—*Time-compression curves—Leda clay* [28].

an average ratio (C_α/C_c) of about 0.025. Poorooshasb et al [41] have developed a model to describe the consolidation of sensitive clays without invoking the concept of primary and secondary, but it requires field verification.

Walker [42] compared laboratory and field rates of secondary consolidation of five sensitive clays and showed that C_α varied with the ratio of effective stress to preconsolidation stress, with the highest rates occurring at stresses higher than

σ'_c. He noted that field and laboratory results were in reasonable agreement and that the higher creep rates occurred in clays with a higher liquidity index. Further evidence of agreement is given by Crawford and Sutherland [43], where the average C_α from oedometer tests is 0.010 and from field observations is 0.012. The variation of C_α with consolidation ratio (σ'/σ'_c) is shown in Fig. 14. A similar relationship has been shown by Crooks [44].

Mesri [45] considered secondary compression to be a continuation of the mechanism of volume change during the primary phase, a mechanism involving the deformation of individual particles and the relative movement of particles with respect to each other. He concluded that soils that are highly compressible in the primary phase would also have a high secondary compressibility and he illustrated this by showing a general relationship between water content and the coefficient of secondary consolidation for several kinds of natural soils.

In subsequent studies, Mesri and Godlewski [46] pursued the relationship between C_c and C_α. They found no reason to suggest any differences in the

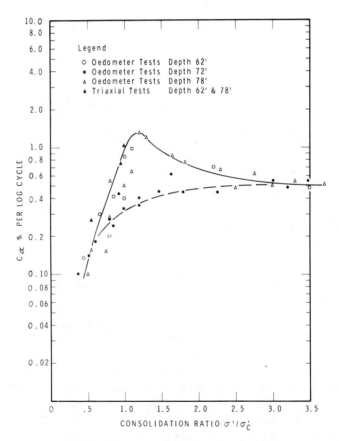

FIG. 14—C_α versus consolidation ratio σ'/σ'_c [43].

mechanism of volume change during the primary and secondary stages of consolidation and concluded that in a particulate system every mechanism of volume change, viscous or nonviscous, is a chain-reaction process and is therefore time dependent. They computed the ratio of C_α/C_c for 22 natural soil deposits and found a range of values from 0.025 to 0.10, with the higher values applying to highly organic soils. It was observed that C_α is not a function of the load increment ratio but is dependent on the applied effective stress and its relation to the preconsolidation stress (σ'_c).

It was shown in oedometer tests that for three undisturbed soil types, Mexico City Clay, Leda Clay, and New Hampshire organic silt, both C_c and C_α increase as the effective stress approaches the critical pressure (σ'_c), reach a maximum at or just beyond the critical pressure, and then decrease at higher effective stresses. Throughout these effective stress changes the ratio C_α/C_c remains reasonably constant. This relationship between C_α and C_c and the rather narrow range of the ratio C_α/C_c is extremely useful for predicting field behavior from laboratory tests.

Comparisons of Field Observations and Laboratory Results

The literature is filled with the results of laboratory tests on the consolidation properties of soils, but field records are scarce. Some useful measurements were reported by Horn and Lambe [47] on the main complex of MIT buildings, which rest on the well-known Boston Blue Clay. Few observations were made in the years immediately after construction in 1915, but there was sufficient information to conclude from time-settlement curves that the primary consolidation in the subsoil was completed on average after about 20 years and that the rate of secondary was greatest at locations undergoing large primary consolidation. The rate of secondary consolidation in the field agreed reasonably well with results of oedometer tests.

In a discussion of the MIT work, Keene [48] gave the results of settlement observations over periods of 7 to 17 years on four Connecticut highway embankments resting on soft organic marine deposits. In general he found a relationship between the coefficient of secondary consolidation and the height of fill (or consolidation pressure). For a 1.5-m fill the value of C_α was about 0.010, for a 3-m fill it was about 0.017, and for high fills (5 and 9 m), C_α was in the range of 0.021 to 0.025. The higher values were attributed to higher shear stresses in the subsoil.

Keene suggested that the coefficient of secondary consolidation might be considered as a percentage of the primary consolidation. From his field observations, the average secondary settlement per log cycle of time is approximately 8% of the preceding primary settlement. From the long-term measurement of settlements of the Empress Hotel, the secondary settlement per cycle is about 12% of primary [43]. Thus there is a direct relationship between secondary and primary consolidation, but the ratio would be expected to vary appreciably between soil types.

Pelletier et al [49] reported results of extensive observations of settlement and piezometric levels under highway embankments near Portland, Maine. Reconstructed laboratory curves of consolidation, taking into account soil disturbance and soil structural effects, agreed well with field observations. The separation of primary and secondary consolidation from the laboratory curves was difficult, especially near the preconsolidation pressure. At greater pressures, the secondary settlement generally represented about 20 to 40% of the total.

Walker and Morgan [50] concluded from a comparison of field and laboratory observations that "where embankment heights are sufficiently large to initiate significant primary consolidation, the rate of settlement in the primary phase is about 200 times faster than that predicted from laboratory test data due to the influence of horizontal silt and sand seams" and "for low embankments, no primary consolidation is observed, and the settlement rate is controlled by secondary consolidation effects. Even for high embankments, secondary consolidation is significant due to rapid completion of the primary phase." It was their experience "that an uncritical application of conventional theory together with standard laboratory testing procedures can lead to gross errors in the prediction of field performance of the Coode Island silt."

These conclusions were based on studies of a bridge approach embankment and an adjacent test embankment in Melbourne, Australia. The subsoil at the site consists of 2 m of fill and 7 m of sand overlying about 15 m of Coode Island silt, a slightly overconsolidated silty clay with thin bands of silt and sand. *In situ* overburden stress and preconsolidation pressure are estimated to be 190 and 280 kPa, respectively. Using the classical consolidation theory, the predicted settlement (to 90% consolidation) was 460 mm in 95 years. From field measurements (Fig. 15) it was deduced that the primary consolidation was completed at Point A (i.e., 170 mm after six months) and secondary consolidation occurred from Point A to Point B, at which point a further load was added, causing the secondary

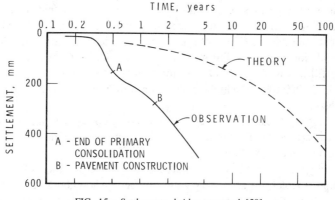

FIG. 15—*Settlements, bridge approach* [50].

rate to increase from 240 to 440 mm per log cycle of time. These observed rates agreed within 30% of laboratory rates.

Recently a very comprehensive investigation of laboratory and field consolidation has been carried out for Hydro Québec in preparation for the further development of James Bay hydroelectric resources. Extensive consolidation studies were used in the analysis of dyke settlement by a subcommittee chaired by Professor G. Mesri of the University of Illinois and reporting to a Committee of Specialists on the Sensitive Clays of the Nottaway, Broadback and Rupert (NBR) Complex, chaired by J.-J. Paré of the Société d'énergie de la Baie James. Laboratory testing was carried out in Canada, the United States, and Norway. Consolidation was monitored in the field under two large test embankments, each nearly 200 m long and more than 100 m wide at the base. The field experiments began in 1971 and the final report was issued in August 1983 [51].

The Committee describes the soft clays of the NBR river region as "highly structured and characterized by an aggregated open fabric with a strong and brittle inter-particle bonding. This structure results in a compressibility in the recompression range drastically smaller than the compressibility in the normally consolidated compression range. Moreover, beyond the pre-consolidation pressure, compression behaviour is highly non-linear." From observed field behavior the Committee developed a methodology for analyzing consolidation rate and settlement based on an assumption of one-dimensional deformation and water flow in the vertical direction. This was considered appropriate for embankments which are wide with respect to the thickness of the compressible clay foundation. It was concluded that the conventional oedometer test is adequate for determination of compressibility characteristics, and the falling-head permeability test in the oedometer provides a reliable procedure for measuring directly the coefficient of permeability.

Because the clay is very brittle, it deforms very little during recompression loading. Immediately beyond the preconsolidation pressure, the compression index is very large, decreasing with increasing pressure. Therefore, it was considered essential to carry out laboratory tests in the appropriate pressure range for settlement calculations. The rate of settlement was evaluated from the compressibility and the permeability of the clay. The Committee concluded that it is possible, on the basis of good-quality laboratory tests, to predict consolidation behavior of these clays with reasonable accuracy.

Figure 16 shows the pressure/void ratio curve for a typical sample of Olga clay, a lacustrine deposit at the south of the NBR complex. The open circles represent void ratios at the end of primary consolidation, and the solid circles, void ratios after 24 h. Figure 17 shows the rate of settlement during the increment near the preconsolidation pressure. Most of the consolidation during this particular load increment is of a secondary nature. Nevertheless, taking into account field stress levels and field performance, the Committee recommended that tests be carried out with 24-h load increments. Some tests are reported for controlled-

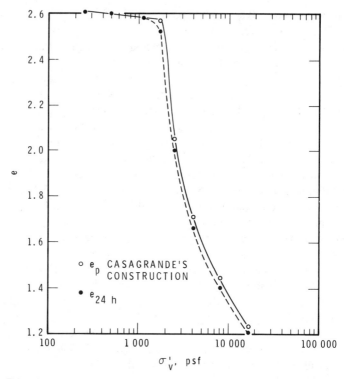

FIG. 16—*Pressure/void ratio curve for a typical specimen of Olga Clay* [51].

rate-of-strain loading, but generally at a rate faster than the average rate by incremental loading. At very slow rates of strain, there appeared to be no further decrease in preconsolidation pressure. Although neither C_c nor C_α is constant for the NBR clays, the ratio of C_α to C_c is reasonably constant at 0.033 (Fig. 18).

The magnitude of primary and secondary settlement was computed routinely using average values for C_c and C_α. For one of the test dykes the secondary settlement does not exceed 10% of the primary consolidation. For low dykes, however, where the final effective stress just exceeds the preconsolidation pressure, secondary settlement could equal or exceed the primary settlement.

The rate of settlement was computed, using the Illicon computer program developed at the University of Illinois, on the basis of the finite-strain theory of one-dimensional consolidation. This program takes into account variation with depth of the initial void ratio, the coefficient of permeability, and effective vertical stresses. It can handle any shape of pressure/void ratio curve, variations in the coefficient of permeability with void ratio, compressibility of the soil structure with time during and after the primary stage, any variation with depth of the vertical stress increase, and time-dependent loading for multistage construction. This procedure gave much better than normal agreement between computed and observed rates of settlement.

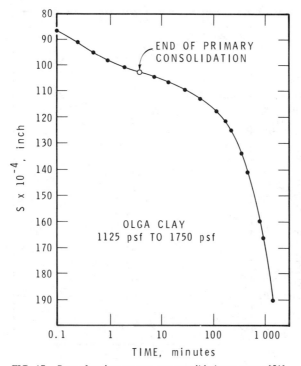

FIG. 17—*Rate of settlement near preconsolidation pressure* [51].

The most comprehensive long-term observations of full-scale field consolidation were begun in Sweden 40 years ago at the suggestion of Terzaghi. The results of these observations are summarized by Chang [52]. The first test fill, with vertical drains, was constructed in 1945 and a second, without drains, was built in 1947. These large fills, 30 m square, 2.5 m high, and resting on 11.5 m of normally consolidated clay, are extensively instrumented for settlement and pore pressure observations. Periodically, strength and water content tests have been made on the subsoil. Nearly 35 years after the fills were installed, the pore water pressures at the middle of the compressible layer were still almost equal to the applied load (Fig. 19), although the water content was significantly decreased. Until 1968 there was little increase in undrained shear strength at the middle of the layer, but by 1979 a significant increase was measured. The total compression of the clay subsoil up to 1980 was about 13%.

Laboratory investigations included oedometer tests on undisturbed specimens obtained by piston sampler. The specimens, 49.9 mm in diameter and 20 mm high, were subjected to 24-h load increment ratios of 0.5 to 1.0. In some cases "end of primary" curves are shown based on interpretations of square-root-of-time plots. On average, the preconsolidation pressure (σ'_c) interpreted from 24-h curves appears to be about 70 to 75% of the value determined at "end of primary."

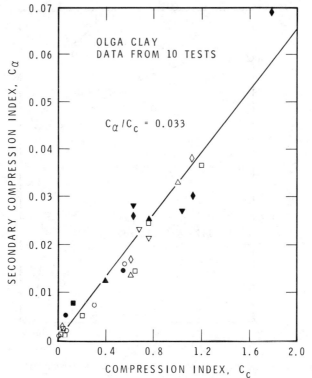

FIG. 18—C_α *to* C_c *relationship for Olga Clay* [51].

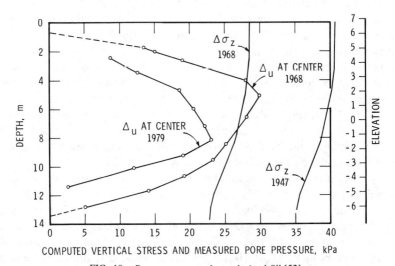

FIG. 19—*Pore pressure under undrained fill* [52].

The actual settlement up to October 1968 of the undrained fill agreed very well with the computed primary consolidation settlement based on the laboratory tests. The settlement is continuing, however, even though the applied load is decreasing as the fill settles into the groundwater, and the computations are rather complicated because as the load decreases some of the subsoils will have been overconsolidated with respect to the final loading. There is evidence that the settlement is approaching its ultimate value.

The delay in the increase of shear strength at the center of the compressible layer is compatible with the high pore water pressures. The observed decreases in water content, which agreed with the measured values of compression, are also compatible. The problem is that the loss of water, according to consolidation theory, should have caused a decrease in pore pressures and an increase in strength. In order to explain these observations, Chang visualizes the consolidation process in three phases. The conventional primary phase is that portion that follows the hydrodynamic laws in dissipating pore pressures created by an applied stress. A self-induced primary consolidation also follows the hydrodynamic laws, but the excess pore pressures have been created by breakdown of the soil structure under constant effective stress. Secondary consolidation is the portion that occurs under negligible excess pore pressures. Chang believed that the three phases probably occurred concurrently and could not therefore be separated clearly on the settlement curve.

A similar field observation of consolidation without gain in strength was reported for a varved clay by Stermac et al [53]. In this case a highway embankment caused the vertical stress in a 3-m layer of subsoil to equal or exceed slightly the estimated preconsolidation stress. After two years the layer had compressed about 12% without any gain in strength. The very slow rate of pore pressure dissipation was attributed to the breakdown of the soil structure, which generated pore pressures almost as quickly as they could dissipate by drainage [54]. After 2½ years the fill was partly removed. The pore pressures immediately dropped but the rate of dissipation appeared to increase, suggesting that the unloading stopped the mechanism that was creating the pore pressures.

A 20-year record of settlements of a Naval Accommodation Block and some earth fills near Ottawa, Canada [55], gave results similar to the Swedish case reported by Chang. Piezometric measurements beneath the building showed that effective stresses in the subsoil reached a threshold level equal to the preconsolidation pressure, while consolidation continued under a constant effective stress. Figure 20 shows the compression of a 3.6-m layer of soil under the Accommodation Block in relation to the effective vertical stresses and compares the field compression with that predicted from laboratory tests. After 2 years and after 20 years, the effective stress in the ground is still approximately equal to the laboratory preconsolidation stress, with the pore water carrying the additional load. It would appear, therefore, that the natural soil cannot accept a load greater than σ'_c until the consolidation approaches the ultimate value, an observation also made by Sällfors [15]. By extrapolation of time-settlement curves, the ul-

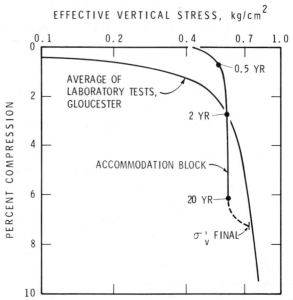

FIG. 20—*Stress-compression in the field and in the laboratory* [55].

timate settlement is expected to occur after about 40 years. Unfortunately, the building has been removed so that precise records are no longer available. From this case, and others in the vicinity, it was concluded that "pore pressures appear to be generated by the collapsing clay structure at a rate equivalent to their dissipation by drainage and this results in a substantial compression under constant effective stress conditions." This building is adjacent to the test fill at Gloucester described in the section on preconsolidation pressure [16].

In an analysis of almost 50 full-scale case records of consolidation, Crooks et al [56] identified eleven cases in which the pore water pressures continued to increase after completion of loading. This behavior, which was attributed to creep/breakdown of the soil structure, resulted in a very slow or negligible rate of *net* pore water pressure dissipation. In the majority of cases, however, the measured rate of dissipation was significantly faster than the predicted rate and this was commonly attributed to inaccurate modeling of field drainage conditions.

Crooks et al found from oedometer tests on a wide variety of soft clays that "the magnitude of c_v decreases significantly as the stress level increases up to and beyond the preconsolidation pressure." For highly sensitive soils, c_v is at a minimum near the σ'_c, due to the breakdown in soil structure. A similar relationship was shown by Janbu et al [22].

Discussion

This review refers to a small portion of the total literature on consolidation testing. It seemed unnecessary to cite every piece of work studied and no doubt much important work has been missed entirely. For example, the highly compressible peat soils are not covered, nor are the unsaturated soils. Attention has been directed primarily at the undisturbed natural inorganic clays and wherever possible at tests that could be related to field observations.

For most natural soils, the standard oedometer test with incremental loading provides a reasonable basis for estimating the magnitude of consolidation settlement in the virgin compression range. Settlement estimates in the recompression range are generally too large, sometimes much too large. Of course sample disturbance and insufficient tests can lead to substantial errors in any calculations. Estimating the rate of consolidation settlement is always difficult. The discrepancy between prediction and performance is attributed primarily to differences between the permeability computed from unidirectional water flow in the laboratory specimen and the three-dimensional flow that occurs in the field. Better results are obtained if permeabilities measured on very large specimens or in the field are used in settlement predictions.

The greatest source of error, in this author's view, is caused by ignoring the time-dependent resistance to compression of the soil structure. For almost all natural clays, if it is assumed at any stage of a normal test that the sole constraint on deformation is the resistance to pore water flow, the computed coefficient of permeability will be too low and the predicted rate of settlement will be too slow. No satisfactory theory has been developed to account for this discrepancy. The structural factor is most pronounced in sensitive clays and special attention has therefore been directed to these clays. This is not because sensitive clays are as common as insensitive ones, but because it is easier to observe their structural influences and to draw conclusions that may have more general application. The division of consolidation into primary and secondary components inhibits understanding more than it helps with calculations. Referring to Fig. 13, at the load increment just below σ'_c, three quarters of the consolidation occurs after pore water pressures have dissipated. Under the next load increment, 40% of the consolidation occurs in the so-called secondary phase. Field evidence has been quoted where these kinds of soil consolidate for years under nearly constant effective stresses.

In both the field and the laboratory, those portions of the soil near drainage boundaries are well into the secondary consolidation phase before the primary phase is half finished at the center of the consolidating layer. Under such circumstances it is not reasonable to divide the consolidation process into these two arbitrary phases. They should not be considered separate phenomena.

The Terzaghi theory for the rate of consolidation is an elegant theory. It provided the basis for developing the principle of effective stresses and led to the understanding of the strength and compressibility of clays. But it was developed using remolded soils and it almost always leads to gross errors in predicting the rate of consolidation in the field. What is needed is a procedure for assessing the effective stress-deformation properties of soil, taking into account the great difference between laboratory and field rates of strain, and a method for estimating the delay in reaching final effective stress values after load application.

Conclusions

The consolidation test is one of the fundamental tests for saturated soils. It is more than 60 years since the first such test was carried out. Much has been learned about consolidation by comparing the results of oedometer tests with consolidation under full-scale field loading, but still there are many uncertainties. The evidence suggests the following conclusions:

1. Soils with large void ratios (high water content) will undergo large consolidation settlements when loaded beyond their preconsolidation pressure. The amount of settlement is usually reasonably predictable from oedometer tests.

2. Soils with small void ratios (low water content) will undergo relatively small consolidation settlements, but the amount of settlement is less predictable from laboratory tests.

3. The laboratory determination of the preconsolidation pressure of soft or sensitive clays is considerably influenced by the method of test, especially by the rate of loading. Nevertheless there is also field evidence that tests using normal incremental loading at 24-h intervals give reasonable pressure-compression relationships for estimating settlements.

4. When highly structured (sensitive) soils are compressed, their structure tends to collapse and this results in increased pore pressure. Field evidence shows that in these cases the soil continues to consolidate, sometimes for decades, under constant effective stress.

5. Soils that exhibit large primary consolidation will generally have proportionally large secondary consolidation. Both C_c and C_α are related to the natural water content of the soil; their ratio (C_α/C_c) is reasonably constant for particular soils and has a rather narrow variation among all soils.

6. Secondary consolidation appears to be of most importance for highly plastic and sensitive soils and greatest at pressures near the preconsolidation pressure.

7. Tests on remolded soils cannot be expected to give reliable values for virgin consolidation. Disturbance of soil samples will cause an overestimate of settlement.

8. The rate of consolidation in the field is almost always faster than predictions based on laboratory tests. There are two main reasons for this. First, the soil is more permeable, especially in the horizontal direction, than the values measured in unidirectional tests in the laboratory; and second, the resistance to deformation of the soil structure masks the hydrodynamic effect, leading to underestimation of the permeability.

9. The understanding of consolidation phenomena will be advanced primarily by field observation. The most useful research is the observation of pore pressures and consolidation settlement of discrete layers of soil *in situ* coupled with laboratory tests on good, undisturbed soil samples.

Acknowledgments

This paper is a contribution of the Division of Building Research, National Research Council Canada.

References

[1] Bjerrum, L., "Problems of Soil Mechanics and Construction on Soft Clays," State-of-the-Art Report to Session IV, in *Proceedings,* 8th International Conference on Soil Mechanics and Foundation Engineering, Moscow, Vol. 3, 1973, pp. 111–159.

[2] Peck, R. B., "Advantages and Limitations of the Observational Method in Applied Soil Mechanics," 9th Rankine Lecture, *Geotechnique,* London, Vol. 19, No. 2, 1969, pp. 171–187.

[3] Terzaghi, K., "Settlement and Consolidation of Clay," *Engineering News-Record,* 26 Nov. 1925, pp. 874–878.

[4] Taylor, D. W., "Research on the Consolidation of Clays," Massachusetts Institute of Technology, Department of Civil and Sanitary Engineering, Serial 82, Cambridge, 1942.

[5] Bjerrum, L., "Engineering Geology of Norwegian Normally-Consolidated Marine Clays as Related to Settlements of Buildings," 7th Rankine Lecture, *Geotechnique,* London, Vol. 17, No. 2, 1967, pp. 81–118.

[6] Berre, T. and Iversen, K., "Oedometer Tests with Different Specimen Heights on a Clay Exhibiting Large Secondary Compression," *Geotechnique,* London, Vol. 22, No. 1, 1972, pp. 53–70.

[7] Garlanger, J. E., "The Consolidation of Soils Exhibiting Creep Under Constant Effective Stress," *Geotechnique,* London, Vol. 22, No. 1, 1972, pp. 71–78.

[8] Foss, I., "Secondary Settlements of Buildings in Drammen, Norway," in *Proceedings,* 7th International Conference on Soil Mechanics and Foundation Engineering, Mexico City, Vol. 2, 1969, pp. 99–106.

[9] Engesgaar, H., "15-Storey Building on Plastic Clay in Drammen, Norway," in *Proceedings,* 8th International Conference on Soil Mechanics and Foundation Engineering, Moscow, Vol. 1.3, 1973, pp. 75–80.

[10] Simons, N. E., "Normally Consolidated and Lightly Over-Consolidated Cohesive Materials," Review Paper: Session 2, in *Proceedings,* Conference on Settlement of Structures, British Geotechnical Society, 1974, pp. 500–530.

[11] Butler, F. G., "Heavily Over-Consolidated Clays," Review Paper: Session 3, in *Proceedings,* Conference on Settlement of Structures, British Geotechnical Society, 1974, pp. 531–578.

[12] Rowe, P. W., "The Relevance of Soil Fabric to Site Investigation Practice," 12th Rankine Lecture, *Geotechnique,* London, Vol. 22, No. 2, 1972, pp. 195–300.

[13] Samuels, S. G., "Some Properties of the Gault Clay from the Ely-Ouse Essex Water Tunnel," *Geotechnique,* London, Vol. 25, No. 2, 1975, pp. 239–264.

[14] Mesri, G., Rokhsar, A., and Bohor, B. F., "Composition and Compressibility of Typical Samples of Mexico City Clay," *Geotechnique,* London, Vol. 25, No. 3, 1975, pp. 527–554.

[15] Sällfors, G., "Preconsolidation Pressure of Soft, High-Plastic Clays," Ph.D. thesis, Chalmers University of Technology, Göteborg, Sweden, 1975.

[16] Leroueil, S., Samson, L., and Bozozuk, M., "Laboratory and Field Determination of Preconsolidation Pressures at Gloucester," *Canadian Geotechnical Journal,* Vol. 20, No. 3, 1983, pp. 477–490.

[17] Hamilton, J. J. and Crawford, C. B., "Improved Determination of Preconsolidation Pressure of a Sensitive Clay," in *Papers on Soils, ASTM STP 254,* American Society for Testing and Materials, Philadelphia, 1959, pp. 254–271.

[18] Smith, R. E. and Wahls, H. E., "Consolidation Under Constant Rates of Strain," *Journal of the Soil Mechanics and Foundations Division,* American Society of Civil Engineers, Vol. 95, No. SM2, 1969, pp. 519–539.

[19] Wissa, A. E. Z., Christian, J. T., Davis, E. H., and Heiberg, S., "Consolidation at Constant Rate of Strain," *Journal of the Soil Mechanics and Foundations Division,* American Society of Civil Engineers, Vol. 97, No. SM10, Oct. 1971, pp. 1393–1413.

[20] Lowe, J., Jonas, E., and Obrician, V., "Controlled Gradient Consolidation Test," *Journal of the Soil Mechanics and Foundations Division,* American Society of Civil Engineers, Vol. 95, No. SM1, 1969, pp. 77–97.

[21] Gorman, C. T., Hopkins, T. C., Deen, R. C., and Drnevich, V. P., "Constant-Rate-of-Strain and Controlled-Gradient Consolidation Testing," Geotechnical Testing Journal, Vol. 1, 1978, pp. 3–15.

[22] Janbu, N., Tokheim, O., and Senneset, K., "Consolidation Tests with Continuous Loading," in Proceedings, 10th International Conference on Soil Mechanics and Foundation Engineering, Stockholm, Vol. 1, 1981, pp. 645–654.

[23] Simons, N. E. and Som, N. N., "The Influence of Lateral Stresses on the Stress Deformation Characteristics of London Clay," in Proceedings, 7th International Conference on Soil Mechanics and Foundation Engineering, Mexico City, 1969, pp. 369–377.

[24] Lambe, T. W,. "The Stress Path Method," Journal of the Soil Mechanics and Foundations Division, American Society of Civil Engineers, Vol. 93, SM6, 1967, pp. 309–331.

[25] Crawford, C. B. and Burn, K. N., "Settlement Studies on the Mt. Sinai Hospital, Toronto," Engineering Institute of Canada, The Engineering Journal, Vol. 45, No. 12, 1962, pp. 31–37.

[26] Casagrande, A., "The Determination of the Preconsolidation Load and its Practical Significance," in Proceedings, First International Conference on Soil Mechanics, Cambridge, MA, Vol. 3, 1936, pp. 60–64.

[27] Schmertmann, J. H., "The Undisturbed Consolidation Behavior of Clay," Transactions, American Society of Civil Engineers, Vol. 120, 1955, pp. 1201–1233.

[28] Crawford, C. B., "Interpretation of the Consolidation Test," Journal of the Soil Mechanics and Foundations Division, American Society of Civil Engineers, Vol. 90, SM5, 1964, pp. 87–102.

[29] Crawford, C. B., "The Resistance of Soil Structure to Consolidation," Canadian Geotechnical Journal, Vol. 2, No. 2, 1965, pp. 90–97.

[30] Jarrett, P. M., "Time-Dependent Consolidation of a Sensitive Clay," Journal of Materials Research and Standards, Vol. 7, No. 7, 1967, pp. 300–304.

[31] Leonards, G. A. and Ramiah, B. K., "Time Effects in the Consolidation of Clays," in Papers on Soils, ASTM STP 254, American Society for Testing and Materials, Philadelphia, 1959, pp. 116–130.

[32] Leroueil, S., Tavenas, F., Samson, L., and Morin, P., "Preconsolidation Pressure of Champlain Clays. Part II. Laboratory Determination," Canadian Geotechnical Journal, Vol. 20, No. 4, 1983, pp. 803–816.

[33] Morin, P., Leroueil, S., and Samson, L., "Preconsolidation Pressure of Champlain Clays. Part I. In-situ Determination," Canadian Geotechnical Journal, Vol. 20, No. 4, 1983, pp. 782–802.

[34] Bozozuk, M. and Leonards, G. A., "The Gloucester Test-Fill," in Proceedings, American Society of Civil Engineers Specialty Conference, Performance of Earth and Earth-Supported Structures, Purdue Univ., Lafayette, Ind., Vol. 1, Part 1, 1972, pp. 299–317.

[35] Mesri, G. and Choi, Y. K., "The Uniqueness of the End-of-Primary (EOP) Void Ratio-Effective Stress Relationship," submitted to the 11th International Conference on Soil Mechanics and Foundation Engineering, San Francisco, 1985.

[36] Mesri, G. and Rokhsar, A., "Theory of Consolidation for Clays," Journal of the Geotechnical Engineering Division, American Society of Civil Engineers, Vol. 100, GT8, 1974, pp. 889–904.

[37] Schiffman, R. L., Pane, V., and Gibson, R. E., "An Overview of Non-Linear Finite Strain Sedimentation and Consolidation," in Proceedings, Symposium on Sedimentation and Consolidation Models: Predictions and Validation, American Society of Civil Engineers, 1984, pp. 1–29.

[38] Leonards, G. A. and Altschaeffl, A. G., "Compressibility of Clay," Journal of the Soil Mechanics and Foundations Division, American Society of Civil Engineers, Vol. 90, No. SM5, 1964, pp. 133–155.

[39] Wahls, H. E., "Analysis of Primary and Secondary Consolidation," Journal of the Soil Mechanics and Foundations Division, American Society of Civil Engineers, Vol. 88, SM6, 1962, pp. 207–231.

[40] Walker, L. K. and Raymond, G. P., "The Prediction of Consolidation Rates in a Cemented Clay," Canadian Geotechnical Journal, Vol. 5, No. 4, 1968, pp. 192–216.

[41] Poorooshasb, H. B., Law, K. T., Bozozuk, M., and Eden, W. J., "Consolidation of Sensitive Clays," in Proceedings, 10th International Conference on Soil Mechanics and Foundation Engineering, Stockholm, Vol. 1, 1981, pp. 219–223.

[42] Walker, L. K., "Secondary Settlement in Sensitive Clays," Canadian Geotechnical Journal, Vol. 6, No. 2, 1969, p. 219–222.

[43] Crawford, C. B. and Sutherland, J. G., "The Empress Hotel, Victoria, British Columbia: Sixty-Five Years of Foundation Settlements," *Canadian Geotechnical Journal*, Vol. 8, No. 1, 1971, pp. 77–93.

[44] Crooks, J. H. A., "General Report for Session 2: Validation and Field Studies," in *Proceedings, Symposium on Sedimentation Consolidation Models: Predictions and Validation*, American Society of Civil Engineers, 1984, pp. 337–343.

[45] Mesri, G., "Coefficient of Secondary Compression," *Journal of the Soil Mechanics and Foundations Division*, American Society of Civil Engineers, Vol. 99, SM1, 1973, pp. 123–137.

[46] Mesri, G. and Godlewski, P. M., "Time- and Stress-Compressibility Interrelationship," *Journal of the Geotechnical Engineering Division*, American Society of Civil Engineers, Vol. 103, GT5, 1977, pp. 417–430.

[47] Horn, H. M. and Lambe, T. W., "Settlement of Buildings on the MIT Campus," *Journal of the Soil Mechanics and Foundations Division*, American Society of Civil Engineers, Vol. 90, SM5, 1964, pp. 181–195.

[48] Keene, P., Discussion to Horn, H. M., and Lambe, T. W., "Settlement of Buildings on the MIT Campus," *Journal of the Soil Mechanics and Foundations Division*, American Society of Civil Engineers, Vol. 91, SM5, 1965, pp. 95–107.

[49] Pelletier, J. H., Olson, R. E., and Rixner, J. J., "Estimation of Consolidation Properties of Clay from Field Observations," *Geotechnical Testing Journal*, Vol. 2, No. 1, March 1979, pp. 34–43.

[50] Walker, L. K. and Morgan, J. R., "Field Performance of a Firm Silty Clay," in *Proceedings, 9th International Conference on Soil Mechanics and Foundation Engineering*, Tokyo, Vol. 1, 1977, pp. 341–346.

[51] "Report of the Committee of Specialists on Sensitive Clays of the NBR Complex: Annex III, Report of the Settlement Subcommittee," Société d'énergie de la Baie James, Geology and Soil Mechanics Department, Montreal, Aug. 1983.

[52] Chang, Y. C. E., "Long Term Consolidation Beneath the Test Fills at Väsby, Sweden," Swedish Geotechnical Institute, Report 13, Linköping, Sweden, 1981.

[53] Stermac, A. G., Lo, K. Y., and Barsvary, A. K., "The Performance of an Embankment on a Deep Deposit of Varved Clay," *Canadian Geotechnical Journal*, Vol. 4, No. 1, Feb. 1967, pp. 45–61.

[54] Crawford, C. B. and Eden, W. J., Discussion of "The Performance of an Embankment on a Deep Deposit of Varved Clay," *Canadian Geotechnical Journal*, Vol. 4, No. 1, 1967, pp. 63–64.

[55] Crawford, C. B. and Burn, K. N., "Long Term Settlements on Sensitive Clay," in *Lauritz Bjerrum Memorial Volume*, Norwegian Geotechnical Institute, Oslo, 1976, pp. 117–124.

[56] Crooks, J. H. A., Becker, D. E., Jefferies, M. G., and McKenzie, K., "Yield Behaviour and Consolidation. 1: Pore Pressure Response," in *Proceedings, Symposium on Sedimentation Consolidation Models: Predictions and Validation*, American Society for Civil Engineers, 1984, pp. 356–381.

General Reports

Vincent P. Drnevich[1]

Report on Laboratory Tests for Consolidation Behavior

REFERENCE: Drnevich, V. P., **"Report on Laboratory Tests for Consolidation Behavior,"** *Consolidation of Soils: Testing and Evaluation, ASTM STP 892*, R. N. Yong and F. C. Townsend, Eds., American Society for Testing and Materials, Philadelphia, 1986, pp. 107–114.

Eight papers are covered in this report. The authors and titles are listed in Table 1. All papers focus on laboratory testing or theory related directly to laboratory testing. Some include field experimental results and field applications. A directory to topics covered in the papers also is given in Table 1.

This report summarizes the salient features of each paper. In the process, an attempt will be made to: (1) formulate questions on topics thay may not be clearly understood (at least by this reporter), (2) identify shortcomings and areas of needed research, (3) look for items of agreement or disagreement, and (4) form some general conclusions.

Summary of Salient Features

The paper by *Juárez-Badillo* (pp. 137–153) presents a general one-dimensional, small-strain theory that considers the variation with stress of: coefficient of compressibility, permeability, and coefficient of consolidation. Three constants are introduced to relate these changes, but only two of the three are independent. The variation of compressibility is inversely proportional to the vertical effective stress. The variation of permeability is related to the initial permeability by the ratio of effective stress to initial effective stress where this ratio is raised to a power (permeability versus ratio of effective stress is a straight line on a log-log plot). A similar ratio to the power, λ, is used to relate the coefficient of consolidation, c_v, to its initial value. Juárez-Badillo shows that for $\lambda = 0$ (constant c_v but permeability and compressibility variable), his solution is identical to that of Raymond published in 1965. The only solution given in the paper is that for $\lambda = 0$. Solutions for some other cases are given in a previously published paper.

[1] Professor of Civil Engineering, Department of Civil Engineering, University of Kentucky, Lexington, KY 40506-0046.

TABLE 1—*Authors, paper titles, and subject areas.*

Author(s)	Title	Theoretical			Experimental		
		General	Laboratory	Field	Laboratory	Field	Field Applications
Juárez-Badillo	General Theory of Consolidation for Clays	X	X				
Fredlund and Rahardjo	Unsaturated Soil Consolidation Theory and Laboratory Experimental Data	X	X		X		
Armour and Drnevich	Improved Techniques for the Constant-Rate-of-Strain Consolidation Test		X		X		
Nwabuokei and Lovell	Compressibility and Settlement of Compacted Fills		X	X	X		X
Sills, Hoare, and Baker	An Experimental Assessment of the Restricted Flow Consolidation Test		X		X		
von Fay, Byers and Kunzer	Desktop Computer Application for Consolidation Testing and Analysis				X		
von Fay and Cotton	Constant-Rate-of-Loading (CRL) Consolidation Test		X		X		
Zeevaert	Consolidation in the Intergranular Viscosity of Highly Compressible Soils	X	X	X	X	X	X

For the solutions, it is shown that the degree of consolidation becomes dependent on the ratio of final to initial thickness of the specimen (see Fig. 1 in the paper). This is significant in that somewhat different degrees of consolidation will occur at a given time depending on whether there is swell or compression, with swell occurring more slowly than compression. The author gives ranges for each of the three parameters that might be expected for typical soils. In the version of the paper available to this reporter, the symbol γ was used for one of the constants. It is easy to confuse this with the unit weight symbol which also occurs frequently in consolidation theories.

Fredlund and Rahardjo (pp. 154–169) address the issue of partially saturated soils which occur far more frequently in practice than most of us want to believe. In the paper, a partial differential equation is given for the water phase and another one given for the air phase. These equations are based on a number of assumptions including: (1) isotropic soil, (2) continuous air phase, (3) infinitesimal strains, (4) linear constitutive relations for all phases, (5) coefficients of permeability of water and air phases are functions of the volume-mass soil properties during the transient process, (6) the effect of air diffusing through water is ignored, (7) the effect of water vapor movement is ignored, and (8) air dissolving in water is ignored. This reported did not understand the meaning of assumption (5) above. No further explanations were given in the paper. The actual derivations of the equations were not given but were referenced in a paper that was presented in 1982 and published in the Proceedings of the NATO Conference on Flow Through Porous Media.

The authors conducted five laboratory tests on compacted kaolin. Degrees of saturation ranged from 63 to 90%. Four of the tests were conducted in a triaxial apparatus where the specimen was enclosed in a rubber membrane. A high entry porous stone was used at one end to measure pore water pressures, and a low entry porous stone was used at the other to measure pore air pressures. Height-to-diameter ratios ranged from approximately 0.5 to 1 for these tests.

One of the tests was performed in an Anteus back-pressure consolidometer where the height-to-diameter ratio was 0.35 and the degree of saturation was 72.5%.

In each of the tests, one of the parameters (applied stress, pore water pressure, or pore air pressure) was changed and the other two were monitored with time to approximately 10 000 min (one week); see Table 2. The paper gives graphs (Figs. 1 to 5) of the measured data, and rational explanations are given for the observed behavior.

The theory requires two "compressibility moduli" for the soil structure, two for the air phase and two for the water phase. From the units of these, (stress)$^{-1}$, they should be referred to as compressibilities rather than compressibility moduli. One of these compressibilities in each case is due to changes in total stress minus pore air pressure; the other is due to changes in the pore air pressure minus the pore water pressure.

There is an additional constraint in that the first of the two compressibilities for the soil structure must be equal to the sum of the first of the two compressibilities for the other two phases. The same holds true for the second of the two compressibilities as well.

Values of compressibilities were determined for each of the tests by measurement of the difference in values between the start of the test and the end of the test.

Next, the two governing partial differential equations were modified by dropping the last two terms in each. This means that the variation in the coefficient of permeability during the test was not considered. These equations were solved numerically, and the remaining coefficients were chosen to best fit the measured behavior. Details of this process were not given; hence it would be difficult for someone else to duplicate the process.

Figures are given for all five tests (Figs. 6 to 10), but only two figures (Figs. 11 and 12) are given which compare the fitted curves with measured results. The fitted curves accurately represent the soil structure and water phase measured data.

It appears to this reporter that the process is complicated and is limited to partially saturated soils that closely fit the assumed conditions. Many nearly saturated soils or ones with very low degrees of saturation would not fit the assumed conditions very well. Also, four of the tests have three-dimensional volume changes with one-dimensional flow (triaxial tests). Parameters from these tests would not be expected to correlate very well with the corresponding parameters for tests with one-dimensional volume changes and one-dimensional flow (oedometer tests). The authors provided compressibility values for all tests (Table 3), but did not provide other parameters or attempt to correlate values of compressibility between the tests.

Armour and Drnevich (pp. 170–183) address the constant-rate-of-strain (CRS) consolidation test. Based on the theoretical work of Wissa, an equation is developed to allow for estimating strain rates for testing. The equation considers both the stiffness of the soil (through the liquidity index) and the permeability.

The authors suggest that the test procedure be modified to allow for performing a permeability test on the specimen in the apparatus before beginning the consolidation loading. The pore water pressure is slightly increased at the fixed end of the specimen, and the flow of water through the specimen is measured with time to establish the permeability. After the permeability is established, axial loading is begun and flow is immediately stopped. Thus the initial conditions for the consolidation test are different from those assumed by Wissa. These initial conditions cause the transient component to be smaller in magnitude; hence steady-state conditions may be assumed to exist more quickly than for conventional CRS testing.

Tests on three different soils with widely varying properties were performed. Each test involved four loading cycles, each to higher effective stresses. Between each cycle, permeability tests were performed on the specimen. Results from the tests indicated that pore pressures generated by axial loading tend to peak in the normally consolidated range of stress when the soil is lightly overconsolidated but peak in the overconsolidated range for soils that are highly preconsolidated. Data from these tests were used to provide the coefficients in the equation for estimating strain rates. It is shown that the strain rates provided for ASTM D4186 may not give excess pore pressures in the desirable range of 3 to 20%.

The paper by *Nwabuokei and Lovell* (pp. 184–202) focuses on the problem of settlement of compacted fills. They examine the behavior of a natural soil, New Haven Clay, that is compacted with three different energies in laboratory compaction molds. The term *compaction prestress* was used to ''represent the fraction of the compaction energy effectively transmitted to the soil matrix due to plastic deformation during the compaction process.''

After compaction, the specimens were subjected to various confining pressures (representing various fill heights) and then back-pressure saturated. Volume changes were monitored. It was assumed that the corresponding fills would eventually become saturated.

An attempt is made to model the process by energy considerations beginning with the nominal compactive energy due to laboratory compaction. A number of assumptions must be made to do this. It is assumed that all input energy is converted into plastic strain energy. The calculated total stress generated in the soil as a result of plastic deformation is the predicted as-compacted prestress. Figure 5 of the paper gives this prestress for three compaction energies and a range of water contents. Statistical procedures were used to determine the relationship of the compaction prestress to compaction energy and water content for this soil.

As-compacted compressibilities also were determined by use of incremental loading tests on the compacted specimens. Additional statistical fits were made to establish volumetric strains. Procedures were outlined for predicting settlements for a hypothetical fill made with this clay.

This reporter found it difficult to follow the development of the theory. It is not totally clear how the as-compacted prestress and the saturated prestress are measured. The former is in terms of total stress and the latter is in terms of effective stress. The statistical equations presented are units dependent on and apply only to this soil. Further refinement of the technique appears warranted before it can be applied.

Sills et al (pp. 203–216) present a relatively new type of consolidation test (at least to this reporter): a restricted flow consolidation (RFC) test. In this test, a special one-dimensional consolidometer (Fig. 3 in the paper) is used where the specimen is completely sealed. The specimen must be saturated and the pore water system must be completely de-aired.

The testing process consists of applying an increment of load, usually at the largest level of stress desired for test results. Because of the boundary conditions, there is little increase in effective stress. Then, flow from one end of the specimen is allowed to occur at a restricted rate. Special devices are required for this, and the amount of restriction necessarily depends on the soil being tested.

These conditions set up a small gradient within the specimen which is assumed uniform. Total stress is measured on one face of the specimen, and pore pressures are measured on the drained and undrained faces. Specimen length changes are also measured.

As flow continues, effective stresses continue to increase, the specimen consolidates, and the deformation is measured. A typical test takes about 3 h. The process can be reversed for unloading as long as pressures are maintained to prevent cavitation.

It is possible to determine the permeability of the soil from the measurement during the test. The coefficient of consolidation may be determined from measured values of permeability and compressibility. Secondary compression behavior can be obtained by monitoring behavior with time after the pore pressures have dissipated.

Results of several tests are provided. The authors discuss the sources of error in the test: undissolved gas in the sample and bedding errors. These errors are cause for unreliable data for low effective stresses (in the early part of the test).

This reporter sees merit in the test. It is very amenable to automation. Some technical difficulties still remain with the restriction device and preselecting the appropriate amount of restriction.

Automation is the topic of the paper by *von Fay et al* (pp. 217–235), who discuss the process used to automate data acquisition and reduction of a bank of conventional, incremental loading consolidation test apparatus. Most of the mechanics of performing the test are still done manually (e.g., load application). Displacement transducers are used to measure displacement, pore pressures are not measured, and load cells are not used to measure force. Force levels are obtained from the loading system calibrations and manually keyed into the computer. For the duration of the load, they are assumed constant.

An interesting feature of this system is that many tests are monitored simultaneously; the only restriction made is that there must be at least a 10-s delay between when each test is started or an individual load increment is applied. Also of interest is the fact that the computer also reduces the data from these tests. One does not appreciate the complexity of the "simple" incremental loading test until trying to write computer codes to reduce the data. This reporter would have liked to have seen more details associated with the algorithms and data reduction process. However, paper length limitations probably precluded this. Details of the algorithms are available in reports put out by the Bureau of Reclamation.

One of the authors of the previous page, *von Fay, and Cotton* (pp. 236–256) share their experiences with the constant-rate-of-loading (CRL) test. This test system was constructed in-house and is configured about a floating ring consolidometer that allows for back pressuring. A double set of O-rings seal the floating ring to the upper and lower platens. (If not properly designed, this system could lead to significant friction especially if nonuniform settlement occurs.)

The authors claim that this system can be used with either single or double drainage. If double drainage were used, very little information would be available about excess pore pressures in the middle of the specimen and, hence, pore pressure gradients. (This reporter recently determined that the authors are using piezometer ports in the ring to circumvent this shortcoming.)

The results of tests on two different soils are presented and compared with results of incremental loading tests on the same soils. Data presentation is in the form of scatter bands (replicate tests were performed); data points for individual tests are not presented. The reducibility of the results and the comparisons with results of incremental loading tests both look good.

As with many of the accelerated rate tests, the selection of the rate at which to test is a problem. Some of the tests were run too slowly with low pore pressures and longer than necessary testing times. A preliminary guide for selecting loading rates is given in Fig. 15 of the paper, but additional work is needed on improving the rate-of-loading selection process.

The final paper discussed in this report is by *Zeevaert* (pp. 257–281). To most of us, the term *intergranular viscosity* is not so familiar as the term *secondary compression* which is synonymous with it. The author points out the existence of a critical stress above which the soil structure suffers a collapse followed by a new structural behavior (Fig. 1). The critical stress is above the overburden stress and in many respects is similar to the preconsolidation stress. For most practice situations, it is desirable to limit stresses to values less than this critical stress.

Results of consolidation tests on numerous specimens of undisturbed, highly compressible soils produce four characteristically different displacement-time curves (Figs. 2 to 5). For stress levels that are low relative to the critical stress, the curves (labelled Type I) look similar to those for conventional soils. For stress levels approaching the critical stress, the secondary compression portion of the curve (labelled Type II) has a convex configuration and then a straight portion. At the critical stress, Type III behavior is exhibited where differentiation between primary and secondary compression is not distinguishable from the deformation behavior. The last type of curve (Type IV) occurs for stress levels above the critical stress. Examples of field measured settlements with time are given to support the fact that these same phenomena exist *in situ* (Figs. 6 to 8).

A rheological model is proposed to account for the observed behavior. It consists of a generalized Kelvin-Voigt unit in series with a generalized ''Z'' unit (Fig. 10). The ''Z'' unit consists of two different types of elements, one for linear Newtonian fluids (dashpots) and one for nonlinear fluids where strain rates are hyperbolically related to stress. This model has sufficient flexibility to nicely model all four types of observed behavior.

Methods of fitting measured behavior to obtain the model constants are detailed. The process is relatively direct and easy to follow. The variation of the model parameters is shown (Fig. 17) to be consistent.

The last portion of Zeevaert's paper discusses some of the problems in testing these soils, especially that of gas formation. He recommends that back pressuring be done to simulate the hydraulic (pore water) pressures *in situ* and that seating stresses on the order of 20% of the overburden stresses be applied to avoid swelling. He also recommends that stress increments in the test be limited to one eighth of the overburden stress so that the clay structure is not damaged by the testing process.

Discussion and Conclusions

The papers reported on herein reaffirm the complex nature of consolidation behavior and the difficulty of accurately representing actual field behavior with laboratory tests.

Some issues that need additional research include:

1. Effects of allowing swell versus preventing swell at the initiation of laboratory consolidation testing.
2. Criteria for maximum pore pressure gradients during consolidation testing.
3. Criteria for maximum strain rate during accelerated loading testing.
4. Preferred procedures for determining secondary compression behavior when accelerated loading tests are performed.
5. Consolidation behavior of compacted soils and partially saturated soils.
6. Consolidation of man-made and process waste materials.

Some of the above items were partially addressed in these papers, others were not.

There is still much work to be done before we can consider writing ASTM standards for partially saturated soils, for soils with large particle sizes, and for laboratory compacted soils to simulate field compaction. On the other hand, some test procedures have become sufficiently matured that they are ripe for standardization. Candidates for standardization include controlled gradient and controlled-rate-of-loading consolidation tests.

V. Silvestri[1]

Report on Theory and Laboratory Testing Requirements: Comparisons

REFERENCE: Silvestri, V., **"Report on Theory and Laboratory Testing Requirements: Comparisons,"** *Consolidation of Soils: Testing and Evaluation, ASTM STP 892,* R. N. Yong and F. C. Townsend, Eds., American Society for Testing and Materials, Philadelphia, 1986, pp. 115–121.

The consolidation test provides basic information on the compressibility characteristics of soft soils needed to estimate the magnitude and rate of settlements resulting from one-dimensional straining. For the past 40 years, extensive use has been made of the conventional consolidation test in which samples are subjected to increasingly large increments of load, and the resulting time/deformation curves are used to obtain the various compressibility parameters. About 20 years ago, several investigators [1–3] found that the experimental results obtained were not unique, rather they were functions of the loading sequence used and the time allowed for each load increment. In other words, as in the case of other geotechnical tests, it was realized that the conventional test was by no means a "standard" test.

To overcome some of the problems associated with the conventional test, several types of continuous consolidation tests have been introduced in recent years. The most common of these tests are the constant rate of strain or deformation test, the constant or controlled hydraulic gradient test, the constant rate of loading test, and the continuous loading test. Even with these tests, however, it soon became apparent that the results are also functions of the specific procedures used [4–6].

In order to obtain more meaningful parameters for the analysis of the consolidation of soft soils, the results obtained in these continuous tests should be compared (1) with those obtained in the conventional tests and (2) with existing field data. Such comparisons represent the bulk of this report.

Ten papers were assigned to this reporter and were grouped in four categories depending on their subject (Table 1). Of these, three papers deal with the various test techniques and procedures used in geotechnical laboratories of international reputation. Three additional papers deal with the interpretation of the consoli-

---

[1] Professor, Department of Civil Engineering, École Polytechnique, Montreal, P. Q., Canada.

TABLE 1—*Authors and papers.*

Category	Authors	Procedures	Theory	Comparisons	Miscellaneous
A	Ducasse et al	x			
	Larsson and Sällfors	x		x	
	Sandbaekken et al	x		x	
B	Dubin and Moulin		x		
	Kabbaj et al		x	x	
	Zen and Umehara		x		
C	Bauer and El-Hakim			x	
	Silvestri et al			x	
D	Peters and Leavell	x			x
	Sarsby and Vickers				x

dation test and introduce new theories. Two papers essentially present comparisons between the experimental results obtained in the various tests. Of the remaining two papers, the first presents a new apparatus for measuring anisotropic properties and the second is devoted to the consolidation testing of peat.

Summaries of Papers

Category A: Testing Techniques

This category includes papers by *Ducasse et al* of the Laboratoires des Ponts et Chaussées in France, *Larsson and Sällfors* of Sweden, and *Sandbaekken et al* of the Norwegian Geotechnical Institute.

1. *Summary of Paper by Ducasse et al*—This paper describes six types of consolidation tests, procedures, and interpretation methods used in the Laboratoires des Ponts et Chaussées of France. In all these tests, the consolidation cell used is back-pressured and pneumatic or hydraulic loadings are used so as to allow complete automation. Apart from the controlled gradient test and the constant rate of deformation test, which are similar to those found in other countries, the authors present four additional types of tests. These are: (1) a modified version of the conventional test, (2) a radial drainage test, (3) a creep test, and (4) a heat accelerated test. Regarding the modified conventional test, the most important point is that in order to better define the stress-strain curve in the neighborhood of the preconsolidation pressure, the load increment ratio varies during the test and an unloading-reloading cycle is performed just after exceeding the preconsolidation pressure. In the radial drainage test, a central sand-filled hole allows for the study of radial consolidation. The creep test is used to study secondary consolidation. Normally, in this test only three loadings are used. These correspond to the vertical *in situ* pressure, the preconsolidation pressure, and the expected final effective stress. Each load increment duration is a minimum of seven days.

The most curious test described in the paper is the heat accelerated test in which the consolidation cell is kept in a water bath whose temperature remains constant at 70°C. The principle of this test is based on the fact that the temperature has two main effects: (1) faster consolidation due to the lowered viscosity of the free water and the simultaneous increase in the permeability of the soil, and (2) the increase of the soil density due to the temperature effect upon the diffuse double layer surrounding the clay particles. It should be noted that the compressibility parameters obtained in this test are corrected for temperature. In addition, the total test duration is equal to about 25% of that of the conventional test. It is the reporter's opinion that, in some clays, this type of test and the high temperature associated with it may induce permanent physico-chemical changes in the soil structure which would modify the compressibility characteristics of the natural soil.

2. *Summary of Paper by Larsson and Sällfors*—This paper mainly describes the automatic constant rate of deformation test used in Sweden. On the basis of the work performed by Sällfors [7], the authors recommend the use of a standard rate of 0.0024 mm/min on clay samples with a height of 20 mm (that is, a rate of 0.72%/h). Because of the good comparison between the results obtained in this test and those inferred from either the conventional test or field studies, this type of test is strongly favored since, firstly, it is a continuous test and, secondly, it takes only one to two days to perform. The authors make use of arithmetic plots between stress and strain to determine settlements. Corrections are applied to the experimentally determined curves in order to better determine the preconsolidation pressure, the moduli of deformation, and the coefficients of consolidation.

Because the strain rate remains constant in this test, the authors prefer it to both the continuous loading [8] and the constant hydraulic gradient tests in which the strain rate varies throughout the loading process.

3. *Summary of Paper by Sandbaekken et al*—This paper describes the consolidation test equipment and procedures used at the Norwegian Geotechnical Institute (NGI). Both incremental and automatic constant rate of strain tests are carried out.

In the incremental loading test, the load increment ratio varies during the consolidation process. Smaller load increments are used to better define the stress-strain curve in the neighborhood of the preconsolidation pressure. As in the case of the paper by Ducasse et al, the test procedure includes an unloading-reloading cycle to account for the effects of sample disturbance. Two overnight loadings under constant effective stress are used to determine the 24-h virgin compression line and thereby determine the 24-h preconsolidation pressure value. The other load increment durations are equal to 2.5 h. In addition, two constant head permeability tests are performed at the end of the two overnight loadings. The values of the coefficients of permeability thus obtained are compared with those calculated for each load increment, on the basis of Taylor's curve fitting method. For the permeability tests, the permeant used has the same salinity as that of the pore fluid.

Concerning the constant rate of strain tests, it has been found that for many :lays tested at NGI, a rate of strain of 0.5 to 1.0%/h is sufficiently low to maintain the ratio of pore pressure to total axial stress between 2 to 7%. As in the case of the incremental test, an unload-reload cycle is also carried out.

Category B: New Theories

This group includes the papers by *Dubin and Moulin, Kabbaj et al,* and *Zen and Umehara.*

1. *Summary of Paper by Dubin and Moulin*—The authors point out the existence of a threshold or critical gradient in the flow of water through clay soils. The flow of water through clay does not follow Darcy's law until the gradient exceeds the critical gradient. In the study reported by the authors, critical gradients having values between 10 and 20 have been determined in some soft French clays.

By modifying Terzaghi consolidation theory with the introduction of a non-Darcian flow law, the authors show that the existence of critical gradients in effect stops the decrease of the pore pressure for a while although the consolidation process is not over.

2. *Summary of Paper by Kabbaj et al*—The authors develop a numerical model for the analysis of the consolidation of natural sensitive clays in the various test methods that are presently used. Starting with the consolidation theory of Berry and Poskitt [9], the deformation behavior of the clay skeleton is expressed in the form of a normalized stress-strain relation and of a preconsolidation pressure–strain rate relation. By modifying some of the principles of Cam-Clay models and combining them with those of elastic-viscoplastic theory, the authors obtain the description of the consolidation process subjected to the hypotheses retained in the analysis.

The results show that each element in a consolidating clay element in the consolidation ring follows a specific stress-strain relationship, related to the strain history of that element. Thus the overall stress-strain relationship obtained from the usual interpretation of the consolidation test represents an average of all the relations followed by each element in the specimen. As a result, the overall stress-strain relation reflects both the testing procedure and the clay behavior itself. Another important result which is mentioned in the paper is the fact that, by taking into account the viscous behavior of the soil, pore pressure isochrones do not show a strong change in shape when the overconsolidated clay just becomes normally consolidated, as compared to the results found on the same clay by means of the CONMULT numerical model [10], which does not consider creep.

3. *Summary of Paper by Zen and Umehara*—On the basis of Mikasa's consolidation theory [11], the authors present a new interpretation of the constant rate of strain test. For the application of their method, the authors present three working diagrams. The various compressibility parameters as well as the average effective stress-strain relationship are determined by means of these diagrams.

The validity of the approach is examined by analyzing the *in situ* consolidation behavior of a model fill built on Osaka bay mud.

Category C: Comparisons

In this group are found the papers by *Bauer and El-Hakim* and *Silvestri et al.*

1. *Summary of Paper by Bauer and El-Hakim*—The authors report the results of a comparative study of the consolidation behavior using two different soils subjected to various different testing techniques. The first series of tests were carried out on undisturbed samples of a Champlain clay from Ottawa (Ontario); the second series of tests were performed on laboratory-prepared kaolin specimens. For each series, four different consolidation equipments were used: a standard oedometer, a modified Rowe cell, an Anteus apparatus, and a K_0-triaxial cell. Several sizes of specimens and different loading techniques were also used.

On the basis of the experimental results, the authors conclude that the controlled gradient test using the Rowe cell with a membrane much softer than the one originally provided is the most suitable technique to establish a well defined stress-strain curve. The Anteus apparatus also yielded good results, but the operational procedures were found quite cumbersome. The K_0-triaxial test was found time consuming and needed an elaborate data recording and controlling system. The conventional test yielded the poorest results.

Concerning the Champlain clay, it was again found that the preconsolidation pressure was time dependent. In addition, the coefficients of consolidation for both soils were found to be independent of strain rates for stresses higher than the preconsolidation pressures.

2. *Summary of Paper by Silvestri et al*—The authors describe a comparative study of the laboratory consolidation behavior of a soft sensitive Champlain clay. Undisturbed specimens of this clay were subjected to three different types of test: incremental loading, constant rate of strain, and controlled hydraulic gradient tests.

The paper indicates that the various tests yield comparable results when the strain rate used is taken into account. Higher strain rates and hydraulic gradients are found to yield higher values of the preconsolidation pressure, modulus of deformation, and coefficient of consolidation.

A simple creep law was found to adequately describe the strain rate dependence of the preconsolidation pressure. When comparing the results with those obtained on other marine clays, a unique relationship was found to exist when the preconsolidation pressures were normalized with respect to a reference value.

Category D: Miscellaneous Papers

In this group are found the papers by *Peters and Leavell* and *Sarsby and Vickers*.

1. *Summary of Paper by Peters and Leavell*—This paper describes a test, the biaxial consolidometer test, developed to measure consolidation properties of anisotropic materials. The consolidometer consists of a modified triaxial cell that allows the measurement of flow from the specimen's circumference and ends separately. In addition to preliminary tests on a laboratory prepared kaolin, consolidation tests were also carried out on an undisturbed sample of a highly anisotropic clay shale.

The device has some interesting features. Firstly, it has the ability to determine both the principal values of the permeability coefficients and the three elastic constants from one loading increment. This allows the evaluation of, for example, the effect of the stress-induced anisotropy on the consolidation properties. Secondly, since the ratio of the volume of flow in axial direction to that in radial direction is a function of the pore pressure distribution, the test offers the possibility of studying the effects of anisotropy and inelasticity on consolidation.

2. *Summary of Paper by Sarsby and Vickers*—This paper describes a 25.4-cm (10-in.)-diameter consolidometer of the Rowe type, used to study the compressibility characteristics of peat. In order to determine the actual vertical stress distribution on the base of the sample, the bottom plate was supported by three load transducers. The difference between the applied pressure at the top of the specimen and the pressure measured by the load transducers was used as a measure of the friction mobilized on the inside wall of the consolidometer.

The magnitude of the side friction was found to be dependent on the sample geometry. In the tests reported in the paper the side friction was observed to account for as much as 50% of the top applied stress.

Closing Comments

The ten papers in this group present a variety of experimental and theoretical results concerning the consolidation behavior of soft clays. It is hoped that this summary will provide a framework for consulting the papers themselves.

Regarding some of the procedures described in the papers, this reporter would like to take the opportunity to add the following comments, with the hope that they will stimulate further studies in this area:

1. Concerning the procedure followed in some of the incremental tests described, it would be preferable that the unloading process of the unload-reload loop should be performed either just when the preconsolidation pressure is exceeded or just before it is reached. Indeed, when dealing with structured clays, if unloading is carried out at stresses higher than the preconsolidation pressure,

the resulting parameters do not apply to the original soil but to a soil whose structure has been destroyed.

2. All consolidation tests should be performed with provisions for measuring the pore water pressure at the undrained end. In doing so, the consolidation phenomenon would be more adequately studied and creep could be accounted for. Thus, when performing incremental consolidation tests, each additional load should be applied only when the pore water pressure would be dissipated to some predetermined degree. For example, a degree varying between 95 and 98% would be acceptable.

3. Even though the consolidation test represents one of the most common and important tests carried out in geotechnical laboratories, the existing state of stress in the sample is very poorly understood. In order to overcome this lack of knowledge, consolidometers should be equipped with lateral pressure measuring devices for the establishment of the true stress paths. This procedure would permit a better determination of the yield curve of the soil.

References

[1] Crawford, C. B., *Journal of the Soil Mechanics and Foundation Division, Transactions of ASCE*, Vol. 90, No. SM5, 1964, pp. 87–102.

[2] Leonards, G. A. and Altschaeffl, *Journal of the Soil Mechanics and Foundations Division, Transactions of ASCE*, Vol. 90, No. SM5, 1964, pp. 133–155.

[3] Leonards, G. A. and Ramiah, B. K. in *Papers on Soils—1959 Meetings, ASTM STP 254*, American Society for Testing and Materials, Philadelphia, 1959, pp. 116–130.

[4] Gorman, C. T., Hopkins, T. C., Deen, R. C., and Drnevich, V. P., *Geotechnical Testing Journal*, Vol. 1, No. 1, 1978, pp. 3–15.

[5] Larsson, R., Report 12, Swedish Geotechnical Institute, 1981, pp. 1–157.

[6] Leroueil, S., Tavenas, F., Samson, L. and Morin, P., *Canadian Geotechnical Journal*, Vol. 20, No. 4, 1983, pp. 803–816.

[7] Sällfors, G., "Preconsolidation Pressure of Soft High Plastic Clays," Ph.D. thesis, Chalmers University of Technology, Gothenburg, Sweden, 1975, p. 231.

[8] Janbu, N., Tokheim, O., and Senneset, K. in *Proceedings, 10th International Conference on Soil Mechanics and Foundations Engineering*, Stockholm, Vol. 4, 1981, pp. 645–654.

[9] Berry, P. L. and Poskitt, T. J., *Geotechnique*, Vol. 22, No. 1, 1972, pp. 27–52.

[10] Tavenas, F., Brucy, M., Magnan, J. P., LaRochelle, P., and Roy, M., *Revue Française de Géotechnique*, Vol. 7, 1979, pp. 29–43.

[11] Umehara, Y. and Zen, K., *Soils and Foundations*, Vol. 20, No. 2, 1980, pp. 79–95.

David Bloomquist[1]

Report on New Experimental Methods and Techniques

REFERENCE: Bloomquist, D., "**Report on New Experimental Methods and Techniques,**" *Consolidation of Soils: Testing and Evaluation, ASTM STP 892*, R. N. Yong and F. C. Townsend, Eds., American Society for Testing and Materials, Philadelphia, 1986, pp. 122–128.

This report provides synopses of eight papers. Seven deal exclusively with the consolidation characteristics of soft clays; the eighth also includes strain characteristics of clays, sands, and peats.

While there are several ways to group the papers, the most convenient is to divide them into three areas:

1. New experimental methods and devices that measure consolidation behavior.
2. Centrifuge and numerical modeling comparisons.
3. New techniques to evaluate and predict field consolidation phenomena.

At the conclusion of this report is a general commentary on the papers, including questions for discussion and several salient aspects that deserve mention.

New Methods and Devices

The paper by *Sethi et al* (pp. 490–499) introduces a novel method of determining solids contents in a dilute suspension. The technique involves measuring the velocity of a drop (20 μL) of clay suspension as it falls through an immiscible organic liquid. This velocity is then compared with a calibration curve of fall rate versus clay concentration. An attribute of the procedure is that only 0.5 mL of slurry is extracted from each elevation in the settling column. This minute amount precludes the serious disruption of subsequent settlement.

Fifteen microsamples are collected simultaneously using a modified Andresean sampler. From each sample, a drop of slurry is micropipetted into and allowed to settle through castrol oil. Its terminal velocity is then correlated to clay concentration.

Solids contents versus time and depth measurements were conducted using two different initial concentrations. The results are shown in Figs. 5 and 6 of

---

[1] Assistant Engineer in Civil Engineering, University of Florida, Gainesville, FL 32601.

their paper. At low initial concentrations (4.1% w/v), no sharp interface was observed until 60 min had elapsed and even then some turbidity (i.e., fine particles) remained in the supernatant. At 10.6% initial concentration, a sharp interface occurred immediately and the material settled as a plug. The authors found from their sampling effort that the increase in concentration with time and depth was attributable not only to the increase in quantity of material present, but also to the size of the particles. Thus they recommended that the consolidation process must be analyzed in concert with the segregation phenomena.

The paper by *Richards* (not in this publication) describes the nondestructive determination of a marine clay's bulk density using gamma radiation.[2] The data are then computer analyzed to automatically compute and plot an e log p curve. The program also differentiates between spurious and anomalous yet valid data. The range of the measuring device is 1 to 2 ± 0.02 Mg/m^2.

A core sample (100 m length is not uncommon) is raised from the seabed. The intact sample is then passed through a gamma ray emitting radioisotope. The amount of rays passing through the sample and collected on the detector are inversely related to the bulk density. The relationship may be expressed as

$$R = R_0 \, (e^{-a\gamma d})$$

where

R = count rate,
R_0 = radiation intensity with no sample present,
a = constant, f (specific gravity and chemical composition),
d = dimension of the sample, and
γ = bulk density.

Since a vast amount (150 points per 1.5 m) of data is collected, a core quality assessment computer program was developed. It basically determines the modal or most frequent value (of ρ sat) for a particular core section, which is then used in the e log p plot. The effective vertical stress, p or σ, is determined by

$$\sigma_{v0_i} = \sigma_{(v0_i - 1)} + (\rho_i - \rho_{sw})\Delta z_i$$
$$\quad\quad\quad\; 2 \quad\quad\; 3 \quad\; 4 \;\; 5$$

where

2 = effective overburden pressure for the core section immediately above,
3 = mean bulk density for the ith interval,
4 = density of seawater, and
5 = incremental increase in depth.

The void ratio is found from

$$e_i = \frac{(\rho_i - \rho_{gi})}{(\rho_{sw} - \rho_i)}$$

[2] Richards, A. F., "Nondestructive Measurement of Consolidation in Marine Soils."

where ρ_{gi} is the specific gravity or grain density.

Figure 6 of Richards' paper illustrates the results of two passes (0 and 90°) through the densitometer; Figs. 7 and 8 show typical e log p plots resulting from using the technique. A major attribute of this system is that heterogeneity of the soil with depth is accounted for. This results in the variable C_c indicated on the plots. Richards states that this knowledge is extremely important, since settlement analysis based on the oedometer test should reflect the correct stratum of interest.

The third paper reviewed involves a new measurement technique employing ultrasonic waves to measure lateral deformations in the triaxial test. As is painfully apparent, it is extremely difficult to measure the horizontal displacements of a triaxial specimen. The most common and semiaccurate method of calculating the horizontal displacement is to measure the change in volume (using a burette) and axial displacement of a saturated specimen. Of course, for dry or partially saturated specimens, this will not work. Another device is to attach LVDTs (using clamps) to the specimen inside the chamber. While this will work, it will at best measure the lateral movements in only two or three locations.

Barański and Wolski (pp. 516–525) developed an ultrasonic device shown in their Fig. 1. Three heads mounted at 120° intervals beam 3.5 MHz waves into the chamber and are reflected off the specimen and picked up by the oscilloscope and computer. The heads can be moved vertically up and down the length of the sample, thus obtaining a complete picture of the deformation geometry. The device was calibrated using plastic rings and found to yield a resolution of 0.1 mm ± 0.2%.

Their research showed some interesting results. The end effects are readily apparent and occur over 15% of the sample's length at both ends. Thus the middle 70% was used for strain computations. On the tests using silty clay (Fig. 7), a comparison of ϵ_1 versus ϵ_3 shows very good agreement with a nonlinear hyperbolic elastic model for v_t. For sandy clay (Fig. 8), however, the results show that the actual strains are quite different from the theoretical prediction.

Isotropic compression tests were performed on a saturated peat. As Fig. 9 indicates, the lateral volumetric strains are greater than indicated by the quantity of water expelled. Wolski and Barański state that this discrepancy is probably due to the compression of the soil skeleton.

Finally, a comparison between the rate of strain, the shear stress, and the shear rate is plotted in Fig. 10. This shows that secondary compression or soil creep may be modeled using the rate process theory.

Centrifuge and Numerical Modeling Comparisons

Shen et al (pp. 593–609) investigated the settlement of a storage tank during filling and emptying which was founded on a compressible clay foundation. Both centrifuge and numerical models were used in the analysis.

Their Fig. 2 illustrates the test setup. Basically, a preconsolidated kaolinite clay stratum 152 mm (6 in.) deep was topped with 25.4 mm (1 in.) of sand on which a model storage tank was placed. A rubber diaphragm served as the flexible tank floor. Once up to test acceleration (60 g), a series of water loadings and unloadings was performed. Surface settlement profiles at six locations and pore pressure responses at two depths were monitored. The resulting data were then compared to a finite element program based on Dafalias' bounding surface plasticity model.

The plasticity model's fundamental premise is the existence in stress space of a bounding surface which limits the range of possible stress states. Eighteen soil parameters were used in the model including classical critical state values, material constants that describe the shape of the bounding surface, the hardening parameter, the location of the projection point, and the size of the elastic zone. These parameters were determined from standard oedometer tests and from extension/compression triaxial tests performed at several overconsolidation ratios (OCRs). Unfortunately, some of the parameters (e.g., the sand's modulus) had to be estimated and the soil bucket interface was assumed to be frictionless.

The results, shown in their Figs. 9 and 10, illustrate the pore pressure and settlement responses, respectively. Very good agreement between the FEM analysis and pore pressure response during the loading/unloading cycles is indicated. The displacements are somewhat larger than predicted and are most probably attributable to the inadequate representation of the Monterey sand layer.

Another paper on FEM/centrifuge model comparisons was presented by *Mitchell and Liang* (pp. 567–592). They investigated the time-dependent pore pressure and deformation behavior of a model embankment founded on a soft clay layer. The centrifuge model is shown in Fig. 1 of their paper. It consists of a sand embankment overlaying a sand drainage layer which is in turn founded on a doubly drained kaolinite clay layer. Foundation pore pressures were monitored with five miniature pressure transducers, while foundation displacements were photographically analyzed using a 25.4 by 25.4 mm (1 by 1 in.) grid of marker beads. This method allowed for bead movement measurements of ± 10 μm, with a maximum error in linear strain of 0.5%.

A new numerical model is currently being formulated that incorporates the combined effects of creep and hydrodynamic consolidation. The Kavazanjian, Mitchell, and Bonaparte (1984) model requires eleven soil parameters obtained from standard triaxial and consolidation tests. Since the program was not fully operational during this research, however, Mitchell and Liang used a two-dimensional FEM, CON2D (developed by Chang and Duncan, 1977), to analyze their centrifuge results. It should be noted that CON2D does not account for creep deformations but, according to the authors, secondary compression and creep effects are small for kaolin clay.

Two series of centrifuge tests were performed. Series I consisted of four "identical" model tests performed at 70 g after a 30 g stress reinitialization period. These tests were used to measure any experimental error and to verify data consistency. Series II was a standard modeling of models program to evaluate model size effects. This procedure accelerates various scaled models to different accelerations, yet maintains an equivalent prototype geometry. The accelerations varied between 75 and 97 g.

The results are shown in their Figs. 4 to 9. In general, for the Series I tests, pore pressures accurately reflected those predicted using CON2D within 15 to 20%. However, the program overpredicted the immediate deformations by 50% beneath the embankment and underpredicted them near or beyond the toe. Long-term settlement was well predicted. For the Series II tests, the researchers found the same general trend exhibited in Series I, and that model size effects are minor.

New Techniques to Evaluate and Predict Field Consolidation

Scott et al (pp. 500–515) performed a series of self-weight consolidation tests on waste slimes using 2 and 10 m high settling columns in conjunction with slurry consolidometer results. The data from these were compared to a finite strain consolidation (FSC) program. The primary goals were to determine if results from a slurry consolidometer test and FSC program could predict a prototype event, and to ultimately develop a method to more rapidly consolidate the slime deposits.

The material used in the research was an oil sand waste slurry, a by-product of the bitumen extraction process. The 2-m cylinder test sequence had two drawbacks: (1) insufficient effective stresses allowed thixotropic strength gains to inhibit consolidation, and (2) removing samples for solids contents disrupted the subsequent settling process.

Two 10-m cylinder tests have been running for over a year. One contains a 31% solids content slime, the other a sand/sludge mix at "substantially higher solids content and density". Sampling ports at 1 m intervals allowed for 55 mL of material to be extracted for various tests. Pore pressures were monitored via the transducer/manometer arrangement shown in their Fig. 4.

The results of the testing program are shown in Figs. 8 and 9. Basically, Scott et al found that consolidation is present in the upper 3 m of the 10-m tests and that volume changes are occurring with little change in pore pressures. After 500 days, the interface has settled 0.5 m. Figure 10 shows the reasonable agreement between the FSC prediction and the 10-m settlement rate. This comparison between the numerical model (as in this case), centrifugal or laboratory model, and the prototype event, is the foremost method to validate predictive tools.

Mikasa and Takada (pp. 526–547) provide further insight into coefficient-of-consolidation computations. Specifically, they present two methods for determining c_v: (1) the curve rule method, and (2) the correction of c_v by the primary consolidation ratio, r. In addition, the effects of finite strain and variable c_v on time-consolidation relationships are explored.

Using Mikasa's general consolidation equation as a starting point, the authors show that even though permeability, k, and compressibility, m_v, change during consolidation (i.e., nonlinear stress-strain relationship), since they decrease proportionally, c_v remains relatively constant. This explains why Terzaghi's theory has worked so well for so long (linear stress-strain and constant k).

The determination of c_v usually is found by either Taylor's $\sqrt{t}$ or Casagrande's log t method. The authors explain that the latter method is incorrect in its assumptions and should not be used. Their new technique, termed the *curve rule method*, replaces the log t method and gives almost the same results as the $\sqrt{t}$ method.

Basically, an oedometer time-settlement curve is superimposed on a series of theoretical curves (Figs. 1 and 2) from which d_0, d_{100}, and t_{50} may be read off directly; C_v is then calculated in the standard manner. The authors note that this value should not be used to predict field settlements because it only reflects primary consolidation, whereas the field predictions are based on total (primary + secondary compression) settlements. Thus they recommend multiplying c_v by the consolidation ratio, r, to obtain a corrected c_v value.

Next, the effects of finite strain and variable c_v are shown. Four normalized time-consolidation curves for various strains and c_v's are presented. These curves are used in the same manner as the previously mentioned curve fitting process. Two sample problems show that errors of 14 to 19% are introduced if these c_v corrections are not performed.

Finally, the authors outline two recommended procedures to follow for c_v determinations, one for research applications and the other for consulting uses. Their Appendix is a short treatise on why Casagrande's method should be replaced by the curve rule method.

As *Takada and Mikasa* (pp. 548–566) point out in their paper, compressibility and permeability of very soft clays are difficult parameters to obtain from conventional oedometer tests. Thus they used the centrifuge to induce self-weight stresses and settlements on several samples of Nanko and Tokuyama clays. From the tests they were able to determine k, f log p, m_v, and c_v.

As a starting point, Mikasa's consolidation theory is reviewed. From this, the pertinent scaling relationships are derived; for example,

$$\frac{\dot{s}_m}{\dot{s}_p} = \frac{s_m/t_m}{s_p/t_p} = \frac{s_m}{s_p} \cdot \frac{t_p}{t_m} = \frac{n^2}{n} = n$$

where

$\dot{s}$ = rate of settlement,
s = settlement,
t = elapsed time,
p = prototype,
m = model, and
n = acceleration of model.

Next, a derivation of permeability is provided. Takada and Mikasa point out that during self-weight consolidation with single drainage conditions, the apparent velocity of pore water flow relative to the soil structure is merely the same in value as, but opposite in direction to, the settlement rate of the clay in stationary water.

Four different clays were tested. The typical sequence involved performing a self-weight consolidation test at 100 to 150 g until settlement ceased. The supernatant was immediately removed to prevent swelling of the consolidated material. An undisturbed coring was then removed from which water contents at 2 to 3 mm intervals were obtained. From this, an f log p curve was then plotted.

Permeability versus volume ratio curves are shown in their Figs. 7 and 8. Basically, k increases with increasing volume ratio (or $1 + e$) but at a decreasing rate of increase.

An important problem that arose dealt with particle segregation. The authors point out that this is a characteristic of centrifuge modeling and not of the prototype event. They have tentatively defined a water content below which segregation does not occur. For the highly plastic clays, 250% at 100 g is mentioned.

Summary

With the rapidly improving numerical techniques available to the geotechnical engineer, more precise and thus complicated testing procedures are concurrently required to provide input parameters. These papers have addressed this aspect by introducing either new measurement techniques or a refinement of those currently in use. The important point is not whether they widely succeed but rather that they help inspire others to continue the research in this vitally important area of engineering.

Two themes or points brought out in several of the papers are worth mentioning: the segregation phenomenon and the thixotropic effect on self-weight sedimentation/consolidation. The segregation of particles during settlement introduces another complication into the engineer's ability to predict field responses. Such questions as the following should be addressed: At what solids content does the process stop? How does this affect the overall dissipation of excess pore pressures?

The thixotropic strength gain during settlement is a very difficult, yet extremely important, aspect to be considered. Perhaps this effect can be used to our advantage by allowing sand surcharges to be placed on a very soft clay to enhance consolidation. Of course, this phenomenon is not accounted for in the mathematical modeling schemes, and may be one reason why many theoretical predictions indicate faster settlement than is actually observed.

J. F. Peters[1]

Report on Consolidation Behavior

REFERENCE: Peters, J. F., **"Report on Consolidation Behavior,"** *Consolidation of Soils: Testing and Evaluation, ASTM STP 892,* R. N. Yong and F. C. Townsend, Eds., American Society for Testing and Materials, Philadelphia, 1986, pp. 129–134.

The papers of the session discussed here deal primarily with the traditional role of consolidation theory in soil mechanics, that of controlling and predicting the volume change of fine-grained soils.[2] Reclamation of soft ground, performance of soil-drain systems, behavior of man-made fill materials and industrial wastes, and problems in characterization of behavior of special soil types are addressed. With one exception, behavior as determined by laboratory-scale investigations is emphasized. This report is organized into three major themes: Compression Behavior, Performance of Drains, and Special Problems. The report concludes with a selection of topics for general discussion.

Compression Behavior

The effective stress-volume change relationship for soil is an important part of determining its overall time-dependent consolidation behavior. For example, the time for 50% consolidation varies depending on (1) the initial effective stress level, (2) whether the soil is on the recompression or virgin portion of the compression curve, and (3) whether the loading change is a loading or unloading increment. Differences in t_{50} stem primarily from variations in compressibility.

The paper by *Bemben* (pp. 610–626) addresses the problem of defining the demarcation between rebound and virgin compression segments of the compression curve for brittle (cemented) soils. For soils which are not cemented, the two legs of the compression curve can be related to the maximum past consolidation (preconsolidation) pressure. Bemben suggests that if a soil becomes cemented, the relevant pressure is not the maximum past pressure but rather the pressure that existed at the time of cementation. He asks, "Could the cementation be preventing the subsequent water content changes in response to subsequent ver-

---

[1] Research Civil Engineer, Geotechnical Laboratory, U.S. Army Engineer Waterways Experiment Station, Vicksburg, MS 39180-0631.
[2] This is in contrast to more comprehensive roles made possible by modern computer-based numerical methods which permit consolidation theory to be used as general means to formulate problems in terms of effective stress.

tical effective stress changes?'' Bemben makes an important contribution by demonstrating that accounting for cementation in the estimate of past consolidation stress (σ'_{cp}) leads to a consistent relationship between past stress and depth, whereas use of the break point in the compression curve (σ'_{cc}) shows no correlation with depth. It is also shown that the condition of the soil fabric relative to the cementation has an important influence on the rate of consolidation.

Fukue et al (pp. 627–641) describe the infuence of the coarse fraction (sand and gravel) on compressibility of sand-clay mixtures. The use of such mixtures is of practical interest because of the potential to reduce the quantity of soft clayey sediments that must be displaced by granular materials obtained off-site. In earthquake-prone regions, sand-clay mixtures offer the additional advantage of being less susceptible to liquefaction than noncohesive granular fills. The authors propose to describe the soil as a four-phase system in which the sand and clay fractions are considered as independent phases. The principal finding is that as the void ratio of the sand phase reaches a critical threshold value, the behavior of the mixture becomes like that of sand. The threshold void ratio is somewhat greater than the maximum void ratio of the sand without clay. The threshold value increases slightly if the clay content is increased, suggesting that the clay acts to stabilize the structure of the sand particles at high void ratios. It was also found that the influence of salt concentration in the pore water is small if the void ratio of the sand component is below the threshold value. Thus, a classification for sand-clay mixtures is developed, based on the void ratio of the sand component, which combines both the clay content and the compacted dry density of the mixture.

The paper by *Alvi and Lewis* (not in this publication) concerns the compressibility characteristics of industrial sludge.[3] The determination of compressibility of sediment waste is of great practical concern because of increasing competition for suitable disposal sites. These properties are needed if such sites are to be economically reclaimed once disposal capacity is reached. For active sites, the compressibility and permeability of the materials are needed to develop efficient disposal procedures for the containment area. Industrial sediment wastes pose a special challenge to the engineer because the constituents tend to be less chemically stable than most natural soils. For example, the authors note that a simple procedure such as water content determination is complicated if the solids contain calcium salts or gypsum because both free water and water of hydration (or chemically bound water) are driven off when dried at the standard temperature (105°C).

Alvi and Lewis present data from eight one-dimensional consolidation tests performed on specimens trimmed from Shelby tube samples. They found that the consolidation behavior of the sludge resembled that of nonplastic silts and fine sands although some pecularities were observed. The preconsolidation pres-

---

[3] Alui, P. M. and Lewis, K. H., "Consolidation Behavior of Industrial Sludge." (For the reader's convenience, the title & author(s) of papers presented at the symposium but not appearing in this volume will be given in full.)

sures determined from the curves appeared high relative to the overburden pressures, suggesting that a "pseudo preconsolidation" pressure "is probably due to chemical alterations caused by cementing agents and/or ion exchange." Two of five direct shear tests performed also showed evidence of psuedo preconsolidation. The authors conclude that traditional physical property tests alone are insufficient to define the behavior of waste materials and that knowledge of chemical composition and its influence on mechanical behavior must also be understood.

Drain Performance

Vertical drains have been used for over a half century to increase the rate of consolidation of compressible strata when preloading (surcharging) is used as a means of stabilizing soft ground. In the past decade there has been an increasing trend towards use of geofabrics for drains in lieu of graded sand. With the advent of geofabrics, several types of prefabricated drains can be designed to provide necessary filter characteristics while also providing large drainage capabilities. Using the geocomposite concept, a fabric filter jacket is wrapped around a hollow plastic core which provides rigidity. Generally, these drains have a flat shape, similar to a kerosene lamp wick, and are referred to as wick drains. Wick drains provide performance comparable to sand drains and are easier and more economical to install. According to Guido and Ludewig, prefabricated drains have replaced sand drains in over 80% of all applications nationwide. At the present time, the major obstacle to design of wick drains is the lack of specific knowledge on individual drain performance, a problem created in part by the variety of designs made possible by the geocomposite concept. Two of the three papers discussed in this section describe laboratory tests for wick drains.

The paper by *Guido and Ludewig* (pp. 642–662) begins with a review of consolidation theory for ground drained by radial flow to wells. The percent consolidation can be expressed as

$$1 - \bar{U}_{r,v} = (1 - \bar{U}_r)(1 - \bar{U}_v)$$

where $\bar{U}_{r,v}$ is degree of consolidation due to both radial flow to the drain and vertical flow to drainage boundaries. The degree of consolidation due to vertical flow $\bar{U}_v$ can be obtained using Terzaghi's equation for one-dimensional consolidation. The consolidation coefficient C_v is based on the properties of the soil without drains. The percent consolidation due to horizontal drainage was found by R. A. Barron to be

$$\bar{U}_r = 1 - \exp(\lambda)$$

where λ is a function of time t and the horizontal coefficient of consolidation C_h, well diameter d_w, diameter of influence d_e, and well resistance. For a particular sand well, the horizontal and vertical consolidation coefficients, C_h and C_v, along with well resistance are needed to estimate consolidation times. The same relationships can be applied to wick drains by defining an equivalent drain diameter:

$$d_w = 2(t + b)/\pi$$

where t and b are the thickness and width of the drain, respectively.

The experimental results described by Guido and Ludewig were obtained from a specially designed wick drain consolidometer. The soil specimen was loaded axially, as in a conventional device, but drainage is permitted both by vertical flow to the specimen bottom and by radial flow to a wick drain installed in the specimen center. Conventional consolidation tests also were performed to determine the properties of the soil. The test was found to be repeatable and appears to be useful in distinguishing performance of various types of wick drains.

Suits et al (pp. 663–683) describe a consolidation test similar in concept to the device used by Guido and Ludewig except vertical drainage was not permitted; thus C_h was measured directly. Two additional tests were developed to measure the flow capacity of the wick drain when subjected to lateral loads and crimping. It was found that lateral pressure, but not crimping, influences the capacity of soft-core wick drains. The performance of rigid-core wick drains is not seriously influenced by lateral pressure or crimping.

The paper by *Adachi et al* (pp. 684–693) describes a case history in which vertical drains are used to speed up consolidation of a preloaded clayey alluvial foundation. The construction procedure consisted of first applying a surcharge of 1.5 m height over the 2 m high permanent fill. The surcharge was then removed and building construction completed. It was expected that the residual settlement that occurred after removal of surcharge would be less than 10 cm. The residual settlement was computed as the difference between settlement at 50% consolidation under surcharge load and full consolidation under the permanent 2 m of fill. To speed up consolidation, sand drains were installed at the edges of the site and rope drains were used in the remaining part.

The field performance failed to meet design predictions in two important aspects: (1) the residual settlement was much greater than anticipated, and (2) there was twice as much settlement under areas where rope drains were used than where sand drains were used. The authors relate the excessive residual settlement to secondary compression on the basis that complete dissipation of excess pore pressure has occurred (i.e., primary consolidation complete). No full explanation of the differential settlement is given but the conclusions imply that it is related to differences in construction time and type of drains used.

Special Problems

The final two papers address quite different aspects of the volume change relationships for soil. Andersland and Al-Khafaji investigate volume balance for soils where the constituents can undergo a change in physical makeup with time. Mowafy, Bauer, and El-Sohby consider the swelling behavior of soil. Both topics lie outside the traditional problems of compressibility discussed previously.

The paper by *Andersland and Al-Khafaji* (not in this publication) describes an experimental study of the volume change of soil having organic constituents which can undergo decomposition and solid-phase volume reduction.[4] Compressibility relationships for saturated inorganic soils are simplified by the fact that the volume of the solids constituent is constant and any volume change is the result of fluid flow from the element. In contrast, organic solids in soil can decompose into soluble solids and gases through a process that depends on the nature of dissolved gases in the pore fluid and the pH. The possibility for decomposition suggests an interesting but complicated extension to the time-consolidation problem. The authors consider a less comprehensive problem of predicting the ultimate volume change due to decomposition. A method is proposed for estimating settlement using an empirically determined relationship between void ratio, pressure, and organic fraction. Data also showed that decomposition significantly reduces the consolidation coefficient C_v.

The paper by *Mowafy et al* (not in this publication) describes an investigation of the influence of consolidometer sidewall friction, percent clay, water content, and salt concentration on the measured swell potential of silt-clay and sand-clay mixtures.[5] Of the items studied, the method used to reduce side friction appear to have the greatest effect on percent swell. It is not clear from the data presented that side friction has a corresponding effect on measured swell pressure. It was also found that the percent swell is inversely proportional to the normal pressure applied prior to flooding. The presence of sodium chloride reduces the percent swell, although no data are given to document the influence of salinity on swell pressure.

The influence of dry unit weight γ_d, clay content C, and initial water content w_n on swell pressure p_s are summarized by the empirical relationship

$$\log p_s = 0.203\gamma_d + 0.017C - 0.026w_n - 4.292$$

It is, of course, expected that $\log p_s$ is proportional to clay content. The inverse proportionality between $\log p_s$ and water content is similar to the often observed logarithmic relationship between suction potential and water content. In all the test data reported, γ_d had the same value of 17.66 kN/m^3, so the statistical significance of γ_d is difficult to assess.

---

[4] Andersland, O. B. and Al-Khafaji, A. W. N., "Soil Volume Change Due To Decomposition of Organic Solids."
[5] Mowafy, Y. M., Bauer, G. E., and El-Sohby, M. A., "A Study of the Swelling Properties of Soils."

General Discussion Topics

Topics for general discussion are:

1. There is an increasing need for nonconventional consolidation testing for design of waste disposal sites. Sediment wastes result from industrial processes and dredging of waterways. Determination of material properties for these materials is made difficult because the wastes sediments are often very soft and compressible and may be susceptible to chemical changes after deposition. Papers in this session addressed problems of industrial sludge and organic soils. Other papers in this symposium have addressed special laboratory techniques for very soft soils. Various industries and the U.S. Army Corps of Engineers have invested heavily in research on these problems. Is there a general need to develop standard guidelines for analysis of consolidation of sediment waste materials?

2. Most papers in this session addressed the compressibility of various soils and little emphasis was placed on time-consolidation behavior. All laboratory investigations were performed using some variation of the oedometer. In view of the extensive literature that exists on general constitutive equations for soil, it is surprising that consolidation research is still concentrated on the behavior of soil subject to one-dimensional loading. Of course, consolidation is a difficult subject and there are considerable theoretical and experimental difficulties in dealing with multi-dimensional problems. Yet, as the case history by Adachi et al makes clear, there is a need to improve the understanding of consolidation behavior under field loading conditions.

Technical Papers

Eulalio Juárez-Badillo[1]

General Theory of
Consolidation for Clays

REFERENCE: Juárez-Badillo, E., "**General Theory of Consolidation for Clays,**" *Consolidation of Soils: Testing and Evaluation, ASTM STP 892*, R. N. Yong and F. C. Townsend, Eds., American Society for Testing and Materials, Philadelphia, 1986, pp. 137–153.

ABSTRACT: A general nonlinear differential equation for the one-dimensional consolidation of saturated plastic soils is presented. The variations of compressibility, of permeability, and of the coefficient of consolidation are taken into account. Application of the theory to the oedometer test, neglecting the submerged unit weight of the soil and assuming the change in thickness to be small, is made. The consolidation curves are found to depend on the ratio of the final to the initial thicknesses H_2/H_1 and on the ratio of the parameters λ/γ. Here $\lambda = 1 - \gamma\kappa$, where γ and κ are the nonlinear coefficient of compressibility and the coefficient of permeachange, respectively. For the special case that the coefficient of consolidation is constant, $\lambda = 0$, the consolidation curves move somewhat (to the left for compression and to the right for swelling), more as the values of H_2/H_1 are smaller for compression or higher for swelling compared to unity, respectively. The effective stresses at any time and depth are smaller than those predicted by the linear Terzaghi's theory. The significance of the degree of consolidation given by Terzaghi's theory, for the case $C_v = $ constant, is given introducing a nonlinear concept for U. For the general case that the coefficient of consolidation is not constant, $\lambda \neq 0$, the corresponding consolidation curves are the subject of a companion paper.

KEY WORDS: consolidation, one-dimensional consolidation, standard consolidation test, oedometer test

Nomenclature

B Function defined by Eq 47
C_v Terzaghi's coefficient of consolidation
C_{v1} Initial C_v
C_{v2} Final C_v
H Half of thickness of specimen drained at both faces
H_1 Initial H
H_2 Final H
ΔH $H_1 - H_2$
k Coefficient of permeability

---

[1] Technical Adviser of the General Director of Technical Services, Ministry of Communications and Transports, Mexico, and Research Professor, Graduate School of Engineering, National University of Mexico, Mexico, D. F., Mexico.

k_1 Initial k
M Constant defined by Eq 45
m_v Terzaghi's linear coefficient of volume decrease
m_{v1} Initial m_v
N Natural integer number
S_t $H_1 - H$
T Reduced variable defined by Eq 30
t Time
U Degree of settlement defined by Eq 54
U_n Natural degree of settlement defined by Eq 64
U'_n Modified natural degree of settlement defined by Eq 67
U_T Terzaghi's degree of consolidation
u Excess pore-water pressure
u_h Hydrostatic water pressure
V Volume
v Variable defined by Eq 22
w Variable defined by Eq 27
z Distance along the vertical direction of flow
z' Reduced variable defined by Eq 31
γ Nonlinear coefficient of compressibility
γ_m Unit weight of soil
γ'_m Submerged unit weight of soil
γ_p Nonlinear coefficient of swelling and recompression
γ_w Unit weight of water
ϵ Natural deformation defined by Eq 62
ϵ_T Final total natural deformation ϵ
ϵ_z Natural deformation at depth z and time t defined by Eq 66
$d\epsilon$ Instantaneous deformation defined by Eq 61
κ Coefficient of permeachange
λ $1 - \gamma\kappa$
ρ Expansion-compressibility ratio
σ Total vertical pressure
σ' Effective vertical pressure
σ'_1 Initial σ'
σ'_2 Final σ'

Since Terzaghi's theory of consolidation for saturated clays [1], many efforts have been made to remove the assumptions on which it is based. Most of them are developed using the void ratio e as the main variable to deal with volume changes [2] instead of $(1 + e)$ as it should be, and many of them also assume the coefficient of consolidation to be constant [3,4]. The main characteristic of the new differential equation is that due account is taken of the variations of compressibility, permeability, and coefficient of consolidation, making use of general equations for these concepts. The new differential equation is applied to the oedometer test to get a better understanding of the consolidation process. Secondary compression is not considered. "General Time Volume Change Equation for Soils" is the title of an unpublished paper by the author, which considers secondary compression. Analysis of laboratory data will improve when primary and secondary compression are better understood. The aim of the present paper is to improve our present understanding of primary consolidation process.

Fundamental Equations

Continuity and Compressibility

Using the law of conservation of matter and Darcy's law for flow through soils, and neglecting the compressibility of water and the compressibility of the solid particles compared with the compressibility of the soil structure, we can readily write, for the one-dimensional consolidation [5]

$$\frac{d\sigma'}{dt} = - \frac{1}{\gamma_w m_v} \frac{\partial}{\partial z} \left(k \frac{\partial u}{\partial z} \right) \tag{1}$$

where

u = excess pore water pressure,
σ' = effective vertical pressure,
z = distance along vertical direction of flow,
t = time,
γ_w = unit weight of water,
k = Darcy's coefficient of permeability, and
m_v = Terzaghi's linear coefficient of volume decrease.

Compressibility

It has already been shown [6,7] that plastic soils obey the law

$$\frac{dV}{V} = -\gamma \frac{d\sigma'}{\sigma'} \tag{2}$$

where γ is the nonlinear coefficient of compressibility. Since m_v is defined [1] by

$$\frac{dV}{V} = -m_v d\sigma' \tag{3}$$

then [8]

$$m_v = \frac{\gamma}{\sigma'} \tag{4}$$

The coefficient γ in Eq 4 should be substituted by the coefficient γ_p in swelling and recompression curves. The nonlinear expansion coefficient γ_p [7] is given by

$$\gamma_p = \rho\gamma \tag{5}$$

where ρ is the expansion-compressibility ratio of the soil.

Permeability

It has already been shown [9] that soils obey, very nicely, the law

$$k = k_1 \left(\frac{V}{V_1}\right)^\kappa \tag{6}$$

where $k = k_1$ for $V = V_1$ and κ is the coefficient of permeachange. Integration of Eq 2 gives

$$\frac{V}{V_1} = \left(\frac{\sigma'}{\sigma'_1}\right)^{-\gamma} \tag{7}$$

where $V = V_1$ for $\sigma' = \sigma'_1$. Introducing Eq 7 into Eq 6 we get

$$k = k_1 \left(\frac{\sigma'}{\sigma'_1}\right)^{-\gamma\kappa} \tag{8}$$

General Differential Equation

Introducing Eqs 4 and 8 into Eq 1, we obtain

$$\frac{\partial\sigma'}{\partial t} = -\frac{k_1\sigma'_1}{\gamma_w\gamma}\frac{\sigma'}{\sigma'_1}\frac{\partial}{\partial z}\left[\left(\frac{\sigma'}{\sigma'_1}\right)^{-\gamma\kappa}\frac{\partial u}{\partial z}\right] \tag{9}$$

which is the "general differential equation" of primary consolidation. It should be observed that

$$\frac{k_1\sigma'_1}{\gamma_w\gamma} = \frac{k_1}{\gamma_w m_{v_1}} = C_{v_1} \tag{10}$$

where C_{v_1} is the Terzaghi's coefficient of consolidation at the start of the consolidation process.

It should also be observed that, in general, using Eqs 4 and 8, C_v is given by

$$C_v = \frac{k}{\gamma_w m_v} = \frac{k_1\sigma'_1}{\gamma_w\gamma}\left(\frac{\sigma'}{\sigma'_1}\right)^{1-\gamma\kappa} = C_{v_1}\left(\frac{\sigma'}{\sigma'_1}\right)^\lambda \tag{11}$$

where

$$\lambda = 1 - \gamma \kappa \tag{12}$$

and, accordingly, C_v is constant only when

$$C_v = \text{constant if } \lambda = 0 \tag{13}$$

For this special case, Eq 9 is identical to the one obtained by Davis and Raymond [3].

In practice, the variation for the parameter λ seems to be not very great. For Mexico City clay $\gamma \sim 0.4$ and $\kappa \sim 4$, therefore $\lambda \doteq -0.6$. For this same clay, in the recompression curve $\gamma_p \sim 0.03$ and using again $\kappa \sim 4$, then $\lambda \doteq 0.88$. For some Norwegian marine clays it has been found that $\lambda = 0.4$ to 0.6 [10]. For other clays, common values of γ are from 0.05 to 0.15, and common values of κ are from 5 to 10; accordingly, common values for λ may be expected to be from 0.8 to -0.5, but the author believes that values of the order of -1.0 may not be infrequent and that values of the order of -2.0 may be rare.

Equation 9 may be written, using Eq 10, as

$$\frac{1}{\sigma'} \frac{\partial \sigma'}{\partial t} = -C_{v_1} \frac{1}{\sigma'_1} \frac{\partial}{\partial z} \left[\left(\frac{\sigma'}{\sigma'_1} \right)^{-\gamma \kappa} \frac{\partial u}{\partial z} \right] \tag{14}$$

Application to Oedometer Test

Differential Equations

In one-dimensional consolidation, if σ is the total vertical pressure and u_h is the hydrostatic pressure in the water, we have

$$\sigma = \sigma' + u_h + u \tag{15}$$

Derivating with respect to z, if γ_m and γ'_m are the total unit weight and submerged unit weight of soil, respectively, we have

$$\gamma_m = \frac{\partial \sigma'}{\partial z} + \gamma_w + \frac{\partial u}{\partial z}$$

that is,

$$\frac{\partial u}{\partial z} = \gamma'_m - \frac{\partial \sigma'}{\partial z} \tag{16}$$

For thin layers of soil for which

$$\frac{\partial \sigma'}{\partial z} >> \gamma'_m \qquad (17)$$

for almost the whole thickness of the sample (disregarding the effect close to the middle plane in a sample drained by the two horizontal faces) during practically the whole process of consolidation (disregarding the effect close to $t = 0$ and $t = \infty$) we can make the common assumption that

$$\frac{\partial u}{\partial z} = -\frac{\partial \sigma'}{\partial z} \qquad (18)$$

and, accordingly, also

$$\frac{\partial \sigma'_1}{\partial z} = \frac{\partial \sigma'_2}{\partial z} = 0 \qquad (19)$$

where σ'_2 is the final σ' after consolidation has taken place.

Introducing Eqs 18 and 19 into Eq 14, we get

$$\frac{1}{\sigma'}\frac{\partial \sigma'}{\partial t} = C_{v_1}\frac{\partial}{\partial z}\left[\left(\frac{\sigma'}{\sigma'_1}\right)^{-\gamma\kappa}\frac{1}{\sigma'_1}\frac{\partial \sigma'}{\partial z}\right] \qquad (20)$$

It should be observed that in Eq 20, σ'_1, may be substituted by any other pressure, say σ'_2, if the corresponding C_v, C_{v_2} for σ'_2, is used. That is,

$$\frac{1}{\sigma'}\frac{\partial \sigma'}{\partial t} = C_{v_2}\frac{\partial}{\partial z}\left[\left(\frac{\sigma'}{\sigma'_2}\right)^{-\gamma\kappa}\frac{1}{\sigma'_2}\frac{\partial \sigma'}{\partial z}\right] \qquad (21)$$

which is found to be somewhat more convenient to deal with than Eq 20.

Equation 21 may be written in more convenient ways. However, two different cases are to be treated separately: C_v variable, $\lambda \neq 0$; and C_v constant, $\lambda = 0$. For the general case for which C_v is variable, $\lambda \neq 0$, introducing the variable v defined by

$$v = \left(\frac{\sigma'}{\sigma'_2}\right)^{\lambda} \qquad (22)$$

we have

$$\frac{1}{v}\frac{\partial v}{\partial t} = \lambda\frac{1}{\sigma'}\frac{\partial \sigma'}{\partial t} \qquad (23)$$

and

$$\frac{\partial v}{\partial z} = \lambda \left(\frac{\sigma'}{\sigma'_2}\right)^{-\gamma\kappa} \frac{1}{\sigma'_2} \frac{\partial \sigma'}{\partial z} \qquad (24)$$

Introducing Eqs 23 and 24 into Eq 21, we obtain

$$\frac{1}{v} \frac{\partial v}{\partial t} = C_{v_2} \frac{\partial^2 v}{\partial z^2} \qquad (25)$$

For the special case for which C_v is constant, $\lambda = 0$ and Eq 21 reduces to

$$\frac{1}{\sigma'} \frac{\partial \sigma'}{\partial t} = C_v \frac{\partial}{\partial z} \left[\frac{1}{\sigma'} \frac{\partial \sigma'}{\partial z}\right] \qquad (26)$$

Introducing the variable w defined by

$$w = \ell n \frac{\sigma'}{\sigma'_2} \qquad (27)$$

we can write Eq 27 as

$$\frac{\partial w}{\partial t} = C_v \frac{\partial^2 w}{\partial z^2} \qquad (28)$$

This linear differential equation with C_v constant but k and m_v variables was first derived and solved by Davis and Raymond [3].

Boundary-Value Problems

The boundary-value problems for the oedometer test will now be stated. Let $2H_1$, $2H$, and $2H_2$ be the initial, current, and final thicknesses of a specimen drained at both horizontal faces. For the statement of the boundary conditions, the usual assumption is made that the change in thickness is small in relation to the thickness and accordingly that

$$H_1 \doteq H_2 \doteq H \qquad (29)$$

It is convenient to use the reduced variables T and z' defined by

$$T = \frac{C_{v_2} t}{H^2} \qquad (30)$$

$$z' = \frac{z}{H} \qquad (31)$$

For the general case, $\lambda \neq 0$, since

$$\frac{\partial v}{\partial t} = \frac{\partial v}{\partial T}\frac{dT}{dt} = \frac{C_{v_2}}{H^2}\frac{\partial v}{\partial T} \qquad (32)$$

and, similarly,

$$\frac{\partial^2 v}{\partial z^2} = \frac{1}{H^2}\frac{\partial^2 v}{\partial z'^2} \qquad (33)$$

Introduction of Eqs 32 and 33 into Eq 25 gives for the general case, $\lambda \neq 0$

$$\frac{1}{v}\frac{\partial v}{\partial T} = \frac{\partial^2 v}{\partial z'^2} \qquad (34)$$

For the special case, $\lambda = 0$, using similar equations into Eq 28

$$\frac{\partial w}{\partial T} = \frac{\partial^2 w}{\partial z'^2} \qquad (35)$$

Therefore, for the general case, C_v variable, the boundary-value problem is

$$\frac{1}{v}\frac{\partial v}{\partial T} = \frac{\partial^2 v}{\partial z'^2} \qquad (34)$$

with the boundary conditions, from Eq 22

$$z' = 0 \qquad 0 \leq T \leq \infty \qquad v = 1 \qquad (36)$$

$$z' = 2 \qquad 0 \leq T \leq \infty \qquad v = 1 \qquad (37)$$

and the initial condition

$$T = 0 \qquad 0 < z' < 2 \qquad v = \left(\frac{\sigma'_1}{\sigma'_2}\right)^{\lambda} \qquad (38)$$

For $T = \infty$ the solution must provide

$$T = \infty \qquad 0 \leq z' \leq 2 \qquad v = 1 \qquad (39)$$

Alternatively, the second boundary condition, Eq 37, may be substituted by the following boundary condition

$$z' = 1 \qquad 0 \leq T \leq \infty \qquad \frac{\partial v}{\partial z'} = 0 \qquad \text{(37 bis)}$$

For the special case, C_v constant, the boundary-value problem is

$$\frac{\partial w}{\partial T} = \frac{\partial^2 w}{\partial z'^2} \qquad (35)$$

with the boundary conditions, from Eq 27

$$z' = 0 \qquad 0 \leq T \leq \infty \qquad w = 0 \qquad (40)$$

$$z' = 2 \qquad 0 \leq T \leq \infty \qquad w = 0 \qquad (41)$$

and the initial condition

$$T = 0 \qquad 0 < z' < 2 \qquad w = \ell n \frac{\sigma'_1}{\sigma'_2} \qquad (42)$$

For $T = \infty$ the solution must provide

$$T = \infty \qquad 0 \leq z' \leq 2 \qquad w = 0 \qquad (43)$$

Solution for the Special Case C_v *Constant*

The solution for the special case was obtained by Davis and Raymond [3] in a similar way to Terzaghi's linear theory since the boundary conditions are similar in terms of u and w. Therefore, for the special case, C_v constant

$$w = \ell n \frac{\sigma'}{\sigma'_2} = \left(\ell n \frac{\sigma'_1}{\sigma'_2} \right) \sum_{N=0}^{\infty} \frac{2}{M} (\sin Mz') \epsilon^{-M^2 T} \qquad (44)$$

where

$$M = (2N + 1) \frac{\pi}{2} \qquad (45)$$

Therefore

$$\frac{\sigma'}{\sigma'_2} = \left(\frac{\sigma'_1}{\sigma'_2} \right)^B \qquad (46)$$

where

$$B = \sum_{N=0}^{\infty} \frac{2}{M} (\sin Mz') \epsilon^{-M^2 T} \qquad (47)$$

Equation 46 may also be written as

$$\frac{\sigma'}{\sigma'_1} = \left(\frac{\sigma'_2}{\sigma'_1}\right)^{1-B} \tag{48}$$

It should be observed that in Terzaghi's theory

$$B = \frac{u}{\sigma'_2 - \sigma'_1} \tag{T-1}$$

and therefore, in Terzaghi's theory

$$\frac{\sigma'}{\sigma'_2} = \frac{\sigma'_2 - u}{\sigma'_2} = 1 - B\left(1 - \frac{\sigma'_1}{\sigma'_2}\right) \tag{T-2}$$

It can easily be checked that the effective stresses given by Eq 46 are always smaller than those predicted by the Terzaghi's theory, Eq T-2, and that their ratio increases when σ'_2/σ'_1 increases, as correctly stated by Davis and Raymond [3]. These authors obtained also a solution for the degree of settlement assuming the traditional e-log σ' relationship and found that the degree of settlement coincides with the degree of consolidation given by Terzaghi's theory. This is not so, however, for the volume-pressure relationship given by Eq 10. The significance of the Terzaghi's degree of consolidation is established in the Appendix.

Degree of Settlement

The settlement due to consolidation may be found using the expressions already published [8]. For simplicity and without any loss of generality, think now about a specimen drained only at one of its faces such that H_1, H, and H_2 are total thicknesses. The infinitesimal element at depth z of initial thickness $(dz)_1$ under effective pressure σ'_1 will, at time t, have a thickness dz under effective pressure σ', given by (from Eq 7)

$$\frac{dz}{(dz)_1} = \left(\frac{\sigma'}{\sigma'_1}\right)^{-\gamma} \tag{49}$$

Integrating for the whole thickness H_1 gives

$$H = \int_0^{H_1} \left(\frac{\sigma'}{\sigma'_1}\right)^{-\gamma} (dz)_1 \tag{50}$$

and the settlement S_t at time t is

$$S_t = H_1 - H = H_1 - \int_0^{H_1} \left(\frac{\sigma'}{\sigma'_1}\right)^{-\gamma} (dz)_1 \tag{51}$$

Similarly, we can write

$$H_2 = \int_0^{H_1} \left(\frac{\sigma'_2}{\sigma'_1}\right)^{-\gamma} (dz)_1 = H_1 \left(\frac{\sigma'_2}{\sigma'_1}\right)^{-\gamma} \tag{52}$$

and the total final settlement is

$$\Delta H = H_1 - H_2 = H_1 \left[1 - \left(\frac{\sigma'_2}{\sigma'_1}\right)^{-\gamma} \right] \tag{53}$$

The concept degree of settlement U may therefore be written as

$$U = \frac{S_t}{\Delta H} = \frac{1 - \dfrac{1}{H} \displaystyle\int_0^{H_1} \left(\frac{\sigma'}{\sigma'_1}\right)^{-\gamma} dz}{1 - \left(\frac{\sigma'_2}{\sigma'_1}\right)^{-\gamma}} \tag{54}$$

where the subindices 1 for H and dz have been dropped for conciseness.

Using the reduced variable z' given by Eq 31, the degree of settlement (Eq 54) may be written as

$$U = \frac{S_t}{\Delta H} = \frac{1 - \displaystyle\int_0^1 \left(\frac{\sigma'}{\sigma'_1}\right)^{-\gamma} dz'}{1 - \left(\frac{\sigma'_2}{\sigma'_1}\right)^{-\gamma}} \tag{55}$$

For the general case, $\lambda \neq 0$, Eq 55 may be written, in terms of the variable v, Eq 22, as

$$U = \frac{1 - \left(\frac{\sigma'_2}{\sigma'_1}\right)^{-\gamma} \displaystyle\int_0^1 v^{-\gamma/\lambda} dz'}{1 - \left(\frac{\sigma'_2}{\sigma'_1}\right)^{-\gamma}} \tag{56}$$

For the special case, $\lambda = 0$, using Eq 48, Eq 55 may be written

$$U = \frac{1 - \left(\dfrac{\sigma'_2}{\sigma'_1}\right)^{-\gamma} \displaystyle\int_0^1 \left(\dfrac{\sigma'_2}{\sigma'_1}\right)^{\gamma B} dz'}{1 - \left(\dfrac{\sigma'_2}{\sigma'_1}\right)^{-\gamma}} \tag{57}$$

where B is given by Eq 47.

Introducing Eq 52 into Eq 57 we can write for the special case

$$U = \frac{1 - \dfrac{H_2}{H_1} \displaystyle\int_0^1 \left(\dfrac{H_1}{H_2}\right)^B dz'}{1 - \dfrac{H_2}{H_1}} \tag{58}$$

Therefore, U is a function of T, through parameter B (see Eq 47), and of the ratio of the final to the initial thicknesses of the specimen. The relationship between U and T for different values of H_2/H_1 will be presented later.

For the general case, C_v variable, introducing Eq 52 into Eq 56 we can also write

$$U = \frac{1 - \dfrac{H_2}{H_1} \displaystyle\int_0^1 v^{-\gamma/\lambda} dz'}{1 - \dfrac{H_2}{H_1}} \tag{59}$$

Furthermore, introducing Eq 52 into the initial condition, Eq 38, of the boundary-value problem, we obtain

$$v = \left(\frac{\sigma'_1}{\sigma'_2}\right)^\lambda = \left(\frac{\sigma'_2}{\sigma'_1}\right)^{-\gamma(\lambda/\gamma)} = \left(\frac{H_2}{H_1}\right)^{\lambda/\gamma} \tag{60}$$

and therefore, for the general case, U is a function of T (v is function of T, Eq 34), of the ratio of the final to the initial thicknesses of the specimen H_2/H_1 and of the ratio of parameters λ/γ. The corresponding consolidation curves require the solution of the boundary-value problem. This is the subject of a companion paper already published [11].

Consolidation Curves

The consolidation curves U versus T for the special case C_v constant for which $\lambda = 0$, for $H_2/H_1 = 0.5$, 0.7, 1, 1.4, and 2.0, were obtained by numerical

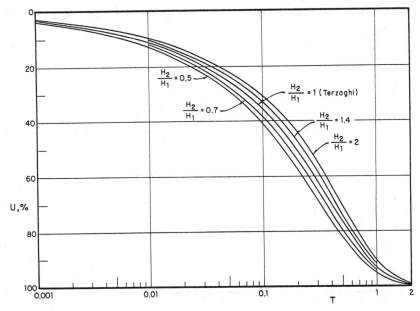

FIG. 1—$T - U\%$ *relationship for* $\lambda = 0$.

methods and are presented in Fig. 1. The case $H_2/H_1 < 1$ corresponds to compression cases, $H_2/H_1 > 1$ to swelling cases, and $H_2/H_1 = 1$ to Terzaghi's theoretical linear solution.

It may be observed that the consolidation curves move to the left for the compression cases and move to the right for the swelling cases. However, the amount of displacement is not very great, especially for the usual values of H_2/H_1 normally used in the laboratory.

In the companion paper it is shown, furthermore, that Terzaghi's theory gives very good results for the degree of settlement, Eq 59, for the case for which $\lambda/\gamma = 1.0$ for the whole interval of H_2/H_1 considered, that is, from $H_2/H_1 = 0.5$ to $H_2/H_1 = 2.0$. Figures 2 and 3 illustrate some of the curves obtained.

Conclusions

The principal conclusions are as follows:

1. A general nonlinear differential equation for the one-dimensional consolidation of saturated plastic soils has been presented (Eq 9 or 14). It takes into account the variations of linear compressibility, of permeability, and of the coefficient of consolidation during the consolidation process.

2. The coefficient of consolidation C_v for a given soil varies with pressure (Eqs 11 and 12).

3. Application to the oedometer test is made neglecting the submerged unit weight of the soil and assuming that the change in thickness is small. The

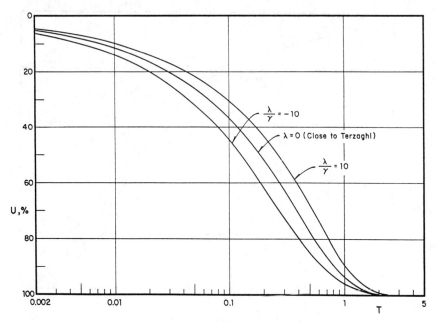

FIG. 2—$T - U\%$ relationship for $H_2/H_1 = 0.9$.

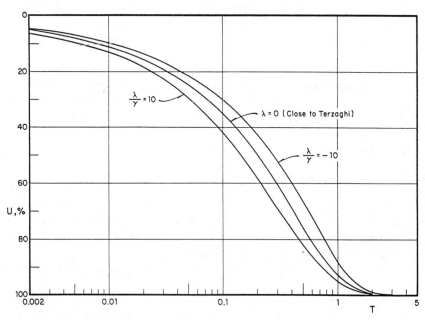

FIG. 3—$T - U\%$ relationship for $H_2/H_1 = 1.1$.

consolidation curves are found to depend on the ratio of the final to the initial thickness H_2/H_1 and of the ratio of the parameters λ/γ.

4. For the special case, $\lambda = 0$, that is, when C_v is constant, the consolidation curves (Fig. 1) move somewhat to the left for the compression cases, and move somewhat to the right for the swelling cases. The effective stresses at any time and depth are given by Eqs 46 and 47, and their values are smaller than those predicted by Terzaghi's theory. The degree of settlement is given by Eq 58. Terzaghi's equation is found to give, for this case, the "modified natural degree of settlement" defined by Eq 67 of the Appendix.

5. For the general case, $\lambda \neq 0$, that is, when C_v is variable, the degree of settlement is given by Eq 59 which requires the solution of the boundary-value problem stated by Eqs 34, 36, 37, and 60. The corresponding consolidation curves are the subject of a companion paper [11].

Acknowledgments

The author is grateful to the National University of Mexico (Graduate School of Engineering and Institute of Engineering) and to the Ministry of Human Settlements and Public Works for their support of this work. Suggestions made by Gonzalo Alduncin-González are gratefully acknowledged. The numerical solution of Eq 58 (Fig. 1) was obtained by Carlos E. Juárez-Badillo R., son of the author.

APPENDIX

Significance of Terzaghi's Equation

The significance of the degree of consolidation given by Terzaghi's theory, for the case C_v constant, may be found introducing a nonlinear concept for U.

Let us first define a "natural degree of settlement U_n" in terms of natural deformation. The instantaneous deformation $d\epsilon$ at time t is

$$d\epsilon = \frac{dH}{H} \tag{61}$$

The natural deformation at time t will therefore be

$$\epsilon = \int_{H_1}^{H} \frac{dH}{H} = \ell n \frac{H}{H_1} \tag{62}$$

The final total natural deformation will be

$$\epsilon_T = \ell n \frac{H_2}{H_1} \tag{63}$$

Let us define U_n by

$$U_n = \frac{\epsilon}{\epsilon_T} = \ell n \frac{H}{H_1} \bigg/ \ell n \frac{H_2}{H_1} \tag{64}$$

Introducing subsequently Eqs 50, 48, 52, and 31 into Eq 64, we can write

$$U_n = \frac{1}{\ell n \ (H_2/H_1)} \ \ell n \left[\frac{1}{H_1} \int_0^{H_1} \left(\frac{\sigma'}{\sigma'_1} \right)^{-\gamma} (dz)_1 \right]$$

$$= \frac{1}{\ell n \ (H_2/H_1)} \ \ell n \left[\frac{1}{H_1} \int_0^{H_1} \left(\frac{H_2}{H_1} \right)^{1-B} (dz)_1 \right]$$

$$= \frac{1}{\ell n \ (H_2/H_1)} \left[\ell n \ \frac{H_2}{H_1} + \ell n \left\{ \frac{1}{H_1} \int_0^{H_1} \left(\frac{H_1}{H_2} \right)^B (dz)_1 \right\} \right]$$

Therefore

$$U_n = 1 - \frac{1}{\ell n \ (H_1/H_2)} \ \ell n \left[\int_0^1 \left(\frac{H_1}{H_2} \right)^B dz' \right] \qquad (65)$$

It should be observed that in Eq 64 the relation between the natural deformations of the whole specimen of soil was used. The deformations, however, are not uniform in the specimen, and an alternative approach would be to take into account the real distribution of natural deformations within the specimen.

At depth z and time t the natural deformation ϵ_z is given by

$$\epsilon_z = \ell n \ \frac{dz}{(dz)_1} \qquad (66)$$

The "area of natural deformations" of the specimen would be the integral of Eq 66 through the total thickness of the specimen. Defining the "modified natural degree of settlement U'_n" by the ratio of this area to the final area of natural deformations we have

$$U'_n = \frac{\int_0^{H_1} \left[\ell n \ \frac{dz}{(dz)_1} \right] (dz)_1}{\int_0^{H_1} \left[\ell n \ \frac{(dz)_2}{(dz)_1} \right] (dz)_1} \qquad (67)$$

Introducing Eqs 49 and 48 into Eq 67 we obtain

$$U'_n = \frac{\int_0^{H_1} \left[\ell n \left(\frac{\sigma'}{\sigma'_1} \right)^{-\gamma} \right] (dz)_1}{\int_0^{H_1} \left[\ell n \left(\frac{\sigma'_2}{\sigma'_1} \right)^{-\gamma} \right] (dz)_1} = \frac{\int_0^{H_1} \left[\ell n \left(\frac{\sigma'_2}{\sigma'_1} \right)^{-\gamma(1-B)} \right] (dz)_1}{\int_0^{H_1} \left[\ell n \left(\frac{\sigma'_2}{\sigma'_1} \right)^{-\gamma} \right] (dz)_1}$$

$$= \frac{H_1 \ \ell n \left(\frac{\sigma'_2}{\sigma'_1} \right)^{-\gamma} - \ell n \left(\frac{\sigma'_2}{\sigma'_1} \right)^{-\gamma} \int_0^{H_1} B(dz)_1}{H_1 \ \ell n \left(\frac{\sigma'_2}{\sigma'_1} \right)^{-\gamma}} = 1 - \frac{1}{H_1} \int_0^{H_1} B(dz)_1 \qquad (68)$$

In Terzaghi's theory, by Eq T-1, we have

$$U'_n = 1 - \frac{1}{H_1} \int_0^{H_1} \frac{u}{\sigma'_2 - \sigma'_1} (dz)_1 \qquad (T\text{-}3)$$

Using U_T for degree of consolidation in Terzaghi's theory, we then have, by Eq T-3,

$$U'_n = U_T \qquad (69)$$

where U_T, as is well known, is given by

$$U_T = 1 - \sum_{N=0}^{\infty} \frac{2}{M^2} \epsilon^{-M^2 T} \qquad (70)$$

where M is given by Eq 45.

Therefore, the significance of the degree of consolidation in Terzaghi's theory is that, for the case when C_v is constant, $\lambda = 0$, it gives the modified natural degree of settlement, defined by Eq 67.

References

[1] Terzaghi, K., *Theoretical Soil Mechanics,* Wiley, New York, 1943, p. 265.
[2] Gibson, R. E., England, G. L., and Hussey, M. J. L., "The Theory of One Dimensional Consolidation of Saturated Clays," *Geotechnique,* Vol. 17, 1967, pp. 261–273.
[3] Davis, E. H. and Raymond, G. P., "A Non-Linear Theory of Consolidation," *Geotechnique,* Vol. 15, No. 2, 1965, pp. 161–173.
[4] Butterfield, R., "A Natural Compression Law for Soils (An Advance on e-log p')," *Geotechnique,* Vol. 29, No. 4, 1979, pp. 469–480.
[5] Jacob, C. E., "Flow of Ground Water," in *Engineering Hydraulics,* Hunter Rouse, Ed., Wiley, New York, 1950, p. 321.
[6] Juárez-Badillo, E., "Compressibility of Soils," in *Proceedings,* Fifth Symposium of the Civil and Hydraulic Engineering Department on Behavior of Soils under Stress, Indian Institute of Science, Bangalore, India, Vol. 1, 1965, pp. A2/1–35.
[7] Juárez-Badillo, E., "Pore Pressure and Compressibility Theory for Saturated Soils," Specialty Session No. 12 on Advances in Consolidation Theories for Clays, University of Waterloo, Ontario, Canada, 1969, pp. 99–116.
[8] Juárez-Badillo, E., "Constitutive Relationships for Soils," in *Proceedings,* Symposium on Recent Developments in the Analysis of Soil Behaviour and Their Application to Geotechnical Structures, University of New South Wales, Kensington, Australia, July 1975, pp. 231–257.
[9] Juárez-Badillo, E., "General Permeability Change Equation for Soils," in *Proceedings,* International Conference on Constitutive Laws for Engineering Materials, University of Arizona, Tucson, Jan. 1983, pp. 205–209.
[10] Janbu, N., "Consolidation of Clay Layers Based on Non-Linear Stress-Strain," in *Proceedings,* Sixth International Conference on Soil Mechanics and Foundation Engineering, Canada, Vol. 2, 1965, pp. 83–87.
[11] Juárez-Badillo, E. and Chen, B., "Consolidation Curves for Clays," *Journal of Geotechnical Engineering,* American Society of Civil Engineers, Vol. 109, No. 10, Oct. 1983, pp. 1303–1312.

Delwyn G. Fredlund[1] and Harianto Rahardjo[1]

Unsaturated Soil Consolidation Theory and Laboratory Experimental Data

REFERENCE: Fredlund, D. G. and Rahardjo, H., **"Unsaturated Soil Consolidation Theory and Laboratory Experimental Data,"** *Consolidation of Soils: Testing and Evaluation, ASTM STP 892*, R. N. Yong and F. C. Townsend, Eds., American Society for Testing and Materials, Philadelphia, 1986, pp. 154–169.

ABSTRACT: The volume change process of an unsaturated soil was considered as a transient flow problem with water and air flowing simultaneously under an applied stress gradient. Laboratory experiments were performed to study the volume change behavior of unsaturated soils. Several compacted kaolin specimens were tested using modified Anteus oedometers and triaxial cells. Theoretical analyses were also made to best-fit the results from laboratory experiments. The fitting was accomplished by approximating the compressibility coefficients and adjusting the coefficients of permeability. The comparisons of the results obtained from the theory and the laboratory yield to a similar behavior of volume changes with respect to time.

KEY WORDS: unsaturated soil, volume change, consolidation, two-phase flow

For saturated soils, a settlement analysis is usually performed using Terzaghi's theory of consolidation. In practice, many geotechnical problems involve unsaturated soils. The construction of earth-fill dams, highways, and airport runways always use unsaturated compacted soils. In addition, large portions of the earth's surface are covered with residual soils that are unsaturated. Heave and settlement problems are commonly encountered with these unsaturated soils.

The constitutive relations for volume change in an unsaturated soil were proposed by Fredlund and Morgenstern in 1976 [1]. Laboratory experiments were also performed to study the volume change behavior of unsaturated soils [2].

This paper presents theoretical analyses of the volume change process in unsaturated soils based on the proposed constitutive relations. Comparisons are made between the theoretical analyses and the laboratory time rate of deformation.

Theory

A stress gradient applied to an unsaturated soil will cause two phases (i.e., soil particle and contractile skin or air-water interface) to come to equilibrium,

[1] Professor and Research Engineer, respectively, Department of Civil Engineering, University of Saskatchewan, Saskatoon, SK, Canada S7N 0W0.

whereas the other two phases (i.e., air and water) will flow. The volume change process of an unsaturated soil can be treated as a transient flow problem with water and air flowing simultaneously under an applied stress gradient.

The increase (swelling) or the decrease (consolidation) in the overall volume of an unsaturated soil can be predicted using the proposed constitutive relations (see Ref 1). Assuming the soil particles are incompressible and the volume change of the contractile skin is internal to the element, the continuity requirement for a referential element can be written as

$$d\epsilon = d\theta_w + d\theta_a \qquad (1)$$

where

$d\epsilon$ = change in unit volumetric strain of the soil structure; that is, $d\epsilon = d\epsilon_x + d\epsilon_y + d\epsilon_z$ where ϵ_x, ϵ_y, and ϵ_z are the normal strains in the x, y, and z directions, respectively,

$d\theta_w$ = net inflow or outflow of water from the unit element per unit volume, and

$d\theta_a$ = net inflow or outflow of air from the unit element per unit volume.

Equation 1 shows that only two of the three possible constitutive relations (i.e., the soil structure, water, and air phases) are independent [3]. Therefore two constitutive relations are required to describe the volume change behavior in an unsaturated soil. The volume change constitutive relations for unsaturated soils were formulated using two independent stress-state variables, $(\sigma - u_a)$ and $(u_a - u_w)$ [3], where

σ = total normal stress,

u_a = pore-air pressure,

u_w = pore-water pressure, and

$(u_a - u_w)$ = matric suction.

The constitutive equations for the soil structure, water, and air phases are, respectively,

$$d\epsilon = m_1^s d(\sigma - u_a) + m_2^s d(u_a - u_w) \qquad (2)$$

$$d\theta_w = m_1^w d(\sigma - u_a) + m_2^w d(u_a - u_w) \qquad (3)$$

$$d\theta_a = m_1^a d(\sigma - u_a) + m_2^a d(u_a - u_w) \qquad (4)$$

where

m_1^s = compressibility of soil structure with respect to a change in $(\sigma - u_a)$,

m_2^s = compressibility of soil structure with respect to a change in $(u_a - u_w)$,

m_1^w = slope of $(\sigma - u_a)$ versus θ_w plot,

m_2^w = slope of $(u_a - u_w)$ versus θ_w plot,

m_1^a = slope of (σ u_a) versus θ_a plot, and
m_2^a = slope of ($u_a - u_w$) versus θ_a plot.

Considering the volumetric continuity requirement shown in Eq 1, the following relationship for the compressibilities must be maintained:

$$m_1^s = m_1^w + m_1^a \tag{5}$$

$$m_2^s = m_2^w + m_2^a \tag{6}$$

The pore-water and pore-air pressures change with time and can be solved simultaneously using two independent partial differential equations. This method is called a two-phase flow approach and has been presented by Fredlund and Hasan [4], Lloret and Alonso [5], and Fredlund [3].

In this paper, a one-dimensional transient flow process will be considered. The one-dimensional flow equations for the water and air phases can be derived by equating the time derivative of the relevant constitutive equation to the divergence of the velocity as described by the flow laws. Darcy's and Fick's laws are applied to the flow of water and air phases, respectively. The time derivative of the constitutive equation represents the amount of deformation that occurs under various stress conditions, while the divergence of velocity describes the rate of flow of the air and water.

Several assumptions are used in the derivations: (1) isotropic soil, (2) continuous air phase, (3) infinitesimal strains, (4) linear constitutive relations, (5) coefficients of permeability of water and air phases are functions of the volume-mass soil properties during the transient process, and (6) the effects of air diffusing through water, air dissolving in water, and the movement of water vapor are ignored.

The partial differential equation for the water phase in the y-direction can be written as [3]

$$\frac{\partial u_w}{\partial t} = -C_w \frac{\partial u_a}{\partial t} + c_v^w \frac{\partial^2 u_w}{\partial y^2} + \frac{c_v^w}{k_w} \frac{\partial k_w}{\partial y} \frac{\partial u_w}{\partial y} + c_g \frac{\partial k_w}{\partial y} \tag{7}$$

where

t = time,
C_w = $(1 - m_2^w/m_1^w)/(m_2^w/m_1^w)$ and is called the interactive constant associated with the water phase equation,
c_v^w = $k_w/(\rho_w \, g \, m_2^w)$ and is called the coefficient of consolidation with respect to the water phase,
k_w = coefficient of permeability with respect to the water phase,
ρ_w = density of water,
g = gravitational acceleration, and
c_g = $1/m_2^w$ and is commonly referred to as the gravity term constant in the soil science literature.

The coefficient of permeability of water can vary significantly with space in the unsaturated soil. This variation is taken into account by the last two terms of Eq 7. The volume-mass properties of the soil can be used to describe the variation in the coefficient of permeability [6,7].

The air phase partial differential equation has the form [3]

$$\frac{\partial u_a}{\partial t} = -C_a \frac{\partial u_w}{\partial t} + c_v^a \frac{\partial^2 u_a}{\partial y^2} + \frac{c_v^a}{D^*} \frac{\partial D^*}{\partial y} \frac{\partial u_a}{\partial y} \qquad (8)$$

where

$$C_a = \frac{m_2^a/m_1^a}{(1 - m_2^a/m_1^a) + \dfrac{(1 - S)n}{(u_a - u_{atm}) m_1^a}}$$

and is called the interactive constant associated with the air phase equation,

S = degree of saturation of the soil,

n = porosity of the soil,

u_{atm} = atmospheric pressure,

$$c_v^a = \frac{D^*R\theta}{\omega} \frac{1}{(1 - m_2^a/m_1^a)(u_a + u_{atm})m_1^a + (1 - S)n}$$

and is called the coefficient of consolidation with respect to the air phase,

D^* = D/g and is called the transmission constant of proportionality for the air phase,

D = a transmission constant for the air phase having the same unit as coefficient of permeability,

R = universal gas constant,

θ = absolute temperature, and

ω = molecular weight of the mass of air.

The last term in Eq 8 accounts for the variation of the transmission constant of proportionality for the air phase with respect to space. The constant of proportionality can be written as a function of the volume-mass properties of the soil.

The excess pore-air and pore-water pressures generated by a change in total stress can be computed using the pore pressure parameters associated with the air and water phases, respectively [8].

Laboratory Tests

Laboratory experiments were performed on several compacted kaolin specimens using modified Anteus oedometers and triaxial cells [2].

Equipment

The modified Anteus oedometers were used to perform one-dimensional consolidation tests. The modified Wykeham Farrance triaxial cells allowed isotropic volume change testing conditions.

A high-air-entry ceramic disk was sealed to the lower pedestal. This disk allows the flow of water but prevents the flow of free air. Therefore the measurement of the pore-air and pore-water pressures can be made independently. A low-air-entry disk was placed on the top of the specimen to facilitate the control of the pore-air pressure. This allowed the translation of the air and the water pressures to positive values in order to prevent cavitation of the water below the high-air-entry disk (i.e., the axis translation technique [9]). The total, pore-air, and pore-water pressure conditions can then be controlled to study the volume change behavior of unsaturated soils.

Two rubber membranes separated by a slotted tin foil were placed around the specimen. This composite membrane was found to be essentially impermeable for a long period of time.

Presentation of Results

Five specimens of kaolin were compacted in accordance with the standard Proctor method. The initial volume-mass properties of each specimen are summarized in Table 1. Each experiment was performed by changing one of the stress-state variable components: the total stress (σ), the pore-water pressure (u_w), or the pore-air pressure (u_a). The stress-state component changes associated with each experiment are given in Table 2. In each case, the volume changes of soil structure and water phase were monitored.

Figure 1 shows a typical plot of volume changes due to an increment of total stress in Test 1. The plot exhibits a decrease in volume of soil structure, air, and water phases during the transient process. A large instantaneous volume decrease occurred at the time when the load was applied.

An increase in the pore-air pressure (Test 2) caused the soil structure to expand temporarily (Fig. 2). However, the increase in air pressure results in an increase in matric suction ($u_a - u_w$), which in turn can cause a decrease in the soil structure at the end of the process.

TABLE 1—*Volume-mass relations for specimens tested.*

Test No.	Diameter, cm	Height, cm	Total Volume, cm³	Water Content, %	Void Ratio	Dry Density, kN/m³	Degree of Saturation, %
1	10.006	11.815	929.09	34.32	1.0696	13.185	78.87
2	9.945	11.703	909.10	33.17	1.0251	13.298	80.61
3	10.543	5.867	503.53	29.62	1.2242	11.529	63.29
4	9.832	5.758	437.16	32.12	0.9310	13.281	90.25
5	6.350	2.283	72.29	31.18	1.1247	12.069	72.51

TABLE 2—*Change in stress-state variable components associated with each test.*[a]

Test No.	Total Stress (σ), kPa		Pore-Water Pressure (u_w), kPa		Pore-Air Pressure (u_a), kPa		Change, kPa
	Initial	Final	Initial	Final	Initial	Final	
1	358.7	560.9	163.8	164.4	214.4	215.6	$\Delta\sigma = +202.2$
2	560.9	559.0	164.4	163.1	215.6	421.1	$\Delta u_a = +205.5$
3	475.1	476.8	41.9	42.2	397.8	206.5	$\Delta u_a = -191.3$
4	611.4	610.1	177.3	379.2	532.0	530.9	$\Delta u_w = +201.9$
5	606.7	605.2	216.5	323.6	413.8	413.8	$\Delta u_w = +107.1$

[a]All tests performed in a modified triaxial apparatus except Test 5, which was performed in a modified oedometer.

A volume change process associated with a meta-stable structure is shown in Fig. 3. The decrease in air pressure (Test 3) reduced the matric suction $(u_a - u_w)$ and allowed more water to flow into the specimen. The intake of water reduced the normal and shear stresses between the soil particles. As a result, the soil structure underwent a decrease in volume (i.e., collapse phenomenon).

An increase in water pressure could cause an increase in volume of the soil structure (Fig. 4 for Test 4) if the soil has a stable structure. On the other hand, when the soil structure was meta-stable, an increase in water pressures caused the soil structure to decrease in volume or collapse (Fig. 5).

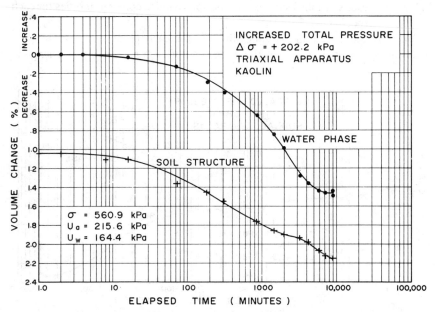

FIG. 1—*Soil structure and water phase volume changes associated with Test 1.*

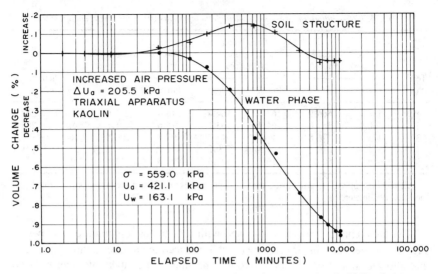

FIG. 2—*Soil structure and water phase volume changes associated with Test 2.*

Theoretical Analyses

Attempts were made to best-fit the theoretical analyses of volume change with the results from laboratory experiments. This was accomplished by approximating the compressibilities of the soil structure, air, and water phases based on the laboratory results. Table 3 summarizes the approximate compressibilities for each of the five specimens.

The magnitudes of the compressibilities for each phase were obtained by dividing the amount of deformation at the end of each process by the change in

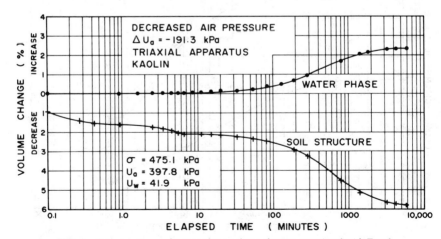

FIG. 3—*Soil structure and water phase volume changes associated with Test 3.*

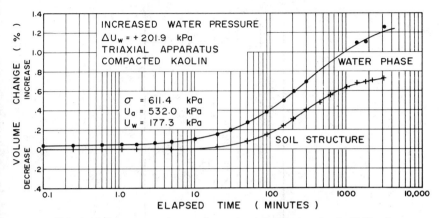

FIG. 4—*Soil structure and water phase volume changes associated with Test 4.*

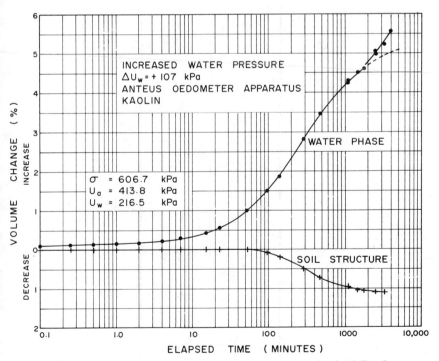

FIG. 5—*Soil structure and water phase volume changes associated with Test 5.*

TABLE 3—Compressibilities for each specimen.

Test No.	Soil Structure		Water-Phase		Air-Phase	
	m_1^s $\times 10^{-4}$ kPa^{-1}	m_2^s $\times 10^{-4}$ kPa^{-1}	m_1^w $\times 10^{-4}$ kPa^{-1}	m_2^w $\times 10^{-4}$ kPa^{-1}	m_1^a $\times 10^{-4}$ kPa^{-1}	m_2^a $\times 10^{-4}$ kPa^{-1}
1	1.17	3.52	0.80	2.41	0.37	1.11
2	0.006	0.032	0.131	0.657	-0.125	-0.625
3	-0.76	-3.80	0.30	1.49	-1.06	-5.29
4	0.09	0.35	0.15	0.61	-0.06	-0.26
5	-0.26	-1.04	1.20	4.81	-1.46	-5.85

the stress-state variable. A sign (positive or negative) is attached to each of the compressibilities based on the direction (increase or decrease) of the volume change associated with each phase and the change of the stress state variables [3]. Table 3 indicates that the soil structure compressibilities have positive signs for a stable structured soil (Tests 1, 2, and 4) and negative signs for a metastable soil (Tests 3 and 5).

The theoretical analysis for each of the laboratory results was based upon the consideration of a one-dimensional transient flow process. The pore-water and pore-air pressures during the transient process were computed by solving simultaneously Eqs 7 and 8. An explicit central difference technique was used for the calculations [10]. The coefficients of permeability of air and water were assumed to be constant during the transient process. Therefore the terms that account for the variation in coefficient of permeability in Eqs 7 and 8 were not used in the computations.

The volume changes associated with the water and the air phases during the transient process were computed according to Eqs 3 and 4. The compressibilities used in the calculations were assumed to be constant throughout the process. The soil structure volume change was obtained by adding the volume changes associated with the water and air phases according to the continuity requirement in Eq 1.

Figures 6 to 10 show the theoretical analyses associated with Tests 1 to 5. The fitting was accomplished by using different combinations for the coefficients of permeability for the water and air phases. The combinations of the water and air

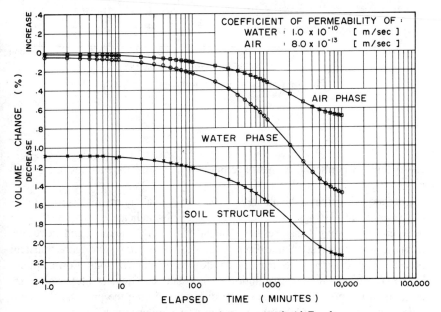

FIG. 6—*Theoretical analysis associated with Test 1.*

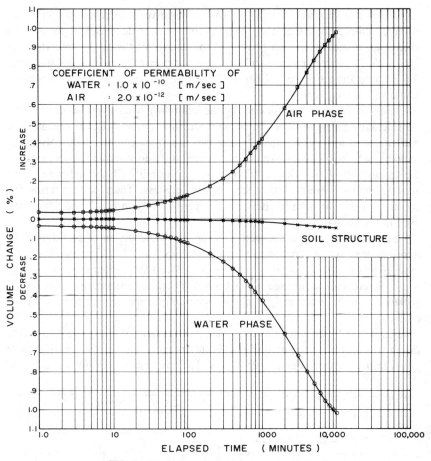

FIG. 7—*Theoretical analysis associated with Test 2.*

coefficients of permeability that gave the best-fit results for each test are shown in each figure.

Comparisons between the theoretical analyses and laboratory results were made for Tests 4 and 3 (Figs. 11 and 12). The results indicate a close agreement between the theoretical analyses based on the constitutive equations and the results from the laboratory tests. Some discrepancies can be observed during the transient process. The disagreements may be due to one or more reasons. For example, they may be due to the assumption of constant coefficients of permeability throughout the process. In spite of the difference during the process, the theoretical analyses and the laboratory results show similar trends.

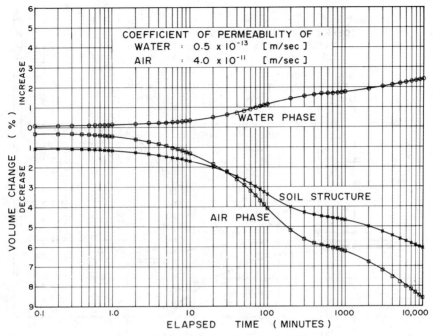

FIG. 8—*Theoretical analysis associated with Test 3.*

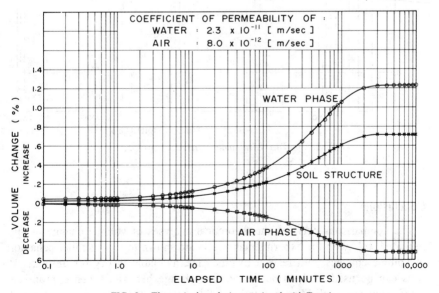

FIG. 9—*Theoretical analysis associated with Test 4.*

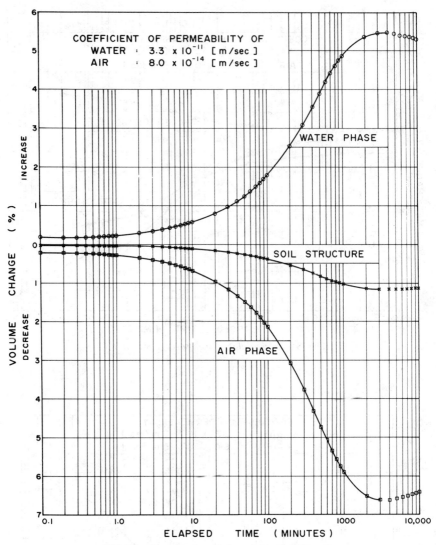

FIG. 10—*Theoretical analysis associated with Test 5.*

Conclusions

The comparisons between the theoretical analyses of volume change in unsaturated soils with the experimental results are shown to yield similar behavior of volume changes with respect to time. The available theory for unsaturated soils can be used to describe the volume change behavior when appropriate coefficients are used. More research should be conducted where each of the soil

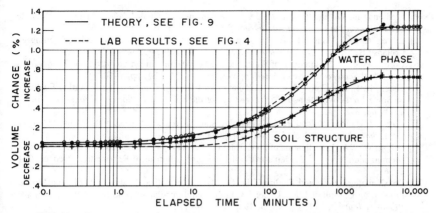

FIG. 11—*Comparisons between theoretical analyses and laboratory results for Test 4.*

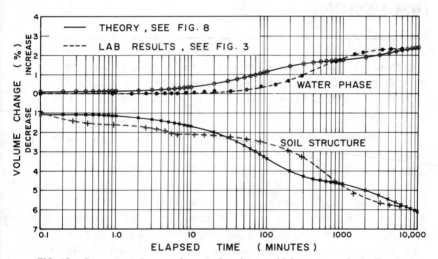

FIG. 12—*Comparisons between theoretical analyses and laboratory results for Test 3.*

properties are independently measured and the time-dependent volume changes are predicted and independently monitored. The relationship between the coefficients of permeability and other soil properties also requires further study.

References

[*1*] Fredlund, D. G. and Morgenstern, N. R., "Constitutive Relations for Volume Change in Unsaturated Soils," *Canadian Geotechnical Journal,* Vol. 13, No. 3, 1976, pp. 261–276.
[*2*] Fredlund, D. G., "Volume Change Behavior of Unsaturated Soils," Ph.D. Dissertation, University of Alberta, Edmonton, Alberta, Canada, 1973.
[*3*] Fredlund, D. G., "Consolidation of Unsaturated Porous Media," Presented to NATO Advance Study Institute Symposium on the Mechanics of Fluids in Porous Media—New Approaches in Research, Newark, Del., 1982.

[4] Fredlund, D. G. and Hasan, J. U., "One-Dimensional Consolidation Theory: Unsaturated Soils," *Canadian Geotechnical Journal,* Vol. 16, No. 3, 1979, pp. 521–531.
[5] Lloret, A. and Alonso, E. E., "Consolidation of Unsaturated Soils Including Swelling and Collapse Behavior," *Geotechnique,* Vol. 30, No. 4, 1980, pp. 449–477.
[6] Corey, A. T., "Measurement of Water and Air Permeability in Unsaturated Soil," *Proceedings of the Soil Science Society of America,* Vol. 21, No. 1, 1957, pp. 7–10.
[7] Green, R. E. and Corey, J. C., "Calculation of Hydraulic Conductivity: A Further Evaluation of Some Predictive Methods," *Proceedings of the Soil Science Society of America,* Vol. 35, 1971, pp. 3–8.
[8] Hasan, J. U. and Fredlund, D. G., "Pore Pressure Parameters for Unsaturated Soils," *Canadian Geotechnical Journal,* Vol. 17, No. 3, 1980, pp. 395–404.
[9] Hilf, J. W., "An Investigation of Pore Water Pressures in Compacted Cohesive Soils," Technical Memorandum 654, U. S. Department of the Interior, Bureau of Reclamation, Denver, Colo. 1956.
[10] Dakshanamurthy, V. and Fredlund, D. G., "A Mathematical Model for Predicting Moisture Flow in an Unsaturated Soil Under Hydraulic and Temperature Gradients," *Water Resources Research Journal,* Vol. 17, No. 3, 1981, pp. 714–722.

DISCUSSION

V. Drnevich[1] *(written discussion)*—(1) What is meant by the statement that permeability is a function of the volume-weight properties of the soil?

(2) Why were the last two terms in Eq 7 omitted when trying to fit the theory and the laboratory results?

D. G. Fredlund and H. Rahardjo (authors' closure)—(1) The coefficient of permeability with respect to the water phase can be written as a function of two of the commonly used volume-weight variables of the soil. It may take any one of the following forms: $k_w = f(S,e)$, $k_w = f(e,w)$, or $k_w = f(w,S)$, where e = void ratio, S = degree of saturation, and w = water content. For an unsaturated soil, the degree of saturation or the water content is taken as the predominant variable and void ratio is assumed to be of secondary importance. Also, permeability is often written as a function of the negative pore-water pressure or matric suction. Any of the above forms are satisfactory.

(2) The laboratory tests were performed by placing specimens in contact with a high-air-entry ceramic disk. This disk has a low permeability and impedes the flow of water out of the specimen. In other words, the rate of volume change may be more an indication of the permeability of the high-air-entry disks than the soil. The last two terms in Eq 7 refer to the spatial variation of permeability of the soil with time. These terms were not meaningful for the type of laboratory test performed.

Finally, we emphasize that the laboratory results presented do not provide a rigorous verification of the consolidation theory for an unsaturated soil. It is difficult to provide such a verification because of the low permeability of the

[1] University of Kentucky, Lexington, Ky.

high-air-entry disk. A rigorous verification would require an independent measurement of all soil properties involved in the formulation and a measurement of the pore-air and pore-water pressure with respect to time and space. Experimentally this is extremely demanding. A lower level of verification involves the independent measurement of total and water volume changes with respect to time rather than the measurement of the pore pressures. The verification provided in this paper simply demonstrates that two independent volume changes (e.g., total volume change and water flux) occur during consolidation and that these can be modeled using two partial differential equations. Certainly there is a challenge for further verification of the consolidation theory.

Donald W. Armour, Jr.[1] and Vincent P. Drnevich[1]

Improved Techniques for the Constant-Rate-of-Strain Consolidation Test

REFERENCE: Armour, D. W., Jr., and Drnevich, V. P., **"Improved Techniques for the Constant-Rate-of-Strain Consolidation Test,"** *Consolidation of Soils: Testing and Evaluation, ASTM STP 892,* R. N. Yong and F. C. Townsend, Eds., American Society for Testing and Materials, Philadelphia, 1986, pp. 170–183.

ABSTRACT: This paper presents several improvements on the constant-rate-of-strain consolidation (CRS) test. The improvements include a rational approach for selecting strain rate and a change in testing procedures to aid in: (1) the selection of strain rate and (2) obtaining more reasonable values of the coefficient of consolidation in initial phases of testing. Based on theoretical work by Wissa, a strain-rate equation is developed that is a function of permeability, liquidity index, and desired maximum pore pressure ratio. The change in procedure involves a method of conducting a permeability test in a back-pressure consolidation apparatus and then initiating consolidation loading without removing the gradient associated with the permeability test. Results of multiple-loading-cycle CRS tests on three clay soils with widely differing characteristics are used to establish the coefficient for the equation. The equation is an improvement over the procedure in ASTM Test for One-Dimensional Consolidation Properties of Soils Using Controlled-Strain Loading (D 4186) which bases the selection only on liquid limit. The proposed equation accounts for specimen permeability and stiffness.

KEY WORDS: consolidation, constant-rate-of-strain consolidation, permeability, liquidity index, strain rate, coefficient of consolidation

The constant-rate-of-strain consolidation (CRS) test was first introduced in 1959 [1] and the basic theory for the present data reduction methods was developed in 1971 by Wissa et al [2]. Since the development of this theory, the CRS test has gained popularity for a number of reasons: (1) the test and data reduction methods are relatively simple and are easily automated; (2) more data points are acquired, allowing for a more accurate determination of settlement parameters; and (3) the CRS test is usually completed in much shorter time than the incremental loading test.

Although the CRS test has many advantages, it is not without disadvantages. Unlike the incremental loading test, secondary compression data cannot easily be distinguished, unless special procedures are instituted. Also, the generated pore pressures are strain-rate dependent; therefore a suitable strain rate must be

[1] Graduate Assistant and Professor of Civil Engineering, respectively, Department of Civil Engineering, University of Kentucky, Lexington, KY 40506.

established before loading commences. Finally, a transient state exists in the initial phases of the test where pore pressures generated by axial loading are low. The small pore pressures are difficult to measure accurately and generally lead to unreasonable values of coefficient of consolidation.

This paper presents a rational method of estimating a suitable strain rate, and a change in testing procedure that aids in (1) estimating strain rate and (2) obtaining more reasonable values of coefficient of consolidation in initial portions of testing.

A nonlinear strain-rate equation was developed by rearranging the equation from Wissa et al for coefficient of consolidation (assuming a log-stress versus linear-strain relation). Strain rate is found to be a function of vertical effective stress, permeability, specimen height, unit weight of water, and pore pressure ratio (ratio of excess pore pressure generated at the base of the specimen by axial loading to the magnitude of total stress). Results of multiple-loading-cycle CRS tests on three clay soils with widely differing characteristics are used to establish the coefficient for the equation.

Strain-Rate Selection Criteria

Previous Work

Several methods have been put forth to select the strain rate, but all of them either require assuming critical parameters or rely solely on empirical correlations. Smith and Wahls [3] derived an equation for strain rate that included the coefficient of consolidation (c_v), compression index (C_c), compressibility (m_v) and pore pressure ratio (u_b/σ_v). In their equation, values for c_v, C_c, and m_v are assumed, the maximum desired pore pressure ratio is set, and strain rate is established. Empirical correlations between the liquid limit of the soil and assumed parameters are used. The current ASTM standard [Test for One-Dimensional Consolidation Properties of Soils Using Controlled-Strain Loading (D 4186)] relies on empirical data published by Gorman [4] and selects strain rate directly from the liquid limit of the soil.

In addition to the problem of strain-rate selection, there is the related problem of the maximum allowable pore pressure ratio, u_b/σ_v. ASTM D 4186 states that the pore pressure ratio be between 3 and 20%, and the chosen rates may be adjusted at discrete intervals if the maximum pore pressure ratio does not fall within these limits. Wissa et al [2] recommend strain rates that result in maximum pore pressure ratios between 20 and 50%.

Strain-Rate Equation

The relationship between strain rate and its controlling parameters is dependent on which stress-strain relation is chosen. Wissa et al [2] presented solutions based on a linear elastic stress-strain behavior and on one where strain was a function of the log-of-stress. It was decided to utilize the latter solution for this work because this theory is more valid for cases where large pore pressure ratios are likely to occur. From Wissa et al [2], with the assumption that strain varies with the log-of-stress, the equation for the coefficient of consolidation is

$$c_v = \frac{-0.434 \, r \, H^2}{2 \, \sigma'_v m_v \log(1 - ppr)} \tag{1}$$

where

c_v = coefficient of consolidation,
H = specimen height,
σ'_v = effective stress,
m_v = compressibility, and
ppr = pore pressure ratio, u_b/σ_v.

Rearranging this equation and solving for the rate term yields

$$r = \frac{-c_v \, 2 \, m_v \, \sigma'_v \log(1 - ppr)}{0.434 \, H^2} \tag{2}$$

Also, from the definition of c_v,

$$c_v = \frac{k}{\gamma_w \, m_v} \tag{3}$$

where k is the coefficient of permeability and γ_w the unit weight of water. Substituting Eq 3 into Eq 2 yields the following equation for strain rate

$$r = \frac{-2 \, \sigma'_v \, k \log(1 - ppr)}{\gamma_w \, 0.434 \, H^2} \tag{4}$$

The parameters to be determined now become permeability, pore pressure ratio, effective stress, and specimen height. For a constant rate of strain, each varies throughout the test. The pore pressure ratio usually passes through a peak or rises to a peak value and stays relatively constant thereafter. The effective stress continually increases during loading and the permeability and specimen thickness continually decrease. Specimen thickness changes are much smaller in magnitude than are changes in permeability or effective vertical stress. Effective stress increases and permeability decreases during loading and, hence, there is a tendency for these parameters to cancel one another. It is desirable to limit the pore pressure ratio during the test to some maximum value. Therefore the values of σ'_v, k, and H should correspond to the values at the maximum pore pressure ratio. Because it is not possible to determine the values of k, and H at the maximum pore pressure ratio before the test is begun, it is proposed to use values of these parameters at the start of the test and then account for their variation between the start of the test and the point where the pore pressure ratio becomes a maximum.

The remaining parameter is the effective stress at the maximum pore pressure ratio. It probably varies with a number of parameters, including soil type, soil stiffness, and degree of disturbance. Rather than trying to consider all of these individually, it is proposed to account for them in a general way by use of an index property. The procedure is outlined subsequently. To keep the equation dimensionally consistent, atmospheric pressure will be used in place of effective vertical stress in Eq 4.

Because of variation in soil behavior and to account for use of initial height and permeability, a soil index property dependent coefficient, C, will be applied to Eq 4. Substituting into Eq 4 yields

$$r = \frac{-C\, P_a k_i}{\gamma_w\, H_i^2}\, \log(1 - \text{ppr}_{\text{max}}) \tag{5}$$

where

C = a soil index property dependent coefficient,
P_a = atmospheric pressure,
k_i = permeability at start of test,
H_i = initial thickness of specimen, and
ppr_{max} = maximum value of pore pressure ratio desired.

Determination of Parameters

The first parameter, permeability, can either be estimated using available methods [6,8] or can be measured directly. The authors recommend measuring the permeability in the consolidation apparatus. A method to accomplish this is outlined subsequently.

Permeabilities of the specimen in the apparatus can range over orders of magnitude. They depend not only on soil type and stiffness, but also are significantly affected by sampling and trimming processes. By Eq 5 the strain rate for a desired maximum pore pressure ratio is directly proportional to permeability. Therefore a measurement of specimen permeability in the apparatus will afford the best opportunity to accurately establish a suitable strain rate. Furthermore, measured values of permeability may indicate a faulty trimming process if they vary significantly from expected values.

The initial specimen thickness can be measured directly and presents no problem.

The desired maximum pore pressure ratio is set by specification or by ASTM D 4186. Pore pressure ratio should be large enough so that adequate pore pressures are built up to determine coefficient of consolidation, but small enough so that large hydraulic gradients are not established. ASTM D 4186 limits this ratio to

20%; others have suggested ratios of 50% [2] or higher [7]. The authors believe that a maximum pore pressure ratio of 20% is too restrictive. A maximum value of 40 or 50% could be allowed so long as nonlinear theory is used to calculate coefficient of consolidation, c_v.

Testing Equipment

CRS Apparatus

A schematic of the CRS testing apparatus is presented in Fig. 1. The equipment is very similar to equipment described by previous authors [4,5]. To accommodate the permeability test, an extra valve was added to the back-pressure line allowing control of waterflow to the base of the specimen from a burette. As with the system described by Gorman [4], a low-friction seal system was used to guide the loading ram into the chamber. To reduce the flexibility of the pore pressure measuring system the pressure transducer was mounted directly under the testing apparatus. In addition to the changes described above, two fastening posts were added to keep the consolidation ring firmly in contact with the base. (In these types of tests where no drainage is permitted from the base of the specimen, it is extremely important to ensure that seals at the base of the specimen between the consolidation ring and the base remain intact.)

Data Acquisition and Control Equipment

One of the advantages of the CRS test is its adaptability to automation. A personal computer with a data acquisition system was used for the tests reported herein. All specimen and test control parameters are entered from the keyboard in response to menu prompts and the computer program computes the required strain rate. At the present time, the corresponding rate of deformation must be manually set on the loading press. Once the test is started, the computer program takes and stores all data, applies calibration factors, and prints out results. The program also controls the loading press for stress reversals and stops the loading press and test when one of several limits (force transducer, pressure transducer, displacement transducer, etc.) is reached. A separate program reduces all data and produces a printed and plotted output.

Test Procedure

The test procedure recommended herein has the following steps:

1. Carefully trim the specimen, measure its dimensions and weight, and determine water content and Atterberg limits from the trimmings.
2. Place the specimen in the apparatus, assemble the apparatus, and apply a small seating load [5 kPa (100 psf)] to the specimen.
3. Fill the chamber with water and back-pressure saturate the specimen from both top and bottom using small increments of pressure to prevent large pressure

(a) SOIL SPECIMEN
(b) POROUS STONES
(c) LOADING PLATEN
(d) CONSOLIDATION RING
(e) LOADING PISTON
(f) LOW FRICTION SEAL
(g) DISPLACEMENT TRANSDUCER
(h) REGULATOR FOR PERMEABILITY TEST
(i) BURETTE
(j) VALVE FOR PERMEABILITY TEST
(k) BACK PRESSURE VALVE
(l) DIFFERENTIAL PRESSURE
    TRANSDUCER
(m) VALVE TO WATER RESERVOIR
(n) FASTENING POST FOR
    CONSOLIDATION RING
(o) CELL PRESSURE REGULATOR

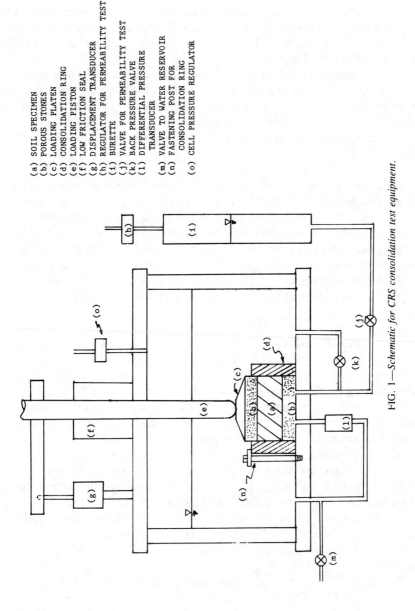

FIG. 1—Schematic for CRS consolidation test equipment.

gradients within the specimen. (During this process, the specimen is not permitted to swell and the swelling pressures are recorded.)

4. Perform a permeability test on the specimen by closing the valve that allows for equilibrating the pressures between the top and bottom of the specimen and applying a small differential pressure [approximately 7 kPa (1 psi)] to the base of the specimen. The water inflow to the base of the specimen is monitored with time by use of a burette. When flow rate becomes constant, determine the coefficient of permeability.

5. Determine the rate of deformation from Eq 8 and set it on the loading press.

6. Begin loading and shut off the valve which allows for flow from or to the base of the specimen. (Note that the specimen will have a pore pressure gradient within it which remains from the permeability test.)

7. Continue with the test according to the procedures established in ASTM D 4186.

In the testing program currently underway, the foregoing procedure is used except that multiple load cycles are being performed on the same specimen to reduce specimen variability. Specimens are being loaded to a stress slightly above the preconsolidation stress and are then unloaded to the seating load and pore pressures are allowed to dissipate. Another permeability test is then conducted followed by loading to a higher stress level than in the previous loading (usually well into the normally consolidated range). This sequence of loading and unloading is continued until the limit of the load transducer is reached.

Comments on Testing Procedure

Early tests indicated that the specimen trimming process was critical; poor trimming procedures resulted in erratic values of permeability. To reduce the effect of the trimming process as a test variable, a trimming guide was designed. The trimming guide fits into the chuck of a drill press and fits over the top of the trimming ring. Trimming is conducted ahead of the ring and the guide holds and advances the cutting ring evenly. The guide and ring are advanced about 3 mm (⅛ in.) at a time, then locked in place during trimming.

After the trimming process was completed the specimen was transferred to a Teflon-coated stainless steel ring. The specimen was adjusted in the ring to make it flush with the bottom edge of the ring and then placed on the bottom porous stone with care to prevent entrapment of air between the specimen and porous stone. The ring was then pushed down over the O-ring seal at the base and two clamping posts were installed to keep it intact with the base. (In some cases, it is possible to use the same ring for trimming and testing and, hence, less disturbance is likely.)

To saturate the specimen a cell and back pressure of 413 kPa (60 psi) was applied in 138 kPa (20 psi) increments while adjusting the load to prevent swell. The cell and back pressure were kept on the specimen for about 24 h. (This length of time was for convenience. Shorter times could have been used.)

The time for the completion of the permeability test was usually less than 1 h, but required several hours or more for highly plastic clays. The time to complete one loading cycle was usually less than one day.

Soil Description

An ongoing testing program involves the testing of a large variety of specimens. At the present time, data from three undisturbed specimens of clays are available. The engineering index properties of the soils are presented in Table 1.

Test Results

Data from a typical series (TECH) are presented in Figs. 2 to 5. Data for the loading portion of each one of the four loading cycles are given. For the first cycle the loading portion took less than 1 h, whereas for the fourth cycle the time for loading was almost 6 h. Examination of the data for the coefficient of consolidation in each of the loading cycles reveals that only the first data point has erratic values. The first point is usually obtained within 3 min after loading begins. The use of the gradient from the permeability test at the start of the consolidation loading phase is helpful in acquiring good coefficient of consolidation data early in the test.

Results from the multiple permeability and multiple loading cycles on the three soils are summarized in Table 2. The value of preconsolidation stress for the first loading cycle was obtained by use of the Casagrande construction on the strain versus log-of-stress plot. The values for the other loading cycles are the maximum effective stress for the previous load cycle.

The values of the stress due to swell are the stresses measured during the backpressuring and permeability testing. Values of preconsolidation stress divided by these values provided the overconsolidation ratios (OCR) in Table 2.

Rates of strain for each loading cycle were determined from a preliminary version of Eq 5. The desired maximum pore pressure ratio used in Eq 5 ranged from 15 to 25%. Data in Table 2 from the test show that actual values of the maximum pore pressure ratio ranged from 13 to 43%.

TABLE 1—*Engineering index properties of tested soils*

Index Property	TECH	MCC20	KTRP21
Liquid limit	31	31	81
Plasticity index	14	14	52
Percent sand	6	26	9
Percent silt	70	...	23
Percent clay	24	...	67
Specific gravity	2.75	2.72	2.71
Unified classification	CL	CL	CH

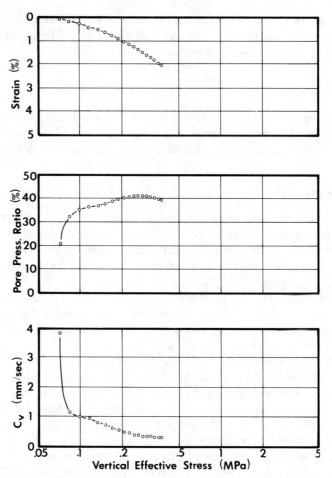

FIG. 2—*First loading cycle of CRS consolidation test on Cumberland River silty clay.* (a) *Axial strain versus effective vertical stress.* (b) *Pore pressure ratio versus effective vertical stress.* (c) *Coefficient of consolidation versus effective vertical stress.*

Equation 5 was solved for the coefficient C, which becomes

$$C = \frac{-r \, \gamma_w \, H_i^2}{P_a k_i \log(1 - \mathrm{ppr}_{max})} \tag{6}$$

The units of the coefficient C are "percent" and all parameters in Eq 6 must be in consistent units.

From earlier discussion, the coefficient C is to account for differences in the parameters used in Eq 5 from the parameters in Eq 4. It was found that the

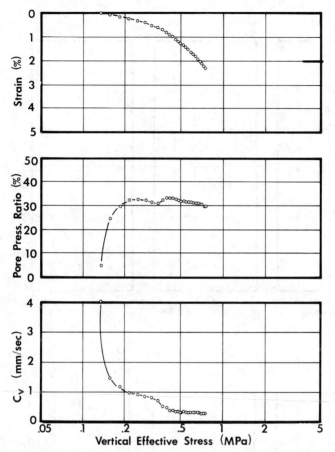

FIG. 3—*Second loading cycle of CRS consolidation test on Cumberland River silty clay.* (a) *Axial strain versus effective vertical stress.* (b) *Pore pressure ratio versus effective vertical stress.* (c) *Coefficient of consolidation versus effective vertical stress.*

coefficient C was a function of the liquidity index (LI). Figure 6 presents these data and shows that for these soils, C can be represented by

$$C \, (\%) = \exp(8 - 3 \, \text{LI}) \tag{7}$$

where "exp" is the base of the natural logarithm. Combining Eqs 5 and 7 yields the proposed rate equation

$$r = \frac{-\exp(8 - 3 \, \text{LI}) \, P_a k_i}{\gamma_w \, H_i^2} \log(1 - \text{ppr}_{max}) \tag{8}$$

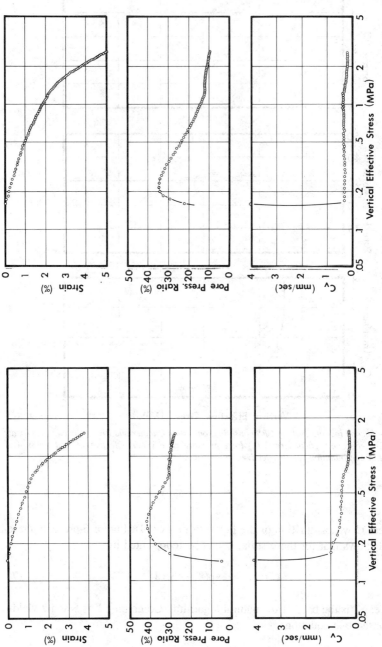

FIG. 5—Fourth loading cycle of CRS consolidation test on Cumberland River silty clay. (a) Axial strain versus effective vertical stress. (b) Pore pressure ratio versus effective vertical stress. (c) Coefficient of consolidation versus effective vertical stress.

FIG. 4—Third loading cycle of CRS consolidation test on Cumberland River silty clay. (a) Axial strain versus effective vertical stress. (b) Pore pressure ratio versus effective vertical stress. (c) Coefficient of consolidation versus effective vertical stress.

TABLE 2—Summary of results for soils tested.

Test No.	Initial Specimen Thickness, mm	Initial Water Content, %	Final Water Content, %	Initial Void Ratio	Final Void Radio	Initial Liquidity Index	Initial Permeability, cm/s	Preconsolidation Stress, MPa	Swell Stress, MPa	OCR	Maximum Pore Pressure Ratio (ppr), %	Actual Rate, %/min	Calculated Rate, %/min
TECH8	25.4	25.1	24.2	0.713	0.688	0.58	5.0E-08	0.18	0.03	5.3	42.5	0.054	0.061
TECH9	24.9	24.2	23.5	0.688	0.668	0.51	3.2E-08	0.38	0.12	3.1	33.7	0.044	0.036
TECH10	24.6	23.5	22.4	0.668	0.638	0.46	1.8E-08	0.78	0.14	5.8	43.0	0.045	0.032
TECH11	24.2	22.4	21.2	0.638	0.602	0.39	1.3E-08	1.69	0.15	10.5	36.5	0.018	0.025
MCC20A	24.9	21.6	20.9	0.646	0.624	0.33	2.4E-08	0.10	0.01	8.1	22.8	0.016	0.029
MCC20B	24.5	20.9	19.9	0.624	0.596	0.28	1.1E-08	0.49	0.06	8.0	27.6	0.018	0.021
MCC20C	24.1	19.9	19.0	0.596	0.568	0.21	6.6E-09	1.25	0.11	11.4	34.9	0.020	0.021
MCC20D	23.7	19.0	16.9	0.568	0.505	0.14	3.0E-09	1.90	0.16	12.1	13.3	0.003	0.004
KTRP21A	25.2	34.7	33.8	1.031	1.005	0.11	1.8E-07	0.16	0.05	3.3	31.9	0.052	0.643
KTRP21B	24.8	33.8	32.7	1.005	0.972	0.09	1.3E-08	0.45	0.09	5.1	14.0	0.016	0.019
KTRP21C	24.3	32.7	31.3	0.972	0.932	0.07	6.6E-09	0.93	0.16	5.8	15.2	0.014	0.012
KTRP21D	23.9	31.3	29.5	0.932	0.877	0.04	3.3E-09	1.54	0.28	5.5	15.6	0.014	0.007

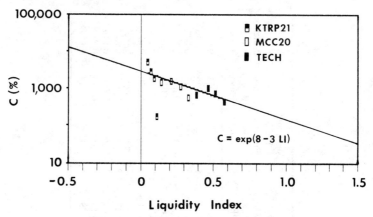

FIG. 6—*Coefficient C versus liquidity index.*

where

r = strain rate (percent/unit of time),
LI = liquidity index with soil saturated,
P_a = atmospheric pressure,
k_i = permeability at start of test,
H_i = initial thickness of specimen, and
ppr_{max} = maximum value of pore pressure ratio desired.

Comparisons of the strain rates by Eq 8 with those applied for the measured maximum pore pressure ratios are given in the last two columns of Table 2. Agreement is good except for the first cycle of test series KTRP21, where either initial permeability measurements were in error or the specimen trimming process allowed flow between the specimen and the ring. Values for the second and subsequent loading cycles all appear reasonable.

Values of strain rates for these tests as prescribed by ASTM D 4186, assuming pore pressure ratios to be less than 20%, are given in Table 3. For the same assumptions, values calculated by use of Eq 8 are also given in Table 3. The ASTM method does not account for any preconsolidation effects. For two of the three soils, the ASTM procedure overpredicts the strain rate (and hence maximum pore pressure ratios would most likely exceed 20%). For the third soil, a highly plastic clay, the ASTM procedure appears to significantly under-predict the strain rate (and hence pore pressure ratios during the test may be immeasurably small for portions of the test).

Conclusions

A rational approach has been developed to estimate the strain rate for use in the constant-rate-of-strain (CRS) consolidation test. It makes use of specimen height, liquidity index, and permeability measured just prior to starting the consolidation test.

TABLE 3—*Comparison of ASTM D 4186 rates with rates from Eq 8 for 20% maximum pore pressure ratio.*

Test No.	ASTM D 4186 Rate, %/min	Rate by Eq 8, %/min
TECH8	0.040	0.025
TECH9	0.040	0.019
TECH10	0.040	0.012
TECH11	0.040	0.012
MCC20A	0.040	0.025
MCC20B	0.040	0.014
MCC20C	0.040	0.010
MCC20D	0.040	0.006
KTRP21A	0.001	0.370
KTRP21B	0.001	0.027
KTRP21C	0.001	0.016
KTRP21D	0.001	0.008

Additional test data are needed to strengthen relationships between the rate coefficient, C, and the liquidity index.

Performing the permeability test on the consolidation specimen in the consolidometer has the additional advantage of applying a pore pressure gradient to the specimen which is close to the gradient associated with CRS testing. As a result, more reliable pore pressure measurements are possible in the early stages of the test and, hence, values of the coefficient of consolidation will be more accurate for these stages.

References

[1] Hamilton, J. J. and Crawford, C. V., "Improved Determination of Pre-consolidation Pressure of a Sensitive Clay," in *Papers on Soils—1959 Meetings ASTM STP 254*, American Society for Testing and Materials, Philadelphia, 1959, pp. 254–271.

[2] Wissa, A. E. Z., Christian, J. T., Davis, E. H., and Heiberg, S., *Journal of the Soil Mechanics and Foundations Division*, American Society of Civil Engineers, Vol. 97, No. SM10, Oct. 1971, pp. 1393–1413.

[3] Smith, R. E. and Wahls, H. E., "Consolidation under Constant Rates of Strain," *Journal of the Soil Mechanics and Foundations Division*, American Society of Civil Engineers, Vol. 95, No. SM2, March 1969, pp. 519–539.

[4] Gorman, C. T., "Strain Rate Selection in the Constant-Rate-of-Strain Consolidation Test," Research Report UKTRP-81-1, Kentucky Transportation Research Program, University of Kentucky, Lexington, 1981.

[5] Deen, R. C., Drnevich, V. P., Gorman, C. T., and Hopkins, T. C., "Constant Rate-of-Strain and Controlled-Gradient Consolidation Testing," *Geotechnical Testing Journal*, Vol. 1, No. 1, March 1978, pp. 3–15.

[6] Samarasinghe, A. M., Huang, Y. H., and Drnevich, V. P., "Permeability and Consolidation of Normally Consolidated Soils," *Journal of the Geotechnical Engineering Division*, American Society of Civil Engineers, Vol. 108, No. GT6, June 1982, pp. 835–850.

[7] Janbu, N., Tokheim, O., and Senneset, K., "Consolidation Tests with Continuous Loading," in *Proceedings*, Tenth International Conference on Soil Mechanics and Foundation Engineering, Vol. 1, 1981, pp. 645–654.

[8] *Design Manual—Soil Mechanics*, NAVFAC DM-7.1, Department of the Navy, Naval Facilities Engineering Command, Washington, DC, May 1982, pp. 7.1-137 to 7.1-139.

Samuel O. Nwabuokei[1] and Charles W. Lovell[1]

Compressibility and Settlement of Compacted Fills

REFERENCE: Nwabuokei, S. O. and Lovell, C. W., **"Compressibility and Settlement of Compacted Fills,"** *Consolidation of Soils: Testing and Evaluation, ASTM STP 892,* R. N. Yong and F. C. Townsend, Eds., American Society for Testing and Materials, Philadelphia, 1986, pp. 184–202.

ABSTRACT: Settlement predictions are necessary for compacted fills under a variety of circumstances (e.g., when fills are high, where major structural loads are to be supported, and when the fill materials are clays or weak rocks). As a result of an extensive testing program on both laboratory- and field-compacted materials, a systematic approach to this prediction has been developed. The recommended procedure involves one-dimensional compression testing in three parts.

The initial vertical strains experienced by the compacted fill are those due to self-weight. The magnitudes of these are highly dependent upon the values of prestress established by the compaction process; however, they occur as rapidly as the fill can be built. The major vertical strains will probably occur as the fill becomes wetter and "softens" in service. The timing of this settlement is uncertain, because it depends upon climatic variables at the site. Materials in the upper portion of the fill will often swell under the low confinement, and the net movement is an appropriate summing of these events with the compressions that occur deeper in the fill. Assuming that the fill becomes essentially saturated in service, further settlements can be produced by loading of the saturated material. The latter response is conventional in nature.

By testing a series of specimens, with a range of confinements which match those which will be imposed by the prototype fill, representative vertical strains are defined. Integration of these strains over the fill height produces the prediction of movement for the upper surface of the fill. The total testing and prediction technique is illustrated in the paper by example.

KEY WORDS: compacted soil, compressibility tests, as-compacted prestress, saturated prestress, regression equations, settlement

Nomenclature

A Cross-sectional area of mold
A_{EP} Area under curve of energy versus plastic deformation
e_L Void ratio obtained by filling mold loosely
e_0 As-compacted void ratio
E Nominal compaction energy

---

[1] Graduate Assistant and Professor, respectively, School of Civil Engineering, Purdue University, West Lafayette, IN 47907.

184

h Height of drop of hammer

H_0 Height of mold

H_L Height of soil corresponding to void ratio, e_L, required to produce a compacted height H_0

H_{sL}, V_{sL} Height and volume of soil solids in a soil of height H_L

H_{s0}, V_{s0} Height and volume of soil solids in a soil of height H_0

H_i, H_j, H_r Thickness of soil layers i, j, and r

H_{vL}, V_{vL} Height and volume of voids in a soil of height H_L

H_{v0}, V_{v0} Height and volume of voids in a soil of height H_0

ΔH_i, ΔH_j, ΔH_r Change in height of soil layers of thicknesses H_i, H_j, and H_r due to self-weight

ΔH_{si} Change in height of soil layer, H_i, due to saturation

k Number of additional layers; varies from $k = 0$ to t, where t = number of iterations

$k1$ Number of lifts above layer r; varies from $k1 = 0$ to $n1$

n Number of soil layers

N_B, N_L Number of blows and layers, respectively

U Strain energy per unit volume

$\Delta V/V_0$ One-dimensional volume change due to saturation, %

w Water content, %

W Weight of hammer

z_i Depth from top of embankment to center of layer, i

γ_d Dry unit weight, kg/m^3

γ_m Wet unit weight, kg/m^3

γ_{mnj}, γ_{mnr} Wet unit weight of layers j and r, respectively, after considering the effects of vertical strains due to self-weight

δ_0, δ_1 Embankment deformations

δ_p Plastic deformation, m

ϵ_{vi} One-dimensional vertical strain due to self-weight for layer, i

ϵ_{vsi} One-dimensional vertical strain due to saturation in layer, i

ϵ_{zp} Plastic strain in z-direction

σ_0 Equivalent fill pressure

σ_s As-compacted prestress

σ'_{s0} Saturated prestress

σ_{0zi} Embankment pressure at any depth z_i

σ_{0nzi} Embankment pressure at any depth z_i, using saturated unit weight

σ_{mnj} Stress at center of layer j after addition of layers due to settlement; j varies from 1 to $n + t$

The improvement of the engineering characteristics of clays or weak rocks used for the construction of fills is most economically accomplished by specifying suitable placement compaction conditions so as to ensure their adequate short- and long-term performances. In order to satisfy these goals, the engineer should be able to quantitatively predict and control the overall performance of the compacted material.

With the increasing demand for higher fills, in which the soil within defined layers is at different "overconsolidation ratios," due to as-compacted prestress, the need arises to produce definite and predetermined compressibility and settlement responses for constructed fills.

Compressibility characteristics and the fabric generated in a compacted material are influenced by the compaction water content, the compaction pressure/energy, and the mode of compaction. These characteristics are modified during the service life of the fill due to the unavoidable changes in the environmental conditions.

Consequently, this paper examines the compressibility behavior of laboratory impact compacted soils, with a view to determining the effects of various compaction variables (compaction water content, compaction pressure/energy, and mode of compaction) and fill pressure on the as-compacted prestress, volume changes due to saturation, saturated prestress, and saturated compressibility.

Using statistical techniques, prediction models were developed for as-compacted prestress, volume changes due to saturation, and saturated prestress.

Also, a systematic procedure for the prediction of the settlement of compacted fills is outlined and is illustrated by an example.

Experimental Apparatus and Procedure

The soil used for this study is a plastic, fine-grained lacustrine clay deposit from near New Haven in Northeastern Indiana. The index properties and classifications of the clay are LL = 47, PL = 20, G_s = 2.75, and classification CL (Unified Soil Classification System, USCS), and A-7-6 (American Association of State Highway Transportation Officials, AASHTO), respectively. The percentage finer than 0.002 mm was found to be 33. X-ray diffraction analyses indicated that it is a polymineral soil containing illite, kaolinite, vermiculite, chlorite, quartz, and feldspar.

The soil was preprocessed by sieving through a No. 4 sieve, mixed with a desired amount of water, and then allowed to cure for a period of five days. The "15-blow" low-energy Proctor, Standard AASHTO and Modified AASHTO [ASTM Test Methods for Moisture-Density Relations of Soils and Soil-Aggregate Mixtures, Using 5.5-lb (2.49-kg) Rammer and 12-in. (304.8-mm) Drop (D 698-78) and . . . Using 10-lb (4.54-kg) Rammer and 18-in. (457-mm) Drop (D 1557-78), respectively] compactions were subsequently carried out at the various water content levels of interest. The relationships between dry density and water content are given in Fig. 1. At high water content levels, 26% and greater, the dry densities obtained under the influence of the various compactive efforts are essentially the same. At these water contents (degrees of saturation S_r greater than 90%), a large portion of the compactive effort is expended in shearing the plastic swollen soil aggregates, and is resisted by the pore water-occluded air pressures. This is in agreement with the compaction mechanism developed by

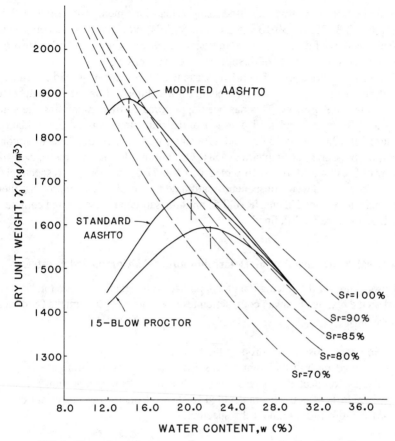

FIG. 1—*Dry unit weight versus moisture content curves for New Haven clay.*

Hodek and Lovell [*1*], where for dry-of-optimum moisture contents, the clay aggregates are shrunken, hard, and brittle. The compaction forces move them about and may even break them, leaving the system with a minimum volume of small pores and a maximum volume of large pores. In contrast, on the wet side of optimum, the clay aggregates are swollen and plastic, with the compaction forces moving the pieces together and deforming them to minimize interaggregate voids. This leaves a system of few large pores, but many small ones.

Following compaction at a desired water content, the soil specimen was transferred to an adjustable Proctor mold from which the sampling, using an oedometer ring, was carried out.

As-compacted compressibility of the soil was determined in a Karol-Warner fixed ring oedometer. The oedometer ring is 63.5 mm (2.5 in.) inside diameter, 101.6 mm outside diameter, and 25.4 mm (1.0 in.) high. Loading was accomplished by a lever arm-weight system. Following a seating pressure of 10 kPa, total applied pressure was increased using a load increment ratio (LIR) of 0.5, to 14.86, 22.3, 33.44, 50.16, 75.24, 112.86 kPa, etc., until the as-compacted prestress was well-defined. The duration of each load increment was 16 min. This time was adequate to define all presecondary effects.

During the service life of a fill, environmental changes can lead to a near-saturation condition, with attendant changes in volume and in the as-compacted prestress in the soil mass. This has been approximated by compressing the soil, using an LIR of 0.5 and load duration of 16 min, until vertical consolidation pressures of 10.0, 69.4, 137.5, and 276.2 kPa were achieved. These total consolidation pressures, at Standard AASHTO optimum density, correspond to fill heights of 0.61, 4.2, 8.4, and 16.9 m (2.0, 13.9, 27.5, and 55.3 ft), respectively. The soil specimens were subsequently saturated by a back-pressure process, then unloaded and reloaded at an LIR of 0.5, until the saturated prestress and compression indices were well-defined.

Analytical Procedure for the Determination of As-Compacted Prestress

A procedure for the calculation of as-compacted prestress (σ_s), as a total stress, based on the results of impact compaction tests was developed, with the following assumptions:

1. No energy loss in the drop of the hammer.
2. The void ratio of the compacted soil in the mold is uniform.
3. The as-compacted prestress (i.e., the fraction of the compaction pressure/ energy which is effectively transmitted to the soil matrix due to plastic deformation) is uniform throughout the specimen.

The plastic deformation (δ_p) which occurs at any energy level for a soil with a loose void ratio (e_L) and water content (w) is derived as follows (Fig. 2):

$$\frac{H_{sL}}{H_L} = \frac{1}{1 + e_L} \tag{1}$$

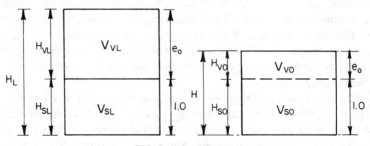

FIG. 2—*Phase diagram.*

$$\frac{H_{s0}}{H_0} = \frac{1}{1 + e_0} \tag{2}$$

but

$$H_{sL} = H_{s0}$$

Hence

$$\frac{H_L}{H_0} = \frac{1 + e_L}{1 + e_0} \tag{3}$$

$$H_L - H_0 = \delta_p = H_0 \frac{e_L - e_0}{1 + e_0} \tag{4}$$

The nominal compaction energy for the impact-type compaction test is given by

$$E = W \times h \times N_B \times N_L \tag{5}$$

Assuming one-dimensional deformation, the strain energy per unit volume (U) stored within the soil is

$$U = \frac{1}{2} \sigma_s \epsilon_{zp} \tag{6}$$

Equating the strain energy per unit volume (U) stored within the soil to the external work done per unit volume:

$$\frac{1}{2} \sigma_s \epsilon_{zp} = \frac{E}{V_L} \tag{7}$$

$$\frac{1}{2} \sigma_s \frac{\delta_p}{H_L} = \frac{E}{AH_L} \tag{8}$$

$$\frac{1}{2} \sigma_s \delta_p A = E \tag{9}$$

Multiplying both sides of Eq 9 by δ_p:

$$\frac{1}{2} \sigma_s \delta_p^2 A = E \delta_p \tag{10}$$

From the results of the laboratory compaction tests (Fig. 3), the relationship

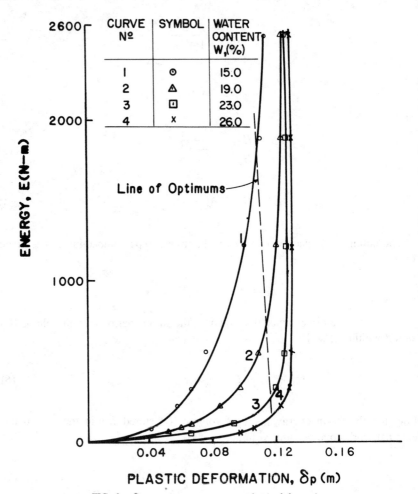

CURVE Nº	SYMBOL	WATER CONTENT, $W_s(\%)$
I	⊙	15.0
2	△	19.0
3	□	23.0
4	x	26.0

FIG. 3—*Compaction energy versus plastic deformation.*

between E and δ_p is seen to be nonlinear. Consequently, $E\delta_p$ can best be evaluated by the area under the curve (A_{Ep}). Observe from Fig. 3 that for water contents wet of optimum, Curves 3 and 4, there are specific energies (less than the externally applied energies) required to yield the maximum plastic deformations. Thus, Eq 10 becomes

$$\frac{1}{2} A\sigma_s H_0^2 \frac{(e_L - e_0)^2}{(1 + e_0)^2} = A_{Ep} \tag{11}$$

and

$$\sigma_s = \frac{2A_{Ep}(1 + e_0)^2}{H_0^2(e_L - e_0)^2 A} \tag{12}$$

The calculated total stress generated in the soil, as a result of plastic deformation, is the predicted as-compacted prestress (σ_s). It was computed for various water contents and energy levels, and values are given in Fig. 4.

Results and Analysis

As-Compacted Prestress

Typical as-compacted compressibility curves are given in Fig. 5. The following codes were adopted for specimen identification: L, S, and M refer to 15-blow (low energy) Proctor, Standard AASHTO, and Modified AASHTO compactions,

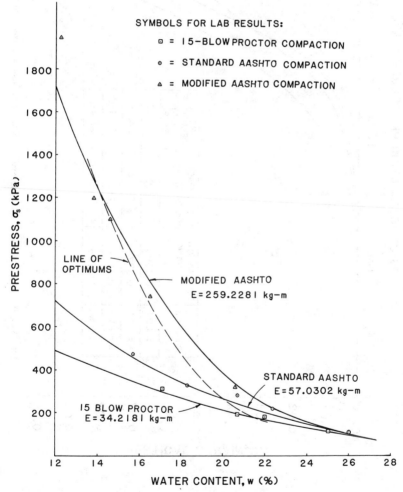

FIG. 4—*As-compacted prestress – water content relationship for New Haven clay.*

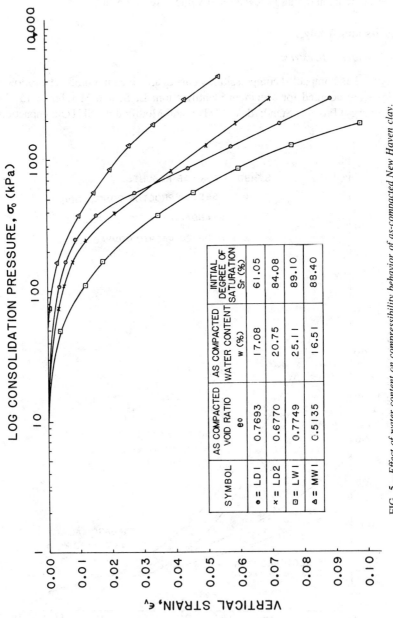

FIG. 5—*Effect of water content on compressibility behavior of as-compacted New Haven clay.*

SYMBOL	AS COMPACTED VOID RATIO e_0	AS COMPACTED WATER CONTENT w (%)	INITIAL DEGREE OF SATURATION Sr (%)
● = LD I	0.7693	17.08	61.05
× = LD2	0.6770	20.75	84.08
▫ = LW I	0.7749	25.11	89.10
◁ = MW I	C.5135	16.51	88.40

respectively. The letters D, O, and W refer to moisture content conditions of dry of optimum, optimum, and wet of optimum. The numbers 1, 2, and 3 are used to differentiate between specimens that fall within identical moisture content zones of dry of optimum, optimum, and wet of optimum.

The relative compression (compression at any given time divided by compression at 16 min) for dry-of-optimum specimens (LD1) was greater than that for wet-of-optimum samples (LW1). The high relative compression for dry-of-optimum specimens is attributable to the more readily achievable outflow of air through their interconnected voids. Note in Fig. 5 the effect of increasing water content and degree of saturation on the compressibility of the low-energy compacted specimens. At the low consolidation pressure levels, the wet side specimens (LW1) are more compressible than the dry side specimens (LD1), while at high pressure ranges (pressures greater than their respective as-compacted prestress) the dry side specimens exhibit a more compressible behavior. Consequently, for embankment design in which as-compacted compressibility is important, the as-compacted prestress should be evaluated or determined.

The laboratory-determined, as-compacted prestress, which represents the fraction of the compaction effort effectively transmitted to the soil skeleton due to plastic deformation, is plotted as points on the predicted curves (Fig. 4). There is a good correspondence between the values measured and those predicted from Eq 12.

Using statistical regression procedures, a prediction equation for as-compacted prestress was developed for the impact laboratory compacted New Haven clay:

$$\sigma_s = -45.9398 + 131337.66 \frac{\sqrt{E}}{w^2}$$
$$- 18982.205 \sqrt{E}/w + 1023.6757 \sqrt{E}$$
$$- 17.80117 \, w \sqrt{E} - 0.12497 \times 10^{-4} w^2 E^2 \qquad (13)$$

A good coefficient of determination (R^2) which indicates the amount of variation explained by the equation is 0.9974. The as-compacted prestress is seen to be a function of the compaction energy and water content. For the New Haven clay, at any water content less than 26%, an increase in compaction effort gives an increase in as-compacted prestress. Also, at a given energy level, an increase in water content leads to a decrease in the as-compacted prestress.

Volume Change and Saturated Prestress

In order to determine the effects of changes in environmental conditions on compressibility and settlement behavior of a compacted fill, specimens were loaded to approximate different levels of fill confining stress, and were saturated in the oedometer by a back-pressure process. The one-dimensional volume changes $(\Delta V/V_0)$ were measured.

The specimens were subsequently unloaded and reloaded at an LIR of 0.5 until the saturated prestress and saturated compressibility index could be well defined. Load increments were applied at the end of 100% primary consolidation, which was determined by the Casagrande logarithm of time-dial reading procedure. The soaked prestress values were determined by the Casagrande logarithm of applied stress (effective)-void ratio construction procedure. Typical data are plotted as logarithm of applied stress-vertical strain in Fig. 6.

Using statistical regression techniques, a prediction equation was developed for the one-dimensional percent volume change, $(\Delta V/V_0)\%$. The percent volume change is described in terms of as-compacted void ratio (e_0), compaction water content (w), and equivalent fill pressure (σ_0):

$$\frac{\Delta V}{V_0}\% = -3.0284903 - 0.39039732 \, e_0^2 \sqrt{\sigma_0} + 0.63841268 \, e_0 \sqrt{\sigma_0}$$

$$- 0.321048\,83 \times 10^{-3} w^2 \sigma_0 + 0.369098\,48 \times 10^{-2} w^2 \quad (14)$$

The coefficient of determination of the above model is 0.835. Similar results were obtained by Abeyesekera and Lovell [2], DiBernardo and Lovell [3], and Lin and Lovell [4] from their studies on compacted shales, laboratory-compacted highly plastic residual clay, and field-compacted plastic residual clay. For this model (Eq 14), a positive value of percent volume change $(\Delta V/V_0)\%$ indicates settlement, while a negative value represents swelling. The percent volume change at zero fill pressure $(\sigma_0 = 0)$ is a function of water content only. Thus, drier specimens will indicate greater swelling tendencies.

In Fig. 7, the effect of void ratio on the estimated volume changes at constant water content is illustrated (contrast Curves 1 and 2). Also shown is the effect of water content at a constant beginning void ratio (contrast Curves 3, 5, and 6). Observe that a decrease in void ratio results in increased swelling tendencies, while an increase in water content yields increased compression. Observe also that for each specimen at a given compaction condition [void ratio (e_0) and water content (w)], there is a critical confining pressure for which there is no volume change.

Specimens compacted dry possess high negative pore pressures. Introduction of water to the compacted clays results in a decrease in their negative pore pressures and consequently a decrease in their effective stresses. The amount of swell also depends on the type of clay minerals present in the soil and the initial compaction moisture content. An expansion of the clay minerals results in a softening of the clay aggregates which, under the effect of confining stresses, will result in settlements. Thus the net volume change experienced by a compacted soil is a combination of several processes which depend on the compaction water content (w), as-compacted void ratio (e_0), and confining pressure (σ_0).

The independent variables used for the prediction equation for saturated prestress (σ'_{s0}) are equivalent pressure (σ_0), compacted void ratio (e_0), compaction

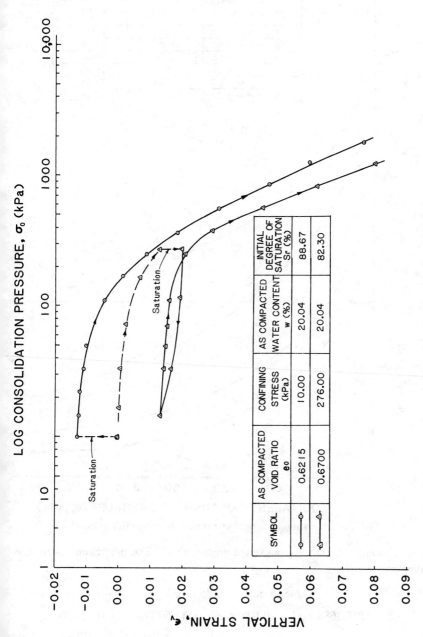

FIG. 6—*Effect of confining pressure on compressibility behavior of standard AASHTO compacted New Haven clay.*

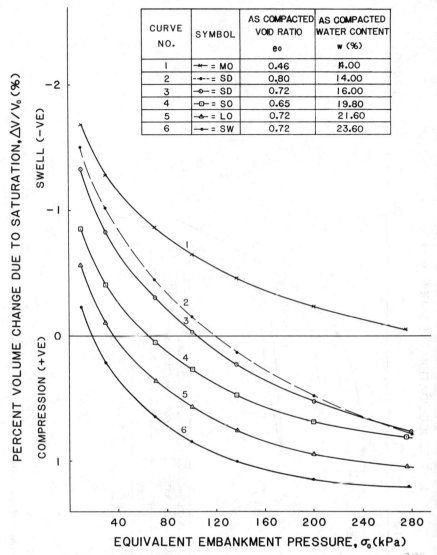

CURVE NO.	SYMBOL	AS COMPACTED VOID RATIO e_0	AS COMPACTED WATER CONTENT w (%)
1	—*— = MO	0.46	11.00
2	—•—— = SD	0.80	14.00
3	—◦— = SD	0.72	16.00
4	—□— = SO	0.65	19.80
5	—△— = LO	0.72	21.60
6	—•— = SW	0.72	23.60

FIG. 7—*Effect of void ratio and water content on percent volume change.*

water content (w), and as-compacted prestress (σ_s). The prediction equation is given as

$$\sigma'_{s0} = 1559.6762 + 2.24866 \, \sigma_0(1 - 0.85007e_0)$$
$$- 4707.3684 \, e_0(1 - 0.63896 \, e_0) - 5.9509 \, \sigma_s(1 - 0.12317w$$
$$+ 0.003621w^2) + 0.831270w^2 \quad (15)$$

The coefficient of determination of the above prediction equation is 0.937.

Dry-of-optimum specimens with their brittle and shrunken aggregates, large interconnected voids, and high negative pore pressures exhibit swelling tendencies upon the addition of water. This is a consequence of the decrease in the negative pore pressure, giving rise to decreased effective stresses. Hence, the saturated prestress values were determined to be much less than the as-compacted prestress.

Consequently, for compacted soils subjected to confining pressures less than the critical confining stresses, the swelling tendencies result in decreased effective stresses, and hence a lower saturated prestress. For confining pressures in excess of their critical values, compression occurs, with the soil aggregates being deformed plastically into a more parallel oriented configuration of particles. Depending on the magnitude of the confining pressure and hence the plastic deformation, the saturated prestress equals or exceeds the as-compacted prestress and confining pressure. This is a consequence of the attendant plastic deformation (primary and secondary compression) which has oriented and arranged the clay particles in a more stable configuration.

The results of this study also indicate:

1. For a given initial compaction condition, the saturated compressibility is greater than the as-compacted compressibility.

2. At a given compaction energy and level of confinement on saturation the compression index decreased with increase in water content.

3. For a given degree of saturation and confining pressure on saturation the compression index decreased with increase in compaction energy.

Settlement Prediction of a Compacted Embankment

A procedure for the estimation of the settlement of an embankment due to self-weight and the effects of changes in environmental conditions was developed. Methods for estimating field responses using laboratory control data have previously been developed by Lin and Lovell [4].

Given a wide embankment, 10.0 m high, constructed of New Haven clay in 0.25-m-thick lifts and compacted to a moisture content (w) and dry unit weight (γ_d) of 20.04% and 1648.0 kg/m^3 ($e_0 = 0.67$), respectively. It is required to estimate the settlement of the embankment due to self-weight and saturation in service.

Two assumptions were made so that the procedure developed previously could be simply applied:

1. When the soil is compacted to given values of moisture and density, the subsequent response is independent of the mode of compaction.

2. One-dimensional compression occurs in the field prototype. A schematic vertical section through the embankment is shown in Fig. 8. Also indicated are some of the symbols used for this analysis. Let γ_d = dry unit weight, and wet

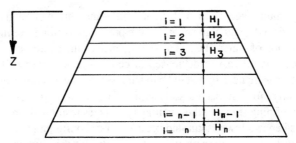

FIG. 8—*Section through a typical embankment.*

unit weight $\gamma_m = \gamma_d(1 + w)$; then, the embankment pressure at any depth, z_i, is

$$\sigma_{0zi} = \gamma_d(1 + w)z_i = \gamma_m z_i \qquad (16)$$

Using the oedometer test results, similar to those indicated in Fig. 5, or the as-compacted consolidation portion of Fig. 6, the vertical strain at the center of layer, i, thickness, H_i, corresponding to an embankment pressure σ_{0zi} is

$$\epsilon_{vi} = \frac{\Delta H_i}{H_i} \qquad (17)$$

$$H_i \epsilon_{vi} = \Delta H_i \qquad (18)$$

Thus the deformation of the embankment due to overburden stresses is

$$\delta_0 = \Sigma_{i=1}{}^n \Delta H_i = \sum_{i=1}^{n} H_i \epsilon_{vi} \qquad (19)$$

Settlement due to self-weight occurs about as rapidly as the fill is constructed. If there is need to control the magnitude of the settlement due to self-weight, the as-compacted prestress should be as large as possible, with due consideration to subsequent side effects on compressibility and strength characteristics, once the material becomes saturated in service.

Determination of the vertical strain due to saturation ($\Delta H_{si}/H_i$) is accomplished by substituting the as-compacted void ratio, as-compacted water content, and the appropriate vertical fill pressure in Eq 14. In the one-dimensional process, ($\Delta V/V_0$)% $= \epsilon_{vsi}$. For very small embankment deformations (due to self-weight), and for conditions in which the embankment is saturated (no buoyancy effect), the appropriate vertical fill pressure (σ_{0nzi}) can be approximated by the use of the saturated unit weight corresponding to the as-compacted void ratio (e_0), given by

$$\sigma_{0nzi} = \gamma_{sat} z_i \tag{20}$$

where

$$\gamma_{sat} = \frac{G_s \gamma_w + \gamma_w e_0}{1 + e_0} \tag{21}$$

However, for a saturated embankment in which a submerged condition exists, the submerged unit weight of the fill material should be used. Thus, since

$$\frac{\Delta H_{si}}{H_i} = \epsilon_{vsi} \tag{22}$$

settlement of the fill due to saturation is given by

$$\sum_{i=1}^{n} \Delta H_{si} = \sum_{i=1}^{n} H_i \epsilon_{vsi} \tag{23}$$

On the other hand, if the deformation (δ_0) determined from Eq 19 is large, then new embankment overburden stresses must be computed, to account for the effect of the added compacted soil required to give the desired embankment height.

The stress at the center of the added layer, thickness δ_0, is

$$\sigma_{mn1} = \frac{1}{2}\delta_0 \gamma_m \tag{24}$$

The overburden stress at the center of each of the previous n-layers is

$$\sigma_{mni} = \gamma_m \delta_0 + \sigma_{0zi} \tag{25}$$

From an as-compacted compressibility curve (Fig. 6), the new vertical strains and deformations corresponding to the stresses given in Eqs 24 and 25 for the various layers are determined. Using Eq 19 for the appropriate number of layers, the new embankment deformation (δ_1) is computed. An iterative process is repeated for each additional layer until the difference between the preceding and present embankment deformations (e.g., $\sigma_1 - \delta_0$) is insignificant or acceptable. The new unit weight for each layer, j, after t iterations is given by

$$\gamma_{mnj} = \frac{\gamma_m}{1 - \sum_{k=0}^{t} \frac{\Delta H_j}{H_j}} \tag{26}$$

Using Eq 26, the void ratios and saturated unit weights for the various layers are computed. Depending on the saturated conditions (saturated with no buoyancy effect or submerged) the appropriate vertical fill pressures are then computed for use in Eq 14. Deformations for each layer and for the total embankment are then computed using Eqs 22 and 23, respectively.

A stage construction approach, in which the deformations are obtained from as-compacted compressibility curves (Fig. 6) for each layer with the addition of each successive lift until the final elevation of the embankment is attained, could be utilized for the determination of self-weight deformations. For a constant water content deformation process, the new unit weight at the center of each layer, r, is given by

$$\gamma_{mnr} = \frac{\gamma_m}{1 - \sum_{k1=0}^{n1} \frac{\Delta H_r}{H_r}} \tag{27}$$

The compressed void ratios and consequently the saturated unit weights are computed using Eq 27. Substitute the as-compacted void ratio, as-compacted water content, and the vertical fill pressure corresponding to the appropriate saturation condition, in Eq 14. This allows the vertical strain at the center of each layer, and consequently the deformation of each layer and of the total embankment, to be computed.

If structural loads are imposed on the saturated fill, the conventional procedure is used for the determination of the additional settlements. Consideration should be given to the variation of saturated prestress and compression index over the depth of the fill.

The numerical solution of an example embankment problem is given in Table 1. For an embankment 10 m in height, the predicted settlement under self-weight is 3.5 cm; additional settlement when saturated is 2.0 cm. The variations of settlement due to saturation, and saturated prestress with depth are given in Fig. 9.

Summary

1. An analytical procedure for the evaluation of as-compacted prestress has been developed and compared with laboratory determined values.

2. The effects of low energy, Standard, and Modified AASHTO compaction on the as-compacted compressibility, volume change due to saturation, and saturated compressibility are discussed.

3. With the aid of statistical regression techniques, prediction equations for as-compacted prestress, volume change on saturation, and saturated prestress were developed.

4. A procedure for the determination of settlements of an embankment due to self-weight and saturation in service is proposed.

TABLE 1—Sample solution for settlements within an embankment.

$H = 10.0$ m; $H_1 = H_2 = H_n = 1.0$; $\gamma_d = 1648.0$ kg/m³; and $w = 20.04\%$

Layer No. i or j	Depth z_i (m)	Embankment Pressure σ_{0zi} (kPa)	Vertical Strain ϵ_{vi} (Fig. 5)	Settlement Due to Self-Weight (m)	Depth z_j (m)	Saturated Unit Weight γ_j (kg/m³)	Embankment Pressure σ_{satj} (kPa)	Vertical Strain Due to Saturation ϵ_{vsj}	Settlement Due to Saturation (m)
1	0.5	9.703	0.000098	0.000098	0.0350	2047.9	0.3516	-0.013987	-0.000490
2	1.5	29.11	0.00050	0.00050	0.5345	2048.8	10.7515	-0.007874	-0.007874
3	2.5	48.52	0.0015	0.0015	1.5347	2049.3	30.8464	-0.003423	-0.003422
4	3.5	67.92	0.0020	0.0020	2.5338	2050.3	50.9346	-0.000719	-0.000718
5	4.5	87.33	0.0030	0.0030	3.5319	2050.8	71.3902	0.001278	0.001276
6	5.5	106.73	0.0035	0.0035	4.5529	2051.9	91.4637	0.002801	0.002792
7	6.5	126.14	0.0045	0.0045	5.5262	2052.4	111.5298	0.004027	0.004013
8	7.5	145.55	0.0055	0.0055	6.5222	2053.5	131.5886	0.005035	0.005013
9	8.5	164.96	0.0065	0.0065	7.5172	2054.5	151.6376	0.005874	0.005840
10	9.5	184.36	0.0080	0.0080	8.5122	2055.6	171.6768	0.006575	0.006532
11					9.5040	2057.2	191.7038	0.007163	0.007106

$$\sum_{i=1}^{n} \Delta H_i = 0.0351 \text{ m} = 3.51 \text{ cm}$$

$$\sum \Delta H_{sj} = 0.020 \text{ m} = 2.0 \text{ cm}$$

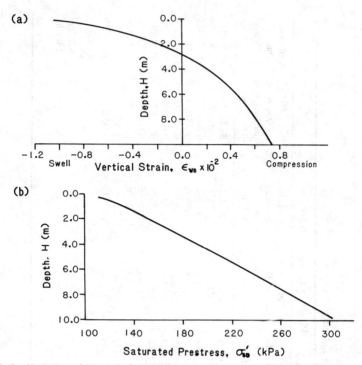

FIG. 9—*Variations of* (a) *vertical strain due to saturation, and* (b) *saturated prestress, in the embankment.*

References

[1] Hodek, R. J. and Lovell, C. W., "A New Look at Compaction Process in Fills," *Bulletin of the Association of Engineering Geologists,* Vol. 106, No. 4, 1979, pp. 487–499.
[2] Abeyesekera, R. A., Lovell, C. W., and Wood, L. E., "Stress Deformation and Strength Characteristics of a Compacted Shale," *Papers of the Conference on Clay Fills,* Institution of Civil Engineers, London, Nov. 1978, pp. 1–14.
[3] DiBernardo, A. and Lovell, C. W., "Compactive Prestress Effects in Clays," *Transportation Research Record 945,* Transportation Research Board, Jan. 1982, pp. 51–58.
[4] Lin, P. S. and Lovell, C. W., "Compressibility of Field Compacted Clay," *Transportation Research Record 897,* Transportation Research Board, May 1983, pp. 51–60.

G. C. Sills,[1] S. D. L. Hoare,[1] and N. Baker[1]

An Experimental Assessment of the Restricted Flow Consolidation Test

REFERENCE: Sills, G. C., Hoare, S. D. L., and Baker, N., **"An Experimental Assessment of the Restricted Flow Consolidation Test,"** *Consolidation of Soils: Testing and Evaluation, ASTM STP 892*, R. N. Yong and F. C. Townsend, Eds., American Society for Testing and Materials, Philadelphia, 1986, pp. 203-216.

ABSTRACT: In the first part of the paper, the restricted flow consolidation test is described. It is simple, requires no consolidation theory for its interpretation, and is very quick, with a complete stress/strain curve obtainable in a matter of hours.

In the second part, the experimental requirements are detailed and an appropriate cell, the Oxford cell, is described.

Finally, some results are presented for reconstituted silty clay samples, to show that the restricted flow test is successful, producing repeatable and reliable results.

KEY WORDS: consolidation, soil properties, test

In many engineering problems it is important to be able to predict the magnitude and rates of soil settlement under given loads. This information is generally obtained from a one-dimensional consolidation test. The standard test requires an incremental loading to be applied, typically over a period of a few days. Information about the stress/strain behavior, as an effective stress/void ratio relationship, is then available at specific effective stress levels and can be obtained elsewhere by interpolation. The permeability is the soil property governing the rate of settlement; this can be obtained from the standard test only by the use of consolidation theory, and the calculation of the coefficient c_v. One of the main drawbacks of the test is the time it takes, another is the discrete nature of the stress/strain results, and a third disadvantage is the indirect nature of the permeability calculation.

Recently, a number of alternative tests have been developed, all of which produce results in a shorter time than the traditional test. These alternatives include the constant rate of strain test (CRS), the controlled gradient test (CG), the continuous loading test (CL), and the restricted flow consolidation (RFC) test. These tests vary in complexity both in the interpretation of results and in the control and feedback systems necessary to undertake them. The simplest of these tests is the restricted flow consolidation test. The purpose of this paper is to describe the details of this test and to present some results obtained in a specially

[1]Department of Engineering Science, Oxford University, England.

designed cell by comparison with those in a Rowe cell and with results obtained from a standard oedometer.

Restricted Flow Consolidation Test

In the traditional standard test, a load is applied to a soil sample which is allowed to drain from one or both faces. The application of the stress causes a nonuniform pore pressure distribution through the sample, with the highest values occurring farthest from the drainage boundary or boundaries. The pore water drains and the pore pressures dissipate, driven by this pressure gradient. The vertical total stress is nearly uniform across the sample (since the total stress due to selfweight is generally very small by comparison with the applied stress), and the vertical effective stress is nonuniform until all the pore pressures have dissipated. Only at this stage can a uniform, known vertical effective stress be linked with a uniform void ratio. The basis of the restricted flow consolidation test is that, if the pore pressure distribution across the sample during consolidation can be kept nearly uniform, then both the effective stress and void ratio will also be close to uniform. In the absence of drainage, and for a saturated sample, the applied loading increases the pore pressures to equal the total stress (without altering the effective stress). If restricted drainage is then permitted from one face of the sample, the drained face pore pressure will drop slightly, causing flow through the soil, increasing the effective stress and causing compression of the sample. The magnitude of the pressure drop across the sample can be controlled by the efficiency of the restriction. Figure 1 illustrates this principle schematically. The total stress is measured on one face of the sample, and the pore pressures on both drained and undrained faces. The vertical effective stress can be calculated from the mean pore pressure and the total stress, and the void ratio can be obtained from the current height of the sample. It is therefore possible to obtain a continuous vertical effective stress/void ratio relationship for the complete range required by applying one total stress increment. Bounds on the accuracy of the effective stress calculation can be obtained by replacing the mean pore pressure by the drained face pore pressure (to give an upper bound) and the undrained face pore pressure (for a lower bound). The idea followed from a theoretical analysis of drainage boundary efficiency described by Lee [1] and was detailed in Lee [2]. A further publication (Lee and Sills) [3] is in preparation, while some earlier results were reported in Hoare [4].

For the pore pressure distribution across the sample to be nearly uniform, the effective permeability of the flow restrictor in the drainage line must be significantly less than the permeability of the soil sample. The majority of the total pressure drop will then occur across the flow restrictor. The permeability of the flow restrictor will affect the time taken for complete dissipation of pore pressures. It has been found that experiments on a normally consolidated silty clay typically require 2 to 3 h. Although low permeability, over-consolidated soils will need a less permeable restrictor, the test duration will still be substantially shorter than the traditional incremental loading test.

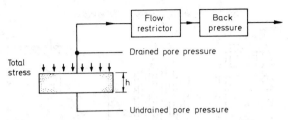

FIG. 1—*Schematic illustration of principle of restricted flow test.*

The same principle may be applied to obtain an unloading stress/strain relationship. It is not generally possible to remove the total stress in one increment, since the pore water pressures cannot achieve values lower than -1bar, or a vacuum. However, if a back pressure nearly equal to the total stress is applied to the sample through the flow restrictor without changing the total stress, the effective stress will decrease uniformly as the pore pressures increase together to the level of the back pressure. The final stage of the test is then the reduction of the total stress and back pressure together to allow the sample to be removed.

The standard test also provides a measure of permeability by analysis of the rate of settlement for each incremental loading, coupled with the gradient of the stress/strain curve. The RFC test has the advantage that a direct calculation of permeability can be made at any stage in the test, assuming only that Darcy's law is valid. Thus, the difference between the pore pressures on the drained and undrained faces, combined with the current sample thickness, provide a measure of the hydraulic gradient across the sample, while the flow rate can be obtained either by direct measurement or by calculation from the sample compression or swelling. However, this paper is concerned with the use of the RFC test for stress/strain information, and no permeability measurements are reported here.

Flow Restrictor

The requirements for the flow restrictor are that it should sufficiently impede the flow of drainage water from the soil sample and should be simple and unaffected by small amounts of air that might occur in the drainage water. The earliest version was a length of hypodermic tubing in the drainage line but, although simple, this was susceptible to blocking completely. The present system uses a sequence of very fine Millipore filter papers compressed together, in a suitable holder. The number of filters can be varied to provide different amounts of restriction. Figure 2 illustrates the features of the flow restrictor. The body is acrylic for ease of checking for air bubbles, and the support disks between each group of 15 to 20 Millipore filters are stainless steel. Different types of flow restrictor could be equally effective (e.g., a suitable flow control valve).

Cell Requirements for RFC Test

The consolidation cell for the RFC test must isolate the sample completely, so that a high pore pressure can be maintained on the drained face as well as on

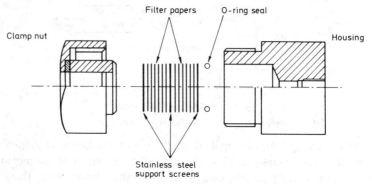

FIG. 2—*Flow restrictor.*

the undrained face. It should have the facility for direct measurement of pore pressure on both faces of the sample, of total stress on one face and of compression of the sample. An important advantage is ease of de-airing, since undissolved gas in the system can cause immediate compression when the total stress is applied. The pore pressures do not reach the level of the total stress so that an effective stress has been arbitrarily applied, with all information relating to lower effective stress levels lost completely. Other measures can be taken to reduce the possibility of this occurrence (and will be described later), but the facility to flush through drainage lines, porous stones, etc., once the cell is assembled is very useful.

Two different cells have been used for the RFC tests to be described: a standard Rowe cell and the Oxford cell, designed specifically for this test. The Rowe cell uses air pressure for the total stress loading, while the Oxford cell can be used either in a standard loading frame or with an attachment that provides the load from a compressed air reservoir like the Rowe cell.

Figure 3 shows the Oxford cell. The sample (1, shown partly consolidated) in a standard 75-mm-diameter cutting ring (2) is placed on a porous sintered stainless steel disk (3) on the base of the cell and the body of the cell (4) is lowered in position to surround the ring. O-ring seals (5) are provided at the top and bottom of the ring to prevent leakage around the outside. The load is provided through a piston (6) which seals against the body of the cell with a quadrant sealing ring (7), minimizing friction forces.

Measurement of the total stress is made by a pressure transducer (8) (Druck PDCR 22) mounted in the base of the cell (9), projecting through a hole in the base porous disk. The transducer has a flush diaphragm face and this is level with the top face of the stainless steel disk. Pore pressures are measured by transducers mounted in the base (10) and on the piston (11), with the active transducer face isolated from the soil by a porous disk (12). Each of these pore pressure mountings contains a valved bleed line (13, valves not shown) so that the system can be flushed through with de-aired water. This also provides the facility for a separate permeability measurement at the beginning or end of the consolidation test, allowing water to enter at one face and flow out at the other. Standard constant or falling head permeability tests could be undertaken or an improved test such as the low gradient flow pump method developed by Olsen. If permeability measurements are required during the RFC test, a differential pressure transducer (not shown) is mounted at the level of the soil sample to measure directly the difference between the drained and undrained face pore pressures. This then allows an accurate measurement of the pore pressure gradient.

The Oxford cell contains various features to assist with de-airing. The bleed lines in the pore pressure transducer mountings have already been noted. In addition, an acrylic trough (14) is provided round the base of the cell to allow assembly under water, and an additional valved bleed hole (15, valve not shown) is provided in the piston to allow this to be easily lowered on to the sample.

The testing and the loading parts of the Oxford cell have been developed to be separately self-contained. There are two main advantages: the sample can be loaded by the traditional lever arm oedometer frame as well as by a pneumatic system using bottled gas, and easy access is maintained on the test section for the instrumentation and bleed lines on the piston. In both loading techniques, a rod bears onto a ball bearing in the center of the piston and a linear variable differential transducer (LVDT) measures the compression of the soil sample continuously during the test. Measurements of the initial height and the weights of wet and dry sample allow the void ratio to be calculated throughout the test.

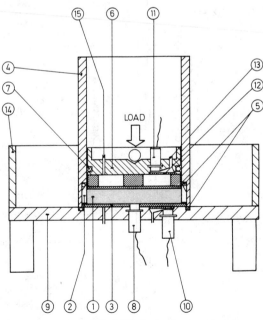

FIG. 3—*Oxford cell.*

Figure 4 shows the pneumatic loading system that can be bolted on to the Oxford cell. The loading rod is moved by the action of compressed gas on a piston which is sealed with a rolling diaphragm to minimize friction.

Some experiments were undertaken also in the Rowe cell, modified to take a standard 75-mm cutting ring. The base of the cell was drilled to take a line to a pore pressure transducer. A pressure transducer was installed in the drainage line between the soil sample and the flow restrictor.

Experimental Details

The stress/strain curve of a given soil sample should clearly be independent of the type of test. A test program was therefore undertaken to test similar soil samples using the standard consolidation test and the restricted flow consolidation test. Differences in the results between the tests would be examined, with particular attention to the consistency of the RFC test and the optimal experimental conditions for it. These experiments were confined to an assessment of the RFC

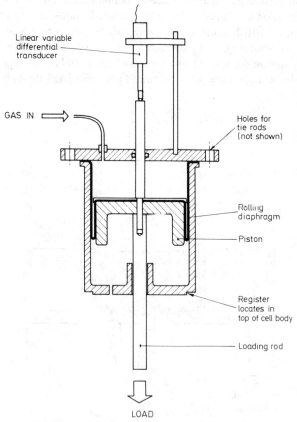

Linear variable
differential
transducer

GAS IN

Holes for
tie rods
(not shown)

Rolling
diaphragm

Piston

Register
locates in
top of cell body

Loading rod

LOAD

FIG. 4—*Pneumatic loading system for the Oxford cell.*

test to obtain stress/strain relationships of a soil; the facility to determine permeability will be examined elsewhere.

This comparative study has been carried out using reconstituted silty clay samples. The clay, referred to as Combwich mud, is a typical estuarine material, about 35% clay size and 65% silt size particles, liquid limit ~ 60%, plastic limit ~ 30%. The soil was mixed to a uniform slurry at a water content of about 100%, then consolidated in a 254-mm (10-in.) Rowe cell to pressures up to 150 kN/m². At the end of consolidation, the sample was unloaded and three cutting rings were pushed together in a guide frame into the 254-mm (10-in.) sample. Comparative measurements of initial water content confirmed that the three samples were very similar to each other. Repeatability between batches of samples may not be as good, since the water content of the initial slurry will be variable. Such differences may affect the whole of the void ratio/effective stress relationship. Nevertheless, the overall pattern should be similar. Another series of tests used soil samples remolded directly into the ring of the test cell without prior consolidation in the Rowe cell. Each of the samples was loaded up to 1000 kN/m² in one of four ways: (1) by the standard test in a standard cell and lever arm loading frame following British Standard BS1377, (2) by the standard test in a Rowe cell, (3) by the RFC test in a Rowe cell, or (4) by the RFC test in the Oxford cell.

The RFC tests were undertaken using the pneumatic loading system rather than the lever arm frame. The main advantage of this system is the ease with which the back pressure can be applied to achieve the swelling stage of the test. This is done by connecting the applied pressure to a gas/water interface in the drainage line while maintaining its connection to the loading system. This is illustrated in Fig. 5. This then provides total stress and back pressures that differ only by

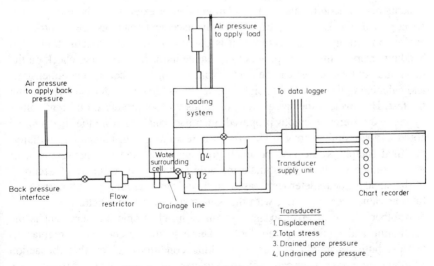

FIG. 5—*Details of the RFC test arrangement.*

whatever small losses (due to friction etc.) exist in the cell. Once the swelling stage is complete, reduction of the applied gas pressure automatically reduces total stress and back pressure together.

The results of each test were recorded in analogue form on a four-pen chart recorder and digitally on a data logger. The analogue output provides a simple check that the test is behaving normally, while the digital results are easier to analyse.

Development of the Test

A comparison between different tests and different equipment can only be as accurate as the individual measurements. Considerable attention was therefore paid to assessing the inaccuracies in all the systems examined, including the standard one, and reducing them as far as possible. Errors can occur in a variety of ways, including undissolved gas in the sample, bedding errors at the start of the test, and strains in the test system. For each of these tests, the solid volume of the sample was calculated from the sample dimensions, dry weight, and specific gravity. The volume of water was similarly calculated (using the difference between wet and dry weights and the sample dimensions). The two volumes should add together to equal the total sample volume. In practice, it was found that this rarely happened and that uncertainties in the height of the solids could be as great as 0.5 mm. The most likely explanation for this difference is undissolved gas, although errors in the specific gravity assumptions are also possible. The figure obtained using the solids weight is likely to be the more reliable.

Another error that can occur is due to bedding errors in the test system. The void ratio calculations are normally made on the basis of the initial dial gage or LVDT readings, but subsequent changes from these values will include components due to bedding and system compression or extension. These errors will be largest at the start of a test. If the displacement readings are recorded in analogue form on a chart recorder, it is possible to identify system strains and bedding errors as instant responses to a change in loading. Another check on the magnitude of these errors can be made by measuring the final sample dimensions and calculating the change in height due to the volume of water expelled during the test. However, if the calculations of solids volume produced different values by the two different methods proposed, then back analysis from the final sample could contain similar errors. That is, if there were gas in the sample initially, the final sample would also be likely to contain gas and the measurement of water volume would underestimate the void volume. The estimate of accuracy of the standard oedometer tests based on this series is around ±0.05 in the void ratio at values near to unity, with the RFC tests being rather better.

Another source of error in any system using electronic measurement is the calibration and stability of transducers. These should be checked at regular intervals, both separately and under working conditions, given that the action of mounting a transducer can alter its characteristics. As a matter of routine, the cell should be filled with water and pressurized without allowing drainage. The three pressure transducers (total stress, and the two pore pressures) should all provide virtually the same calibrated value of pressure. Any discrepancy will limit the accuracy of the consolidation test results. In the present series, friction losses in the standard oedometer and transducer errors were probably comparable in magnitude, on the order of a few kN/m^2. This is a high percentage uncertainty for the low effective stress values, and should be capable of considerable improvement for the RFC test with re-designed transducer mountings.

Early restricted flow tests showed up a major problem with soil samples at less than full saturation. The principle of the test is that no effective stress is developed during the undrained loading stage, and that, when the drainage is opened, the effective stress increases from zero at a rate controlled by the flow restriction. However, in the presence of undissolved gas, the undrained loading is transferred only partly to the pore pressures and there is an immediate increase in effective stress. This reaches a level governed by the amount of undissolved gas, the magnitude of the load, and the stiffness of the soil. Unless pore pressures, total stress, and displacement are monitored during this undrained loading, the early part of the void ratio/effective stress curve will be lost. This monitoring can be achieved either by increasing the undrained loading steadily or incrementally over a period of 5 to 10 min as illustrated in Fig. 6 for two different types of soil sample. The first is fully saturated, so that pore water pressures and total stress are equal and no effective stresses develop. The second, typical of most ''undisturbed'' samples, contains some undissolved gas, and the pore pressures are less than the total stress, giving rise to effective stress. The incremental loading illustrated in Fig. 6 allows a greater accuracy of assessment of system

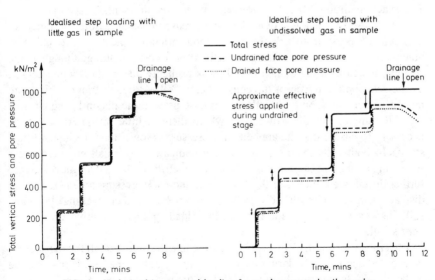

FIG. 6—*Undrained incremental loading for partly saturated soil samples.*

strains or bedding errors from a continuous displacement recording if high accuracies are required. The alternative of a steadily increased loading would, nevertheless, generally give acceptable accuracy. Measurement of the concurrent sample compression allows the corresponding void ratio changes to be calculated.

The Rowe cell was not suitable for this style of low effective stress testing, as it was found that the piston movement was not consistently related to the sample strains. This appears to be due to the loading arrangement whereby a flexible convoluted diaphragm is pressurized to apply a load. Initially the increase of air pressure in the diaphragm can force water from between the diaphragm and the cell wall so that it moves underneath the piston, lifting the face so that it is no longer in contact with the soil. As a result, the position of the sample is lost and in the RFC test, it is not possible to identify when the piston again comes into contact with the sample. This is not a problem when the Rowe cell is used for the standard incremental consolidation test, since the piston *will* be in contact with the soil sample before a given load is applied and again at the end of the consolidation due to that load. For the RFC test, this problem can be overcome to some extent by back analyzing the compression from the final sample thickness, coupled with the LVDT measurements once drainage of the sample is established, but this cannot be used for the undrained loading stage for the low effective stresses.

Once the total load has reached the required value, the drainage line is opened, and the pore pressures begin to dissipate. Figure 7 shows typical behavior during this consolidation stage. To distinguish between the pore pressures for this illustration, the flow restrictor has been set to give a maximum difference of around 40 kN/m^2 between drained and undrained face pore pressures. In practice, a greater flow restriction, giving closer values of pore pressure, would be normal. The exact conditions would be governed by the requirements of accuracy.

The swell-back stage is provided by an increase in back pressure. In principle, it should be possible to achieve this without affecting the total stress, but in practice, for tests in the Oxford cell, there is a corresponding, though much smaller, increase in the total stress. This is probably due to the friction caused by the seals on the piston of the consolidation cell. Although friction losses are not, in general, welcome, they do in this case prevent the piston lifting clear of the soil sample. For the Rowe cell, where there is little piston friction, it is necessary to ensure that the applied back pressure is lower than the total vertical stress. In both cells there is, therefore, a comparatively small effective stress remaining on the soil sample at the end of the swell-back stage. This may decrease further during the removal of the total stress and back pressure together, and the displacement reading will indicate the occurrence of further swelling. For most soils, however, the void ratio changes in unloading small effective stresses are very small.

The crucial part of the RFC test is to ensure that the output of all the transducers is recorded continuously during the test, beginning before the load is first applied to the sample. In general, the initial LVDT readings should be used as the reference

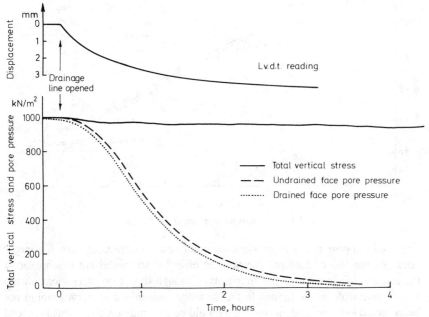

FIG. 7—*Typical results of an RFC test.*

values, corrected where necessary for strains in the system, but a check should be made on the results obtained by back analysis. If there is a significant difference, the cause should be found.

Results

The calculation of the effective stress requires an estimate of the mean pore pressure from the two extreme values of drained and undrained face pore pressures. The distribution of pore pressure through the sample will be closer to parabolic than linear (from the boundary conditions and consolidation theory), so that the mean pore pressure is equal to two thirds of the undrained value plus one third of the drained value.

In order to inspire confidence in the principles of the RFC test, it must be demonstrated that repeated tests on similar samples give similar results. Figure 8 shows the results of four tests on an estuarine silty clay mixed at a water content at around 95% and then hand molded into the consolidation ring. Inevitably, this method produces some variability in the samples, but the loading curves can be seen to be very similar, the differences between them being little more than the order of accuracy expected from the calculation of void ratio (i.e., around ±0.05 near values of void ratio of one and correspondingly higher at higher values). Also, since the effective stress is traditionally plotted, as in this figure, to a logarithmic scale, the small differences in the value of the effective

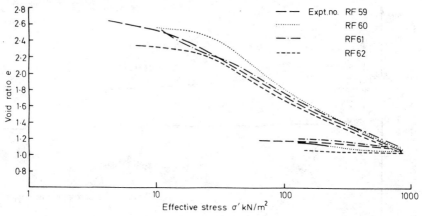

FIG. 8—*Results for soft samples in RFC test.*

stress will appear much more significant for small magnitudes than for larger ones. At the end of loading, the vertical effective stress did not reach exactly the same value in each of the four tests, although this is not very clear on the logarithmic scale of the figure. In this situation, the unloading curves would not be expected to be coincident, but they should be generally parallel. This is indeed the case, and the results of these four tests and of a number of other similar tests confirm that the RFC test provides consistent, repeatable results.

Figure 9 shows some of the results of a series of experiments undertaken to make direct comparison between standard oedometer and RFC tests. Large samples prepared in the 254-mm (10 in.) Rowe cell (as described earlier) at con-

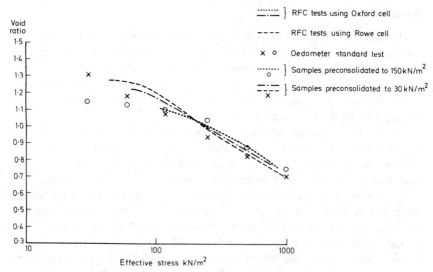

FIG. 9—*Comparative results from standard and RFC tests.*

solidation pressures of 30 or 150 kN/m^2 provided three subsamples for comparison. In these particular tests, the initial undrained loading was not applied gradually, so that some of these results illustrate the loss of low effective stress data that can occur as a consequence of undissolved air in the system. These, and the other tests, confirm that the RFC test and the standard test produce consistent results within the bounds of accuracy of the standard test.

Conclusions

The test program has demonstrated the success of the RFC test in producing quickly and simply a one-dimensional stress/strain (effective stress/void ratio) relationship for a soil sample. The test has been successfully undertaken in both the Oxford cell and the Rowe cell, although the former has considerable advantages for low effective stresses and ease of de-airing. The RFC test has the advantage of speed over the standard test, an advantage which it shares with many of the recently proposed alternative consolidation tests. However, it has two further advantages which make it unique. The first is the simplicity of the test, both in concept and in practice, with no requirements for electronic control or feedback. The second advantage is that the interpretation of the results require no theoretical assumptions to be made about the consolidation behavior of the soil. The stress/strain correlation requires only the assumption of the effective stress law. Although a better approximation to the mean effective stress has been obtained by assuming that the pore pressure distribution is parabolic, this is not essential and the assumption of a linear distribution would be adequate.

The RFC test has further potential. One is the calculation of soil permeability, with the only theoretical assumption being that of Darcy's law. The experimental requirements are slightly different from those of the stress/strain measurement, and pose some additional problems. The main difficulty is the accurate measurement of the difference between the pore pressures on the undrained and drained faces. Initially, the two pore pressures are large (the same order of magnitude as the total vertical stress), and pressure transducers that can monitor these will not provide an accurate measurement of the small difference between the pore pressures. A differential pressure transducer is therefore necessary, to measure the difference directly. However, the transducer is then necessarily remote from at least one of the sample faces, and some losses will occur which must be allowed for in calibration. There is also the danger that if the system should develop a leak anywhere between the undrained face and the downstream side of the flow restrictor, the differential transducer could be subjected to an overpressure that would damage it. There may well be a case for an old-fashioned instrument such as a manometer for this differential measurement!

The RFC test also has applications in specialized testing to examine the possible dependence of soil properties on time and pressure level. The rate of change of effective stress or strain could be varied in different tests or at any stage in a single test by altering the flow restriction. A given effective stress level could be achieved by different combinations of total stress and pore pressure using a back pressure throughout the loading stage.

Acknowledgments

The authors gratefully acknowledge the assistance of members of the soil mechanics group in the Engineering Department at Oxford: Mr. D.McG. Elder, for discussion on the details of the test and interpretation of the results; Mr. S. Oldham, for support on the electronic instrumentation; and Mr. R. Sawala, Mr. C. Waddup, and Mr. M. Smith, for the mechanical construction of the equipment. The original idea for the restricted flow consolidation test came from Dr. K. Lee, then at Oxford, now at Nanyang Technical Institute, Singapore.

The authors are grateful for the generous support of the Wolfson Foundation in developing the cell and undertaking the testing program, and also thank ELE International for the gift of two 75-mm-diameter Rowe cells.

References

[1] Lee, K., "Consolidation with Imperfectly Permeable Drainage Boundary," Cooling Prize submission, Internal Report, Oxford University Engineering Dept., OUEL 1349/81, SM001/KL, 1977.
[2] Lee, K., "An Analytical and Experimental Study of Large Strain Consolidation," D.Phil. thesis, Oxford University, 1979.
[3] Lee, K. and Sills, G. C., "One Dimensional Consolidation with Restricted Drainage," report in preparation.
[4] Hoare, S. D. H., Cooling Prize submission, Internal Report, Oxford University Soil Mechanics Group, SM014/ELE/80, 1980.

Kurt F. von Fay,[1] Jack G. Byers,[2] and Betsy A. Kunzer[3]

Desktop Computer Application for Consolidation Testing and Analysis

REFERENCE: von Fay, K. F., Byers, J. G., and Kunzer, B. A., **"Desktop Computer Application for Consolidation Testing and Analysis,"** *Consolidation of Soils: Testing and Evaluation, ASTM STP 892,* R. N. Yong and F. C. Townsend, Eds., American Society for Testing and Materials, Philadelphia, 1986, pp. 217–235.

ABSTRACT: The use of desktop microcomputers for automatic data acquisition of consolidation test data and data analyses is discussed. The various electrical sensing devices used to measure several test parameters are briefly described. In addition to using the desktop computer for automatic data acquisition, a program was written to analyze the data using several classical empirical procedures. The Casagrande logarithm of time fitting method and the Cour inflection point method for computing C_v are used to analyze the time-dependent data. The Casagrande and Schmertmann construction techniques are used to analyze the time-independent data. Additionally, the density, water content, and degree of saturation of the specimen are determined throughout the test.

KEY WORDS: consolidation test, data acquisition, computers, measurements, automated system, microcomputer, consolidation data analysis, linear variable differential transformer (LVDT)

The phenomenon of soil settlement has long been a concern of and problem for civil engineers. The Leaning Tower of Pisa is a classic example of what can happen when a soil's settlement characteristics are not correctly determined. Before 1925, only parts of the settlement process were understood. Then Karl Terzaghi [1] proposed a theory in *Erdbaumechanik* that related pressure, volume change, and time, and that described the process associated with the compression of a mass of discrete irregular particles into a denser material. This process became known as consolidation. Terzaghi also developed a test to study consolidation characteristics, which consisted of placing a cylindrical, saturated specimen of a particular material in a ring (to prevent lateral displacements), loading the specimen, and letting the pore water drain through porous stones placed on the top or bottom or both. Terzaghi and others developed various

---

[1] Civil Engineer, Soil Testing Section, Geotechnical Branch, U.S. Bureau of Reclamation, Denver, CO 80225.

[2] Supervisory Civil Engineer, Soil Testing Section, Geotechnical Branch, U.S. Bureau of Reclamation, Denver, CO 80225.

[3] Civil Engineering Technician, Soil Testing Section, Geotechnical Branch, U.S. Bureau of Reclamation, Denver, CO 80225.

analysis techniques whereby deformations of the sample were plotted against various functions of time and load. With these data in hand, the engineer is better able to determine the settlement characteristics of a soil stratum and, hopefully, to avoid a modern-day equivalent of the Leaning Tower of Pisa.

With the advent of computer technology in the 1960s, more comprehensive and thorough consolidation analysis was made possible. In the early 1980s, after advances in microcomputer technology, it became feasible to develop very specialized interactive systems using these computers. It is relatively easy, practical, and economical to tailor automated microcomputer systems to specific data acquisition requirements. Along with this is the possibility of using these same computers to utilize various analysis techniques on an interactive basis that is relatively flexible and fast. Results from the analyses can be obtained almost immediately after the user initiates them, instead of the normal one-day turnaround time associated with some larger computer systems.

This paper discusses the program development, equipment interface, and use of a microcomputer for consolidation testing and analysis at the U.S. Bureau of Reclamation, Engineering and Research Center, Geotechnical Laboratory.

Concept

In recent years some of the most rapid advances in geotechnical laboratories have been in the area of automatic data acquisition and analysis, the most common application being a single computer attached to one or two test units. This type of application is good; however, we have found that significant efficiency and productivity increases occur when several test units are monitored, the data immediately analyzed, and test results presented in the form of report quality plots and data tables.

This is the concept used to develop the consolidation data acquisition and analysis programs. The Bureau has 18 pneumatic consolidometers, with each unit in use approximately 60% of the year. The consolidating normal loads are applied manually. The data recording, storage, processing, and presentation are completely automated, utilizing a combination of load cells and LVDTs (linear variable differential transformers), a data acquisition complex, microcomputer, thermal printer, and multipen plotter.

Hardware

The consolidometer used by the Bureau of Reclamation is a Karol-Warner dual range (low and high) consolidometer (Fig. 1). The consolidation load is applied to the 108-mm (4.25-in)-diameter or 50.8-mm (2.00-in.)-diameter specimens and measured by bourdon tube gages. The vertical displacement is measured by alternating-current (2-c) LVDTs, and loads are measured by Bourdon tube pressure gages.

The core of the automatic data acquisition and analysis system is a microcomputer with 1 megabyte of random access memory (RAM), two built-in disk drives

FIG. 1—*Consolidation test apparatus.*

containing 276 kilobytes per disk, a 1-megabyte Winchester hard disk, and a 16-bit microprocessor. This capability allows the computer to store all equipment calibration curves, specimen data, and test data for each of the 18 systems, both in RAM and on disk. This results in faster access for data calculations and graph production. The computer also has an internal clock and an IEEE parallel, interface buss. The data acquisition complex, thermal printer, and a two-pen, 216 mm ($8\frac{1}{2}$ in.) by 279 mm (11 in.) flatbed plotter all connect to the IEEE interface (Fig. 2).

The computer is programmed in extended BASIC utilizing structured programming techniques and subroutines. Automatic LVDT height readings are recorded at programmed intervals (4, 10, 20, 40, 80 s, etc) for each load of each consolidation test. User-initiated random specimen height readings can also be obtained and, if desired, stored as part of the test data. Each specimen data record includes the system and apparatus identification number, LVDT and specimen height reading, the consolidation load on the specimen, the elapsed time in seconds from the moment the present load was applied, the number of readings since the beginning of the test, and the date and time the reading was taken. The

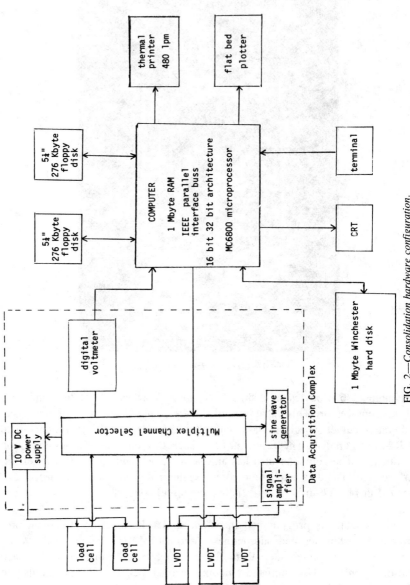

FIG. 2—Consolidation hardware configuration.

computer uses the stored specimen and test data to calculate a number of test values and to generate various plots which can be shown on the CRT or dumped to a thermal printer or multipen plotter.

System Calibration

Before a material is tested, calibrated load cells are used to generate a calibration equation for each Bourdon pressure gage of the pneumatic consolidometers. Because the load cells require a 10-V direct-current (d-c) power source, the computer is programmed to automatically switch the power source and the digital voltmeter to direct current when a pressure gage calibration is initiated from the terminal, and to switch back to alternating current at the end of the calibration process. Force applied to the load cell alters the d-c signal sent from the load cell through the multiplexer to the digital voltmeter. The voltmeter interprets the changed signal as volts and sends the voltage reading to the computer. The computer compares the voltage reading from the load cell to the stored pressure-voltage equation, determines the load on the load cell, and generates a Bourdon tube pressure gage reading versus applied load equation for that particular pressure gage.

During a consolidation test, the required normal load is entered into the computer, which then calculates and displays the necessary Bourdon tube pressure gage reading for the load requested. The Bourdon tube gage is then set to the proper gage reading and the load is instantaneously applied to the specimen through a toggle-operated shutoff valve. The computer uses this load as the specimen load for each LVDT reading taken during this consolidation load. Load cells or pressure transducers are not used to make load measurements during a test because of the unit cost and mechanical configuration.

The system uses LVDTs, electronic displacement transducers, with 2.5 cm (1 in.) of travel to measure the height of the specimen to 0.0025 mm (0.0001 in.) (Fig. 1). Each consolidometer has an LVDT connected to a specific computer channel in the data acquisition unit. Because of the large number of LVDTs being monitored, the signal produced by the 2.5 kHz signal wave generator is boosted by an amplifier to ensure 2.5 kHz to each of the 18 LVDTs. The signal is transmitted through the multiplexer to the digital voltmeter where it is interpreted as a voltage reading and read by the computer. The LVDT is calibrated and the calibration equation stored both in ram and on disk.

Testing

When the consolidation test is initiated, a reference LVDT reading is taken on a gage block of a known height. This reference reading is then stored in the computer. The computer compares the first recorded LVDT reading on the specimen with the LVDT reading of the gage block and the given height of the gage block. During the test, the computer calculates the amount an LVDT core is moved by comparing the voltage reading of the LVDT to the stored LVDT calibration equation. The computer then calculates the position of the LVDT core within the barrel and determining the specimen height to the nearest 0.0025 mm (0.0001 in.). The height of the gage block in inches, the LVDT reading on the gage block, and the first LVDT reading on the specimen are stored in the computer. Each channel is programmed to record LVDT readings at set intervals from the application of a load until a stop order is given through the terminal. These readings take priority over any other computer use. The computer synchronizes the start of each test such that no two LVDT readings from different tests are required at the same instant.

Software

The program "EngCon, Desk Top Computer Analysis of Consolidation Data" calculates various soil phase relationship conditions, determines polynominal best fit equations for the time-dependent and time-independent data and with the equations, determining time-dependent and time-independent parameters of a soil specimen tested according to the Incremental Stress, One-Dimensional Consolidation Soil Theory. It must be remembered, however, that the methods and controlling theories concerning consolidation testing and data analysis were developed for certain specific conditions and simplifying assumptions. Among them were: (1) the coefficient of compressibility is constant, (2) the coefficient of permeability is constant, (3) the specimen is saturated by an incompressible fluid, (4) the mineral grains are incompressible, and (5) the specimen is homogeneous throughout. When those conditions are not met, sound engineering judgment must be used to correctly utilize the computer-generated results.

In 1936, Casagrande [2] developed an empirical, graphical technique to determine the preconsolidation pressure from the semilogrithmic representation of specimen deformation versus logarithm of effective stress. This method has become known as the "Casagrande" construction. Figure 3 illustrates the essential features of this construction.

With the aid of the Casagrande method, the most probable preconsolidation stress is determined as follows:

1. Determine the point of maximum curvature (minimum radius). This can be done by eye, graphically, or analytically. See Point 1 in Fig. 3. To determine the point graphically, extend the straight line portion of the top part of the curve until it intersects the line described in Step 5. To determine the point analytically, the equation describing the curve must be known. The point of minimum radius, R, is found using the relationship [3]

$$R = \frac{1 + [(dy/dx)^2]^{3/2}}{(d^2y/dx^2)} \tag{1}$$

2. Draw a horizontal line from Point 1.

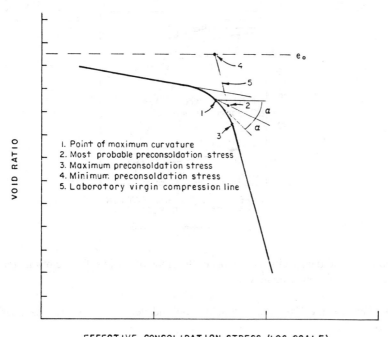

EFFECTIVE CONSOLIDATION STRESS (LOG SCALE)

FIG. 3—*Casagrande construction.*

3. Draw a line tangent to the curve at Point 1.

4. Bisect the angle made by Steps 2 and 3.

5. Extend the straight line portion of the laboratory virgin compression curve (called the laboratory virgin compression line) up to where it meets the bisector line obtained in Step 4. The point of intersection of those two lines is the most probable preconsolidation stress (Point 2 in Fig. 3).

Additionally, the maximum possible preconsolidation stress is at Point 3 where the laboratory virgin compression line leaves the curve. The minimum possible preconsolidation stress is at Point 4, the intersection of the laboratory virgin compression line with a horizontal line drawn from e_0, the initial void ratio.

The Casagrande construction allows one to determine important soil deformation parameters, but fails to take into account the effects of sample disturbance or to determine the slope of the field virgin compression line. In 1955, Schmertmann [4] developed a graphical method to analyze time-independent settlement data which accounted for the effect of sample disturbance and estimated the slope of the field virgin compression line. Figure 4 depicts the important elements of the Schmertmann procedure and is used as described below: (Note: Figure 4 depicts an unload-reload segment in the compression curve which is used to determine the recompression index, C_r. However, since the Bureau laboratories

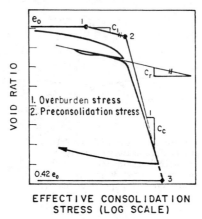

EFFECTIVE CONSOLIDATION
STRESS (LOG SCALE)

FIG. 4—*Schmertmann procedure.*

usually unload the specimen only once, the slope of the best fit line of the unload data is used to approximate C_r.)

1. Perform the Casagrande construction and determine the preconsolidation stress.

2. Calculate the initial void ratio, e_0. Draw a horizontal line from e_0 to the overburden stress. This defines Point 1 in Fig. 4.

3. From Point 1, draw a line parallel to the best fit line of the unload-reload curve to the preconsolidation stress. This establishes Point 2 in Fig. 4. (Note: If a soil is normally consolidated, the overburden stress equals the preconsolidation stress; therefore Points 1 and 2 will be the same.)

4. From a point on the void ratio axis equal to $0.42\ e_0$, draw a horizontal line, and where that line meets the extension of the laboratory virgin compression curve, another point (Point 3) is established.

5. Connect Points 2 and 3 with a straight line. The slope of this line defines the field virgin compression index, C_c.

From the graphical representations created using the one-dimensional consolidation test data, the program determines the laboratory virgin compression line, field virgin compression line, overconsolidation ratio, laboratory virgin compression index, field virgin compression line, preconsolidation stress and corresponding void ratio, compression ratio, and recompression index.

Time-Dependent Analysis

This part of the program is based on Terzaghi's one-dimensional consolidation theory developed in 1925 [1] and is discussed in most soil mechanics texts. The program determines the coefficient of consolidation, the coefficient of permeability, and axial strain values at various average degrees of consolidation (percent consolidation) using two different curve fitting techniques. These empirical procedures were developed to approximately fit the observed laboratory test data to the curves generated using Terzaghi's theory of consolidation. This is valid since actual curves often have very similar shapes to the theoretical percent consolidation (U) versus time factor (T) curves. Figure 5 shows the theoretical curves for two different functions of T.

Casagrande's Logarithm of Time Fitting Method

Casagrande [5] developed a logarithm of time fitting method, called the "graphic method" on the computer generated plots, where deformation values (dial readings, axial strain, or void ratio) are plotted versus the logarithm of time as shown in Fig. 5. For this technique, R_{50} (deformation value at 50% consolidation) and t_{50} (time at 50% consolidation) are determined from R_0 and R_{100} which are obtained using empirically derived methods.

R_0 is determined using the fact that T is proportional to U^2 (time factor is proportional to the percent consolidation squared) up to about 60% consolidation on the theoretical U–log T plot (Fig. 5). Using this, any two times in the ratio of 4 to 1 (t_1 and t_2, where $t_2 = 4t_1$) are chosen and their corresponding deformation values (R_1 and R_2) are determined. Next, the difference between these values is determined ($R_2 - R_1$) and subtracted from the first deformation value [$R_0 = R_1 - (R_2 - R_1)$]. This procedure defines R_0 and is shown in Fig. 6.

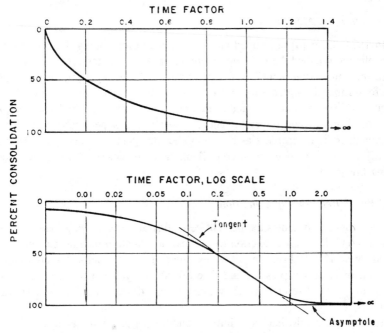

FIG. 5—*Theoretical time-consolidation curves.*

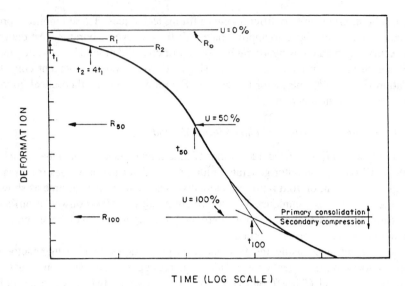

TIME (LOG SCALE)

FIG. 6—*Casagrande logarithm of time-fitting method.*

The R_{100} (deformation value at 100% consolidation) is defined as the intersection of the tangent to the straight line portion of the curve and the tangent to the end of the curve (Fig. 6).

Inflection Point Method

In 1971, Cour [6] developed another technique for finding the coefficient of consolidation, called the *inflection method*, on the computer-generated plots. He used the "inflection point" of the deformation versus logarithm of time curve. He found that the inflection point of theoretical U versus logarithm of T curve occurs at 70% consolidation and a time factor of 0.405 (Fig. 7). From this, R_{70}, the coefficient of consolidation, and the coefficient of permeability can be determined. R_0 is determined as before. Other deformation values corresponding to different average degrees of consolidation can be determined from R_0 and R_{70} using linear relationships.

Soil Phase Relationship Conditions

The program computes the percent vertical strain from seating load conditions (initial conditions) as a function of vertical deformation reading changes monitored by an LVDT. From these values, the soil phase relationship conditions at the end of each test pressure are determined. They are specimen height, void ratio, dry density (unit weight), water content, degree of saturation, and axial strain.

The program calculates the initial conditions needed for analysis from the initial (seating load) height readings, specimen size, specific gravity, and weight.

The program then calculates the appropriate strain values and void ratio values. These values are used where needed by different analysis techniques. Since the specimen diameter is constant, air, water, and soil volumes are determined in terms of the specimens.

Printed Output

A formated copy of the data file is printed, which includes general specimen identification and placement conditions as well as a computer listing of all test measurements. Because of the flexibility of this program, several different output options are available, depending on the analysis procedure used. Point-to-point plots of axial strain versus logarithm of time, axial strain versus load, and void ratio versus logarithm of load are available. In addition, the void ratio and axial strain versus square root of time consolidation plot can be obtained for each load. A table of loads and the corresponding values of height, void ratio, dry density (unit weight), water content, degree of saturation, and axial strain is printed. The values used in the table correspond to the last LVDT reading taken for a specific load. The values for percent rebound, dry weight determined from the initial moisture, dry weight determined from the final weights, and water content determined from the final weight and initial moisture are printed in conjunction with the table above. A table of values for the dry density (unit weight), water content, and degree of saturation corresponding to the initial load and to the maximum load is also printed. A curve fitted void ratio versus logarithm of pressure plot can be generated. The plot shows results of the Casagrande and Schmertmann construction techniques using the analytical and graphical analysis methods or a manual analysis method, depending upon which option was chosen. Among the results shown are the minimum, most probable, and maximum preconsolidation stress, laboratory virgin compression line, field virgin compression line, point(s) of maximum curvature, rebound line, and e_0. A table showing the results from the analysis of the time-independent data is also available. The table has values for the various preconsolidation stresses, void ratios at the different preconsolidation stresses, overconsolidation ratio, compression index, rebound index, and field compression index.

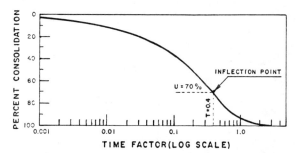

FIG. 7—*Cour inflection point method.*

Curve-fitted axial strain versus logarithm of time plots for specific loads can also be generated. The plots show the results of the Casagrande logarithm of time fitting method ("graphic method") and the Cour inflection point method for determining c_v ("inflection method"). The coefficient of consolidation, coefficient of permeability, and the strain values and time at 0, 50, 70, and 100% consolidation are shown.

The Appendix presents a full range of results in their typical printed form.

Conclusions

We have described the development and use of a microcomputer for data acquisition and analysis of consolidation test data. This system has been extremely reliable in providing excellent report quality products with virtually no downtime.

The principles of this data acquisition and computer analysis system can be applied to several materials tests and data analyses. The Bureau has also used this configuration in automating the triaxial, direct, repeated direct, rotational shear, and back pressure permeability tests. The flexibility of the interactive program allows the user to obtain design and analysis information during and immediately after testing.

Acknowledgments

Henry Hoff, electronics engineer, was instrumental in the test apparatus computer interface and program development.

APPENDIX

The following pages present the results in their printed form.

SAMPLE NUM SPEC NO.	DRILL HOLE DEPTH	CLASS SYM	SPEC GRAV	SPEC TYPE
EXAMPLE 51	DH-282 1.5-5.0	CL	2.69	1
DATE PLACE LOAD NO.	CONT. NO. CONT. HT.	CONT. DIA.	P.I. %	L.L. %
3-01-84 7	208 1.25	4.25	.20	.32
WET SPEC•C CONT•PLATE	DIAL BLOCK DIAL SPEC	DATE REMOV	WATER•CO	C WT TRIM
1825.3 1213.8	.2295	3-29-84	RP•11	CL-230
S.L. %	C IN WATER IN WATER %	F WATER %	.8639	.HT
	CL-241 .13.2	F H519		1.2378
F WET•CONT	F DRY SP•C C FOR SP	F DRY W•C	F•C FOR WS	ADD WATER#
817.4	732.9 200.3	208.9	207.3	2
EFF OVER B	FEATURE CONT			
45	CONSOLIDATION			

LVDT ins	LOAD lb/in2	TIME secs	SYS/LINE	DATE	TIME
.2517	6.0	4	1	3	12:50:50
.2518	6.0	10	2	3	13:04:12
.2518	6.0	10	3	1	13:04:19
.2518	6.0	20	4	3	13:04:30
.2518	6.0	40	5	3	13:04:50
.2518	6.0	80	6	3	13:07:31
.2519	6.0	200	7	3	13:07:31
.2518	6.0	400	8	3	13:10:41
.2518	6.0	800	9	3	13:10:46
.2519	6.0	2000	10	1	13:11:18
.2519	6.0	4000	11	3	13:11:18
.2519	6.0	8000	12	3	13:13:52
.2519	6.0	8640	13	2	13:15:32
.2519	12.5	79960	14	1	11:15:32
.2520	12.5	4	15	2	11:18:27
.2520	12.5	10	16	3	11:18:38
.2521	12.5	20	17	3	11:18:58
.2522	12.5	40	18	3	11:19:39
.2528	12.5	80	19	3	11:21:19
.2533	12.5	200	20	2	11:52:19
.2534	12.5	400	21	3	13:24:49
.2536	12.5	800	22	3	12:25:49
.2538	12.5	2000	23	3	16:52:49
.2539	12.5	4000	24	6	07:02:15
.2550	12.5	8000	25	3	07:26:08
.2550	12.5	71038	26	6	07:26:14
.2595	25.0	4	27	6	07:26:25
.2595	25.0	10	28	3	07:26:40
.2600	25.0	20	29	6	07:26:46
.2605	25.0	40	30	3	07:27:26
.2616	25.0	80	31	6	07:29:26
.2622	25.0	200	32	6	07:34:11
.2628	25.0	400	33	6	07:05:12
.2633	25.0	800	34	6	08:39:12
.2637	25.0	2000	35	6	09:46:12
.2644	25.0	4000	36	6	13:06:42
.2645	25.0	85942	37	7	07:18:26
.2705	50.0	4	38	7	08:00:28
.2718	50.0	10	39	7	08:00:34
.2728	50.0	20	41	7	08:00:44
.2737	50.0	40	43	7	08:09:05
.2756	50.0	80	45	7	08:09:45
.2756	50.0	200	47	7	08:16:40
.2776	50.0	800	49	7	08:23:23
.2786	50.0	2000	50	7	08:23:46
.2795	50.0	4000		7	09:17:12

spec	load	set	press	time
.2795	8000	10	50.0	7 10:24:12
.2804	20000	10	50.0	7 13:44:17
.2808	85929	10	100.0	8 09:16:03
.2808	4	10	100.0	8 09:16:04
.2896	10	10	100.0	8 09:16:21
.2916	20	10	100.0	8 09:16:42
.2930	40	10	100.0	8 09:16:44
.2956	80	10	100.0	8 09:17:23
.2981	200	10	100.0	8 09:19:23
.2989	400	10	100.0	8 09:20:54
.3002	800	10	100.0	8 09:24:30
.3015	2000	10	100.0	8 09:35:30
.3027	13606	10	100.0	8 10:29:08
.3032	20000	10	100.0	8 13:02:46
.3041	77178	10	100.0	8 14:56:08
.3042	4	10	200.0	9 08:44:18
.3129	10	10	200.0	9 08:44:57
.3149	20	10	200.0	9 08:45:08
.3162	40	10	200.0	9 08:46:09
.3174	80	10	200.0	9 08:48:09
.3192	200	10	200.0	9 09:04:18
.3260	400	10	200.0	9 08:48:09
.3280	800	10	200.0	9 09:16:27
.3298	2000	10	200.0	9 10:26:47
.3317	8000	10	200.0	9 11:17:35
.3336	86400	10	200.0	9 14:37:49
.3350	172800	10	200.0	9 09:04:47
.3351	252245	10	200.0	11 09:05:15
.3382	4	10	400.0	12 06:48:51
.3416	10	10	400.0	12 10:10:59
.3426	20	10	400.0	12 08:11:10
.3434	40	10	400.0	12 08:11:31
.3458	80	10	400.0	12 08:12:11
.3495	200	10	400.0	12 08:14:11
.3506	400	10	400.0	12 08:20:15
.3550	800	10	400.0	12 08:30:22
.3592	2000	10	400.0	12 09:24:09
.3628	8000	10	400.0	12 10:31:09
.3654	8000	10	600.0	12 12:51:22
.3663	80971	10	600.0	13 07:37:17
.3664	4	10	600.0	13 07:37:24
.3702	10	10	600.0	13 07:37:34
.3709	20	10	600.0	13 07:37:55
.3714	80	10	600.0	13 07:38:36
.3721	200	10	600.0	13 07:40:36
.3734	400	10	600.0	13 07:46:07
.3748	800	10	600.0	13 07:54:22
.3770	2000	10	600.0	14 08:14:33
.3792	8000	10	400.0	14 08:14:18
.3813	8000	10	400.0	13 06:15:31
.3835	8000	10	400.0	14 06:51:31
.3854	36658	10	400.0	14 07:55:06
.3952	4	10	400.0	14 07:55:13
.3951	10	10	400.0	14 07:55:23
.3850	20	10	400.0	14 07:55:44
.3848	80	10	400.0	14 07:56:25
.3845	200	10	400.0	14 07:58:25
.3842	400	10	400.0	14 08:01:50
.3840	800	10	400.0	14 08:11:22

11 Dec 1984   09:27:41

LOAD lbf/in2	HEIGHT ins	VOID RATIO	DRY UNIT WT lbf/ft3	MOISTURE CONTENT %	DEGREE OF SATURATION %	AXIAL STRAIN %
1.00	1.2578	.4559	115.33	13.20	77.88	0.00
6.00	1.2576	.4557	115.35	16.94	100.01	.02 WETTED
12.50	1.2545	.4521	115.64	16.81	100.01	.26
25.00	1.2450	.4411	116.52	16.40	100.01	1.02
50.00	1.2287	.4223	118.06	15.70	100.01	2.31
100.00	1.2054	.3953	120.35	14.70	100.01	4.17
200.00	1.1744	.3594	123.52	13.36	100.01	6.63
400.00	1.1432	.3233	126.89	12.82	100.01	9.11
600.00	1.1241	.3012	129.05	11.20	100.01	10.63
400.00	1.1263	.3037	129.00	11.29	100.01	10.45
100.00	1.1302	.3175	127.45	11.80	100.01	9.51
50.00	1.1420	.3219	127.03	11.97	100.01	9.21
12.50	1.1528	.3344	125.84	12.43	100.01	8.35
1.00	1.1696	.3538	124.03	13.16	100.01	7.01

REBOUND   34.03 %

DRY MASS(FROM INITIAL MOISTURE) =   540.2 gms

DRY MASS(FROM FINAL MASS) =   534.2 gms

FINAL MOISTURE CONTENT(FROM FINAL MASS AND INITIAL MOISTURE) =   15.3 %

	INITIAL	MAX LOAD
DRY UNIT WEIGHT(lbf/ft3)	115.332	129.050
MOISTURE CONTENT	13.200 %	11.198 %
DEGREE OF SATURATION	77.879 %	100.012 %

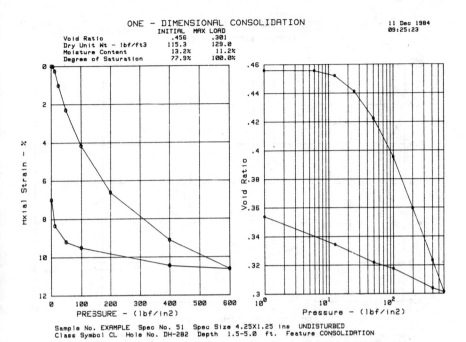

Sample No. EXAMPLE Spec No. 51 Spec Size 4.25X1.25 ins UNDISTURBED
Class Symbol CL Hole No. DH-282 Depth 1.5-5.0 ft. Feature CONSOLIDATION

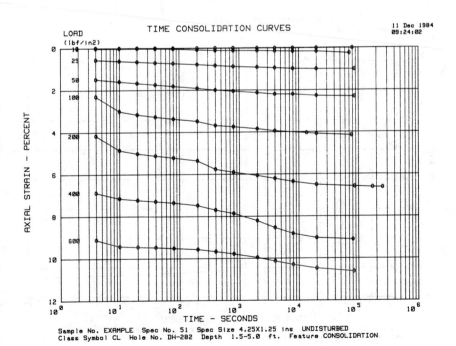

Sample No. EXAMPLE Spec No. 51 Spec Size 4.25X1.25 ins UNDISTURBED
Class Symbol CL Hole No. DH-282 Depth 1.5-5.0 ft. Feature CONSOLIDATION

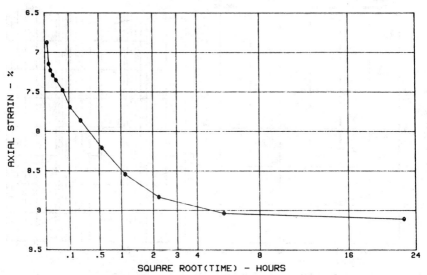

Sample No. EXAMPLE Spec No. 51 Spec Size 4.25X1.25 ins UNDISTURBED
Class Symbol CL Hole No. DH-282 Depth 1.5-5.0 ft. Feature CONSOLIDATION

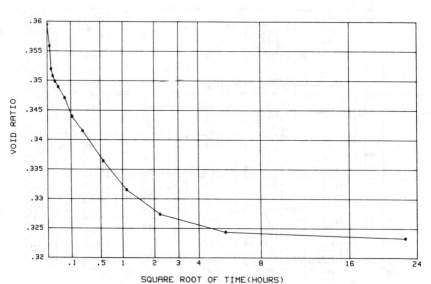

Sample No. EXAMPLE Spec No. 51 Spec Size 4.25X1.25 ins UNDISTURBED
Class Symbol CL Hole No. DH-282 Depth 1.5-5.0 ft. Feature CONSOLIDATION

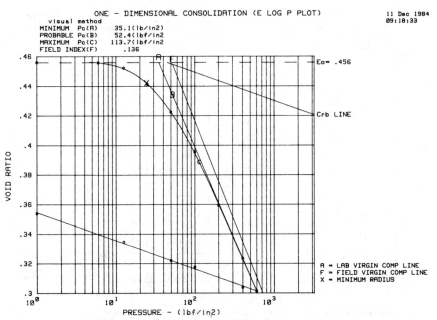

ONE - DIMENSIONAL CONSOLIDATION (E LOG P PLOT)   11 Dec 1984
09:18:33
visual method
MINIMUM Pc(A)   35.1(lb/in2)
PROBABLE Pc(B)   52.4(lbf/in2
MAXIMUM Pc(C)   113.7(lbf/in2
FIELD INDEX(F)   .136

Sample No. EXAMPLE Spec No. 51 Spec Size 4.25X1.25 ins UNDISTURBED
Class Symbol CL Hole No. DH-282 Depth 1.5-5.0 ft. Feature CONSOLIDATION

11 Dec 1984   09:19:20

RESULTS(E LOG P)

SAMPLE NO. EXAMPLE

VISUAL METHOD

	MINIMUM	PROBABLE	MAXIMUM
PRECONSOLIDATION STRESS(lbf/in2)	35.1	52.4	113.7
VOID RATIO AT PRECONSOLIDATION	.456	.434	.389
OVERCONSOLIDATION RATIO	.780	1.165	2.527

LAB COMPRESSION INDEX   =   .126

COMPRESSION RATIO   =   .086

REBOUND INDEX   =   .019

FIELD COMPRESSION INDEX =   .136

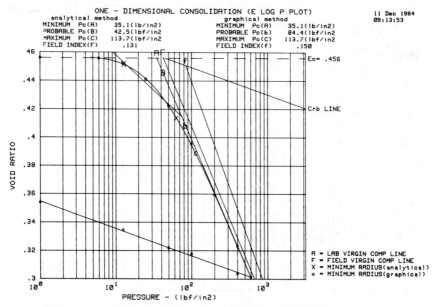

ONE - DIMENSIONAL CONSOLIDATION (E LOG P PLOT)

analytical method		graphical method	
MINIMUM Pc(A)	35.1(lb/in2)	MINIMUM Pc(A)	35.1(lb/in2)
PROBABLE Pc(B)	42.5(lbf/in2	PROBABLE Pc(b)	84.4(lbf/in2
MAXIMUM Pc(C)	113.7(lbf/in2	MAXIMUM Pc(C)	113.7(lbf/in2
FIELD INDEX(F)	.131	FIELD INDEX(f)	.150

11 Dec 1984
09:13:53

Eo= .456

Crb LINE

A = LAB VIRGIN COMP LINE
F = FIELD VIRGIN COMP LINE
X = MINIMUM RADIUS(analytical)
x = MINIMUM RADIUS(graphical)

Sample No. EXAMPLE Spec No. 51 Spec Size 4.25X1.25 ins UNDISTURBED
Class Symbol CL Hole No. DH-282 Depth 1.5-5.0 ft. Feature CONSOLIDATION

11 Dec 1984   09:14:41

RESULTS(E LOG P)

SAMPLE NO. EXAMPLE

ANALYTICAL METHOD

	MINIMUM	PROBABLE	MAXIMUM
PRECONSOLIDATION STRESS(lbf/in2)	35.1	42.5	113.7
VOID RATIO AT PRECONSOLIDATION	.456	.446	.389
OVERCONSOLIDATION RATIO	.780	.944	2.527

LAB COMPRESSION INDEX	=	.126
COMPRESSION RATIO	=	.086
REBOUND INDEX	=	.019
FIELD COMPRESSION INDEX	=	.131

GRAPHICAL METHOD

	MINIMUM	PROBABLE	MAXIMUM
PRECONSOLIDATION STRESS(lbf/in2)	35.1	84.4	113.7
VOID RATIO AT PRECONSOLIDATION	.456	.408	.389
OVERCONSOLIDATION RATIO	.780	1.876	2.527

LAB COMPRESSION INDEX	=	.126
COMPRESSION RATIO	=	.086
REBOUND INDEX	=	.019
FIELD COMPRESSION INDEX	=	.150

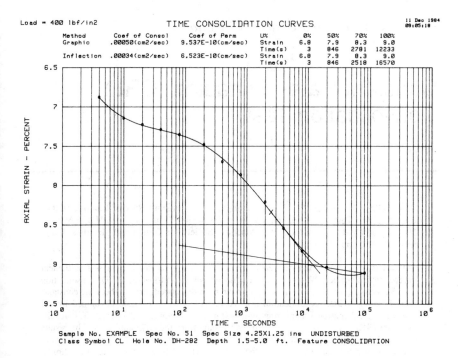

Load = 400 lbf/in2   TIME CONSOLIDATION CURVES   11 Dec 1984 09:05:18

Method	Coef of Consol	Coef of Perm	U%	0%	50%	70%	100%
Graphic	.00050(cm2/sec)	9.537E-10(cm/sec)	Strain	6.8	7.9	8.3	9.0
			Time(s)	3	846	2781	12233
Inflection	.00034(cm2/sec)	6.523E-10(cm/sec)	Strain	6.8	7.9	8.3	9.0
			Time(s)	3	846	2518	16570

AXIAL STRAIN - PERCENT

TIME - SECONDS

Sample No. EXAMPLE Spec No. 51 Spec Size 4.25X1.25 ins UNDISTURBED
Class Symbol CL Hole No. DH-282 Depth 1.5-5.0 ft. Feature CONSOLIDATION

References

[1] Terzaghi, K., *Erdbaumechanik auf Bodenphysikalischer Grundlage*, Franz Deuticke, Leipzig and Vienna, 1925, p. 399.

[2] Casagrande, A., "The Determination of the Preconsolidation Load and its Practical Significance," in *Proceedings*, First International Conference on Soil Mechanics and Foundation Engineering, Vol. 3, 1963, pp. 60–64.

[3] McNulty, E. G., "Computerized Analysis of Stress-Strain Consolidation Data," Kentucky Bureau of Highway, Research Report 468, Lexington, KY, 1977.

[4] Schmertmann, J. H., "The Undisturbed Consolidation Behavior of Clay," *Transactions of American Society of Civil Engineers*, Vol. 120, Paper 2775, 1955.

[5] Casagrande, A., "Notes on Soil Mechanics—First Semester," Harvard University (unpublished), Cambridge, MA, 1938, p. 129.

[6] Cour, F. R., "Inflection Point Method for Computing C," *Journal of the Soil Mechanics and Foundations Division, Transactions of American Society of Civil Engineers*, Vol. 97, No. SM5, 1971, pp. 827–831.

Kurt F. von Fay[1] and Charles E. Cotton[1]

Constant-Rate-of-Loading (CRL) Consolidation Test

REFERENCE: von Fay, K. F. and Cotton, C. E., "**Constant-Rate-of-Loading (CRL) Consolidation Test**," *Consolidation of Soils: Testing and Evaluation, ASTM STP 892,* R. N. Yong and F. C. Townsend, Eds., American Society for Testing and Materials, Philadelphia, 1986, pp. 236–256.

ABSTRACT: For the past 50 years, the information used to estimate soil settlement characteristics has been obtained using a standard-incremental-loading (STD) consolidation test. Recently, several other test methods have been developed, among them the constant-rate-of-loading (CRL) consolidation test. A versatile CRL apparatus, developed by the Bureau of Reclamation Engineering and Research Laboratory, is described. Test results from the CRL apparatus are compared with results from the STD apparatus. The results show that test completion times for the CRL test are much shorter than for the STD test, that agreement between the results of the CRL and STD tests are within reasonable limits, and that the CRL apparatus and test procedure offer an alternative to the STD test.

KEY WORDS: consolidation, conventional-incremental-loading, constant-rate-of-loading (CRL), constant-rate-of-strain, pore pressure, suction

Over the years, several different specialized soil testing apparatuses have been developed in the Bureau of Reclamation Engineering and Research Center laboratory to meet the need for unusual testing requirements. Recently, two such devices, a modified floating confining ring and a constant-rate-of-loading (CRL) drive system, were combined with an air-operated consolidation frame. This combination created a versatile CRL consolidation apparatus.

This paper presents a description of the CRL apparatus, its capabilities, and results of tests on remolded specimens. These results are compared with results obtained by testing similar specimens in a standard-incremental-loading device.

Development of the CRL Apparatus

There were two primary reasons for the development of the CRL apparatus. The first was to better simulate certain construction loading conditions such as those imposed by embankments, dams, and large structures. Since different methods of testing may yield different values for soil properties, it is desirable to simulate field conditions as closely as possible when testing soils in the

[1] Bureau of Reclamation, Denver, CO 80225.

236

laboratory. The second reason was to overcome some of the negative aspects of the conventional method, such as: (1) the long period of time required to obtain test results, from two to as much as five weeks; (2) assumed high hydraulic gradients imposed on the sample; (3) the variability of the effective stress throughout the sample; and (4) innate limitations of the conventional testing equipment itself, for example, no capability for most apparatuses to measure pore pressure (which determines the state of consolidation) and to apply back pressure (which ensures saturation).

The CRL apparatus has the capability to measure initial matrix suction (negative pore pressure), measure pore pressure, load the sample at a constant loading[2] rate, stop and hold a predetermined load on the sample at any time during the loading sequence, apply back pressure, and saturate the specimen at any point during the loading sequence. The specimen may be unloaded at the same rate it was loaded or at a different rate, if desired. For the purpose of this report, only the capability of this CRL apparatus to generate usable consolidation data in a relatively short time was addressed. The other capabilities will be investigated and the results documented in a future report.

Several other testing methods have been developed to determine the consolidation characteristics of a material and to refine the analysis procedure. Lowe et al [1] developed the analysis and procedure for controlled gradient testing. The constant rate of strain (deformation) test was first suggested by Hamilton and Crawford [2], but the proper procedure for analysis was not developed since information about the permeability of the material was not obtained. Two different analysis procedures for this test were developed by Smith and Wahls [3] and Wissa et al [4]. Imai [5] suggested a seepage test to determine the consolidation properties of soils.

Analysis of CRL Test

The testing and analysis techniques for the CRL test were first presented by Aboshi et al [6]. The procedure is based on the theoretical work of Schiffman [7], who developed the governing equation and found the solution for general time-dependent load applications.

The process of consolidation under a constant rate of loading, assuming constant permeability and constant coefficient of consolidation, is governed by

$$\frac{\partial u}{\partial t} = C_v \frac{\partial^2 u}{\partial z^2} + \frac{dp}{dt} \tag{1}$$

where

u = pore pressure,
t = time,

---

[2] Even though the term *stress* is more appropriate than *loading*, the authors use the latter term since it is consistent with the name of the test.

z = location,

C_v = coefficient of consolidation, and

dp/dt = the rate of loading, a constant.

Aboshi et al [6] developed the following solution for this case:

$$\frac{u}{p} = \frac{16}{\pi^3 T} \sum_{n=1,3,5...}^{\infty} \frac{1}{n^3} \sin \frac{n\pi z}{2H} (1 - e^x)$$

(2)

where

$x = -(n^2\pi^2) \, T/4$,

u = pore pressure,

p = applied load,

T = time factor = $C_v t/H^2$,

e = transcendental constant equal to 2.7182818. . . , and

H = drainage path length.

If the pore water pressure is measured at the impermeable base, the pore water pressure ratio equation is

$$\frac{u_d}{p} = \frac{16}{\pi^3 T} \sum_{m=0}^{\infty} (-1)^m \frac{1}{(2m + 1)^3} (1 - e^y)$$

(3)

where

$y = (2m + 1)^2 \, \pi^2 \, T/4$, and

u_d = pore pressure at the impermeable base.

The above expressions show a unique relationship between the ratio of u/p and the time factor T. Results from the CRL test yield the ratio u/p, making it possible to determine the corresponding time factor T. From this, the coefficient of consolidation can be determined as follows:

$$C_v = \frac{T H^2}{t}$$

(4)

Aboshi et al [6] also developed the means to determine the average effective stress within the sample. To accomplish this, a relationship between the average degree of consolidation and the time factor was derived:

$$U = 1 - \frac{1}{pH} \int_0^H u dz$$

$$= 1 - \frac{32}{\pi^4 T} \sum_{m=0}^{\infty} \frac{1}{(2m + 1)^4} (1 - e^y)$$

(5)

where U is the average degree of consolidation. Therefore, for any time factor, there is a corresponding average degree of consolidation. From this, the average effective stress can be calculated using $\bar{p}' = pU$, where $\bar{p}'$ is the average effective stress.

Description of Apparatus

CRL Apparatus

Figure 1 shows the CRL apparatus used to obtain test data for this report. The floating ring consolidometer used to perform the CRL testing is shown in Fig. 2. The apparatus is capable of testing cylindrical specimens 108 mm (4.25 in.) in diameter, of various heights up to a maximum of 64 mm (2.50 in.) An air-operated consolidation frame and a loading drive system were used to load the specimen at a constant rate. The constant-rate-of-loading drive system consisted

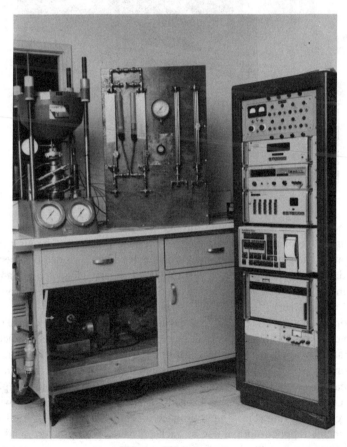

FIG. 1—*CRL apparatus.*

FIG. 2—*Floating ring consolidometer.*

of an electric motor which, acting through a series of clutches and gears, controlled the valve of a plunger operated air pressure regulator. The air regulator applied air to the consolidation frame, which subsequently loaded the specimen. The load was measured with a 22 kN (5000 lb) capacity load cell. Changes in the height of the specimen were measured with an LVDT (linear variable differential transformer).

The top and bottom end plates were machined stainless steel with coarse porous carborundum disks inlaid and held in place with epoxy. The soil specimens can be saturated by applying back pressure through either or both end plates. For this report, the specimens were drained through the top plate, and pore pressures were measured with a pressure transducer connected to the bottom end plate.

Two O-rings set in grooves in each of the end plates form a seal with the floating confining ring (Fig. 3) and produce a negligible amount of friction. All measurements (applied normal stress, back pressure, pore pressure, and deformation) were taken with electronic instruments at predetermined intervals and recorded on paper tape by a microprocessor-controlled data acquisition system (Fig. 4).

STD Apparatus

The STD consolidation tests were performed using an air-operated consolidation frame (Fig. 5). This apparatus is capable of testing specimens according to the requirements of ASTM Test for One-Dimensional Consolidation Properties of Soils (D 2435). The changes in height of the specimen were measured with a dial indicator graduated to 0.0025 mm (0.0001 in.).

The specimens were drained on the top and bottom by coarse porous carborundum disks. A falling head system was used to wet the specimen by connecting a 13-mm ($^1/_2$-in.)-diameter by 305-mm (12-in.)-long glass standpipe to the bottom end plate and filling it with water. As water entered the specimen, more was added into the glass standpipe. This system, by itself, probably does not saturate the specimen. However, as the loads increase and the voids decrease, the specimen becomes saturated. The STD equipment used in the E&R Center geotechnical laboratory has no provision for the application of back pressure or for the measurement of pore pressure.

Testing Program

Soil Description

The testing program was conducted using remolded specimens. The specimens were placed at a predetermined moisture content and unit weight. Each specimen

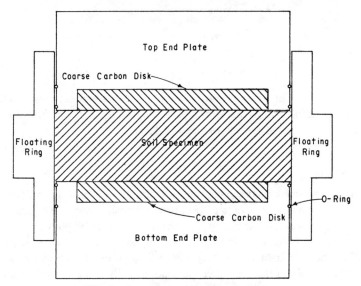

FIG. 3—*Cross section of specimen container.*

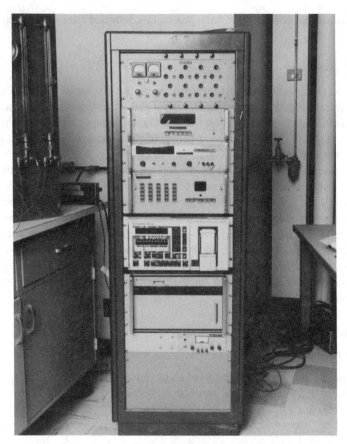

FIG. 4—*Microprocessor-controlled data acquisition system.*

was 108 mm (4.25 in.) diameter by 32 mm (1.25 in.) high and was packed in three lifts, ensuring a fairly uniform unit weight throughout. The physical properties of the soils tested are summarized in Table 1.

CRL Test

Each test specimen was recompacted to 108 mm (4.25 in.) in diameter by 32 mm (1.25 in.) high in a separate specimen container. After the specimen was recompacted, it was removed from the original container and placed in a floating ring specimen container for testing. A seating stress of 6.9 kPa (1.0 psi) was applied to the specimen to properly seat and align all the components of the system. The height of the specimen was automatically recorded at the time. The specimens were saturated using back pressure with the seating stress applied.

FIG. 5—*Air-operated consolidation frame used for the STD tests.*

TABLE 1—*Physical properties.*

Sample No.	Soil Classification [10]	G[a]	LL[b]	PI[c]	Percent Sand	Percent Fines
1	CL[d]	2.66	32	18	9	91
2	CH[e]	2.72	59	40	3	97

[a]G = specific gravity.
[b]LL = liquid limit.
[c]PI = plasticity index.
[d]CL = inorganic clays of low to medium plasticity.
[e]CH = inorganic clays of high plasticity.

The top end plate was placed dry and the bottom end plate and connecting tubing were filled with de-aired water.

To help remove some of the air within the specimen, a small back pressure was applied through the bottom end plate, forcing the water up through the specimen, top end plate, and into a burette which was open to the atmosphere. After the water was visible in the burette, a valve between the burette and specimen was closed and back pressure of 414 kPa (60 psi) was applied through the top and bottom end plates to ensure saturation. The back pressure of 414 kPa (60 psi) remained on the sample throughout the test.

After the back pressure was applied, the degree of saturation was checked using the pore pressure parameter C [8,9]. The loading sequence was not started until the value of C reached at least 0.95 (C is similar to pore pressure parameter B except that C is often used for one-dimensional normal loading).

The loading rate was determined using the PI (plasticity index) and LL (liquid limit) of the soil (see Appendix for discussion of selection of loading rate). One-way drainage (at the top of the specimen) was permitted throughout the test.

The specimens were unloaded at the same rate they were loaded. The normal stress, pore water pressure, and height of the specimen were electrically sensed and automatically recorded every 30 min during loading and unloading. When unloading was complete, the soil specimens were removed from the apparatus and the moisture content of each specimen was determined.

STD Test

Each test specimen was recompacted in the specimen container to 108 mm (4.25 in.) diameter by 32 mm (1.25 in.) high. The specimens were tested in accordance with ASTM D 2435. Two-way drainage (top and bottom of the specimen) was allowed. Each load was held on the specimen for minimum of 24 h, since pore pressures could not be measured before the next load was applied. If holidays or weekends interfered, the next load was placed as soon after the 24-h period as practical.

Results and Discussion

Two soil types, a Cl (inorganic clay of low to medium plasticity) and a CH, (inorganic clay of high plasticity) were tested. The specimens were remolded to near their Proctor laboratory maximum dry unit weight and optimum water content. Table 2 gives the conditions at which the various specimens were placed, the loading rates of the CRL tests, and the type of test run on each specimen.

At least two specimens from each sample were tested using each method and the void ratio versus log effective stress plots were compared. Three specimens of the CH material were tested in the CRL apparatus at two different loading rates, as a preliminary indication of any possible effects of loading rate on void ratio versus log effective stress results.

TABLE 2—*Test conditions.*

Specimen	Classification	Type of Test	Placement Void Ratio	Placement Dry Unit Weight, kN/m^3	Placement Moisture Content	Loading Rate, kPa/h
1A	CL	CRL	0.580	16.49	15.8	55.2
1B	CL	CRL	0.581	16.49	15.8	55.2
1C	CL	STD	0.594	16.37	15.8	...
1D	CL	STD	0.588	16.43	15.8	...
2A	CH	CRL	0.887	14.14	15.0	34.5
2B	CH	CRL	0.887	14.14	15.0	20.7
2C	CH	CRL	0.887	14.14	15.0	20.7
2D	CH	STD	0.952	13.67	15.0	...
2E	CH	STD	0.896	14.08	15.0	...

All the test specimens showed expansive characteristics as they were saturated. However, the normal stress on individual specimens were increased to keep them from expanding past their initial height. After the saturation process was completed, the loading cycle was started.

Figures 6 through 9 show time consolidation curves for the specimens tested in the STD apparatus. Generally, the end of primary consolidation for higher load applications did not occur until after 10 000 s. Also, for the CH material at higher load applications, the end of primary consolidation may not have been reached, according to the plots, even at 80 000 s.

Figures 10 and 11 show coefficient of consolidation versus logarithm of effective stress plots for the CL and CH materials, respectively. In Fig. 10, the CRL results compare relatively well, whereas the results from the STD method vary considerably. Additionally, only a few test loads (the highest three or four depending on the test) from the STD method produced results that could be analyzed properly using classical techniques.

Figure 11 shows the aforementioned plots for the CH material. The CRL results compare well, as do the STD results. Only a few test loads (the highest three) from the STD method produced results that could be analyzed properly using classical techniques, but results obtained were almost identical with those from the CRL test method.

Table 3 gives the compression index (Cc) and the rebound index (Crb, sometimes used to approximate the recompression index) for each soil specimen. As can be seen from the table, there is relatively good agreement between test results for the Cc index, with the STD results being somewhat larger overall. However, these differences are not large, and may not be significant.

There is considerable difference between Crb results from the STD and the CRL test methods. These differences are most likely due to the test methods used. The reasons for and significance of these differences should be investigated.

All specimens were loaded to approximately 1380 kPa (200 psi) and then unloaded to obtain rebound curves. Figure 12 shows approximately how much

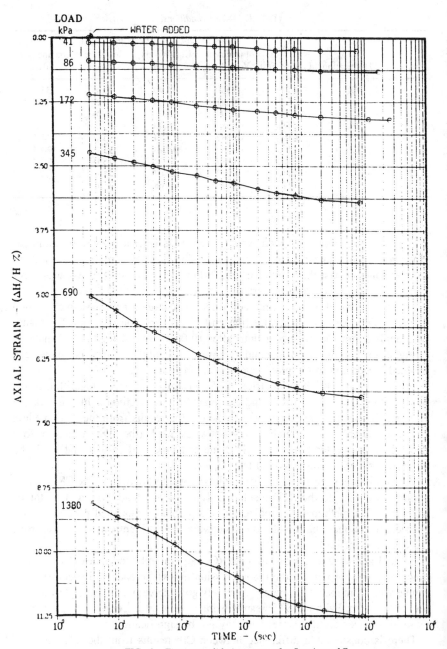

FIG. 6—*Time consolidation curves for Specimen 1C.*

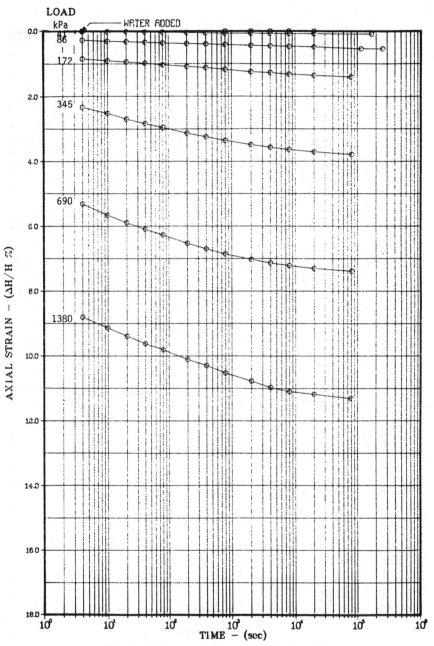

FIG. 7—*Time consolidation curves for Specimen 1D.*

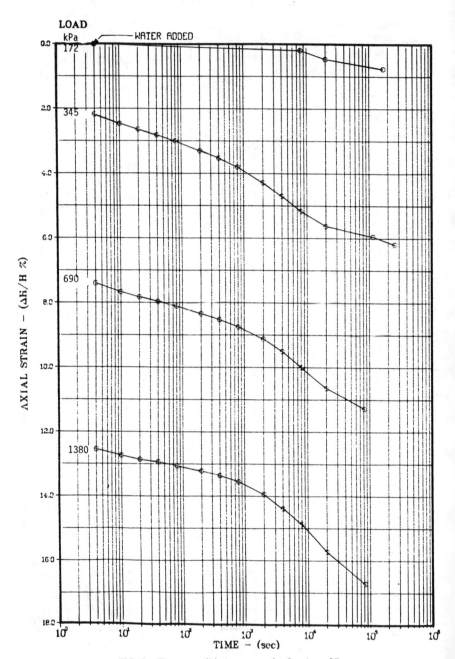

FIG. 8—*Time consolidation curves for Specimen 2D.*

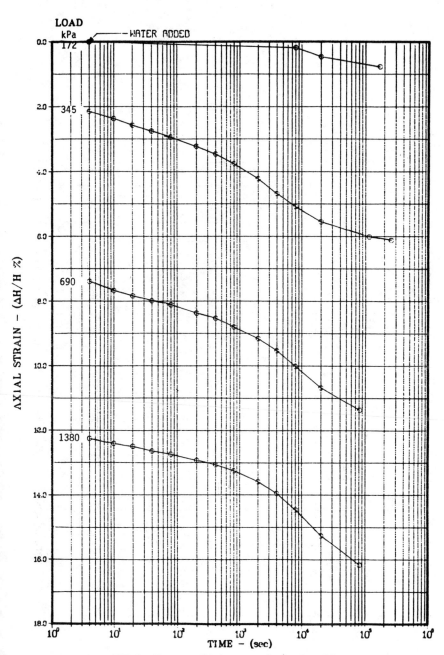

FIG. 9—*Time consolidation curves for Specimen 2E.*

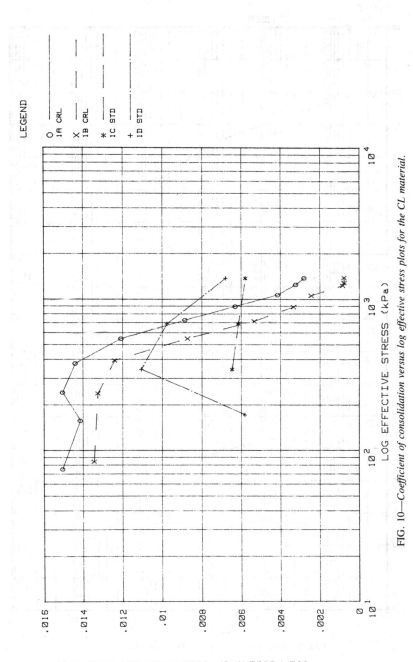

FIG. 10—Coefficient of consolidation versus log effective stress plots for the CL material.

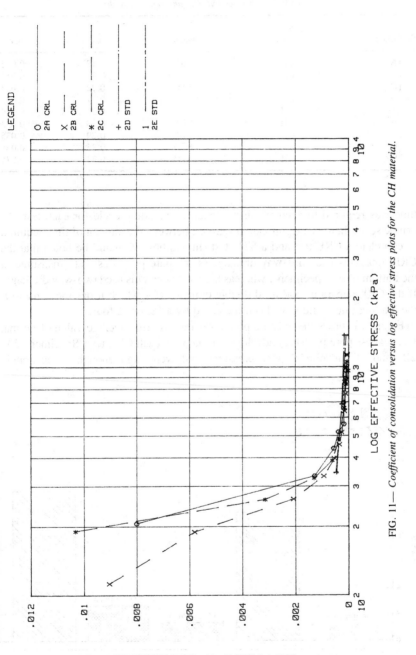

FIG. 11— *Coefficient of consolidation versus log effective stress plots for the CH material.*

TABLE 3—*Compression (Cc) and rebound (Crb).*

Specimen	Soil Classification	Test Method	Cc	Crb
1A	CL	CRL	0.18	0.012
1B	CL	CRL	0.21	0.013
1C	CL	STD	0.22	0.022
1D	CL	STD	0.21	0.022
2A	CH	CRL	0.27	0.029
2B	CH	CRL	0.28	0.033
2C	CH	CRL	0.29	0.035
2D	CH	STD	0.35	0.065
2E	CH	STD	0.33	0.075

time was required to accomplish the loading/unloading cycle for each test. As can be seen, there is a significant difference between the amount of time required to complete a CRL test and a STD test. In addition, it should be noted that the CRL apparatus used one-way drainage (since pore pressures were measured at the bottom of the specimen), whereas the STD apparatus used two-way drainage. If the CRL apparatus had used double drainage (for which it has the capacity), the loading rate could have been increased by a factor of four.

Figure 13 shows the average pore pressure ratio (the average value of normal load/excess pore pressure) developed throughout each CRL test. Specimens 2A, 2B, and 2C developed pore pressures that were high enough to accurately

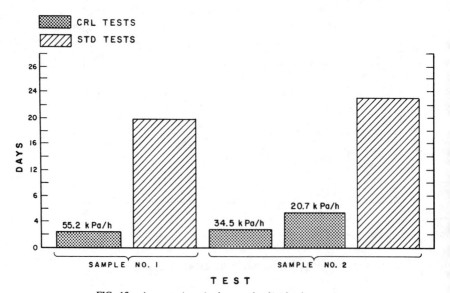

FIG. 12—*Average time, in days, to load/unload a specimen.*

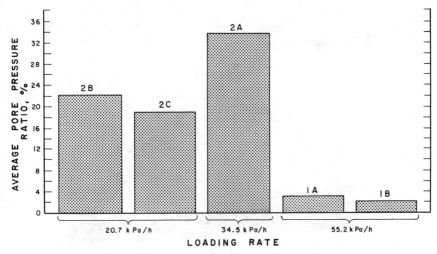

FIG. 13—*Average pore pressure ratio developed during a test.*

measure, but not high enough to disrupt the effective stress distribution throughout the sample. Other investigators have produced satisfactory test results with pore pressure ratios up to about 50% [3,4,6,11, and Appendix]. The excess pore pressures developed while testing Specimen 1A and 1B at the loading rate of 55.2 kPa (8 psi) per hour, were not very high. These specimens could not have been tested at a much faster stress rate and the tests completed even sooner.

Figure 14a compares the void ratio versus log effective stress curves generated from the results of each test method for Sample 1. The shapes of the curves and the overall range of values are very similar. The CRL test results, however, are somewhat lower than the results from the STD test, which may be due to the slightly higher density (lower void ratio) the CRL specimens (Table 2).

Figure 14b compares the void ratio versus log effective stress curves generated from the results of each method for Sample 2. The variation of results from the STD and CRL tests is mostly due to the different placement conditions (Table 2) of the specimens. The range of results from the CRL tests is smaller and is likely due to the specimen being placed at almost identical conditions. Because of the physical limitations of the CRL apparatus, Specimens 2A, 2B, and 2C could only be loaded to an effective stress of approximately 1100 kPa (160 psi), and Fig. 14 reflects this. However, since the pore pressures generated during the test of Specimens 1A and 1B were small, it was possible to load these specimens to an effective stress of approximately 1380 kPa (200 psi).

Figures 14a and 14b show that the variation between results of tests on the same material using different test methods is not large and is within reason. With this in mind, and considering the reduced time to complete a CRL test, the CRL

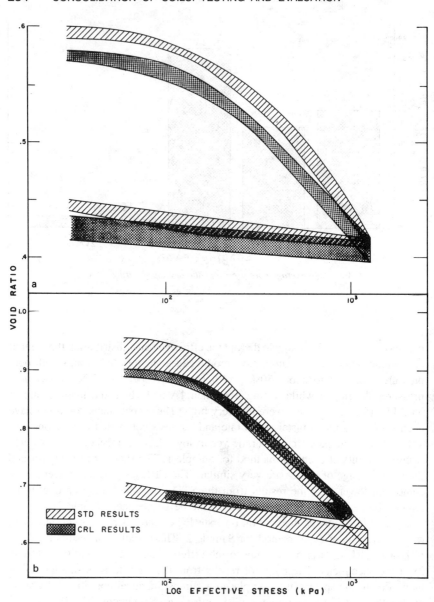

FIG. 14—*Comparison of range of results.*

apparatus and test procedure described in this report appear to offer an attractive alternative to the STD test.

Conclusions

There are significant differences in the time required to complete the consolidation test, depending on the type of test and apparatus used. The time required to complete a STD test is from two to five weeks, depending on the number of load increments and soil type.

The completion time for the CRL test depends on the selected stress rate and the compressibility and permeability of the soil. The CRL tests reported herein required approximately two to five days to complete, offering a great time savings. Additionally, if the CRL apparatus had used double drainage (for which it has the capacity), test completion times would have been even shorter.

The significance of and reasons for the difference between Crb values from the CRL and STD test methods should be evaluated.

Since agreement between the fundamental results of the CRL and the STD tests were within reasonable limits, the CRL apparatus and test procedure described in this report offer an alternative to the STD test.

Acknowledgments

Acknowledgment is due R. Butts, B. Wilson, and D. H. Novotny for their assistance in data acquisition and presentation, and W. Lambert for taking the photographs. Recognition is given H. L. Hoff and S. N. Gavlick for their work writing the computer programs that were used to reduce the data.

APPENDIX

Selection of a Loading Rate

When considering the selection of a loading rate to obtain good results with the CRL apparatus, it is important that the rate selected generate pore pressures of an appropriate magnitude. If the rate of loading is too high, the pore pressure at the impermeable base becomes so large that the proper distribution of the effective stress in the specimen is not maintained. If, on the other hand, the loading rate is too low, the excess pore pressure generated at the impermeable base of the sample may be too small to measure accurately, leading to errors in the results.

Gorman et al [11] suggested that, for the CRS (constant rate of strain test), the pore pressure should be at least 7 kPa (1 psi) and should not exceed 30 to 50% of the applied stress any time during the test. Smith and Wahls [3] obtained good results with their CRS tests until the pore pressures exceeded 50% of the applied load. Wissa [4] reported good results in his paper on the CRS test if pore pressures were in the range of 2 to 5% of the applied load. In their discussion of the CRL test, Aboshi et al [6] stated that the pore pressures should be large enough to accurately measure, but not so large that they disrupt the uniformity of the effective stresses in the specimen.

These criteria give a fair amount of latitude in the selection of a satisfactory loading rate for the CRL test. Figure 15, obtained from our experience with the CRL test, shows one means of selecting an appropriate loading rate. Test data obtained for this report show good comparison between CRL and STD test results, with pore pressures between about 3 and 34%, and it appears, from the references cited above, that ratios as high as 50% may produce acceptable results.

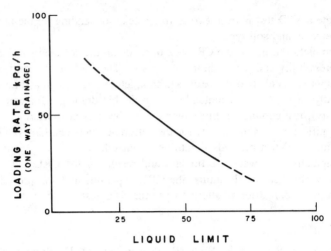

FIG. 15—*Loading rate versus liquid limit.*

References

[1] Lowe, J., III, Jones, E., and Obrician, V., "Controlled Gradient Consolidation Test," *Journal of the Soil Mechanics and Foundations Division*, American Society of Civil Engineers, Vol. 95. No. SM1, Proceedings Paper 6327, Jan. 1969, pp. 77–97.

[2] Hamilton, J. J. and Crawford, C. B., "Improved Determination of Preconsolidation Pressure of a Sensitive Clay" in *Papers on Soils—1959 Meetings ASTM STP 254*, American Society for Testing and Materials, Philadelphia, 1959, pp. 254–271.

[3] Smith, R. E. and Wahls, H. E., "Consolidation Under Constant Rates of Strain," *Journal of the Soil Mechanics and Foundations Division*, American Society of Civil Engineers, Vol. 95, No. SM2, Proceedings Paper 6452, March 1969, pp. 519–539.

[4] Wissa, A. E. Z., Christian, J. T., Davis, E. H., and Heiberg, S., "Analysis of Consolidation at Constant Strain Rate," *Journal of the Soil Mechanics and Foundations Division*, American Society of Civil Engineers, Vol. 97, No. SM10, Proceedings Paper 8447, Oct. 1971, pp. 1393–1413.

[5] Imai, G.,"Development of a New Consolidation Test Procedure Using Seepage Force," *Soils and Foundations*, Vol. 19, No. 3, Sept. 1979, pp. 45–60.

[6] Aboshi, H., Yoshikuni, H., and Mariyama, S., "Constant Loading Rate Consolidation Test," *Soils and Foundations*, Vol. 10, No. 1 1970, pp. 43–56.

[7] Schiffman, R. L., "Consolidation of Soil Under Time Dependent Loading and Varying Permeability," *Proceedings*, 37th Annual Meeting, Highway Research Board, Vol. 37, Washington, DC, 1958, pp. 584–618.

[8] Skempton, A. W., "The Pore Pressure Coefficients A and B," *Geotechnique*, Vol. 4, 1954, pp. 143–147.

[9] Lambe, T. W. and Whitman, R. V., *Soil Mechanics*, Wiley, New York, 1969.

[10] *Laboratory Classification of Soils*, Earth Sciences Training Manual No. 4, Bureau of Reclamation, E&R Center, Denver, CO, Dec. 1977.

[11] Gorman, C. T., Hopkins, T. C., Deen, R. C., and Drnevich,V. P., "Constant-Rate-of-Strain and Controlled-Gradient Consolidation Testing," *Geotechnical Testing Journal*, Vol. 1, No. 1, March 1978, pp. 3–15.

Leonardo Zeevaert[1]

Consolidation in the Intergranular Viscosity of Highly Compressible Soils

REFERENCE: Zeevaert, L., "Consolidation in the Intergranular Viscosity of Highly Compressible Soils," *Consolidation of Soils: Testing and Evaluation, ASTM STP 892*, R. N. Yong and F. C. Townsend, Eds., American Society for Testing and Materials, Philadelphia, 1986, pp. 257–281.

ABSTRACT: The author has been interested for several decades in finding better correlations, from the practical engineering point of view, between theory and the phenomenological behavior of highly compressible soils as observed in oedometer tests and in the field.

The author has developed a theory on secondary compression or intergranular viscosity in the consolidation process as he has gotten more experience in the field. The theory is based on two rheological models: the Terzaghi model and the Z-unit developed by the author. The theory has been revised, and currently offers good correlations in the interpretation of the phenomenon observed in practice.

The theory calls for the ability to develop fitting methods to obtain the parameters necessary for the formulae. Two fitting methods are used to cover the consolidation curves observed in the tests performed in the oedometer with undisturbed specimens and in the range of the recompression close to the critical stress. The method proposed also may be used after the critical stress zone has been passed and the soil structure is stabilized at a new stress level. Several examples of fitting calculations are given in the paper to acquaint the reader with the method for practical use.

The author discusses problems with the oedometer conventional test, and ends with recommendations and an analysis of the material presented.

KEY WORDS: consolidation of soils, secondary compression of soils, oedometer test, consolidation parameters of soils

In foundation engineering one encounters many cases where foundations may have to be built on soil sediments of high and very high compressibility. These sediments may be encountered in marginal lagoons, silty peat deposits, lacustrine areas, and marine environments. The mechanical behavior of these soils show high *intergranular viscosity,* commonly known as secondary compression. The subsoil in these areas shows usually highly stratified conditions. Compressible sediments are interbedded with fine sand strata, permitting a fast primary compression. The so-called secondary compression, therefore, becomes a primary phenomenon that the foundation engineer cannot overlook [1,2]. These conditions

[1] Professor, Graduate School at the Faculty of Engineering, Universidad Nacional Autonoma de México, México, D.F., México.

are encountered in Mexico City clay, where the author has worked for several decades trying to better understand the soil deformation behavior observed in the field and laboratory. This paper contains the latest findings of the author in his attempt to forecast settlements more accurately from the results of the oedometer test [3–8].

The phenomenological behavior of very fine and highly compressible soils obtained in oedometer tests needs to be understood by the engineers who find practical application for the data collected from these tests. The author has established a simple theory based on Terzaghi's rheological model for primary compression and a rheological model devised by the author representing the intergranular viscosity [3]. The purpose of these two models is to correlate the oedometer soils tests with field conditions when the horizontal displacements are restricted.

The application of the theory here proposed is explained using tests on typical Mexico City undisturbed silty clay specimens taken with 12.7 cm (5 in.) diameter Shelby tubes. The specimens used in the oedometer were taken 7.62 cm (3 in.) from the core of the 12.7 cm (5 in.) diameter Shelby tube specimens. The results of oedometer tests on this type of soil give the consolidation curves shown in Figs. 2 to 5.

A typical compressibility curve is shown in Fig. 1 and plotted in arithmetic scales. The flat portion of the curve represents the recompression of the soil up to the break taking place at a stress level σ_b, which the author refers to as the *critical confining stress* or simply the *critical stress*. At this stress level, the soil structure suffers a collapse followed by a new structural behavior [9]. The critical stress σ_b is usually located to the right of the overburden effective stress σ_o, thus giving the soil the characteristics of a pre-consolidated type soil. The author has

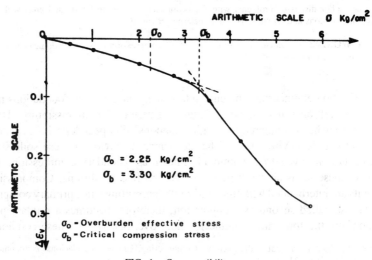

FIG. 1—*Compressibility curve.*

found that, in places where the reduction of the piezometric water levels have increased the overburden effective stress, a corresponding increase of the critical stress is observed, remaining always σ_b over the overburden effective stress. This interesting phenomenon takes place because of the development of higher bonds in the clay structure given by the active clay mineral grains as they are forced to become closer together at a very slow rate.

In sensitive soils the compressibility behavior after the critical stress is not used to avoid strong compressions. Therefore, for practical engineering purposes only the compressional behavior of the clay during recompression, corresponding to the flat portion of the compressibility curve and up to the overburden effective stress (Fig. 1), need be investigated. In deep excavations large elastic heave is observed, and settlement takes place upon load reapplication. The vertical displacements should be forecasted [9,10].

The typical configuration of the consolidation curves obtained in oedometer tests show the following characteristics: For low stress levels of recompression the consolidation curves take the configuration shown in Fig. 2 and referred to as Type I. For higher stress levels on the recompression branch of the compressibility curve and close to the critical stress σ_b (Fig. 1), the configuration of the consolidation curves is as shown in Fig. 3. The secondary or intergranular viscosity compression becomes more evident and assumes a convex configuration after the primary compression is completed. This configuration we call curve Type II (Fig. 3).

For a stress level very close to the critical stress, the configuration of the consolidation curve shows no break merging into a straight logarithmic line, as shown in Fig. 4; this is Type III. After the critical stress, a new clay structure is forced to form under the confined conditions giving a typical configuration curve, Type IV (Fig. 5).

The most common consolidation curves in the field in highly compressible soils for stress levels before the critical stress correspond to configurations of

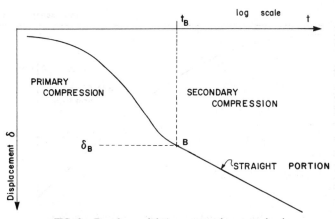

FIG. 2—*Type I consolidation curve at low stress level.*

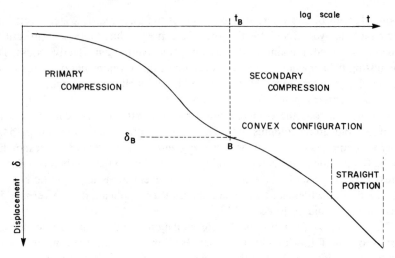

FIG. 3—*Type II consolidation curve before critical stress level.*

Type II, as may be seen in Figs. 6, 7, and 8 for three highly stratified subsoil sites where settlements were recorded. The cases in Figs. 6 and 7 correspond to compensated foundations (see Ref *8*, Chapter 7), and those in Fig. 8 to earth fill. The critical stress in these cases was higher than the overburden effective stress at any depth.

Thus may be seen the importance of developing a theory from the practical engineering point of view to correlate the phenomenological behavior observed in the oedometer with that observed in the field. The parameters should be determined in oedometer tests, and adjusted for application under field state of stress conditions to forecast time settlements [*10*].

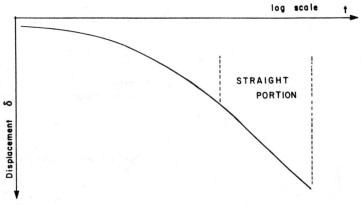

FIG. 4—*Type III consolidation curve on the critical stress zone.*

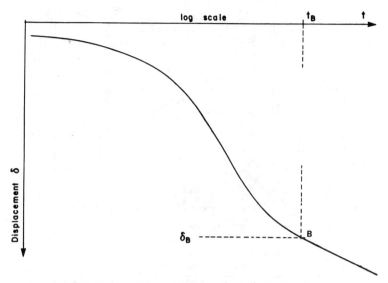

FIG. 5—*Type IV consolidation curve after critical stress.*

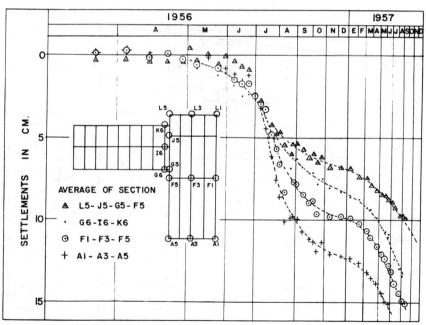

FIG. 6—*Settlement of a compensated foundation on highly stratified silty clay deposit in Mexico City.*

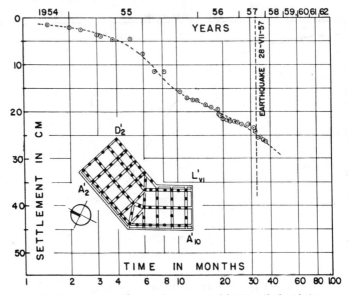

FIG. 7—*Average settlement of compensated friction pile foundation.*

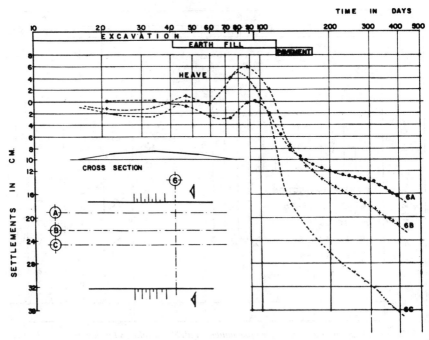

FIG. 8—*Settlement of an embankment.*

Theoretical Considerations

The theory presented is based on two rheological models: the Terzaghi model, and a model developed by the author called the Z-Unit [10], both of which attempt to represent soil behavior under confined conditions. The working hypotheses are:

1. The soil is formed of two structures with different rheological properties, one representing the primary structure and the other the secondary structure (Fig. 9).

2. The primary structure is built of coarser grains forming a continuous skeleton structure capable of taking effective stresses. The volumetric strain of the structure so conceived has the tendency to end upon dissipation of the pore water pressure. Terzaghi's model is assumed to apply. The large pores of the primary structure are saturated with air-free gravitational water [11], and volumetric strain behavior is only of elasto-plastic nature for an applied stress increment.

3. The secondary structure is formed by very fine and ultra-fine soil, forming clusters between the large grains and furring them continuously throughout the primary structure (Fig. 9). The secondary structure so visualized has pores filled with water of different viscosity, like the one assumed for the primary structure. This water has to be drained away from the pores because of the applied pressure during the process of consolidation. Since the secondary structure is built mainly of clay mineral grains, the relative displacement between them is considered to be a phenomenon of high viscous characteristics due to the adsorbed water films surrounding the clay mineral grains [12].

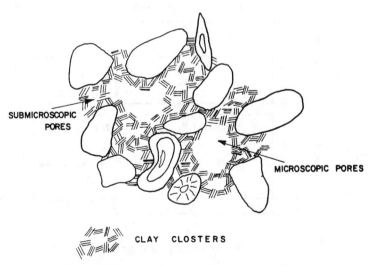

FIG. 9—*Artistic concept of clay structure.*

From these assumptions we can conclude that the total volumetric strain of the soil is the addition of the primary $\Delta\epsilon_{v1}$ and the secondary $\Delta\epsilon_{v2}$ volumetric strains

$$\Delta\epsilon_v = \Delta\epsilon_{v1} + \Delta\epsilon_{v2} \tag{1}$$

In the theory we will analyze the Kelvin unit and the Z-unit in series (Fig. 10). The Kelvin unit contains a resisting element in parallel with a dashpot of linear fluidity ϕ_1, representing the pore water fluidity in the primary structure.

The Z-unit is formed of a highly viscous element increasing its viscosity with time in parallel with a dashpot of linear fluidity ϕ_2. The highly viscous element represents the compression of the clay mineral grain clusters upon shear stresses applied to them. The dashpot represents the hydrodynamic retardation of the volumetric strain $\Delta\epsilon_{v2}$ due to the linear fluidity ϕ_2 of the water in the secondary structure pore space.

The action of these two units requires that the pressure in the dashpot of the Z-unit always be larger or equal to the pressure in the dashpot of the Kelvin unit, in order that the water of the clay clusters in the secondary structure may flow to the drainage surfaces during the primary compression. The two models are analyzed separately, based on the working hypotheses just described.

The Kelvin Unit

In this unit we establish the following relations under an applied sustained increment of unit load Δp at a certain stress level (Fig. 10):

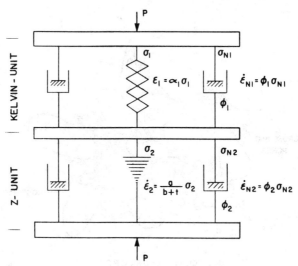

FIG. 10—*Rheological model.*

for static equilibrium: $\Delta p = \Delta \sigma_1 + \Delta \sigma_{N1}$ (2)

for volumetric strains: $\Delta \epsilon_{v1} = \Delta \epsilon_1 = \Delta \epsilon_{N1}$ (3)

stress-strain condition of the resisting element: $\Delta \epsilon_1 = \alpha_1 \Delta \sigma_1$ (4)

where α_1 represents a secant parameter for the stress-strain behavior. For the dashpot we consider the Newtonian liquid with constant fluidity ϕ_1:

$$\Delta \dot{\epsilon}_{N1} = \phi_1 \sigma_{N1} \qquad (5)$$

From these equations we obtain the following differential equation:

$$\Delta \dot{\epsilon}_{v1} + \frac{\phi_1}{\alpha_1} \Delta \epsilon_{v1} = \phi_1 \Delta p$$

Upon integration we obtain

$$\Delta \epsilon_{v1} = \Delta p \cdot \alpha_1 (1 - e^{-\phi_1 t / \alpha_1}) \qquad (6)$$

When a large number of units are considered we write

$$\Delta \epsilon_{v1} = \alpha \cdot \Delta p (1 - \Sigma \frac{\alpha_1}{\alpha} e^{-\phi_1 t / \alpha_1}) \qquad (7)$$

in which $\Sigma \alpha_1 = \alpha$ represents the compressibility of the soil.

On the other hand, the average degree of consolidation from Terzaghi's theory reads

$$\Delta \epsilon_{v1} = m_v \Delta_p (1 - \Sigma \frac{2}{M} e^{-MT_v}) \qquad (8)$$

in which $T_v = c_v t / H^2$ is the primary time factor, and $M = (2m - 1)^2 \pi^2 / 4$. When comparing Eqs 7 and 8 we obtain

$$\alpha = m_v, \quad \frac{\alpha_1}{\alpha} = \frac{2}{M}, \text{ and } \frac{\phi_1}{\alpha_1} = M \frac{c_v}{H^2} \qquad (9)$$

Equations 7 and 8 represent the same phenomenon and they correlate with Eq 9. Hence, we can write for the primary structure volumetric strain,

$$\Delta \epsilon_{v1} = (m_v \Delta p) \cdot F (T_v) \qquad (10)$$

in which F (T_v) is the Terzaghi's time function retarding the deformation by the hydrodynamic process of consolidation.

The Z-Unit

In this unit we establish the following conditions (Fig. 10). For the equilibrium of the elements

$$\Delta p = \Delta\sigma_2 + \Delta\sigma_{N2} \tag{11}$$

The volumetric strains

$$\Delta\epsilon_{v2} = \Delta\epsilon_2 = \Delta\epsilon_{N2} \tag{12}$$

and the stress-strain conditions [3] of the Newtonian liquids with nonlinear and linear fluidities respectively are

$$\Delta\dot{\epsilon}_2 = \frac{a}{b + t} \Delta\sigma_2$$
$$\Delta\dot{\epsilon}_{N2} = \phi_2 \Delta\sigma_{N2} \tag{13}$$

In the equations in 13, a, b, and ϕ_2 are constant parameters and t is the time element. Combining above mentioned equations we obtain the differential equation for the Z-unit

$$\Delta\dot{\epsilon}_{v2} = \frac{a}{b + a/\phi_2 + t} \cdot \Delta p \tag{14}$$

Under a sustained increment of unit load Δp and after integration of Eq 14 we obtain

$$\Delta\epsilon_{v2} = a\Delta p \cdot \ln \frac{b + a/\phi_2 + t}{b + a/\phi_2} \tag{15}$$

Furthermore, the pressure taken by the fluid in the dashpot

$$\Delta\sigma_{N2} = \frac{1}{\phi_2} \Delta\dot{\epsilon}_{v2}$$

Therefore

$$\Delta\sigma_{N2} = \frac{a/\phi_2}{b + a/\phi_2 + t} \cdot \Delta p \tag{16}$$

and for $t = 0$, $\Delta\sigma_{N2} = \Delta p$, obtaining $b = 0$ and

$$\Delta\epsilon_{v2} = 2.31a\Delta p \, \log \left(1 + \frac{\phi_2}{a} \cdot t\right) \tag{17}$$

Considering the sum of all the Z-units we may write Eq 17 in the following convenient form:

$$\epsilon_{v2} = 2.3\bar{a}\Delta p \, \log \left\{1 + \frac{\bar{\phi}_2}{\bar{a}} \frac{\alpha}{\phi_1} \left(\frac{\phi_1}{\alpha} \cdot t\right)\right\} \tag{18}$$

From Terzaghi-Kelvin's correlations we find

$$\alpha = m_v \text{ and } \frac{\phi_1}{\alpha} = 2c_v/H^2$$

and calling $2.3\bar{a} = m_t$, by substitution in Eq 18 we obtain

$$\Delta\epsilon_{v2} = m_t\Delta p \, \log \left\{1 + 4.61 \frac{m_v}{m_t} \frac{\bar{\phi}_2}{\phi_1}(c_v t/H^2)\right\}$$

in which $c_v t/H^2 = T_v$ is the Terzaghi's primary time factor, and calling $m_t/m_v = \beta$, then

$$\Delta\epsilon_{v2} = m_t \cdot \Delta p \, \log \left\{1 + \frac{4.61}{\beta} \frac{\bar{\phi}_2}{\phi_1} T_v\right\} \tag{19}$$

Equation 19 represents the volumetric strain due to the intergranular viscosity phenomenon. We consider β, $\bar{\phi}_2$ and ϕ_1 constant average values for a specific stress level $(p + \Delta p/2)$, and call

$$\frac{4.61}{\beta} \cdot \frac{\bar{\phi}_2}{\phi_1} = \xi \tag{20}$$

The value of ξ is a dimensionless parameter, and may be determined from the consolidation curves obtained from the oedometer test. According to Eq 1 we write in dimensionless form:

$$\frac{\Delta\epsilon_v}{m_v \cdot \Delta p} = \Omega \text{ and } \Omega = F(T_v) + \beta \, \log (1 + \xi T_v) \tag{21}$$

Equation 21 represents the volumetric strain-time behavior in the case of the oedometer test and approximately similar conditions of lateral restraint in the field. Nevertheless, when this condition is not met in the triaxial test or in the field for small loaded areas in comparison with the thickness of the clay stratum, then one should add to Eq 21 a time term representing a *steady creep* [10].

The configuration Ω of the theoretical consolidation curves expressed by the dimensionless Eq 21 for different values of ξ may be found plotted for a value of $\beta = 0.6$ in Fig. 11. Similar curves may be obtained for other values of β. We may notice from them a limiting value of ξ for which the curves show a straight logarithmic behavior after the break in the consolidation curve obtaining a Type I curve (Fig. 2). The limiting theoretical value of ξ is found to be 5. The Type I curves will conform to the theoretical equation

$$\Omega = F(T_v) + \beta \log (1 + 5T_v) \tag{22}$$

Notice in Fig. 11 that the intergranular viscosity phenomenon becomes less important during the primary compression for values of $\xi < 5$. However, after the primary compression has taken place, this phenomenon becomes more evident on a long time basis as the value of ξ becomes smaller. To further visualize the configuration of the consolidation curves, we find curves plotted with $\xi = 0.5$ for different values of β in Fig. 12, where the phenomenon of intergranular viscosity can be readily recognized after the effect of the primary compression has taken place.

Fitting Methods

Type I Curves

In terms of δ versus t the theoretical equation for the oedometer consolidation curves may be written according to Eq 22 as follows:

$$\delta = \delta_v F(T_v) + C_t \log (1 + 5T_v) \tag{23}$$

in which δ_v, C_t, and the coefficient of consolidation c_v are determined at every stress level and increment of pressure Δp.

To explain the fitting method, we use a typical Type I laboratory consolidation curve. We first determine the value of C_t, selecting two points for large values of t on the logarithmic straight line portion of the curve (Fig. 13); hence

$$\delta_2 - \delta_1 = C_t \log T_{v2}/T_{v1}$$

Taking one cycle in the logarithmic scale log $T_{v2}/T_{v1} = 1$, we obtain $C_t = (\delta_2 - \delta_1)$. The value of δ_v is determined by selecting on the consolidation curve a Point B immediately after the break with coordinates δ_B, t_B, at which we may assume $F(T_v) \cong 1$ and $T_v \cong 2$ (Fig. 12). Therefore

$$\delta_B = \delta_v + C_t \log (1 + 5 \cdot 2)$$

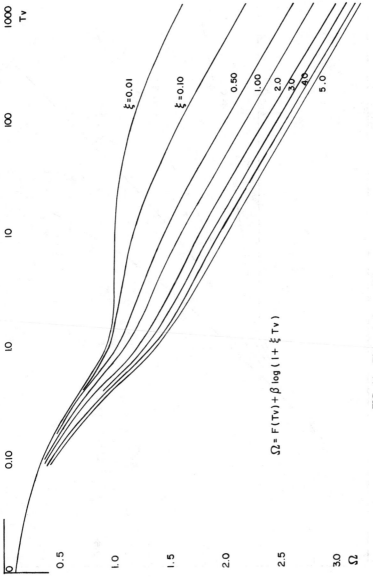

FIG. 11—*Theoretical consolidation curves for* $\beta = 0.6$.

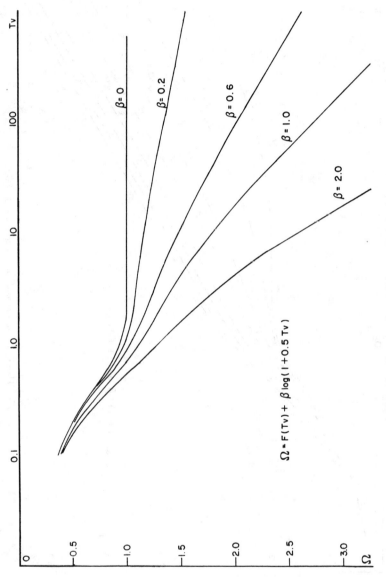

FIG. 12—*Theoretical consolidation curves for* $\xi = 0.50$.

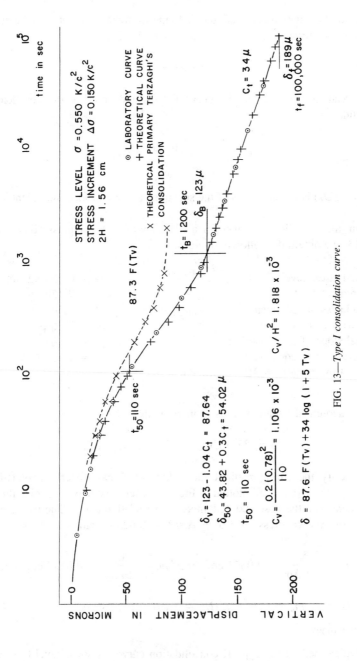

FIG. 13—*Type I consolidation curve.*

Hence $$\delta_v \cong \delta_B - 1.04C_t \qquad (24)$$

The coefficient of consolidation c_v is found from the intercept with the primary compression curve at $\delta_v/2$ for $T_v \cong 0.2$, from Eq 23,

$$\delta_{50} = \delta_v/2 + 0.3C_t \qquad (25)$$

We read the time t_{50} on the curve for δ_{50} and compute the coefficient of consolidation:

$$c_v = \frac{0.2H^2}{t_{50}}$$

in which $(2H)$ is the thickness of the specimen at the unit load increment applied Δp.

From the consolidation curve for a silty clay of high compressibility given in Fig. 13, we obtain the following values:

Initial stress level: $\sigma = 0.550$ kg/cm^2
Increment of stress: $\Delta\sigma = 0.150$ kg/cm^2
Thickness of specimen: $2H = 1.56$ cm
Primary compression: $\delta_v = 87.6$ μ
Final slope of log behavior: $C_t = 34.0$ μ
Coefficient of consolidation: $c_v = 1.106 \cdot 10^{-3}$ cm^2/s
$\mu = 1$ micrometre $= 10^{-4}$ centimetres

The theoretical equation reads in micrometres:

$$\delta = 87.6\, F(T_v) + 34 \log(1 + 5T_v) \qquad (26)$$

To verify the fitting method the values of δ have been calculated with $T_v = (1.818 \cdot 10^{-3})t$ and plotted in Fig. 13, where one may recognize the close agreement with the observed oedometer consolidation curve. The parameters of Eq 22 for average stress level $\sigma + \Delta\sigma/2 = 0.625$ kg/cm^2 are

$$m_v = \frac{\delta_v}{2H \cdot \Delta\sigma} = 0.0373 \text{ cm}^2/\text{kg}, \quad m_t = \frac{C_t}{2H\Delta\sigma} = 0.0145 \text{ cm}^2/\text{kg},$$

$$\beta = m_t/m_v = 0.39, \quad c_v/H^2 = 1.818 \cdot 10^{-3}/\text{s}$$

Type II Curves

A typical oedometer Type II consolidation curve for stress level 0.5 kg/cm^2 and stress increment of 0.50 kg/cm^2 is shown in Fig. 14. The theoretical curve reads (from Eq 23)

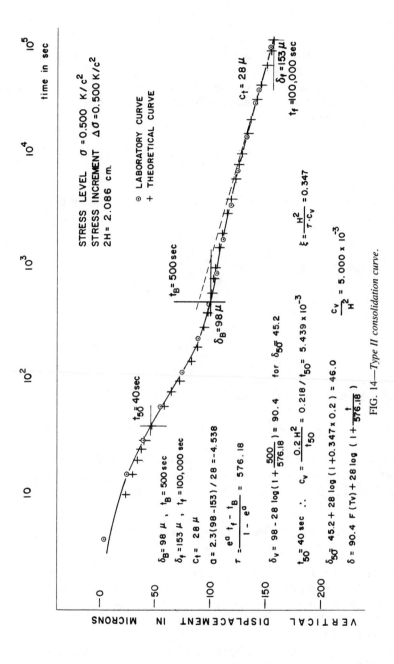

FIG. 14—*Type II consolidation curve.*

$$\delta = \delta_v \, F(T_v) + C_t \log (1 + \xi T_v) \tag{27}$$

The value of C_t is determined as previously described for curves of Type I, where the logarithmic law becomes a straight line. The value of ξ is determined selecting a Point B just after the break of the consolidation curve where $F(T_v) \cong 1$ (Fig. 12). The coordinates at this point we call δ_B, t_B (Fig. 14). Another point on the logarithmic portion of the curve is selected for the maximum observed time, with coordinates δ_F, t_F. Notice that the value $\xi \cdot c_v/H^2$ represents the inverse of a time τ; therefore

$$\frac{1}{\tau} = \xi \frac{c_v}{H^2} \text{ and } \xi = \frac{H^2}{\tau c_v} \tag{28}$$

The problem is to find the value of τ for Eq 28. We write for Points B and F, respectively,

$$\delta_B - \delta_F = C_t \log \frac{\tau + t_B}{\tau + t_F}$$

and solving for τ we obtain

$$\tau = \frac{e^a t_F - t_B}{1 - e^a} \tag{29}$$

in which $a = 2.303 \, (\delta_B - \delta_F)/C_t$.

After the value of τ is determined we calculate

$$\delta_v = \delta_B - C_t \log (1 + t_B/\tau) \tag{30}$$

With $\delta_v/2$ we find on the consolidation curve the time t_{50} as a first approximation for 50% of primary compression, and compute c_v. For an improved value of the coefficient of consolidation we use the following equation:

$$\delta_{50} = \frac{1}{2}\delta_v + C_t \log (1 + 0.2 \, H^2/\tau c_v) \tag{31}$$

With this value we determine in the curve a new value for t_{50} and consequently c_v and ξ. Substituting this value in Eq 31 we compute a new δ_{50} and find in the laboratory curve a better approximation of t_{50}. We repeat the procedure until c_v and ξ do not change substantially, therefore obtaining a final value for ξ satisfying theoretical Eq 27.

The fitting method just described applied in the laboratory consolidation curve of Fig. 14 yields the following values:

Initial stress level $= 0.5$ kg/cm^2 Stress increment $= 0.5$ kg/cm^2
$\delta_B = 98$ μ, $t_B = 500$ s $\delta_F = 153$ μ, $t_F = 100\ 000$ s
$C_t = 28$ μ, $a = -4.538$ $\tau = 576.18$ s

from which

$$\delta_v = 98 - 28 \log (1 + 500/576.18) = 90.4 \ \mu$$

From $\delta_{50} = 45.2$ μ, we obtain on the curve $t_{50} = 38$ s. The sample height at initial stress level of 0.5 kg/cm^2 is $2H = 2.086$ cm; therefore, the coefficient of consolidation has the value

$$c_v = \frac{0.2(1.043)^2}{38} = 5.73 \cdot 10^{-3} \text{ cm}^2/\text{s}, \ \xi = \frac{H^2}{c_v \tau} = 0.330$$

To obtain an improved value of c_v and ξ we enter Eq 31 and find

$$\delta_{50} = 45.2 + 28 \log (1 + 0.2 \cdot 0.330) = 46 \ \mu$$

From the laboratory curve we read $t_{50} = 40$ s and compute $c_v = 5.439 \cdot 10^{-3}$ cm^2/s and $\xi = 0.347$.

Entering $\delta_{50} = 45.2 + 28 \log (1 + 0.2 \cdot 0.347)$, we find $\delta_{50} = 46$ μ. Hence, we consider the values given above final. Therefore, the theoretical oedometer consolidation equation including the intergranular viscosity phenomenon is in micrometres:

$$\delta = 90.4 \ F(T_v) + 28 \log (1 + t/576.18) \tag{32}$$

in which $T_v = (5.0 \cdot 10^{-3})t$.

The curve is calculated with Eq 32 and the plot in Fig. 14 to compare the fitting method with the laboratory soil configuration.

The parameters for dimensionless Eq 21 will be

$$m_v = \frac{90.4 \cdot 10^{-4}}{2.086 \cdot 0.50} = 0.00867 \text{ cm}^2/\text{kg}$$

$$m_t = \frac{28 \cdot 10^{-4}}{2.086 \cdot 0.50} = 0.00268 \text{ cm}^2/\text{kg}$$

$$\beta = \frac{m_t}{m_v} = 0.309, \ \xi = \frac{H^2}{c_v \tau} = 0.347$$

Hence the dimensionless configuration (Eq 21) reads

$$\Omega = F(T_v) + 0.309 \log (1 + 0.347T_v) \tag{33}$$

The results of the method just described have been applied to a set of laboratory consolidation curves. The theoretical curves are shown plotted in Figs. 15 and 16. The fitting may be found satisfactory.

The variation of the parameters against the average stress level for which they were determined may be found plotted in Fig. 17.

Problems with Conventional Oedometer Test [13]

The soil specimen in the field is subjected to a specific state of stress, hydraulic pressure, and degree of saturation. The last-mentioned index property in organic soils may be less than 100%. When the soil is recovered from the ground and placed in the oedometer under water, the original stresses and hydraulic pressures are reduced to zero. Two effects should be considered, namely: the expansion of the gas in the pore space reducing the degree of saturation, taking place according to Mariotte's law; and the release of gas dissolved in the water when relieved from the field hydraulic water pressure, taking place according to Henry's law. The last action develops microscopic bubbles in the water, modifying the permeability of the soil as compared with the permeability for water absent of microscopic gas bubbles, also giving the water some compressibility [11].

The influence of the entrapped gas in the pore space may be recognized in the oedometer consolidation curves in the first phase of the compression (see Figs. 15 and 16). The absorption of the entrapped gas appears not to take place instantaneously in the test, and not until enough compression of the soil has taken place to reduce the void space. By the same token, the degree of saturation in the field should be known [11].

On the other hand, the content of gas in the soil implies an immediate compression upon load application [11]. Hence the basic assumptions made in the theory for primary consolidation do not apply fully in the oedometer test under the conditions stated before, and corrections should be applied.

An important error in the first phase of the primary consolidation also affecting the compressibility determined in the test is the remolded soil membrane enclosing the specimen when trimmed in the consolidation ring. This effect should be carefully investigated. Finally, one should also evaluate the friction against the ring to correct for the applied stress. The friction increases as the pore pressure reduces and becomes more important at the end of the primary phase of consolidation and thereafter.

In spite of all these problems in the conventional oedometer test, one finds in well-conducted tests approximate parameters which are useful to the engineering profession. Nevertheless, the author feels that the oedometer test technique may

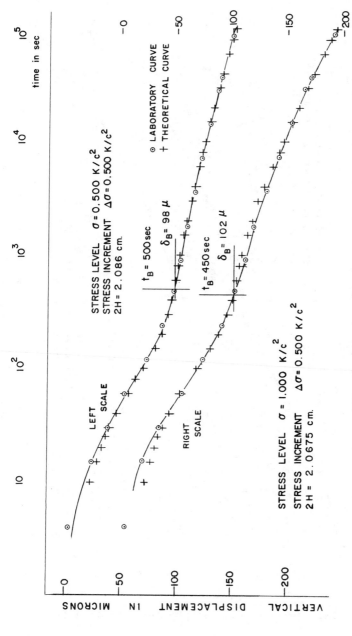

FIG. 15—*Oedometer consolidation curves.*

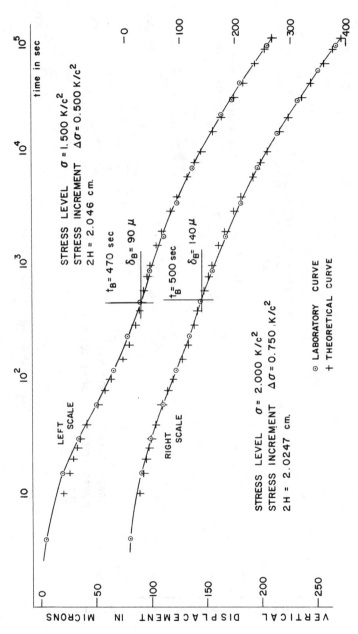

FIG. 16—Oedometer consolidation curves.

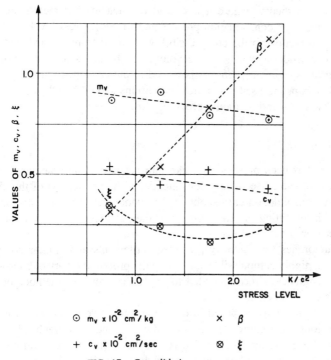

FIG. 17—*Consolidation parameters.*

be improved to obtain better results. The following recommendations may improve the oedometer tests and their interpretation:

1. Plot the consolidation curves, allowing sufficient time during the test to obtain the straight portion of the secondary compression.

2. The effect of friction against the consolidation ring may be important to consider in the final stages of the intergranular viscosity compression. It can be minimized using floating and larger diameter rings and soil repellent agents at the soil-ring interface or if possible by measuring the friction during the test.

3. The consolidation test should be done under an hydraulic pressure equal to what the soil was subjected to in the field. One should allow sufficient time under this pressure to permit the microscopic gas bubbles to dissolve in the pore water. Before applying water pressure to the oedometer, the soil should be loaded with about 20% of the overburden effective stress at which the soil was subjected in the field. This practice will avoid swelling of the clay minerals. Full consolidation should be allowed under the above-mentioned stress.

4. The increment of stresses in the oedometer test should be of equal magnitude and on the order of ⅛ or less of the effective overburden stress, in order not to damage the clay structure and to obtain average values compatible with the

assumptions in theory, where it is considered that the parameters are average values during each consolidation process and for each stress level.

5. The correction on the compressibility obtained from the oedometer test is performed considering the probable volume of the non-saturated remolded soil enclosing the specimen when this is placed in the consolidation ring.

6. All parameters used under field conditions should be adjusted to changes of effective stress and hydraulic pressure [10].

Conclusions

A theory has been presented to take into account "intergranular viscosity", commonly known as secondary consolidation. The application of this theory to laboratory consolidation curves obtained in oedometer gives satisfactory agreement. With the fitting methods proposed, the parameters ruling the theory may be obtained in a practical way.

The author has satisfactorily predicted settlements and similar compression problems in highly compressible soils using the proposed theory of consolidation. The soils engineer is recommended to apply the proposed methodology to his own soils environment.

The fitting methods proposed in this paper may give better than usual agreements on long-time compressions for sediments and stratigraphical conditions similar to those considered here. However, the oedometer tests on undisturbed specimens must be carefully conducted and the corrections mentioned performed.

References

[1] Tan, T.-K., *Secondary Time Effects and Consolidation of Clays*, Academia Sinica, Harbin, China, 1951.

[2] Wahls, H. E., (1962), "Analysis of Primary and Secondary Consolidation," *Journal of Soil Mechanics*, American Society of Civil Engineers, Vol. 88, No. SM6, 1962, pp. 207–231.

[3] Zeevaert, L., "Ecuación Completa de Consolidación para Depósitos de Arcilla que Exhiben Fuerte Compresión Secundaria," *Revista de Ingeniería*, Nos. 6–8, México, April–Aug. 1951, pp. 57–72.

[4] Zeevaert, L., "Discussion on Secondary Consolidation," in *Proceedings*, 3rd International Conference on Soil Mechanics and Foundation Engineering, Switzerland, Vol. 3, Aug. 1953, pp. 130–132.

[5] Zeevaert, L., "Consolidation of Mexico City Volcanic Clay," in *Proceedings*, Joint Meeting of ASTM and SMMS, Dec. 1957, p. 28 (Spanish and English).

[6] Zeevaert, L., "Consolidation Theory for Materials Showing Intergranular Viscosity," in *Proceedings*, 3rd Panamerican Conference on Soil Mechanics and Foundation Engineering, Vol. 1, Caracas, Venezuela, 1967, p. 89.

[7] Zeevaert, L., *Foundation Engineering for Difficult Subsoil Conditions*, 1st ed., Van Nostrand Reinhold, New York, Chapter 2, 1972.

[8] Zeevaert, L., *Foundation Engineering for Difficult Subsoil Conditions*, 2nd ed., Van Nostrand Reinhold, New York, Chapter 2, 1982, pp. 85–113.

[9] Zeevaert, L., "The Outline of a Mat Foundation Design on Mexico City Clay," in *Proceedings*, 7th Texas Conference on Soil Mechanics and Foundation Engineering, Jan. 1947, reprinted 1982.

[10] Reference 8, Chapter VII-5, pp. 300–310 and Chapter II-3, 13, pp. 85–96.

[11] Zeevaert, L., "Descompresión en Depósitos de Suelos Impermeables," División de Estudios de Posgrado, Facultad de Ingeniería, Universidad Nacional Autonoma de México, 1982 (English and Spanish).

[12] Zeevaert, L., "Viscosidad Intergranular en Suelos Finos Saturados," División de Estudios de Posgrado, Facultad de Ingeniería, Universidad Nacional Autonoma de México, 1984.

[13] Van Zelst, T. W., "An Investigation of the Factors Affecting Laboratory Consolidation of Clay," in *Proceedings*, Second International Conference on Soil Mechanics and Foundation Engineering, Vol. 7, 1981, pp. 52–61.

Pierse Ducasse,[1] *Claude Mieussens,*[1] *Michel Moreau,*[2] *and Bertrand Soyez*[2]

Oedometric Testing in the Laboratoires des Ponts et Chaussées, France

REFERENCE: Ducasse, P., Mieussens, C., Moreau, M., and Soyez, B., **"Oedometric Testing in the Laboratoires des Ponts et Chaussées, France,"** *Consolidation of Soils: Testing and Evaluation, ASTM STP 892,* R. N. Yong and F. C. Townsend, Eds., American Society for Testing and Materials, Philadelphia, 1986, pp. 282–298.

ABSTRACT: The Laboratoires des Ponts et Chaussées (LPC) of France are about to publish six new standards on oedometric testing. Two of these tests (controlled gradient test and controlled rate of strain test) are similar to those used in other countries. The other four have been under development at the LPC and other French research centers for the past 15 years, and exhibit special features which distinguish them from traditional standards used abroad. Briefly, the unique features are that: the incremental test was adapted for better accuracy and easier computerization of its data; the radial drainage test is very similar to the standard test; the creep test was simplified to run for less time; and the heat accelerated test was developed in France to obtain the compressibility curve of the soil in less than a week. The paper deals with the main characteristics of the testing process, the material developed in France, and interpretation techniques.

KEY WORDS: soils, oedometer tests, laboratory testing equipment, clay, standard

Since the publication of the works of Terzaghi [1] about 60 years ago, consolidation theory and laboratory testing techniques have been the subject of several hundred articles, theses, and papers. Among the works that have been of special value to us, we make particular mention of Taylor's research on consolidation of clays [2] and the model of the behavior of clays proposed by Bjerrum [3].

Historically, the French Laboratoires des Ponts et Chaussées (LPC) have had to deal with the problems of the compressibility of soils since 1962, when the French motorway construction program was initiated. A compressibility test procedure was published in 1965 [4] and was updated 15 years later in an internal document of the LPC network. In 1984, the new procedure proposes standards for the commonest tests, without ruling out further evolution of the techniques and the development of new procedures.

---

[1] Engineer, Laboratoire des Ponts et Chaussées de Toulouse, 1 Avenue du Colonel Loche, 31400 Toulouse, France.
[2] Engineer, Laboratoire Central des Ponts et Chaussées, 58 Boulevard Lefèbvre, 75732 Paris CEDEX 15, France.

The purpose of this paper, which is not a detailed procedure, is to make the chosen methods better known, with emphasis on the original features that distinguish them from the older, more "classical", procedures.

There are several reasons for the evolution of the testing techniques. First, a procedure is generally a compromise between the demands of the theoretical model of the behavior it is desired to observe and quantify and the normal pace of work in a laboratory. The oedometric test with a load increment every 24 h was perfectly suited to the working day in the past. However, since most laboratories now have recording systems, and even automatic test control systems, it becomes possible to take this new context into account and improve the quality of the routine tests.

At the same time, the development of numerical computation methods now makes it possible to use laws of behavior that are more complex and closer to the actual phenomena. It is possible, for example, to take into account the variation of k_v and c_v during consolidation and to introduce a creep term during the primary stage; routine laboratory tests had to be adapted to allow for the characterization of these laws.

New tests offering theoretical advantages, such as the controlled gradient test and the controlled rate of strain test, now can be used more widely thanks to the generalization of acquisition and servocontrol systems. Before it can become a routine test, however, a new procedure must be subjected to critical analysis. In a first stage, comparisons with more traditional tests are required by the utilization of the customary design methods. In a second stage, the new procedure may become the basis for the utilization of other structural design methodologies.

Finally, testing techniques, including this procedure, must be a compromise between the rigid control demanded for a research test and the economic requirements of a routine investigation.

Evolution of Testing Equipment

The LPC procedures have been designed so that the corresponding tests can be performed using the greatest possible variety of equipment; in particular, the incremental compressibility test is still often carried out using Terzaghi-type cells. For the other tests (radial drainage, controlled gradient, etc.), special apparatus have been developed in the LPC network over the past 15 years as needs dictated (see Refs 5–7). To improve the quality of the classical tests and to allow for the development of new tests, a multipurpose cell, described below, has been designed at the Toulouse Laboratory.

LPC Back-Pressure Cell

Figure 1 shows the principle of the LPC back-pressure cell in the pneumatic or hydraulic loading configuration with the press loading option. For the radial drainage test, the top porous stone is replaced by a tight disk and the bottom

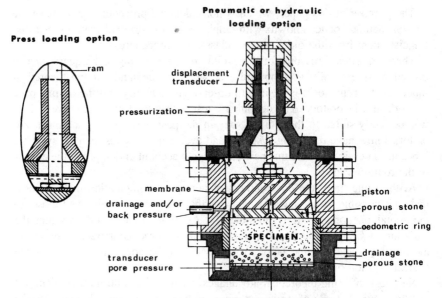

FIG. 1—*Cross-section of the LPC back-pressure cell.*

stone is limited to the dimensions of the center drain, as shown by the diagram in Fig. 6.

There are two main features of this cell:

1. The oedometric ring allows for the use of a cutting edge to cut the specimen.
2. Tightness between the specimen and the top pressurization chamber is ensured by a fabric-reinforced membrane that is tight and inextensible. As the piston is displaced, this membrane simultaneously rolls up and unrolls on the surfaces of the oedometer and piston. During a radial drainage test, the top drain port of the cell is closed. The internal geometry of the oedometer is such that, as the piston and membrane are displaced, the volume of the water chamber above the specimen remains constant.

Saturation of Specimen

It is wise to recall that Terzaghi's vertical consolidation theory assumes that the soil is saturated with an incompressible liquid. For incremental loading tests carried out with classical equipment, saturation is generally effected by immersion of the specimen or by percolation. It is well known that these procedures are inadequate when pore pressure generation measurement is wanted, as in the case of the controlled gradient test (CGT) and the controlled rate of strain test (CRST). It has also been shown that, even in the case of the incremental test, imperfect saturation may affect the results obtained. For example, the values of c_v at low load values may be from one to ten times smaller when saturation is by back-pressure than when the specimen is simply immersed. On the other hand, the compressibility characteristics (C_s and C_c) measured at the same time remain the same order of magnitude.

Saturation by Back Pressure

Currently, the most satisfactory way of saturating a soil specimen is back-pressure. It must, however, be used with care, since in some soils it may have the following drawbacks: (1) swelling of the specimen, for which an attempt will be made to compensate as much as possible; or (2) on the other hand, settlement that may go as far as structural collapse (in the case of a loess).

With most highly compressible clayey soils, a back-pressure of the same order as the *in situ* hydrostatic pressure will generally suffice. Even so, the following procedure may be used in all cases to determine the back-pressure necessary for saturation:

• A paraffin volume gage is inserted in the drainage circuit(s) connected to the back-pressure.
• The back-pressure is gradually increased and the volume of (de-aerated) water entering the soil specimen is noted.
• When the volume of water entering the specimen as the back-pressure is increased becomes negligible, saturation is regarded as satisfactory and the same back-pressure value is maintained for the duration of the test.

Once the specimen is saturated, a few hours are allowed before starting the test so as to allow the pore gases to dissolve more completely in the pore water.

Incremental Compressibility Test

The incremental compressibility test has been the standard oedometric test for many years. Because of its established position and simplicity of application, it is likely to remain the most widely used laboratory technique for investigating the compressibility of clayey soils for a long time to come. Despite its long history and the many publications dealing with it, we thought it might be worthwhile to make use of certain procedures and methods of interpretation that take fuller advantage of modern equipment and new computing techniques.

Choice of Equipment

The main advantage of the incremental compressibility test is the simplicity of the equipment that can be used. The quality of the test will be higher or lower according to the type of equipment used. We put forward the following recommendations:

• The apparatus should provide for the application of a back-pressure so as to ensure perfect saturation of the specimen.

• A system for recording settlement versus time is desirable. (We shall see that it can improve the interpretation of the test results.)

• A pneumatic or hydraulic loading system should be used whenever possible; they are more convenient than weights and allow for complete automation of the test.

Saturation

For the material and technique proposed for the control of saturation, see the section entitled "Saturation of Specimen." As for the other tests, saturation, with or without back-pressure, must be effected in such a way as to prevent any swelling. This procedure is obviously unsuitable for soils having a marked tendency to swell, for which a special test is required.

Test Procedure

Loading Law—Classically, the test consists of subjecting the specimen to a series of loads σ'_n in accordance with a loading law defined by the incremental ratio

$$i = \frac{\sigma'_{n+1} - \sigma'_n}{\sigma'_n}$$

and observing the settlement of the soil during the time each load is maintained.

The first load σ'_1 applied to the oedometric specimen should be small with respect to overburden pressure σ'_{vo}. Preference is given to 5 kPa or a nearby value according to the equipment used. This first load serves to correct specimen surface and piston insertion defects; it also gives the first point of the compressibility curve. For soils subject to swelling, this first load will be chosen equal to the swelling pressure. Up to the overburden pressure σ'_{vo}, the applied loads are chosen equal, respectively, to $\sigma'_{vo}/2$, $3\sigma'_{vo}/4$, and σ'_{vo} (Fig. 3). If the specimen is overconsolidated, loading is continued with an incremental ratio of 0.5 up to pressure σ'_p.

Since this pressure σ'_p is most often unknown at the start of the test, the curve of relative settlement versus effective applied stress $\Delta H/H$ must be plotted as the test proceeds. Preconsolidation pressure σ'_p is located in the vicinity of the maximum curvature of the curve. It is assumed that σ'_p has been passed as soon as a marked curvature appears.

At this point, the load is reduced to the first pressure σ'_1 with a pause at an intermediate load σ' such that the distance between σ' and σ'_p is approximately equal to that between σ' and σ'_1 on a logarithmic scale, then increased again along the same stress path (Fig. 3). The test is then continued with an incremental ratio of 1 up to the final load. This load incrementation procedure defines the compressibility curve more accurately in the vicinity of σ'_p without exceeding ten increments with the soils commonly encountered in France. In general, the loading program should be chosen so that there are about four increments between σ'_1 and σ'_p.

Increment Duration—Each load is applied for a period of 24 h. This duration may be shortened for the increments of the unloading-reloading cycle (generally two a day for a manual test or every 6 h for an automated test).

The duration of the loading increments is a conventional value justified by more than 50 years of common use of this type of test. It may be that this duration is too short with some relatively impermeable soils and too long for others. Nevertheless, it seems to be wiser not to change the procedure of this incremental test, of which we have a long experience, but rather turn to other types of test (such as those described in the section entitled "Controlled Gradient and Controlled Rate of Strain Tests") if we wish to make our oedometric tests shorter.

In any case, we must not change the duration of the increments in the course of a single test, since the appearance of more or less creep according to the duration of the increment would affect the shape of the compressibility curve and the value of C_c. This remark is based on the model of behavior proposed by Bjerrum, shown schematically in Fig. 2.

Classically, we propose making settlement measurements at the following times: 0, 15 s, 30 s, 1 min, 2 min, 4 min, 8 min, 15 min, 30 min, 60 min, 2 h, 4 h, 8 h, and 24 h. The important thing is to record the exact time of the settlement measurement; a recording system can make this easier.

Interpretation of Tests

Compressibility Parameters—The compressibility curve is plotted in the usual way by entering the value of the void ratio at the end of each loading increment in a $(e, \log \sigma')$ diagram. Figure 3 shows the graphic constructions by which the various soil compressibility parameters are determined:

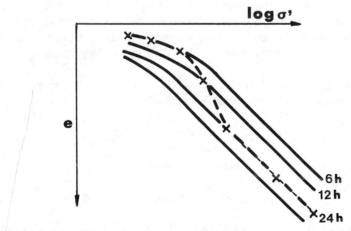

FIG. 2—*Compressibility curve during a test with an increasing duration of the loading steps, according to Bjerrum's model.*

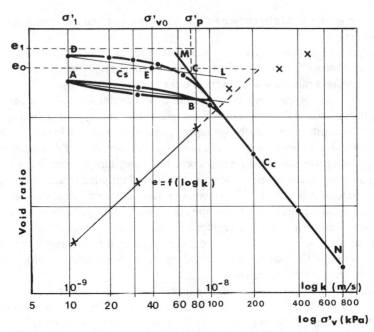

FIG. 3—*Compressibility curve* (e, *log* σ′) *and* (e, *log* k_v) *diagram. Incremental test.*

• Compression index C_c corresponds to the slope of the quasi-linear portion of the diagram for high values of σ′ (straight line *MN*).
• Recompression (or swelling) index C_s is taken as equal to the mean slope of the unloading-reloading loop (straight line *AB*).
• Parallel *DL* to *AB* through the point representing the first loading increment intersects *MN* at Point C, the abscissa of which is taken as equal to σ′$_p$, the preconsolidation pressure. The advantage of this simple method over the classical methods of Casagrande, Burmister, and Schmertmann is that it does not depend on the operator and is easy to program on a computer. On the other hand, it has the drawback of being dependent on the choice of first loading increment.

The purpose of the unloading-reloading cycle carried out in the laboratory is to simulate the one the soil has already undergone:

σ′$_p$ to σ′$_{vo}$: if the soil is overconsolidated

σ′$_{vo}$ to 0 : taking of specimen

0 to σ′$_p$: loading in laboratory

The choice of σ′$_1$ is arbitrary and has little effect on the result. It is based on convenience of construction on a semi-logarithmic diagram.

The result also depends on the load value at which the unloading-reloading cycle is begun. This is why it is preferable to carry out this cycle from a value close to preconsolidation pressure σ'_p.

Finally, it should be emphasized that while all the methods of determining σ'_p are imperfect, the result is probably affected more by such factors as the duration of the test and the choice of load increment than by the principles on which the methods are based.

In this way, from Point E (σ'_{vo}, e_o), we determine the compressibility of the soil by means of two straight lines, EC for the overconsolidated domain and CM for the normally consolidated domain. At large values of σ', in the case of soils that are highly compressible or else characterized by high quasi-preconsolidation, the compressibility curve is no longer linear, and it then becomes necessary to interpret the test in the range of stress values corresponding to the anticipated project service loadings.

Definition of a Remolding Criterion

The mean slope of the unloading-reloading cycle is also a remolding criterion. If it is very different from the slope of the compressibility curve between σ'_{vo} and σ'_p, it may be deduced that the specimen has been remolded. When the straight line passing through the first point of the curve and parallel to the unloading-reloading loop intersects the compressibility curve (Fig. 4), the specimen is said to be intact.

Settlement Versus Time Curves: Determination of c_v and k_v

It is well-known that the methods of Taylor (settlement versus $\sqrt{t}$) and Casagrande (settlement versus $\log t$) give understandably different results, especially

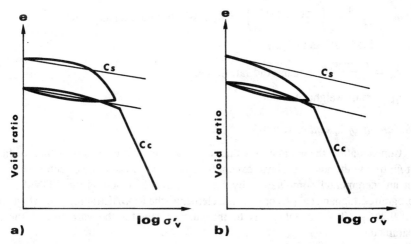

FIG. 4—*Remolding criterion for the tested specimen, using the compressibility curve; (a) Unremolded specimen. (b) Remolded specimen.*

if the soil is characterized by considerable creep (secondary compression). In the LPC procedure, we recommend using only Taylor's method, which is less affected by creep because it is based on the beginning of the settlement curve.

For the principles of this method, we refer the reader to general works on soil mechanics, in particular that of Taylor [8]. However, the proper application of this construction calls for an iterative computation (easy to program) to check that the linearized portion of the curve actually corresponds to 60% consolidation, according to Terzaghi's theory. Using Taylor's classical notation, this gives

$$\frac{d_{60} - d_s}{d_{90} - d_s} = \frac{6}{9}$$

If not, the linearization must be repeated on a different number of experimental points.

Finally, we point out that this method requires good precision, especially on the first measurements, making the use of a recording system advantageous.

The coefficients of consolidation c_v and of permeability k_v are given for each loading increment by the classical relations

$$c_v = \frac{0.848 \; d^2}{t_{90}}$$

$$k_v = c_v \, m_v \, \gamma_w$$

where

d = longest drainage path,

t_{90} = time to 90% consolidation,

$m_v = \dfrac{1}{1 + e_m}\left(\dfrac{e_n - e_{n+1}}{\sigma'_n - \sigma'_{n+1}}\right)$ (compressibility coefficient between loads σ'_n and σ'_{n+1}),

$e_m = \dfrac{e_n + e_{n+1}}{2}$ (mean void ratio), and

γ_w = unit weight of water.

Variations of k_v *and* c_v *with* σ'

Consolidation theory assumes that k_v, c_v and m_v are constants, which is far from true with most soft clays, except for small variations of the effective stress. In an incremental computation by finite differences (such as the CONMULT program, Magnan et al [9]) or by finite elements (the ROSALIE program, Magnan et al [10]), it is advantageous to introduce a law for the variation of these parameters.

The law of variation of the compressibility coefficient is already contained implicitly in the compression indexes C_s and C_c defined above.

For the variation of $k_v = f(\sigma')$ or $k = f(e)$, we suggest simply entering the couples of $(e, \log k_v)$ values on a semi-logarithmic diagram. It is generally possible to linearize and translate the experimental law of variation of k_v into the form: $e = e_k + c_k \log k_v$.

Figure 3 gives an example of this experimental law. With small values of σ', especially in the overconsolidated domain, the dispersion is rather large, probably because of the effect of remolding, but also because of the difficulty of determining c_v in this domain; these experimental points will not be used; the points representing very large values of σ', for which the experimental law ceases to be linear, and which generally correspond to loads much larger than the anticipated service loads, will also be eliminated.

From the relation $e = e_k + c_k \log k_v$, the "corrected values" of c_v can be calculated using two equations

$$c_v = \frac{k_v(1 + e)}{0.434\, C_s\, \gamma_w}\, \sigma' \text{ for } \sigma'_{vo} < \sigma' < \sigma'_p$$

$$c_v = \frac{k_v(1 + e)}{0.434\, C_c\, \gamma_w}\, \sigma' \text{ for } \sigma'_p < \sigma'$$

where e is determined as a function of σ' on the straight line segments determining the compressibility of the soil (Fig. 3), and k_v is determined from the experimental law $e = F(\log k)$.

Figure 5 gives an example of a law of variation of c_v corrected for σ'.

Heat-Accelerated Test

The heat-accelerated test was developed in France by the Centre Experimental de Recherches et d'Etudes du Bâtiment et des Travaux Publics (CEBTP) to make it possible to predict the total settlements of a structure in much less time than is required for the classical incremental test [11]. It is based on the principle that an increase in temperature has two main effects on saturated clays: (1) faster consolidation through the lowered viscosity of the "free" pore water and simultaneous increased permeability of the soil, and (2) an increased soil density at constant load that may be ascribed to the properties of the "double layer" surrounding the soil particles.

An increase in temperature has a large effect on the speed of consolidation and a very small one on the settlement of the soil. This means that a consolidation test can be accelerated without vitiating the determination of settlement amplitudes

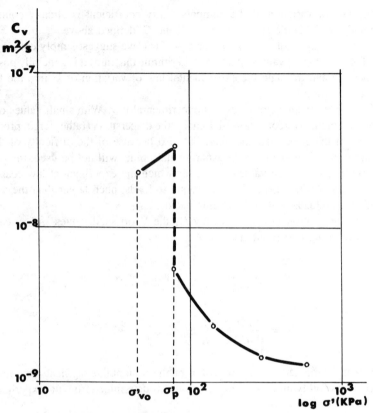

FIG. 5—*Variation of* c_v, *"corrected", versus* σ'_v.

by raising the soil to a temperature above 20°C, the usual temperature in an air-conditioned testing room.

Choice of Equipment

The heat-accelerated test has led to the development by the CEBTP of special apparatus for the running of five simultaneous tests. It consists in fact of five standard oedometric cells immersed in a bath held at a temperature between 68 and 70°C. Loading is by compressed-air cylinders, and the settlement of the specimens is continuously recorded; the system can easily detect movements of 20 μm. Loading and recording devices are fully automated. Tests of this type can be carried out in any laboratory, however, using standard oedometers placed in thermostated baths at a suitable temperature, close to 70°C.

Saturation

Saturation of the specimen before the test is unnecessary in as much as the sole initial objective of the heat-accelerated test is to determine the compressibility curve of the soil. For an investigation of the evolution of the consolidation rate, however, the specimen would have to be saturated.

Test Procedure, Loading Law

The test start-up procedure with the CEBTP equipment is as follows:

• The specimen is placed in the oedometric cell and the whole put into the tank that is to contain the bath water. The weight of the piston of the cell and of the cylinder then applies a pressure of 10.5 kPa to the specimen.

• The cylinder is immobilized to prevent any subsequent swelling of the specimen; the tank is filled with water and the thermostat set to 68°C. When the temperature has stabilized, the actually loading increments can be applied.

If traditional immersed frames are used and the soil tends to swell in the presence of water as soon as the tank is filled, loads will be applied gradually until settlement occurs.

Except for the first load, the loading law applied to the specimen will be the same used in the conventional test, at least as regards the values and staging of the stresses applied to the soil. The temperature effect will be taken into account by reducing the duration of the loading increments from 24 h to 8 h. The duration of the unloading increments, for its part, will be 4 h.

Interpreting the Test

Determining the Compressibility Characteristics of the Soil—The techniques of interpretation explained in the section entitled "Incremental Compressibility Test" are used.

Determining the Coefficients of Consolidation c_v and Permeability k_v of the Soil—Although the heat-accelerated test was originally developed only to determine soil compressibility parameters, it is possible, by including a stage of saturation of the specimens, to calculate the coefficients of consolidation and permeability of the soil for each stage of loading. This is done simply by applying the analysis procedures of the incremental test (see section entitled "Incremental Compressibility Test") to the settlement-versus-time curves.

Allowance will be made for the effect of temperature (70°C) by dividing the values yielded by these calculations by 3 to obtain the usual values of c_v and k_v corresponding to a temperature of 20°C. The corrected values of k_v and c_v will be determined as in the conventional test. This factor 3 was deduced experimentally by comparing the results of numerous incremental tests, performed at 20 and 70°C. This value is consequent with the ratio of 2.6, between the water viscosities at these two temperatures, as emphasized by Philipponat [11]. However, in the opinion of the CEBTP's own experts, it is still preferable to use the "classical" incremental test to obtain reliable values of the coefficient of consolidation c_v.

Radial Drainage Test

Choice of Equipment and Formation of Central Drain

The oedometric test with central drainage, which may be used to determine the coefficient of radial consolidation c_r, is preferred by the French LPC Network over the test with peripheral drainage, considered much less reliable because of the technological difficulties it involves. It will be recalled that coefficient c_r combines the vertical compressibility of the soil and its horizontal permeability.

Except for the actual formation of the vertical drain, the radial drainage test, shown schematically in Fig. 6, depends upon a good seal between the loading piston and the oedometric ring. The dimensions of the specimen must be as large as possible to favor radial flow over the vertical flow that may persist at the perimeter. The design of the LPC back-pressure cell described in the Evolution of Testing Equipment section meets these requirements perfectly. The dimensions of the soil specimen it uses (diameter 65.2 mm; H_0 (initial height of soil sample in the oedometer) = 25 mm) have been judged sufficient for testing a volume of soil that is as representative as possible.

The drain diameter used is one tenth of the inside diameter of the oedometric ring used, as a general rule, or 6.5 mm with the recommended equipment. The preparatory hole of the drain can be made in practice using a small auger or a jetting device. The LPC have developed a jetting device that can dig the preparatory hole for a drain in 30 s, using a fluid pressure of 200 kPa.

The preparation of the specimen for the test ends with the filling of the preparatory hole with clean sand having a suitable particle size distribution. The criteria applied here by the LPC are that no particles be smaller than 0.1 mm, and that the diameter of the largest particles be approximately one twentieth that of the drain, or 0.3 mm with the LPC back-pressure cell.

Saturation

Before the test proper is started, it is necessary to saturate the various parts of the system: (1) the drain and the water evacuation circuit in the cell, and (2)

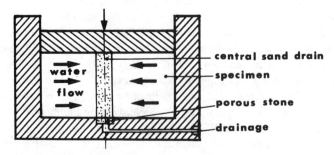

FIG. 6—*Principle of an oedometer for the radial drainage test.*

the soil itself, since Barron's theory [*12*], on which the analysis of the test is based, applies only to soils saturated with an incompressible fluid.

It is possible to saturate separately the two subassemblies mentioned above. However, it is preferable to saturate both together by applying a back-pressure, following the rules stated in the Saturation of Specimen section.

Test Procedure

The radial drainage test used by the LPC is an incremental test that proceeds along the same lines as the standard test discussed in Saturation of Specimen. The loads increase in a geometrical progression and are held constant for periods of 24 or 48 h or longer, depending on the soil. Thus, duration of the increments is the major difference between the loading laws of the radial drainage test and those of the standard test. The interpretation of the radial drainage test, as regards the settlement versus time curves for each loading increment, is again based on the use of Taylor's method. This technique of analysis presupposes that each increment hold time is long enough; that is, longer than the time $(t_{90})_r$ required to reach 90% consolidation—radial consolidation in this particular case. The theory of radial consolidation gives the order of magnitude of this time. For an oedometer with an inside diameter of 65.2 mm and a drain diameter of 6.5 mm, we find that for

$$c_r = 10^{-7} \text{ m}^2/\text{s}, (t_{90})_r \text{ is approximately } 5\frac{1}{2} \text{ h}$$

$$c_r = 5.10^{-8} \text{ m}^2/\text{s}, (t_{90})_r \text{ is } 39 \text{ h}$$

$$c_r = 10^{-8} \text{ m}^2/\text{s}, (t_{90})_r \text{ is } 54 \text{ h}$$

Interpreting the Test

Because of the great similarity between their procedures, the calculations required for the interpretation of radial drainage tests are similar to those used for incremental compressibility tests. There are two exceptions:

1. The calculation of the initial void ratio, in which allowance must be made for the presence of the drain. 2. The formulas for calculating the coefficient of radial consolidation which are based on Barron's theory of radial consolidation rather than on Terzaghi's one-dimensional consolidation theory.

Determining the Compressibility Characteristics of the Soil—As we have already stated, the results of the radial drainage test are analyzed in the same way as those of the incremental test.

Processing of Settlement Versus Time Curves—The settlement versus time curves will be analyzed only at loading increments corresponding to an effective stress σ' greater than the effective stress σ'_{vo} corresponding to the *in situ* overburden pressure. As in the incremental test, Taylor's method will be used, with 1.06 used as the ratio of the slopes of straight lines, rather than 1.15.

The radial coefficient c_r will be calculated using the formula

$$c_r = 0.29 \frac{D^2 \, F(n)}{(t_{90})_r}$$

where

D = diameter of the specimen,
d = diameter of the drain, and
$n = D/d$.

$$F(n) = \frac{n^2}{n^2 - 1} \ln (n) - \frac{3n^2 - 1}{4n^2}$$

The coefficient of horizontal permeability k_r will be calculated using the equation

$$k_r = c_r \, m_v \, \gamma_w$$

where m_v is defined in the same way as for the incremental test. The corrected coefficients of permeability and consolidation k_r and c_r are determined using the same calculations explained for k_v and c_v in the incremental test.

Remarks

The radial drainage test lasts from two weeks to two months, depending on the number and duration of the loading increments. Here again, it is possible to imagine various ways of accelerating this test for the determination of the coefficient of radial consolidation (or permeability): carrying out tests of the "controlled gradient" or "controlled rate of strain" type or using heat to accelerate the consolidation process. However, only the incremental type test is regularly used in France.

Long-Term Creep Test

The long-term creep test is used to characterize the strain rate of a soil specimen subjected to a constant effective stress. A first-draft procedure made it possible to determine the parameters needed for the application either of Koppejan's method [13] or of the method of Gibson and Lo [14]. The creep test so performed called for about six loading increments lasting ten days each.

Experience has shown that the model of Gibson and Lo is not representative of the actual behavior of soils *in situ*, as soon as the thickness of the layer exceeds 1 m.

The theoretical drawback of Koppejan's method is that it separates the consolidation stage from the creep stage, whereas *in situ* the two phenomena are simultaneous at first.

The CONMULT finite-difference computing program used by the LPC [9] makes it possible to introduce the creep phenomenon for each effective stress increment as consolidation progresses. An expression of the form

$$\frac{\Delta h}{h} = c_\alpha \left| log\ t_2/t_1 \right|$$

similar to the one proposed by Koppejan, is used to characterize the law. The aim of the test is accordingly to determine parameter c_α versus effective stress.

The testing procedure simply consists of putting the soil specimen through three stress increments, for example σ'_{vo}, σ'_p and σ'_{vF} (final service stress), each for at least seven days. If the soil is normally consolidated, the stress σ'_p will be replaced by $\sigma'_{vF}/2$. Interpreting the test reduces to computing the slope of the final quasi-linear part of each diagram $H = f\ (log\ t)$.

Controlled Gradient and Controlled Rate of Strain Tests

The two tests, although now well known around the world, previously were reserved for research projects in French universities and among the LPC network. Due to their potential advantages, the LPC issued provisional standards to aid their introduction as routine tests in a few years time. These standards are the close application of the testing fundamentals already published by Lowe et al [15] and Wissa and Heiberg [16], respectively. They will be open to improvement as experience increases. Here again, the LPC back-pressure cell proves itself as a first-rate tool.

Conclusion

This paper offers a panorama of the LPCs thinking both about improving the quality of the classical tests and about the use of new tests. Current trends suggest, moreover, that the controlled rate of strain test, and to a lesser extent the controlled gradient test, will in the medium term become routine in preliminary design calculations for structures on soft soils.

We have emphasized that the soil specimen tested must be as nearly perfectly saturated as possible. A back-pressure oedometric cell has been developed for this purpose, and its design is such that it may be used for all the tests described here.

One of the concerns that underlay the writing of the LPC procedures was adapting the routine tests to the requirements of current progress in rheology. In particular, these tests now make it possible to characterize variations in the consolidation parameters with effective stress and to take creep into account from the start of settlement.

References

[1] Terzaghi, K., *Erdbaumechanik auf bodenphysikalischer Grundlage*, Franz Deuticke, Leipzig and Vienna, 1925.

[2] Taylor, D. W., "Research on Consolidation of Clays," Department of Civil and Sanitary Engineering, Series 82, Massachusetts Institute of Technology, Cambridge, MA, Aug. 1942.

[3] Bjerrum, L., "Engineering Geology of Normally Consolidated Marine Clays as Related to the Settlements of Buildings," *Geotechnique*, Vol. 17, No. 2, 1967, pp. 83–119.

[4] Laboratoire Central des Ponts et Chaussées, "Essai de compressibilité à l'oedomètre—Modes opératoires du Laboratoire Central des Ponts et Chaussées," SMS 2, Dunod, Paris, 1965.

[5] Peignaud, M., "Compressibilité à l'oedomètre sous charge variable," *Bulletin de Liaison des Laboratoires des Ponts et Chaussées*, Spécial T, May 1973, pp. 306–321.

[6] Paute, J. L., "Essai oedométrique à drain central," *Bulletin de Liaison des Laboratoires des Ponts et Chaussées*, Spécial T, May 1973, pp. 322–334.

[7] Thomann, G., "Essai oedométrique avec mesure de pressions interstitielles," *Bulletin de Liaison des Laboratoires des Ponts et Chaussées*, Spécial T. May 1973, pp. 335–345.

[8] Taylor, D. W., in "Fundamentals of Soil Mechanics," 12th ed., Chapter 10, Wiley, New York, 1962, pp. 208–249.

[9] Magnan, J. P., Baghery, S., Brucy, M., and Tavenas, F., "Etude numérique de la consolidation unidimensionnelle en tenant compte des variations de la perméabilité et de la compressibilité du sol, du fluage et de la non-saturation," *Bulletin de Liaison des Laboratoires des Ponts et Chaussées*, No. 103, Sept.-Oct. 1979, pp. 83–94.

[10] Magnan, J. P., Humbert, P., Belkeziz, A., and Mouratidis, A., "Finite Element Analysis of Soil Consolidation, with Special Reference to the Case of Strain Hardening Elastoplastic Stress-Strain Model," in *Proceedings*, 4th International Conference on Numerical Methods in Geomechanics, Edmonton, Alberta, Canada, Vol. 1, May 1982, pp. 327–336.

[11] Philipponat, G., "Mesure de la compressibilité des sols par un essai oedométrique accéléré," *Annales I.T.B.T.P.*, Paris, No. 347, Feb. 1977, pp. 119–132.

[12] Barron, R. A., "Consolidation of Fine-Grained Soils by Drain Wells," *Journal of Soil Mechanics and Foundation Engineering*, Vol. 73, No. SM 6, American Society of Civil Engineers, June 1947, pp. 811–835.

[13] Koppejan, A. W., "A Formula Combining the Terzaghi Load Compression Relationship and the Buisman Secular Time Effect," in *Proceedings*, 2nd International Conference on Soil Mechanics and Foundation Engineering, Rotterdam, Vol. 3, 1948, pp. 32–34.

[14] Gibson, R. E. and Lo, K. Y., "A Theory of Consolidation for Soils Exhibiting Secondary Compression" in *Proceedings*, Norwegian Geotechnical Institute, No. 41, Oslo, 1961.

[15] Lowe, J., III, Jonas, E., and Obrician, V., "Controlled Gradient Consolidation Test," *Journal of Soil Mechanics and Foundation Engineering*, Vol. 95, No. SM 1, American Society of Civil Engineers, Jan. 1969, pp. 77–97.

[16] Wissa, A. E. Z. and Heiberg, S., "A New One-Dimensional Consolidation Test," Research Report 69-9, Soils Publication 229, Department of Civil Engineering, Massachusetts Institute of Technology, Cambridge, MA, March 1969.

Rolf Larsson[1] and Göran Sällfors[2]

Automatic Continuous Consolidation Testing in Sweden

REFERENCE: Larsson, R. and Sällfors, G., **"Automatic Continuous Consolidation Testing in Sweden,"** *Consolidation of Soils: Testing and Evaluation, ASTM STP 892,* R. N. Yong and F. C. Townsend, Eds., American Society for Testing and Materials, Philadelphia, 1986, pp. 299–328.

ABSTRACT: The constant rate of strain test has been successfully used as a routine test in Sweden for more than seven years. The paper describes in detail the experience gained so far. Methods of interpretation of test results and calculation of settlements are given. A critical comparison with other continuous loading tests and incremental loading tests is also presented.

KEY WORDS: clay, compressibility consolidation, preconsolidation pressure, settlement

The testing procedure for the consolidation testing of clays outlined by Terzaghi in 1936 [1] is still used in many laboratories all over the world. The stress/strain characteristics are determined from a test where the load is applied in increments, with a duration of 24 h, each increment being equal to the previous consolidation load. However, in order to obtain more reliable data, smaller load increments have been proposed by Leonards [2] and Bjerrum [3]. A natural development of the consolidation test would be to load the specimen continuously and measure the applied stress along with the strain. The first such test was reported by Crawford [4], where the specimen was loaded at a constant rate of strain. Five years later, Smith and Wahls [5] reported a thorough investigation concerning the constant rate of strain test (CRS test). Wissa and Heiberg [6] reported similar results, although the mathematical derivations were based on somewhat different assumptions. In the same year, Lowe et al [7] published a paper on the constant gradient test (CGT test). In this test the specimen is loaded in such a way that the pore pressure at the undrained bottom of the specimen is kept constant. Yet another test, constant rate of loading (CRL), has been used by some researchers.

All the tests mentioned above gave very interesting results but presented one major difficulty: The faster the test was run, the higher the preconsolidation

---

[1] Research Engineer, Swedish Geotechnical Institute, S-581 01 Linkoeping, Sweden.

[2] Associate Professor, Department of Geotechnical Engineering, Chalmers University of Technology, S-412 96 Gothenburg, Sweden.

pressure seemed to be and the further the oedometer curve was displaced in the diagram. As early as 1964, Crawford stated: "An important question requiring further study is the maximum laboratory rate of compression that can be used to compute the correct field settlement" [4].

Investigations on the applicability of the various types of continuous consolidation tests to Swedish soft clays were started in 1971, following the pioneering work carried out in Canada and the United States. At Chalmers University of Technology, and later also at the Swedish Geotechnical Institute (SGI), intensive studies were done into how the tests should be performed and interpreted, and correlations were made with field tests and with the old, well-established, incremental oedometer test.

In 1975, recommendations were made [8] on how to use the CRS test, how the test should be run, and how to evaluate the preconsolidation pressure. The CRS test became standard at the SGI in 1977, and today all major consultants in Sweden use the CRS test almost exclusively. Larsson and Sällfors [9] have outlined how the compression characteristics should be evaluated from the test results and have also given equations for the calculation of settlements.

The CRS Test

Apparatus

In Sweden, a 50-mm-diameter piston sampler is used for obtaining undisturbed specimens of clay, and therefore the oedometer ring is also 50 mm in diameter. The specimen is 20 mm high and is mounted in a Teflon ring lubricated with high-vacuum silicon grease. The Teflon ring is inserted into a thick casing ring which also serves as a guide for the top piece, containing the porous stone (Fig. 1). The casing ring, containing the Teflon ring with the specimen, is mounted on the oedometer base which at that time should be covered with water. An O-ring seals against the Teflon ring so that after the closing of the bottom valve, drainage is possible only through the top of the specimen. The pore pressure is measured at the undrained bottom by a stiff pressure transducer. Before the specimen is mounted, the system can be saturated by simply flushing the system with deaired water. The oedometer is placed in a compression machine and the test is run by deforming the specimen at a constant rate of strain. The applied vertical force is measured by a force transducer and the deformation by a displacement transducer.

At certain intervals, readings are taken of applied vertical force, pore pressure at the undrained bottom, and deformation. A data logging system or a microcomputer can be used to collect the data, and thus the test does not need any further attention provided there is some kind of automatic stop when critical pressures or deformations are measured. The calculation of the effective stress and the permeability presume a fully saturated specimen. The Swedish soils, for which the simple apparatus just described are used, have a natural degree of saturation of 99 to 100%. For other soils, saturation of the specimen by back-

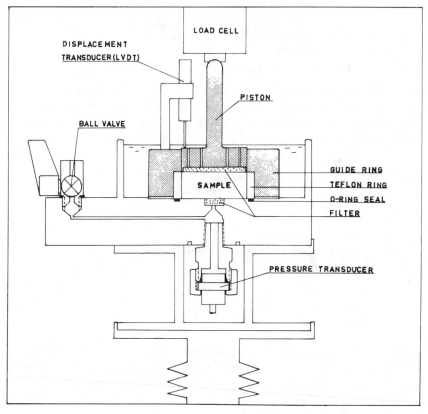

FIG. 1—*Oedometer for automatic consolidation tests.*

pressure can be achieved by submerging the oedometer inside a pressurized cell. This does not alter anything in the following text except that pore pressure becomes excess pore pressure.

Test Procedure

The crucial issue regarding the test procedure is how fast the test can be run and still yield results from which the long-term behavior of the clay can be evaluated. Sällfors recommended, for various reasons which will be discussed later, that for a 20-mm-high specimen, a rate of deformation not exceeding 4×10^{-5} mm/s should normally be used for clays (see Ref 8).

The average effective stress in the specimen is calculated according to Eq 1, under the assumption that the pore pressure throughout the specimen is parabolic

$$\sigma' = P/A - \frac{2}{3} u_b \qquad (1)$$

where

P = applied vertical force,
A = cross-sectional area of specimen, and
u_b = pore pressure at undrained bottom.

An additional requirement for the test is that the pore pressure does not exceed 15% of the applied vertical stress; this limits the error resulting from Eq 1. The permeability, k, is calculated, under the same assumptions, according to Eq 2, and the coefficient of consolidation, c_v, according to Eq 3

$$k = \frac{g\rho_w \cdot H}{2u_b} \cdot \frac{d\epsilon}{dt} \tag{2}$$

where

H = height of specimen,
u_b = pore pressure at undrained bottom, and
$d\epsilon/dt$ = rate of deformation

$$c_v = \frac{k \cdot M}{g\rho_w} \tag{3}$$

where k is the coefficient of permeability and M is the compression modulus.

With the procedure described above, a test is normally completed in 24 to 48 h, all data are obtained automatically, and at the end of the test the microcomputer produces plots of σ' versus ϵ, M versus σ', k versus ϵ, and c_v versus σ', all as continuous curves.

Results from CRS Tests and Evaluation of Consolidation Parameters

From all the CRS tests performed on undisturbed clay specimens unaffected by dry crust effects, a consistent pattern for compression characteristics has emerged. Typical results are given in Fig. 2.

The stress-strain relation follows the same pattern for most clays except for cry crust, boulder clays, and clays disturbed during sampling. Initially, there is a small irregularity due to imperfections in the contact between the specimen and the porous stone. This is best observed on the modulus-pressure curve.

The stress-strain curve then becomes a straight line until the preconsolidation pressure is approached. For stresses close to the preconsolidation pressure the modulus drops until it again becomes constant and the stress-strain curve again becomes a straight line. The modulus then remains constant until a limiting stress is reached; for stresses above this limiting stress, the modulus increases linearly with increasing effective stress.

The permeability of the soil decreases with increasing compression and thus decreasing void ratio. After some initial irregularities the permeability versus compression becomes almost a straight line in a semi-log plot. The coefficient of consolidation, c_v, is a function of the modulus and the permeability (see Eq 3). The dominating parameter is the modulus, and the coefficient of consolidation-stress curve very much resembles the modulus-stress curve. When the preconsolidation pressure is passed and the deformations become large, the shape of the coefficient of consolidation-stress curve depends on the relation between the increase in modulus and the decrease in permeability. As long as the modulus remains constant, the coefficient of consolidation decreases in all soils. After the limiting pressure for the lower modulus is reached and the modulus starts to increase, the coefficient of consolidation also increases in mineral clays while it remains almost constant in organic clays and may even go on decreasing in very organic soils. The experience gained so far has led to the following interpretation of the CRS test (see Fig. 3).

The preconsolidation pressure is evaluated according to Sällfors [8], where the two straight parts of the stress-strain curve are extended until they intersect. An isosceles triangle is then inscribed between the lines and the stress-strain curve. The point of intersection between the base of the triangle and the upper line represents the preconsolidation pressure σ'_c (Fig. 3). This construction is somewhat dependent on the scales and is therefore always made in a plot where the scales are such that the length representing 10 kPa on the stress axis corresponds to the length representing 1% of the strain axis.

After determination of the preconsolidation pressure, the stress-strain curve for higher stresses is moved horizontally a distance c to pass through the point where σ'_c was evaluated (Fig. 3). With the low testing rates used according to Swedish practice, the value of c is usually small. As will be shown later, the adjusted stress-strain curve so obtained corresponds very well to the curve obtained from standard incremental oedometer tests.

The modulus-stress plot is now modified. The initial constant modulus M_0 is extended to σ'_c. At σ'_c the modulus is assumed to drop instantaneously to the second constant modulus M_L. The part of the curve where the modulus increases linearly with effective stress is moved c kPa horizontally to the left and extended to the baseline where M equals 0. The stresses at the intersection with the baseline a and at the intersection with the constant modulus σ'_L are evaluated and the modulus number M' is evaluated as $\Delta M/\Delta\sigma'$ for the part of the curve where M increases linearly with effective stress.

Thus the curve is divided into three parts:

1. The part in the stress interval $\sigma'_0 - \sigma'_c$ where $M = M_0$.
2. The part in the stress interval $\sigma'_c - \sigma'_L$ where $M = M_L$.
3. The part in the stress region where $\sigma' > \sigma'_L$ and where $M = M'(\sigma' - a)$.

The initial modulus M_0 from the first loading of a natural "undisturbed" specimen in the oedometer is never used. It is always too low compared with the in situ initial modulus due to specimen disturbance, swelling, and imperfect fit in the

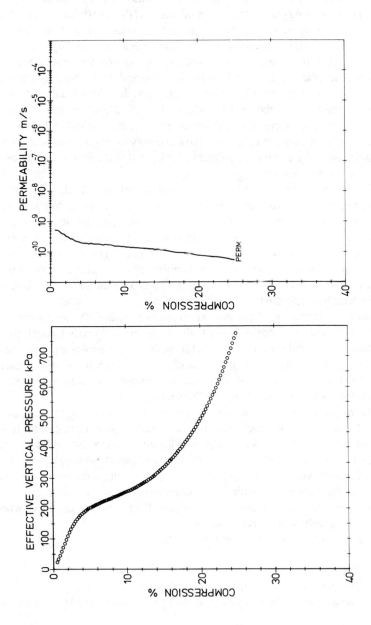

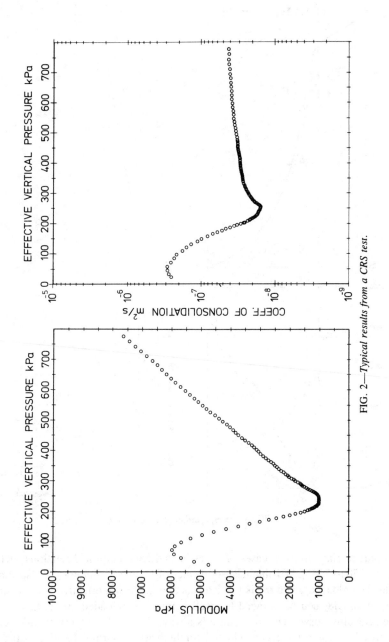

FIG. 2—*Typical results from a CRS test.*

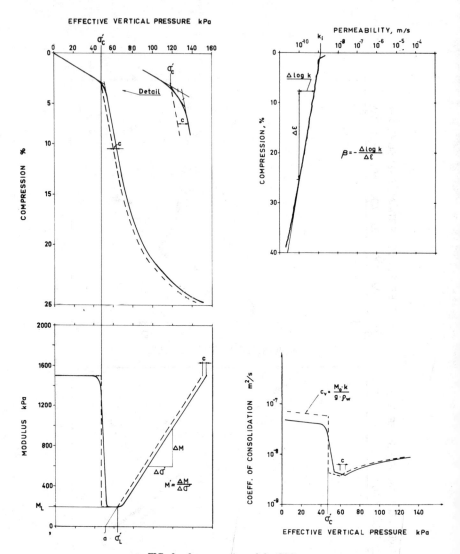

FIG. 3—*Interpretation of the CRS test.*

oedometer ring. In most cases M_0 is estimated from empirical relations such as $M_0 \approx 250$ to $500 \, \tau_{fu}$. To obtain a useful value of M_0 in the laboratory, the specimen must be unloaded when σ'_c is just exceeded to the *in situ* effective vertical stress σ'_0. It should then be allowed to swell before it is reloaded. M_0 is taken from the reloading curve. This procedure is always followed for those boulder clays in Sweden which have consolidated for high loads during the glacial period. These clays are initially loaded to about 1000 kPa (estimated ice pressure) and then unloaded to *in situ* vertical stress before the actual test starts.

If it is known from the loading history of the soil that the soil *in situ* is in a reloading state, the specimen should correspondingly be unloaded to the estimated lowest unloading pressure.

The permeability is evaluated by simplifying the log permeability-strain curve to a straight line. The initial permeability k_i is evaluated at the intersection of the straight line and the horizontal line $\epsilon = 0$ and the decrease in permeability with compression is expressed by the parameter $\beta = -\Delta \log k / \Delta \epsilon$ (Fig. 3).

The coefficient of consolidation c_v is calculated from Eq 3 and if the mathematical expression is used, c_v automatically becomes corrected by the correction of M. If a simple correction is to be made on the plotted c_v-curve, then first the c_v-values for stresses up to the preconsolidation pressure should be calculated from

$$c_v = M_0 \cdot k/g \cdot \rho_w$$

Then the c_v-curve for higher stresses should be moved c kPa horizontally to the left. Just as the modulus c_v drops instantaneously to the lower values at σ'_c. Test on dry crusts, silts, and remolded clays can be evaluated using the same parameters but the patterns often differ [10].

Calculation of Settlements

As a direct consequence of the shape of the stress-strain curves, a simple calculation of the final settlements, which disregards secondary consolidation, can be made using

$$\epsilon = \frac{\sigma' - \sigma'_0}{M_0} \text{ when } \sigma' < \sigma'_c \qquad (4a)$$

$$\epsilon = \frac{\sigma'_c - \sigma'_0}{M_0} + \frac{\sigma' - \sigma'_c}{M_L} \text{ when } \sigma'_c < \sigma' < \sigma'_L \qquad (4b)$$

$$\epsilon = \frac{\sigma'_c - \sigma'_0}{M_0} + \frac{\sigma'_L - \sigma'_c}{M_L} + \frac{1}{M'} \ln \frac{(\sigma' - a)M'}{M_L} \text{ when } \sigma' > \sigma'_L \qquad (4c)$$

where σ'_0 is the *in situ* effective vertical stress and where $\sigma' = \sigma'_0 + \Delta \sigma'$.

Sources of Error in the Tests

The main sources of error in the CRS test are the rate effects and uneven stress distribution in the specimen. The rate effects occur in all types of test even after the excess pore pressures have disappeared due to secondary compression.

Suklje [11] and Bjerrum [12,13] showed that no unique stress-strain relation exists independent of time but that there is a set of relations, each one of them valid for a certain strain rate. The main questions are: What strain rate should be considered appropriate for engineering purposes? and How should the test results be corrected to be valid for this rate? A special advantage with the CRS test in relation to other test procedures in this respect is that it is the only consolidation test with a nearly constant rate of deformation throughout the test. Also, incremental loading tests are evaluated at different rates of deformation. The correction used in Sweden for the CRS tests is based on the empirical observation that using this correction, all results become essentially equal, independent of the rate, provided that a limiting rate of deformation is not exceeded.

Provided that the limiting rate is not exceeded, the correction is comparatively small. There is no strong experimental evidence on whether the rate effects should be taken into account by moving the curve in a linear plot, in a semi-log plot, or in some other way. Provided that the correction is small enough, the practical difference is insignificant. The corrected curves have been thoroughly correlated with field measurements and standard incremental tests.

In a CRS test the distribution of effective stresses over the specimen is uneven. Drainage is provided at the upper boundary where the excess pore pressure is always zero, whereas no drainage is allowed at the bottom where the pore pressure is measured (Fig. 4). The pore pressure distribution is then assumed to be parabolic, and the average effective stress is calculated using

$$\sigma' = \sigma - \frac{2}{3} u_b \qquad (5)$$

In Eq 5 the assumption is made that the void ratio is almost constant in the specimen [5]. More rigorous solutions have been elaborated by Tokheim and Janbu [14] taking variations in void ratio into account.

All calculations, however, assume a gradual change in modulus and a more or less parabolic shape of the pore pressure distribution. As shown by Tavenas et al [15], the abrupt change in modulus at the preconsolidation pressure will

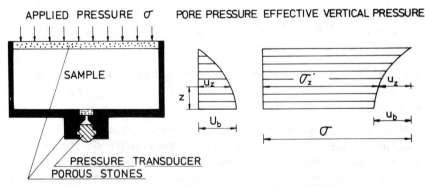

FIG. 4—*Assumed stress distribution in automatic consolidation tests.*

invalidate this assumption. For a stress interval around the preconsolidation pressure, the pore pressure distribution (and thereby also the distribution of effective stresses and void ratio) will deviate significantly from the assumed parabolic shape. The size of the stress interval, affected by errors due to the abrupt change in modulus at the preconsolidation pressure, depends on the pore pressure at the bottom which, in turn, is dependent on the rate of deformation. The magnitude of the error depends on the pore pressure and the relative change in modulus at the preconsolidation pressure. These effects further emphasize the importance of keeping the rate of deformation, and hence also the pore pressure, low.

Experiences from CRS Tests—Preconsolidation Pressures

The evaluation of the preconsolidation pressure from CRS tests [8] was based on the observation in the laboratory that this evaluation eliminated rate effects for slow tests, and on comprehensive field tests for correlation with *in situ* preconsolidation pressures. These studies were made on a soft clay, on a very soft organic clay, and on a relatively stiff clay (see Fig. 5).

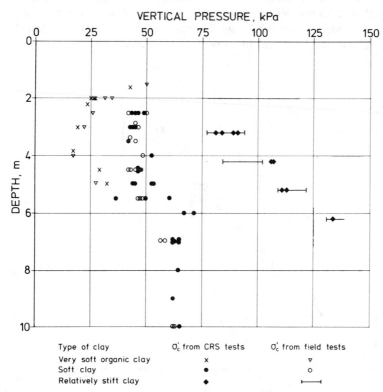

FIG. 5—*Correlations between preconsolidation pressure from CRS tests with full-scale field tests in:* (a) *soft clay,* (b) *very soft organic clay, and* (c) *relatively stiff clay.*

Ever since the test came into use at the Swedish Geotechnical Institute in 1976, the evaluated preconsolidation pressures have continuously been compared with the stress-profile *in situ* and to what is known about the stress history of the soil. In some cases supplementary tests have been made on specimens from the institute's test fields, where the preconsolidation pressures have previously been established by field tests as well as by comprehensive series of standard incremental oedometer tests. During the introduction of the CRS test as a standard test, a large number of standard incremental tests were also run, parallel to the CRS tests. The results of these investigations have resulted in a great confidence in the reliability of the preconsolidation pressures evaluated from CRS tests, according to Sällfors. All the CRS tests have been run at the recommended rate of strain of 4×10^{-5} mm/s on specimens with a height of 20 mm.

A comprehensive study on the preconsolidation pressure in sensitive Canadian clays in the laboratory as well as in the field has recently been presented by Samson et al [16]. From this study, Leroueil [17] concludes that the preconsolidation pressure obtained in standard incremental tests give preconsolidation pressures approximately equal to the preconsolidation pressure *in situ*. In the study by Samson et al a large number of CRS and also CGT tests were performed. When these CRS tests are evaluated according to Swedish practice and compared with the standard incremental tests, it becomes obvious that the evaluated preconsolidation pressures are the same also for these clays irrespective of test procedure (Fig. 6).

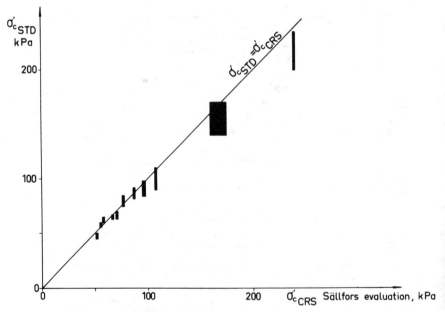

FIG. 6—*Correlation between preconsolidation pressure in Canadian clays from standard incremental tests and CRS tests evaluated according to Swedish practice* [16].

The standard rate of strain used in Sweden originates from the investigations of Sällfors [8]. Then the number of soils investigated were limited. To check also the general validity for more compressible organic soils and soils with lower permeability, new investigations on the rate effects have been performed at SGI. These test series consist of CRS tests at different rate of strain as well as standard incremental tests. Comparative continuous loading tests have also been performed. Mainly in order to study the CL tests, new test series were also made on the previously investigated Bäckebol clay (for data, see Ref 8).

The stress-strain curves from CRS tests on Bäckebol clay at 4-m depth are shown in Fig. 7. From these curves the parameters in Table 1 have been evaluated according to the Swedish method of evaluation previously described. The results do show some scatter but, as can be seen from Table 1 and Fig. 8, the standard rate is well within the lower range where there is no further rate effect on the preconsolidation pressure.

A number of similar series of tests have been performed on a clayey gyttja[3] from Vallda. Basic soil properties are $w_N \approx 170\%$, $w_L \approx 195\%$, $w_p \approx 90\%$, $I_p \approx 105\%$, $\tau_{fu} \approx 10$ kPa, and organic content $\approx 6\%$. The effect of rate on the evaluated preconsolidation pressures is shown in Fig. 8. Each point in the diagram

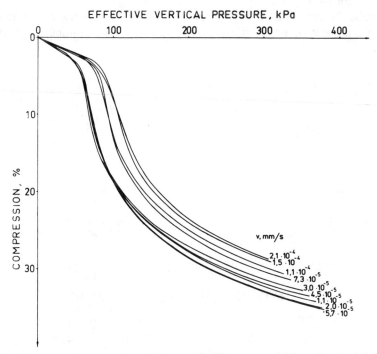

FIG. 7—*Stress-strain curves from CRS tests with different rates of deformation, Bäckebol clay.*

---

[3] Clayey gyttja is a soil with an organic content of 6% to 20%. (The Swedish word *gyttja* is used internationally.)

TABLE 1—*Consolidation parameters from CRS tests. Bäckebol, 4 m.*

Rate of Deformation, mm/s	σ'_c, kPa	M_L, kPa	σ'_L, kPa	M'	a, kPa	k, m/s	β
1.1×10^{-5}	44	175	55	12.18	41	5.0×10^{-10}	3.19
2.0×10^{-5}	48	178	60	12.46	46	6.0×10^{-10}	2.90
3.0×10^{-5}	44	198	57	12.73	41	4.7×10^{-10}	3.00
4.5×10^{-5}	43	190	57	12.78	42	4.1×10^{-10}	2.79
5.7×10^{-5}	46	180	58	12.12	43	5.7×10^{-10}	3.00
7.3×10^{-5}	69	190	80	12.91	65	3.0×10^{-10}	2.56
1.1×10^{-4}	62	273	85	12.69	63	3.8×10^{-10}	2.81
1.5×10^{-4}	75	280	98	12.99	76	3.9×10^{-10}	3.10
2.1×10^{-4}	74	375	109	12.75	79	4.0×10^{-10}	3.00

represents the average of five to six tests. This is why the scatter is moderate in spite of the problem of accurately measuring preconsolidation pressures as low as about 30 kPa. For this very organic soil it is found that the standard testing rate of 4×10^{-5} mm/s is close to the critical rate and it might be advisable to run tests on this type of soil somewhat slower than 4×10^{-5} mm/s, which is common for clays.

Also in the investigations reported by Samson et al [*16*] CRS tests were run at different rates of strain. If these tests are evaluated according to Swedish

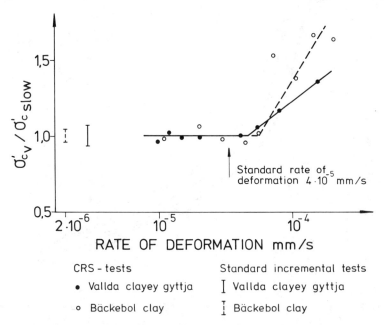

FIG. 8—*Effect of rate of deformation on the evaluated preconsolidation pressure (CRS tests evaluated according to Swedish method).*

practice and compared with standard incremental tests, the same pattern as for Swedish clay is obtained. Provided that the rate of strain is slow enough, the preconsolidation pressures will be about the same for standard incremental tests and CRS tests (see Fig. 9). The standard rate of deformation of 4×10^{-5} mm/s corresponds to a rate of strain of 2×10^{-6} s^{-1}.

Constant Modulus M_L

The modulus M_L, which is a constant, occurs in a stress interval directly after passing the preconsolidation pressure. As previously mentioned, the calculated effective stresses in this stress region may be erroneous if the pore pressure is too high due to a stress distribution deviating from that assumed. Another aspect is that the modulus is calculated for the whole sample using average stress and average deformation. To enable the measurement of the minimum modulus there must be a stage when all elements of the sample are subjected to stresses within the limits for this modulus. This puts another restriction on the allowable pore pressure and thereby rate of strain.

A variation similar to that for the preconsolidation pressure is found when investigating the effect of rate of strain on the measured constant modulus M_L. For rates of strain below a critical value the measured modulus seems to be constant, but if this critical rate is exceeded the modulus increases with increasing rate of strain. The critical rates seem to be about equal to the critical rates for the evaluation of the preconsolidation pressure (Fig. 10).

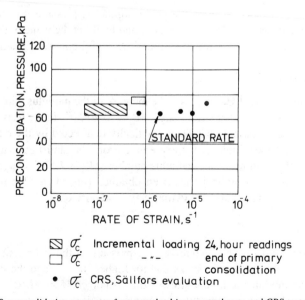

FIG. 9—*Preconsolidation pressures from standard incremental tests and CRS tests with different rates of strain. Gloucester clay 3.4 to 3.9 m. (Data from Refs 16 and 17.)*

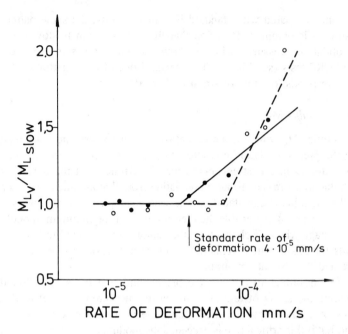

● Vallda clayey gyttja
○ Bäckebol clay

FIG. 10—*Effect of rate of deformation on modulus* M_L *for Bäckebol clay and Vallda clay.*

A number of factors, such as the preconsolidation pressure, the consistency limits, and the liquidity index, have been found to affect the value of the constant modulus M_L. So far the authors have found no useful empirical relation.

Modulus Number M'

The modulus number M' expresses the increase of the modulus with increasing effective stress in the higher stress range, $M' = \Delta M / \Delta \sigma'$. In the tests performed at SGI the modulus number has been practically unaffected by the rate of strain used in the tests (see Table 1, for example). From the statistical data gathered at SGI a simple relation for the modulus number, $M' \approx 4.5 + 6/w_N$, where w_N is expressed as a real number, has been obtained (Fig. 11). This relation has been found to be fairly valid for gyttja, clay, and silty clay.

Parameter a

The parameter a is the effective vertical pressure at the intersection between the straight modulus-effective stress line for higher stresses and the line $M = 0$ (Fig. 3). The modulus at higher stresses thus becomes $M = M'(\sigma' - a)$.

The value of a for an undisturbed specimen, unaffected by dry crust effects, has consistently been found to be approximately equal to the preconsolidation pressure (Fig. 12). In the investigations on the effects of the rate of strain the value of a has been found to follow the value of σ'_c and the rate dependency seems to be identical.

Limiting Pressure, σ'_L

The limiting pressure, σ'_L, is the effective vertical stress where the linear portion of the stress-strain curve after the preconsolidation pressure ends and the modulus starts to increase with increasing effective stress. The value σ'_L is related to the other compression parameters so that $\sigma'_L = a + M_L/M'$. As no empirical relation has been found for M_L, there is none for σ'_L, either. However, the stress interval $\sigma'_L - \sigma'_c$ is often so large that in most settlement calculations for moderate loadings only the constant modulus M_L is used.

The parameters a and M_L are rate dependent and consequently the parameter σ'_L becomes rate dependent and the same critical rate of strain as for the other two parameters can be found (see Fig. 13).

Permeability

The permeability of a soil is calculated using the pore pressure at the bottom of the specimen and assuming a parabolic distribution of the pore pressure through

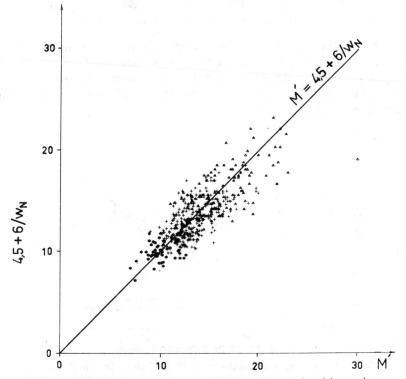

FIG. 11—*Simple equation for modulus number versus measured modulus number.*

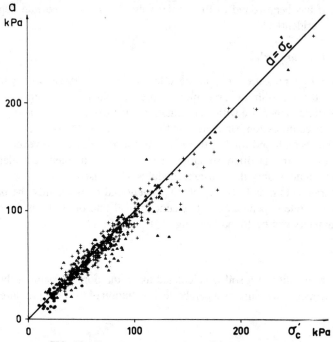

FIG. 12—*Pressure* a *versus preconsolidation pressure.*

the specimen. The permeability is then related to the average deformation of the specimen. Most tests are performed at room temperature ~20°C and the permeability so calculated is therefore corrected for the change in viscosity of water from room temperature to ground temperature. The ground temperature in Sweden is usually assumed to be 7°C.

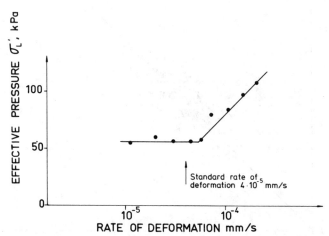

FIG. 13—*Effect of rate of deformation on the stress* σ'_L *(Bäckebol clay, 4-m depth).*

There are numerous sources of errors in the procedure described above, in the test itself, and in its application to field conditions. As pointed out by Tavenas et al [15], the pore pressure distribution may deviate considerably from that assumed when the preconsolidation pressure is passed, which would cause irregularities in the permeability-compression relations at the beginning of the test. These possible faults are at least partly erased by simplifying the log permeability-compression relation to a straight line.

Another source of error is the plotting of the permeability versus the average compression when the actual permeability is measured in the upper part of the specimen, where the compression is greater than the average compression of the specimen. The magnitudes of these errors depend on the magnitude of the pore pressure and the compression characteristics and thus also on the rate of strain. With the restrictions for the rate of strain and maximum pore pressures used in Sweden, the error in permeability would typically be of the order of 1 to 5% but could be larger for extremely brittle clays, especially if higher pore pressures are allowed.

A third source of error is the correction for temperature. Comparative tests at different temperatures in the laboratory have shown that the correction is reasonably correct if the temperature in the laboratory can be held constant. The temperature in the ground, however, is rarely measured. This temperature is related to the average yearly temperature in the air, but very local differences may exist owing to vegetation, average local snow cover during the coldest season, and, of course, urbanization. A difference of one degree Celsius corresponds to a change in permeability of about 3%.

These errors, however, are negligible when compared with the effect of specimen size and natural variations in the soil. In many cases the soil is layered which, apart from variations in vertical permeability, also creates large differences between vertical and horizontal permeabilities. This may also be the case in some nonlayered soils where particle orientation creates the same effects. Leonards [18] pointed out that even the most homogeneous soils have natural variations in water content. Thus there is also a natural variation in vertical permeability in "homogeneous" clays. For the "very homogeneous" Bäckebol clay, Sällfors [8] found variations of water content of ±4% within specimens with a length of 100 mm. With an average water content of about 80% this results in a natural variation in permeability of ±25%. In most clays the variation is larger, even much larger, while the permeabilities in a few gyttjas may be more even.

This means that a quantitative measurement of the permeability for a natural clay which can be applied to the field conditions cannot be obtained from small specimens in the laboratory, irrespective of the type of test. These tests should be seen as qualitative tests only. If accurate measurements of the permeability are required, they have to involve a large volume of the soil, and preferably be made *in situ*.

The permeabilities obtained at different rates of deformation for Bäckebol clay are shown in Fig. 14.

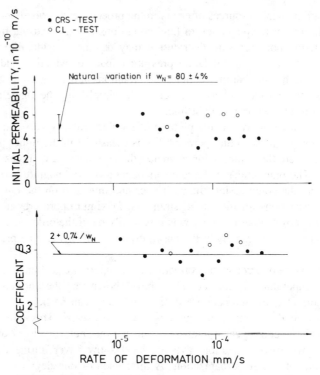

FIG. 14—*Effect of rate of deformation on permeability parameters.*

There might be a small tendency for a decrease in initial permeability with increasing rate of deformation, but all the results fall within the limits for natural variation. It can also be seen that there is a tendency for relatively high permeability to be accompanied by a relatively high value of the parameter β which reflects the possibility of a subjective influence when the slightly curved log permeability-compression relation is simplified to a straight line.

In the investigation reported by Tavenas et al [15] on permeability it was found that log permeability had no linear relation with either compression or log compression. For practical purposes, however, they concluded that the permeability, as in Swedish practice, could be simplified to a straight line in the log permeability-compression plot. They used the permeability change index $c_k = \Delta e/\Delta \log k$ to describe the change in permeability with compression. The initial permeability was found to be a complex function of a number of soil properties, while the permeability change index was found to be a simple function of initial void ratio:

$$c_k = 0.5e_0$$

This equation can also be written as

$$\frac{-\Delta \log k}{\Delta \epsilon} = \beta = 2 + \frac{2}{\rho_d \cdot w_N}$$

which for typical Swedish clay becomes

$$\beta = 2 + 0.74/w_N$$

A check on this relation for Swedish clays has shown it to be a generally good relation for inorganic clays.

The permeabilities obtained in CRS tests have in a number of cases been compared with various kinds of other laboratory permeability test and the results have always been about equal. In the study of permeability by Tavenas et al [15], results from tests on specimens from Bäckebol, Sweden, were included.

Permeability was determined in special oedometer tests where "falling head" tests were performed at various stages of compression. When the results from these tests and 15 CRS tests run on clay taken from the same place and at the same level are compared, the mean values almost coincide. The void ratio for the CRS tests are calculated from the average natural water content, 80%. The scatter in the permeability falls within the expected range due to variations in natural water content (see Fig. 15).

Even with all the shortcomings mentioned, especially those related to specimen

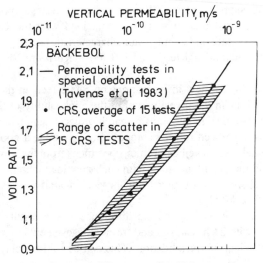

FIG. 15—*Correlation between permeability measured in CRS tests and permeability measured in special oedometer.*

size, the CRS test seems to be a relatively simple and reliable test to obtain a quantitative measure of the permeability of the soil, provided that the rate restrictions for the test are observed.

Coefficient of Consolidation c_v

The coefficient of consolidation is a function of the modulus and the permeability:

$$c_v = \frac{k \cdot M}{g \cdot \rho_w}$$

Besides being rate dependent as M is rate dependent, all the shortcomings involved with the measurement of k are inherited in measuring the coefficient of consolidation. It should therefore be taken into consideration that c_v values from all types of small size oedometer tests are qualitative rather than quantitative, even if we disregard what are perhaps even greater sources of error, such as horizontal water flow and creep effects.

Comparison Between CRS Tests and Standard Incremental Tests

As previously mentioned, a number of standard incremental tests were run parallel to the CRS tests during the introduction of the new test procedure. The results were that the points for the 24-h readings of the deformation for the different load increments practically fell on the corrected CRS curve. In the same way, the supplementary CRS tests on specimens from the test fields gave results almost identical to the previous standard incremental tests. In cases where an older investigation was supplemented with new and more tests, although the points from the older standard incremental tests coincided with the CRS curve, the preconsolidation pressure and compression characteristics had to be re-evaluated, because the older incremental tests had left too much room for subjectivity in the evaluation.

In connection with the new investigations on rate effects in the CRS test and the CL test as well as in the investigation of the applicability of the CRS test to organic soil and also in some previous research projects, supplementary standard incremental tests have been run for comparison. To enable a good comparison, several incremental tests have been run on specimens from the same sample with varying first load increment and then with each new load increment equal to the previous load, $\Delta p/p = 1$. Results from these comparative tests are shown in Fig. 16 (pp. 322 and 323).

As can be seen, there is very little difference in the results from standard incremental tests with 24-h readings and from CRS tests, either as to stress-strain relations or coefficient of consolidation. This can be said for clay as well as for gyttja.

An investigation on the applicability of the Swedish CRS technique to stiffer clays has also been performed on clays from Iraq [19]. This investigation showed no further differences between the CRS tests and the standard incremental tests, except that the continuous CRS tests made it possible to discern whether or not there was a preconsolidation pressure. This was not possible on the basis of results from the standard incremental tests.

The CRS test evaluated according to Swedish practice thus gives results practically identical to the results from standard incremental tests. The results are far superior in most aspects because the curves are continuous and leave very little room for subjective interpretation.

On the other hand, the CRS test does not give any information on the rate of secondary consolidation. If that information is required, the CRS test has to be supplemented with incremental loading tests, or alternatively, the rate of secondary consolidation may be roughly appreciated by empirical relations based on the compression index or the natural water content [10,20].

It has been suggested that the rate of secondary compression might be appreciated from CRS tests run at different rates of deformation, but considering all other rate effects in the test itself and the natural variation of the soil, this hardly seems to be a practical solution.

Comparison Between CRS Tests and Real Field Behavior

The CRS test is a very young test; it has only been in use for eight years as a standard test. Considering the usual time between the testing of the soil and the completion of the construction to the measurement of the "final settlement," this is a relatively short time. The first preliminary results from constructions where vertical drains have been used only recently have become available. These results so far seem very encouraging, as is also the old experience of standard incremental tests from this type of loading.

On the other hand, the comparisons with time-settlements and pore pressures in old test fills on thick homogeneous clay deposits without drains show the same discrepancies as to predicted rates of settlements and pore pressure dissipations as well as "final settlements" as the old tests [21]. In these cases it is obvious that new calculation methods have to be used, taking secondary consolidation and variation in modulus and permeability into account. Because the CRS tests give results almost identical to those from the standard incremental tests, all the gathered experience, good and bad, from the old type of test can be applied to the CRS test. This is provided that the experience is based on well-conducted tests performed in such quantity that there has been very little room for subjective interpretation of the standard incremental tests.

Other Types of Continuous Consolidation Test

During the first investigations as to what type of continuous consolidation test would be suitable for soft Swedish clays, the CGT test was also employed. In the CGT, the governing criterion is that the pore pressure at the bottom of the specimen should be kept constant throughout the test. This implies that the rate

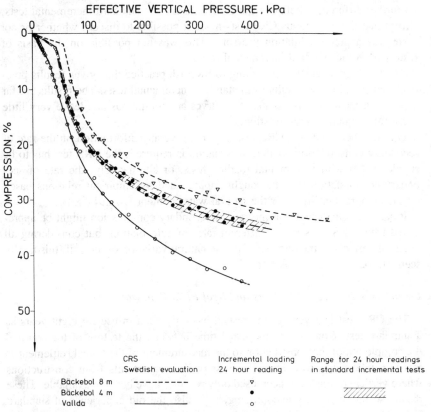

FIG. 16—*Correlation between standard incremental tests and results from CRS tests. Standard incremental tests comprise several tests on specimens from the same sample with varying first-load increments and then with each new load increment equal to the previous load.*

of deformation is governed by the coefficient of consolidation. The rate of deformation is thus relatively rapid until the preconsolidation pressure is reached, but then drops abruptly to a very slow rate of deformation. The further development of rate of deformation with stress depends on the type of soil because the relation coefficient of consolidation-stress depends on the type of soil. In the tests on Swedish clays it soon became evident that the test was rate (gradient)-dependent and that with the very small gradients that could be allowed in soft, normally consolidated clays with fairly low preconsolidation pressures, the CGT would be very time-consuming. This finding and the more complicated equipment required to perform the CGT resulted in its being abandoned and the research concentrated on the CRS test.

Another type of test where the relation between the pore pressure at the bottom of the specimen and the applied pressure is kept constant, the CL test, has been developed at the Norwegian Technical University [22]. The rate of deformation in this test is dependent on the coefficient of consolidation, as in the CGT, but

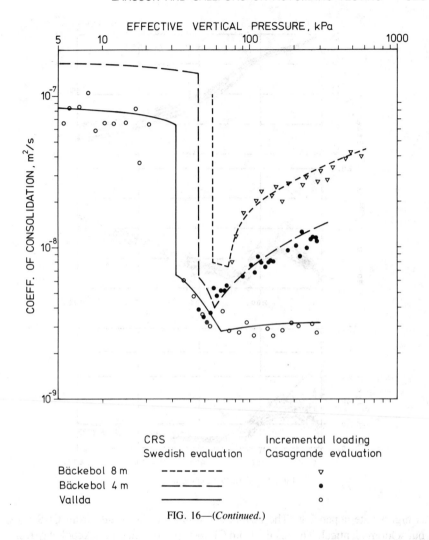

CRS   Incremental loading
Swedish evaluation   Casagrande evaluation

Bäckebol 8 m   – – – – – – –   ▽
Bäckebol 4 m   —— —— ——   •
Vallda   ——————————   ○

FIG. 16—(*Continued.*)

as the pore pressure and the gradient are allowed to increase with increasing stress, the CL test is much faster.

Janbu et al [22] have presented curves from CL tests that are very similar to the curves obtained from CRS tests (Fig. 17). The CL test has the advantage of automatically adjusting the rate of deformation to the properties of the tested soil. The test equipment has also become fairly simple with the recent development in electronics and electronically geared compression test machines. The mathematical solution for the interpretation of the test was also given by Janbu et al [22].

The test has therefore been investigated at SGI and compared with the CRS test. These investigations have shown that the CL test, like all oedometer tests,

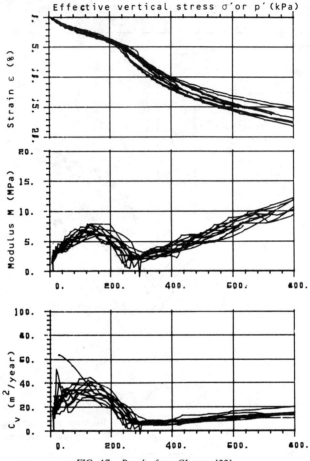

FIG. 17—*Results from CL tests* [22].

is highly rate-dependent. The results are similar to the results from CRS tests but seldom identical. The results from CL tests performed on Bäckebol clay from the same sample as the tests in Table 1 are given in Table 2 and Fig. 18 (the Swedish method of interpretation has been used). As can be seen from the results, the CL tests give modulus numbers M', permeabilities and values of the parameter β that are practically unaffected by the rate of deformation and very close to the results from the CRS tests.

As to the rate-dependent parameters, the slowest of the CL tests ($u_b/\sigma' = 0.1$) gave parameters compatible with the slow CRS tests. The rate of deformation was also close to the standard rate at and just after the preconsolidation pressure. At higher stresses the rate in the slow CL test increased and, consequently, the curves deviated somewhat. This can also be observed in the evaluated parameters where the values of σ'_L and a are comparatively low. This implies that the CL

TABLE 2—*Consolidation parameters from CL tests. Bäckebol, 4 m.*

u_b/σ'	$\sigma'_c,$ kPa	$M_L,$ kPa	$\sigma'_L,$ kPa	M'	$a,$ kPa	$k,$ m/s	β
0.1	48	163	55	12.27	42	4.9×10^{-10}	2.94
0.2	56	235	73	12.65	54	5.9×10^{-10}	3.09
0.3	55	255	72	12.31	51	6.0×10^{-10}	3.28
0.4	58	230	75	12.67	57	5.9×10^{-10}	3.15

test leaves the straight-line portion of the curve earlier than the CRS test and the modulus starts to increase at lower stresses due to the increasing rate of strain.

Similar results have also been obtained for other types of clay, although there is one exception. In gyttja and highly organic clay where the coefficient of consolidation is almost constant, after the preconsolidation pressure is passed, the CL tests became tests with an almost constant rate of strain with the rate perfectly matched to the consolidation properties of the soil. The results from CL tests with low ratios u_b/σ' (≈ 0.1) also became equal to the results from the slow CRS tests in these soils.

Samson et al [16] investigated the relation between the CGT and CRS tests and found certain differences. When results from the three types of tests (CGT, CRS, and CL) are compared, the following picture emerges. All tests have to be run in the low gradient-slow rate-slow ratio u_b/σ' ranges to enable a proper

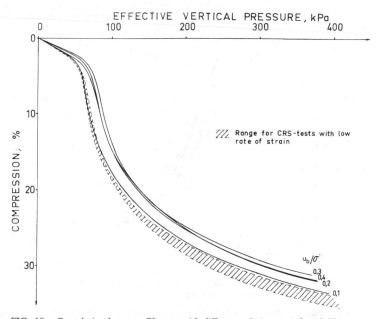

FIG. 18—*Correlation between CL tests with different relations u_b/σ' and CRS tests.*

evaluation of the settlement properties. The typical differences then remaining between the types of test are effects of the varying rates of strain. These differences are illustrated in Fig. 19.

The CRS test has a constant rate of strain throughout the test, but the rate of strain varies in a CGT test. After the preconsolidation pressure is passed, the rate of strain decreases dramatically and the curve will follow a stress-strain relation corresponding to very low strain-rates, usually lower than the rate in the CRS test. The rate of strain will then again increase in the CGT test as the c_v-value increases, and the stress-strain curve will coincide with relations for higher rates of strain.

In the CL test the rate of deformation will increase throughout the test after the preconsolidation pressure is well passed. The stress-strain curve will thus climb to new stress-strain relations valid for the actual rate as the test goes on.

While there is a well-established procedure available to evaluate the CRS test, there is at present no similar method for evaluation of the CGT or CLT tests which accounts for the varying rates of strain in the tests.

Conclusions

The CRS test is an automatic continuous consolidation test that will provide continuous relations of stress-strain and permeability-strain and thus coefficients of consolidation-stress or strain. The results are directly compatible with the

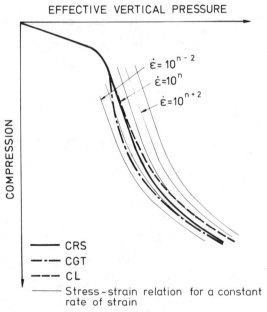

FIG. 19—*Schematic relation between a CRS test at low rate of strain, a CGT test with a low gradient, and a CL test with low relation u_b/σ'.*

results from old standard incremental tests but, because it is continuous, the possibility of subjective evaluation is avoided.

To obtain a completely defined stress-strain curve with incremental tests requires several parallel tests. The mathematical expression for the stress-strain curve has also been shown to be somewhat different from that used previously. The duration of the test is reduced from a normal time of over a week to one to two days. Cases where the standard rate might have to be lowered are the same as those cases where the duration of the load increments might have to be increased, and the time reduction is then even greater. The time for the test setup is about the same in a CRS test as in a standard incremental test, but all the readings are automatically recorded and the calculation and plotting of the results are also automatic. The final evaluation is a very simple and fast procedure.

On the other hand, no information on the rate of secondary compression is obtained in a CRS test. If this information is required, either empirical relations have to suffice or supplementary incremental tests have to be performed.

The CRS test, like all other oedometer tests, suffers from all the shortcomings involved in testing a small volume of a soil that normally has considerable natural variation.

Acknowledgments

The research on continuous consolidation tests has been sponsored by the Swedish Council for Building Research, the Swedish Road Administration, and internal funds at the Swedish Geotechnical Institute and Chalmers University of Technology.

References

[1] Terzaghi, K. and Frölich, O., *Theorie der Setzung von Tonschichten,* F. Deutike, Vienna, 1936.
[2] Leonards, G. A., "Engineering Properties of Soils," in *Foundation Engineering,* McGraw-Hill, New York, Chapter 2, 1962.
[3] Bjerrum, L., "Problems of Soil Mechanics and Construction on Soft Clays and Structurally Unstable Soils (Collapsible, Expansive and Others)," in *Proceedings,* 8th International Conference on Soil Mechanics and Foundation Engineering, Moscow, Vol. 3, 1973, pp. 111–159.
[4] Crawford, C. B., "Interpretation of the Consolidation Test," *Journal of the Soil Mechanics and Foundations Division,* American Society of Civil Engineers, Vol. 90, No. SM5, 1964, pp. 87–102.
[5] Smith, R. E. and Wahls, H. E., "Consolidation Under Constant Rates of Strain," *Journal of the Soil Mechanics and Foundations Division,* American Society of Civil Engineers, Vol. 95, No. SM2, 1969, pp. 519–539.
[6] Wissa, A. E. Z. and Heiberg, S., "A New One-Dimensional Consolidation Test," Massachusetts Institute of Technology Soils Publication 229, Cambridge, MA, 1969.
[7] Lowe, J., III, Jonas, E., and Obrician, V., "New Concepts in Consolidation and Settlement Analysis," *Journal of the Geotechnical Engineering Division,* American Society of Civil Engineers, Vol. 100, No. GT6, 1974, pp. 571–613.
[8] Sällfors, G., "Preconsolidation Pressure of Soft High Plastic Clays," Ph.D. thesis, Chalmers University of Technology, Gothenburg, Sweden, 1975.
[9] Larsson, R. and Sällfors, G., "Beräkning av sättningar i lera," *Väg- och Vattenbyggaren,* No. 3, 1981, pp. 39–42.

[*10*] Larsson, R., "Drained Behaviour of Swedish Clays," Report 12, Swedish Geotechnical Institute, Linköping, Sweden, 1981.

[*11*] Suklje, L., "The Analysis of the Consolidation Process by the Isotache Method," in *Proceedings*, 4th International Conference on Soil Mechanics and Foundation Engineering, London, Vol. 1, 1957, pp. 200–206 and Vol. 3, 1957, pp. 107–109.

[*12*] Bjerrum, L., "Seventh Rankine Lecture; Engineering Geology of Norwegian Normally-Consolidated Marine Clays as Related to Settlements of Buildings," *Geotechnique*, Vol. 17, No. 2, 1967, pp. 81–118.

[*13*] Bjerrum, L., "The Effect of Rate of Loading on an p_c-Value Observed in Consolidation Tests on Soft Clays," written discussion in Session III, *Proceedings*, American Society of Civil Engineers Special Conference, Purdue University, Lafayette, IN, Vol. 3, 1972, pp. 167–168.

[*14*] Tokheim, O. and Janbu, N., "A Continuous Consolidation Test," Internal Report, Norwegian Institute of Technology, Trondheim, 1976.

[*15*] Tavenas, F., Leblond, P., Jean, P., and Leroueil, S., "The Permeability of Natural Soft Clays," *Canadian Geotechnical Journal*, Vol. 20, No. 4, 1983, pp. 629–660.

[*16*] Samson, L., Leroueil, S., Morin, P., and Le Bihan, J. P., "Pression de préconsolidation des argiles sensibles," D.S.S. Contract 15X79-00026, Report prepared for the Division of Building Research of the National Research Council of Canada, Ottawa, 1981.

[*17*] Leroueil, S., "Discussion in *Proceedings*, 10th International Conference on Soil Mechanics and Foundation Engineering, Stockholm, Vol. 4, 1981.

[*18*] Leonards, G. A., Personal communication, 1975.

[*19*] Svensson, B., "Geotekniska egenskaper hos mycket fasta utländska leror," Report, Swedish Council for Building Research, Stockholm, 1983.

[*20*] Mesri, G., "Coefficient of Secondary Compression," *Journal of the Soil Mechanics and Foundations Division*, American Society of Civil Engineers, Vol. 99, No. SM1, 1973, pp. 123–137.

[*21*] Chang, Y. C. E., "Long Term Consolidation Beneath the Test Fills at Väsby, Sweden," Report 13, Swedish Geotechnical Institute, Linköping, Sweden, 1981.

[*22*] Janbu, N., Tokheim, O. and Senneset, K., "Consolidation Test with Continuous Loading," in *Proceedings*, 10th International Conference on Soil Mechanics and Foundation Engineering, Stockholm. Vol. 1, 1981, pp. 645–654.

G. Sandbaekken,[1] T. Berre, [1] and S. Lacasse[1]

Oedometer Testing at the Norwegian Geotechnical Institute

REFERENCE: Sandbaekken, G., Berre, T., and Lacasse, S., **"Oedometer Testing at the Norwegian Geotechnical Institute,"** *Consolidation of Soils: Testing and Evaluation, ASTM STP 892*, R. N. Yong and F. C. Townsend, Eds., American Society for Testing and Materials, Philadelphia, 1986, pp. 329–353.

ABSTRACT: The paper describes the test equipment, the testing procedures, and the interpretation of the results from oedometer tests as practiced today by the Norwegian Geotechnical Institute. Both the incremental loading and the constant-rate-of-strain methods are used. The test procedures and the interpretation include steps to account for the effects of sample disturbance. A new method for determination of the coefficient of consolidation is presented.

KEY WORDS: oedometer test, clay, sample disturbance, procedures, coefficient of consolidation

Oedometer tests were originally performed with an incremental loading procedure. Continuous loading tests were introduced and reported by Wissa and Heiberg [1], Wissa et al [2], Lowe et al [3], Smith and Wahls [4], Aboshi et al [5], and Janbu et al [6] among others. The continuous loading method includes constant rate of strain, controlled gradient, and constant rate of loading. In 1980 the Norwegian Geotechnical Institute (NGI) developed a fully automated continuous loading oedometer test device. The constant-rate-of-strain loading method was preferred to a controlled gradient because it was found mechanically simpler and easier to run.

Oedometer testing at NGI includes special trimming and mounting procedures and an unloading-reloading loop to correct the stress-strain curve for the sample disturbance affecting the behavior of a soil, especially at stresses lower than the preconsolidation stress.

This paper describes the oedometer cells, the specimen preparation techniques, the test procedures, and the interpretation of the tests as done at NGI. Typical results and the soil parameters inferred from the test are given.

---

[1] Norwegian Geotechnical Institute, Oslo, Norway.

Oedometer Cells

Three oedometer cells with soil specimen cross-sections of 20, 35, and 50 cm^2 are in use (Figs. 1 and 2). The height of the specimens is typically 20 mm, but 30-mm-high specimens are sometimes used with 50-cm^2 cross sections. The stainless steel oedometer ring has a highly polished inner surface. The 20-cm^2 ring has its own cutting edge. With the 35- and 50-cm^2 rings, an additional cutting ring is used to trim and mount the specimen. The top cap and the base plate, both made of chromed brass, are provided with 5-mm-thick porous stones to which two drainage tubes are connected. The diameter of the top cap is 0.15 to 0.25 mm less than the inner diameter of the oedometer ring. This type of oedometer allows one to:

1. Flush the porous stones with water after mounting of the specimen. In addition, evaporation from the specimen is prevented before flushing of the stones, which is important in the case of specimens mounted with dry filter stones.
2. Perform permeability tests and, in constant-rate-of-strain tests, measure pore pressure at the bottom stone.

The oedometer cell has also the possibility for back-pressuring if the drainage tubes from the top cap are removed and a chamber is placed around the specimen instead of the Perspex guiding ring (Figs. 3 and 4). The back-pressuring capacity of the chamber is 700 kN/m^2. A bellofram seals the chamber around the piston.

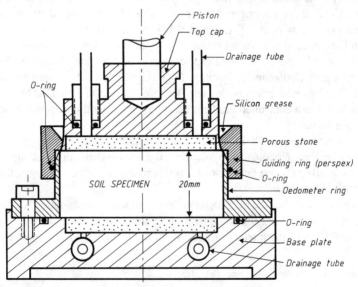

FIG. 1—*Schematic drawing of 20-cm^2 oedometer cell.*

FIG. 2—*20- and 50-cm² oedometer cells.*

Trimming Equipment and Procedure

Two different methods of mounting the specimen into the oedometer ring exist. For very soft clays (undrained shear strength less than about 12 kN/m²), the soil is pushed vertically from the sample tube directly into the oedometer ring. In this case, the extrusion is done continuously and at a rate of about 5 mm/min. For stiffer clay, the sample is first extruded and trimmed down to a diameter

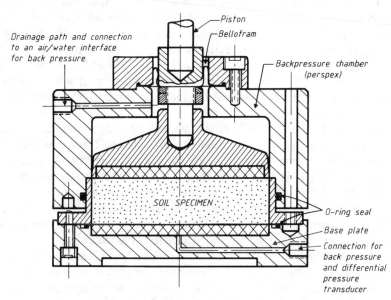

FIG. 3—*Back-pressure chamber for oedometer cell.*

FIG. 4—*20- and 50-cm² oedometer cells with back-pressure chamber.*

slightly larger than the oedometer ring, and then pushed into the ring. To minimize evaporation from the soil specimen, the samples are trimmed and mounted in a room with humidity around 90%.

Very Soft Clay

Figure 5 illustrates the vertical extrusion of a sample directly into the oedometer ring. The oedometer ring is fastened upside down onto a yoke which is fixed to two guiding rods. The ring is coated with liquid paraffin oil to minimize wall friction. To provide the clay with as much lateral support as possible during the trimming operation, the distance between the top of the sample tube and the cutting ring is kept at a minimum (less than 10 mm). The extrusion rate is kept constant and slow enough to allow cutting away of the excess material with a spatula. Extrusion is stopped when the soil specimen protrudes 5 to 10 mm above the oedometer ring.

A wire is used to cut the clay several times under the edge of the cutting ring. Two very thin bronze plates are pushed into the cut previously made by the wire. The plates penetrate from each side of the specimen. A single bronze plate may also be used, but the clay above the plate may then tilt due to the insertion of the plate. The clay is several times below the brass plates with a wire, and while the wire is drawn quickly through the clay, the yoke holding the oedometer ring and the bronze plates is raised. The clay inside the oedometer ring is now completely separated from the rest of the sample in the tube.

The clay protruding from the top of the oedometer ring is cut away by means of a wire saw and a steel straight-edge before the base plate is mounted. The yoke is unfastened from the guide rods and rotated so that the cutting plates come on top. The cutting plates are removed and the top of the specimen is trimmed level.

For 35- and 50-cm² cells, the cutting ring on top of the oedometer ring is removed before a final trimming of the top surface of the specimen is done. The Perspex guiding ring and top cap are then put in place. The groove between the top cap and the Perspex guiding ring is filled with silicon grease to prevent evaporation from the specimen.

Medium-to-Stiff Clay

A 40- to 50-mm-high sample is extruded from the tube and trimmed down to a diameter about 5 mm larger than the inner diameter of the oedometer ring. The ring is coated with paraffin. The sample is pushed incrementally into the ring held by the yoke which is fixed to the two guiding rods. The excess material is cut away by means of a spatula. The remaining steps in the procedure are the same as described earlier for very soft clay except that the plate on which the specimen is placed replaces the two bronze cutting plates.

Mounting with Dry Filter Stones

At NGI, oedometer specimens are always mounted with dry filter stones to prevent swelling of the unloaded specimen. Figure 6 illustrates the difference observed in the behavior of a stiff clay mounted with dry and saturated filters. The main reason for the difference between the two curves is believed to be that the specimen with wet stones absorbs water from the stones and thereby softens. One may argue that dry stones may absorb water from the specimen and thereby cause an extra compression, which again could explain the higher p'_c-value for this specimen. However, when clay specimens are left overnight at a vertical stress equal to $0.25\ p'_0$, almost no compression is observed.

FIG. 5—*Specimen extruded from sample tube directly into the oedometer ring.*

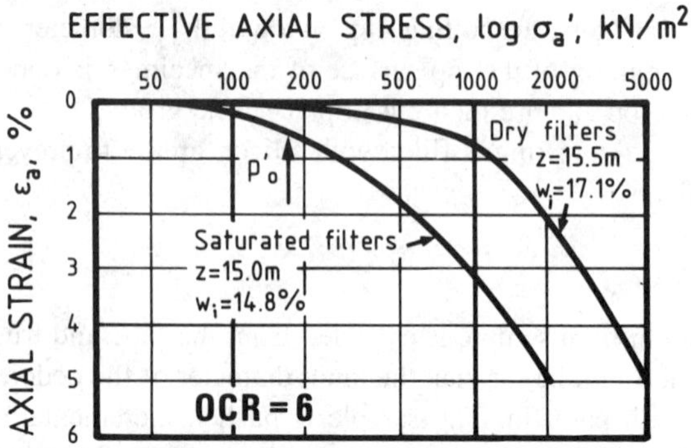

EFFECTIVE AXIAL STRESS, log σ_a', kN/m²

FIG. 6—*Results of incremental loading oedometer tests on stiff clay specimens mounted with dry and saturated filters* [7].

For quick clays, a wet filter paper (Whatman No. 54) is placed on the top surface of the specimen to prevent squeezing of the soil up in between the top cap and the oedometer ring. Otherwise filter paper is not used.

Incremental Loading (IL)

Apparatus

The loading apparatus (Fig. 7) consists of a standard Geonor loading frame with capacity of 10 kN and a 10:1 lever arm. The vertical compression of the specimen is measured by a dial gage with divisions in 0.01 mm.

Loading Procedure

Figure 8 illustrates the incremental loading procedure. The first load increment is usually equal to 25% of the *in situ* vertical effective stress, p'_0. For heavily overconsolidated clays, one may apply p'_0. A load increment ratio, $\Delta p/p$, equal to 1.0 (i.e., doubling the load for each increment) is used for loading up to p'_0. The stress at which unloading starts (called p'_1) is equal to either once or twice the preconsolidation stress p'_c (see later section on Sample Disturbance). When loading from p'_0 to p'_1, the load increment ratio is reduced to 0.5 to better define the stress-strain (load-compression) curve around the preconsolidation stress. The increment duration is usually 2.5 h (three increments in a workday). For most clays tested at NGI, this duration allows the specimen to reach end-of-primary consolidation. However, if the time compression curves (with Taylor's method) indicate that more time is required to reach end-of-primary, the increment duration is increased.

At p'_1, the stress is kept constant overnight. The compression reading after one night with constant stress is assumed to come close to the virgin curve for a test where all the increments have lasted 24 h. If it is assumed that the 2.5-h and the overnight virgin curves are parallel on a semilogarithmic plot [8], the overnight reading may be used to estimate the 24-h virgin curve and the 24-h p'_c. For tests where oedometer creep parameters are required, the stress may be held constant a week or even longer. At the end of the p'_1-increment, the porous stones are saturated with water of the same salinity as the pore water of the clay, and a constant-head permeability test is carried out (described below). The specimen is unloaded to p'_0, usually in two steps, and the stress is again kept constant overnight.

From p'_0 to $(p'_0 + p'_1)/2$ on the reloading curve, the load increment ratio is 1.0. Thereafter the load increment ratio is 0.5 until the stress has reached the virgin curve. The rest of the oedometer test is performed with a load increment ratio of 1.0. The specimen is loaded to about $9\,p'_c$ or to the loading capacity of the apparatus. The maximum load is kept constant overnight (16 to 24 h) before a second constant head permeability test is performed. The whole loading procedure is performed with free drainage from the top and bottom filter stones.

FIG. 7—*Incremental loading apparatus and constant-head permeability test setup.*

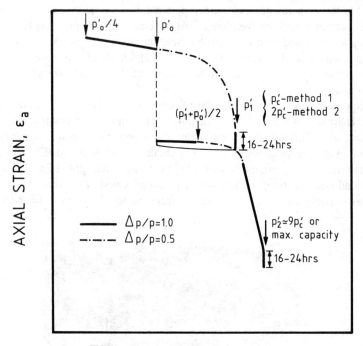

FIG. 8—*Incremental loading procedure.*

Correction for False Deformation

The initial deformation, as determined from a plot of deformation versus square root of time for the first load increment, is considered as false (although this is not quite true). For the rest of the test, false deformation is found from a calibration curve which has been determined with a steel dummy specimen between the porous stones.

Determination of Coefficient of Consolidation

The coefficient of consolidation, c_v, is computed from the coefficient of permeability, k, and the tangent constrained modulus, M (see later section on Determination of c_v). The relationship between k and axial strain, ϵ_a, called the k-line, is determined from time compression plots and from the two constant-head permeability tests run at p'_1 and p'_2. Axial strain and the logarithm of the permeability coefficient are assumed to plot on a straight line.

Time-Compression Plots and Coefficient of Permeability

From at least three load increments on the virgin stress-strain curve between p'_c and p'_2, the coefficient of permeability is calculated by means of Taylor's construction (Fig. 9).

Constant-Head Permeability Tests

The coefficient of permeability is measured twice during an incremental oedometer test. A schematic drawing of the portable apparatus used at NGI is shown on Fig. 10. Water of the same salinity as the pore water in the clay is pressed through the specimen, from bottom to top, by a mercury column in a U-shaped saran tubing. The amount of water flowing in and out of the specimen is measured separately. The test is continued until the water inflow and outflow are approximately equal. Almost no deformation of the specimen takes place during the permeability test, first because the test is carried out after the effective vertical stress has been kept constant overnight, and second because the applied gradient causes only a minor reduction in the effective stress from the stress on the virgin compression line.

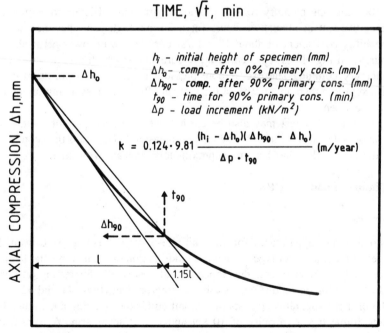

FIG. 9—*Coefficient of permeability k from time-compression curve.*

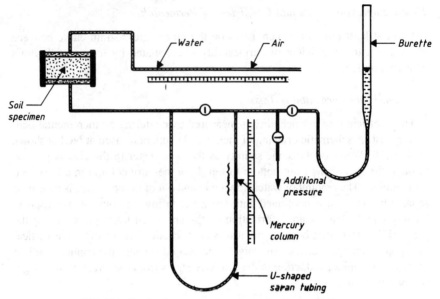

FIG. 10—*Sketch of apparatus for constant-head permeability test.*

A 100-mm-long mercury column provides a gradient sufficient to complete a permeability test in about two hours for a 20-mm-high clay specimen with a permeability coefficient of about 10^{-7}cm/s. For clays with lower permeability, the burette in Fig. 10 is replaced by a pressure source which acts in series with the mercury column. The excess pore pressure in the bottom porous stone is always kept less than about 10% of the vertical effective stress acting on the top of the specimen.

The permeability test may also be carried out with back pressure. Only the amount of water flowing into the specimen is measured. The test is then continued until the rate of flow has become approximately constant with time.

Continuous Loading (CRS)

Apparatus

Two different apparatuses for constant-rate-of-strain loading have been developed at NGI: an older type, which is a modification of the incremental loading device, and a new apparatus developed in 1984 especially for testing on board the drilling ships in the difficult North Sea environment (Figs. 11 and 12).

The onshore apparatus consists of a modified Geonor loading apparatus. The loading frame has a capacity of 10 kN and a 10:1 lever arm. A rectangular proving ring or a load cell for measuring vertical load is mounted in the vertical deadweight hanger which can be secured to a screw-thread jack (Fig. 11). The motor (12 V dc), placed on the front of the apparatus, drives a shaft connected to the screw-thread jack. The standard motor can apply a constant axial strain rate between almost 0 and 4% per hour on a 20-mm-high specimen.

The offshore apparatus consists of a loading frame and a ball screw assembly controlled by a stepping motor (24 V dc) (Fig. 12). The load capacity for the apparatus is 10 kN. An initial load up to 100 N can be applied on the specimen by dead weights. This new apparatus facilitates mounting of the oedometer cell into the loading apparatus under storm conditions, and minimizes the effects on the test from sea waves and vibrations in the ship. The new apparatus has proved much better than the older offshore CRS device in both respects (see next section).

For both apparatuses, the vertical load is measured with electronic load measuring devices. Vertical displacement is measured by a linear variable differential transducer (LVDT) mounted on the back side of a dial gage. Pore pressure is usually measured with a total pressure transducer. If back pressure is used, the difference between the pore pressure at the bottom of the specimen and the back pressure at the top is measured with a differential pressure transducer.

FIG. 11—*Constant-rate-of-strain oedometer used onshore.*

FIG. 12—*Constant-rate-of-strain oedometer used offshore.*

Data Acquisition and Regulation System

A microcomputer logs the test data at preselected time intervals. The data are recorded on a magnetic disk and then transferred to the main NGI computer for storage, computations, and plotting of standard diagrams. Offshore, the test data are stored on tapes or floppy disks and the test results are processed by the microcomputer. The interfacing box between the microcomputer and the loading apparatus was developed at NGI. The interfacing box provides the excitation voltage (6.0 V) to the transducers and the adjustable voltage to the motor. It also switches the motor on or off in one of two directions, depending on the orders from the computer.

The computer program gives the operator a choice of loading procedures. The operator inputs the following parameters:

- Time interval between readings.
- Stress before unloading (p'_1).

- Maximum stress after reloading (p'_2).
- Time interval under constant stress at p'_1 and/or p'_2.
- Number of unload-reload loops.

The loading procedure calls for very little manual intervention. Except for the first loading to 0.25 p'_0, saturation of the porous stones, and connection of the pore pressure transducer to the base plate, the entire test is performed by the microcomputer.

Loading Procedure

The loading procedure is very similar to the incremental loading procedure except that the load is applied at a constant rate of strain. The rate is chosen manually such that the excess pore pressure at the undrained base of the specimen never exceeds 5 to 10% of the total vertical stress.

Prior to the start of the continuous loading, a vertical stress approximately equal to $p'_0/4$ is applied by dead weights in one step. The vertical stress is thereafter increased continuously at a constant rate of strain. Saturation of the porous stones is done at the overburden stress p'_0, or above the assumed initial negative pore pressure in the specimen, if higher than p'_0. The porous stones are flushed with CO_2 for 5 min before saturation with water of the same salt concentration as the pore water of the specimen. For a test with back pressure, the latter is applied immediately after saturation of the porous stones. If the specimen tends to swell after saturation of the stones, the load is manually increased to prevent swelling. The remainder of the test is completely automatic.

As previously described, the specimen is continuously loaded to p'_1, where the load is held constant overnight (16 to 24 h). The excess pore pressure and the vertical displacement are recorded during this period. The specimen is then unloaded to p'_0. Some negative pore pressures develop during this unloading, but they are allowed to dissipate before reloading. Finally, the specimen is loaded to the maximum stress p'_2. Constant-head permeability tests are sometimes carried out to check the values back-calculated from the measured pore pressures. The permeability tests are done in the same manner as for the incremental loading tests.

The procedure described above is used onshore. Offshore, the periods with constant stress are considerably reduced in order to increase the number of tests per apparatus.

For many clays tested at NGI a rate of axial strain of 0.5 to 1.0% per hour seems adequate and maintains the ratio of pore pressure to total axial stress between 2 and 7% throughout the test. Readings of total axial load, axial displacement, and excess pore pressure are taken every 5 to 10 min. The stress-strain curve is plotted simultaneously on the monitor of the microcomputer.

Correction for false deformation is done exactly as described earlier for the incremental loading tests.

Test Results

Incremental Loading

Figure 13 presents the stress-strain curve and permeability line (the k-line) from an incremental loading oedometer test on a lean North Sea clay. The data are generally plotted both on a semilog scale and on a natural scale (the latter recommended by Janbu [9] and Janbu et al [6]. In addition to the stress-strain curve, the plots provide coefficient of consolidation, coefficient of permeability, and constrained modulus. The interpretation of the test is discussed later.

Continuous Loading

Figure 14 gives an example of the results obtained from a CRS oedometer test on a soft plastic marine clay. The variation of the excess pore pressure during loading and reloading is also shown. Appendix I presents a few details of the method of interpretation used for the constant-rate-of-strain oedometer test.

The first CRS oedometer tests carried out offshore (in 1983) were performed with the device shown in Fig. 11. Marginal weather conditions and vibrations on the ship resulted in wavy stress-strain curves as illustrated in Fig. 15a. The uniaxial type of loading frame now used offshore (Fig. 12) solved this difficulty, as shown by the comparative plots in Fig. 15a.

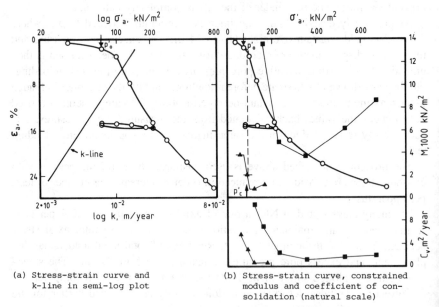

(a) Stress-strain curve and
    k-line in semi-log plot

(b) Stress-strain curve, constrained
    modulus and coefficient of con-
    solidation (natural scale)

FIG. 13—*Results of incremental loading oedometer test on lean North Sea clay.*

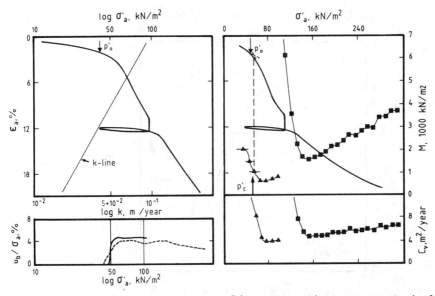

(a) Stress-strain curve,
    k-line and excess pore
    pressure in semi-log plot

(b) Stress-strain curve, constrained
    modulus and coefficient of con-
    solidation (natural scale)

FIG. 14—*Results of constant-rate-of-strain oedometer test on soft plastic clay.*

Sample Disturbance and Correction of Stress-Strain Curve

Figure 15*b* shows the stress-strain curves for (1) a slightly disturbed clay specimen, and (2) a perfect specimen of the same clay. Both specimens of undisturbed plastic clay from Drammen were initially artificially preconsolidated in oedometer cells. (This part of the loading is not shown in Fig. 15*b*.) One of the specimens was completed unloaded, mounted into a new oedometer cell with a smaller diameter and reloaded. This reloading curve is the "slightly disturbed specimen" curve in Fig. 15*b*. The other specimen was reloaded without doing anything which could disturb it. This reloading curve is the "perfect specimen" curve in Fig. 15*b*. The shape of the curve for the "slightly disturbed specimen" is not too bad with respect to disturbance. In fact, if the curve for the perfect specimen had not been there, the test would have been classified as very good. Still, the difference between the two curves is quite important. The curve for the slightly disturbed specimen has been corrected for disturbance according to the method proposed by Schmertmann [9]. It is seen that the corrected curve agrees very well with the curve for the perfect specimen.

When an oedometer curve is corrected for sample disturbance, the corrected curve usually deviates quite a lot from the uncorrected one as demonstrated in Fig. 15*b*. Therefore, correcting properly for sampling disturbance seems to be much more important than whether the load increments should be large or small,

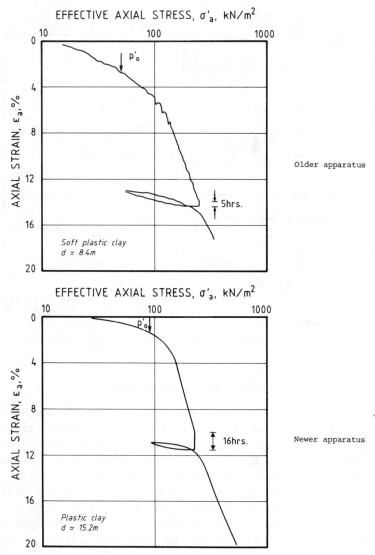

FIG. 15a—*Comparison of two types of CRS apparatus used offshore.*

whether the increment duration should be short or long, or whether the loading should be incremental or continuous. Consequently, extreme care should be taken:

- To obtain samples of high quality.
- To avoid, as much as possible, mechanical disturbance during extrusion, trimming, and mounting of the specimen into the oedometer cell.
- To avoid gaps between the specimen and the oedometer ring during mounting.

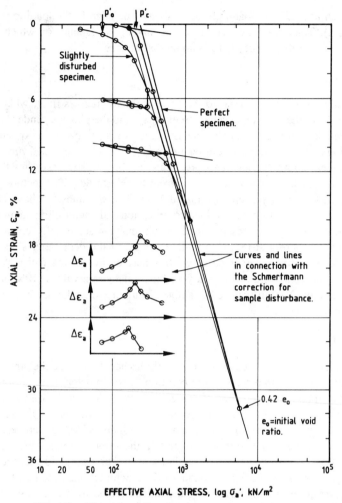

FIG. 15b—*Effect of sampling disturbance on stress-strain curve.*

- To prevent swelling of the specimen in the early stages of the test by using initially dry porous stones.

The disturbance is usually more severe for less-plastic materials. For disturbed samples, at stresses below the preconsolidation stress, the measured moduli tend to be too low, and at stresses above p'_c, the moduli tend to be too high (Fig. 15b).

The derived preconsolidation stress, p'_c, usually tends to become too low for disturbed specimens (Fig. 15b). However, p'_c may also be too high. The relative importance of sample disturbance is more pronounced at lower stresses.

To correct the stress-strain curve for the effects of sample disturbance at low stresses, an unloading-reloading loop is always included in the loading program (see Fig. 8). The stress at which unloading starts (p'_1) depends on which of two methods is used to correct p'_c for sample disturbance.

Method 1

The value of p'_1 is set equal to the preconsolidation stress (p'_c), where p'_c is estimated from the undrained shear strength, s_u, and the plasticity index, I_p [11].

The detailed procedure for this determination of p'_c is given in Appendix II. The oedometer reloading curve after unloading to p'_0 is then used *directly* for the settlement computations. This procedure has proved successful for the non-fissured heavily overconsolidated clays from the North Sea. The method should not be used for slightly overconsolidated clays. The s_u-values for the evaluation of p'_c are determined from unconfined compression and unconsolidated undrained triaxial tests. The p'_c-values obtained in this manner are usually considerably higher than the stress at the maximum curvature of the oedometer stress-strain curves. At least for uncemented clays, this may be due to the fact that the s_u-values for such clays are reached at very high axial strain values (10% or more) and are therefore less influenced by sample disturbance than the oedometer stress-strain curve around p'_c.

Method 2

The value of p'_1 is set equal to or slightly higher than twice the stress corresponding to the maximum curvature point on the oedometer stress-strain curve. The stress-strain curve may then be corrected for sample disturbance by Schmertmann's procedure [10], which requires an initial determination of p'_c. The value of p'_c is usually determined by Casagrande's [12], Janbu's [13], or Schmertmann's [10] methods, which all give approximately the same estimate. With Janbu's method, both the stress-strain and modulus-strain curves should be used to determine p'_c, as shown in Figs. 13 and 14. The Schmertmann method to correct the stress-strain curve includes using the unload-reload loop to account for disturbance effects during the initial loading to p'_c.

For the plastic, slightly overconsolidated Norwegian clays, the *in situ* p'_c-values may be considerably lower than even the 24-h oedometer p'_c. For plastic Drammen clay ($I_p \approx 30\%$) for example, the 100-year *in situ* p'_c seems to be about 25 to 30% lower than the 24-h oedometer p'_c. These figures have been estimated from data presented by Bjerrum [14] and Foss [15]. The indications are that the difference between the *in situ* p'_c and the oedometer p'_c increases with increasing plasticity.

Determination of c_v by the k-Line Method

For both incremental and continuous oedometer tests, the coefficient of consolidation c_v is computed from

$$c_v = k \cdot M \frac{1}{\gamma_w}$$

where k is the coefficient of permeability and M the tangent constrained modulus on the stress-strain curve.

For incremental tests, k is computed by the Taylor curve-fitting method for at least three load increments. Constant-head permeability tests are run at p'_1 and at the end of the test. The permeability values are plotted on a semilogarithmic scale versus vertical strain. For clays with a plasticity index below 20%, the two sets of k-values usually agree reasonably well, and plot on a straight line, the k-line. For more plastic clays, the k-values based on the time-compression curves tend to be lower than the directly measured values (see Fig. 16), more so for smaller load increments. In the case of such disagreement, the directly measured k-values are used to compute c_v. The k-values from the time-compression curves are believed to be too low because "delayed compression" in the soil skeleton retards the rate of compression. Delayed compression is a reduction in volume at unchanged effective stress if there were no water in the pores of the specimen (no hydrodynamic time lag) [14].

For sandy clays and silts, a difference between the two sets of k-values is also often seen (Fig. 17). The shape of the time-compression curve for this soil type is believed to be also largely governed by "delayed compression", because although the delayed compression in the soil skeleton is not as pronounced as for plastic clays, delayed compression still predominates the hydrodynamic time lag, because of the very high permeability of the material. In addition, one finds a large scatter in the permeability values back-figured from the time-compression curves because primary consolidation takes place so rapidly that it is difficult to

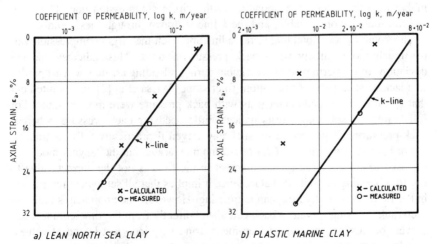

a) *LEAN NORTH SEA CLAY*   b) *PLASTIC MARINE CLAY*

FIG. 16—*Comparison of* k-*line for lean and plastic clays (incremental loading test).*

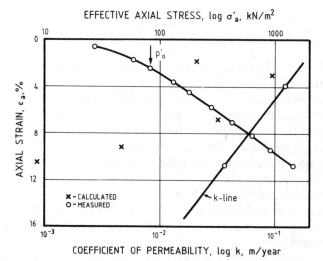

FIG. 17—*Oedometer curve for a clayey sandy silt (incremental loading test).*

interpret the time curves. For such soils, the directly measured k-values are also used to compute c_v.

For continuous tests, the value of the coefficient of permeability is computed from the excess pore pressure measured at the bottom of the specimen. The experience so far indicates that these k-values agree with constant-head test values over a wider range of soil types than the values back-calculated from time-compression curves in incremental tests. Constant-head permeability tests are carried out now and then to ensure that the k-values based on the excess pore pressure are realistic. As shown on Fig. 18, the k-values at the initial stage of the CRS test and during reloading (in the unloading-reloading loop) are much higher than the values corresponding to the virgin compression line. The latter are considered more reliable, and the k-line is drawn through these points.

The k-values corresponding to reloading approach the virgin compression values much more quickly when back pressure is used. This indicates that the deviations of the permeability coefficient during reloading are due, at least partly, to a lack of saturation of the bottom filter stone. Wissa et al [2] also pointed out that c_v-values from oedometer tests with back pressure were more reliable. For fully saturated soft clays, tests indicate little need for back pressure, whereas back-pressuring is preferable for stiff clays, even if practically fully saturated.

For both *incremental* and *CRS-tests,* one may argue that the tangent modulus, M, used to compute the total settlement in the field is not the correct value to use for obtaining the field c_v-value, since it implies that all settlement retardation in the field is due to hydrodynamic time lag. However, the approach is believed preferable to using c_v-values back-calculated directly from the time-compression curves, because these c_v-values can mean using very incorrect k- and M-values. The M-values may be incorrect because the ratio between primary and secondary compression may be much higher in the field than in the laboratory because of the usually much longer drainage path in the field. Another advantage with the k-line method is that the c_v-values may be based on modulus values corrected for sample disturbance and k-values from the k-line at the same void ratio as *in situ*. For loads between p'_0 and p'_c, this procedure can lead to c_v-values about three times larger than uncorrected c_v-values, which in turn may be two or three times higher than c_v-values derived directly from time-compression curves. The difference between the uncorrected c_v-values from the k-line method and those back-figured directly from the time-compression curves varies between 0 for clays with low secondary compression to as high as 3 for clays with pronounced secondary compression. The difference and scatter could well be more significant for clayey silts.

Summary

Both incremental and automatic continuous loading oedometer tests are run at the Norwegian Geotechnical Institute. Recently a special uniaxial automatic continuous oedometer has been developed for offshore use.

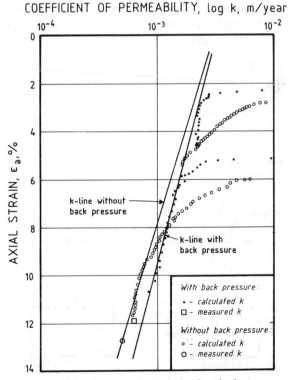

FIG. 18—*CRS oedometer test with and without back pressure.*

The commonly used test procedure includes an unload-reload loop to account for the effects of sampling disturbance, and two overnight loadings under constant effective stress to determine points on the 24-h virgin compression line and thereby the 24-h preconsolidation stress value. The durations of the other load increments are usually 2.5 h. Constant-head permeability tests are carried out before unloading, as a routine step of the incremental loading oedometer test. For constant-rate-of-strain tests, direct permeability measurements are made now and then to check the values back-calculated from the measured pore pressures.

The coefficient of consolidation is determined from the coefficient of permeability and the tangent constrained modulus on the stress-strain curve both for incremental and CRS tests. For incremental tests, the coefficient of permeability is based on direct measurements and the time compression curves.

It is essential in oedometer tests to correct properly for the effects of sampling disturbance. The procedures used to account for these are presented.

Acknowledgments

The authors wish to express their gratitude to their NGI colleagues who have contributed with ideas and careful work in the oedometer test section of the laboratory: K. Iversen, P. Morstad, E. Heier, R. Nordsether, and S. Bakke. K. Iversen, E. Heier, and the first author developed and implemented the automatic CRS apparatus. The helpful reviews by K. Høeg, G. Aas, and O. Gregersen, of NGI, are gratefully acknowledged.

APPENDIX I

Interpretation of CRS Oedometer Test

The theoretical solution for computing M, c_v, and k is based on the equations developed by Wissa et al [2]. The parameters M and c_v are assumed constant over an increment of time. The pore pressure distribution in the specimen is assumed parabolic (Fig. 19). The following test data are recorded:

- Δt time interval between readings,
- σ_a total stress at top of specimen (assumed uniform in specimen),
- u_b excess pore pressure at undrained specimen bottom,
- ϵ_a axial strain.

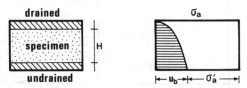

FIG. 19—*Stress distribution in CRS oedometer specimen.*

The test parameters are obtained as follows:

Average effective stress on the specimen:

$$\sigma'_{a_{av}} = \sigma_a - \frac{2}{3} u_b$$

Oedometer modulus:

$$M = \frac{\Delta \sigma'_a}{r \, \Delta t}$$

where r is the average rate of strain and $\Delta\sigma'_a$ the change in effective stress over an increment of Δt.

Modulus of compressibility:

$$m_v = \frac{1}{M}$$

Coefficient of permeability:

$$k = \frac{1}{2} \frac{r}{u_b} H^2 \cdot \gamma_w$$

where H is the height of the sample and γ_w the unit weight of water.

Coefficient of consolidation:

$$c_v = \frac{M \, k}{\gamma_w}$$

APPENDIX II

Determination of p'_c of Stiff Clay from Plasticity Index and Undrained Shear Strength

For very stiff clays, the value of p'_c is often estimated from the undrained shear strength, s_u, and the plasticity index, I_p, as previously described by Berre [11] and Andresen et al [7]. In this method, the value of s_u is determined from unconfined compression, unconsolidated undrained, or pocket penetrometer tests. The preconsolidation stress, p'_c, is obtained from

$$p'_c = s_u/(s_u/p'_c)$$

where the ratio (s_u/p'_c) is taken from the curve in Fig. 20 for an estimated value of p'_c/p'_0. If the first estimate of p'_c/p'_0 proves to be too far off, the computation of p'_c is repeated once or twice, with new values of p'_c/p'_0. This method has been used successfully for very stiff nonfissured North Sea clays.

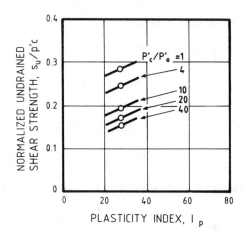

FIG. 20—*Normalized undrained shear strength as a function of* p'_c/p'_0 *ratio and plasticity index (data from Drammen clay).*

References

[1] Wissa, A. E. Z. and Heiberg, S., "A New One-Dimensional Consolidation Test," Department of Civil Engineering Research Report 69-9, Soil Publication, No. 229, Massachusetts Institute of Technology, Cambridge, MA, 1969.

[2] Wissa, A. E. Z., Christian, J. T., Davis, E. H., and Heiberg, S., "Consolidation at Constant Rate of Strain," *Journal of the Soil Mechanics and Foundation Engineering Division,* American Society of Civil Engineers, Vol. 97, No. SM10, 1971, pp. 1393–1413.

[3] Lowe, J., Jonas, E., and Obrician, V., "Controlled Gradient Consolidation Test," *Journal of the Soil Mechanics and Foundation Engineering Division,* American Society of Civil Engineers, Vol. 95, No. SM1, *Proceedings* Paper 6327, 1969, pp. 77–97.

[4] Smith, R. E. and Wahls, H. E., "Consolidation Under Constant Rates of Strain," *Journal of the Soil Mechanics and Foundation Engineering Division,* American Society of Civil Engineers, Vol. 95, No. SM2, 1969, pp. 519–539.

[5] Aboshi, H., Yoshikumi, H., and Maruyama, S., "Constant Loading Rate Consolidation Test," *Soil and Foundations,* Vol. 10, No. 1, 1970, pp. 43–56.

[6] Janbu, N., Tokheim, O., and Senneset, K., "Consolidation Tests with Continuous Loading" in *Proceedings,* 10th International Conference on Soil Mechanics and Foundation Engineering, Stockholm, Vol. 1, 1981, pp. 645–654.

[7] Andresen, A., Berre, T., Kleven, A., and Lunne, T., "Procedures Used to Obtain Soil Parameters for Foundation Engineering in the North Sea," *Marine Geotechnology,* Vol. 3, No. 3, 1976, pp. 201–266.

[8] Taylor, D.W., *Fundamentals of Soil Mechanics,* Wiley, New York, 1948.

[9] Janbu, N., *Grunnlag i Geoteknikk (Elementary Soil Mechanics),* Tapir, Trondheim, Norway, 1970.

[10] Schmertmann, J. S., "The Undisturbed Consolidation of a Clay," *Transactions,* American Society of Civil Engineers, Vol. 120, 1955, p. 1201.

[11] Berre, T., "Triaxial Testing at the Norwegian Geotechnical Institute," NGI Publication No. 134, Oslo, 1981; also published in *Geotechnical Testing Journal,* ASTM, Vol. 5, No. 1/2, 1983, pp. 3–17.

[12] Casagrande, A., "The Determination of the Preconsolidation Load and Its Practical Influence" in *Proceedings,* 1st International Conference on Soil Mechanics and Foundation Engineering, Boston, Discussion D-34, Vol. 3, 1936, pp. 60–64.

[*13*] Janbu, N., "The Resistance Concept Applied to Deformation of Soils" in *Proceedings,* 7th International Conference on Soil Mechanics and Foundation Engineering, Mexico City, Vol. 1, 1969, pp. 191–196.
[*14*] Bjerrum, L., "Engineering Geology of Norwegian Normally-Consolidated Marine Clays as Related to Settlements of Buildings," *Geotechnique,* Vol. 17, No. 2, 1967, pp. 81–118.
[*15*] Foss, I., "Secondary Settlements of Buildings in Drammen, Norway" in *Proceedings,* 7th International Conference on Soil Mechanics and Foundation Engineering, Mexico City, Vol. 2, pp. 99–106.

Bernard Dubin[1] and Gérard Moulin[2]

Influence of a Critical Gradient on the Consolidation of Clays

REFERENCE: Dubin, B. and Moulin, G., **"Influence of a Critical Gradient on the Consolidation of Clays,"** *Consolidation of Soils: Testing and Evaluation, ASTM STP 892*, R. N. Yong and F. C. Townsend, Eds., American Society for Testing and Materials, 1986, pp. 354–377.

ABSTRACT: Several studies in the laboratory have shown the existence of a permeability law which is different from Darcy's law and which defines the flow of the pore water in certain natural clays.

Consolidation tests in the laboratory, including the measurement of permeability at different stages of loading, that were carried out on Saint Herblain clay at the E.N.S.M. Soil Mechanics Laboratory prove the existence of major critical gradients ($12 < i_0 < 20$). According to this observation, a consolidation model worked out with a non-Darcian permeability law has been produced by means of a finite-difference computer program.

On the one hand, this model is compared with Terzaghi's consolidation model and, on the other hand, the law that has been established is tested next to the laboratory tests.

KEY WORDS: permeability, clay, laboratory tests, consolidation law

Evaluation of settlement velocity duration for the study and design of civil structures built on clay foundations requires the use of the usual one-dimensional consolidation theory by Terzaghi (1925). As the predictions to which this theory used to lead were not always representative, other models have been proposed; however, it seems they are not much used. In fact, two major parameters in Terzaghi's theory govern consolidation; that is, on the one hand, the permeability which is supposed to be constant for the duration of the consolidation (k = constant) and, on the other hand, the drainage which obeys Darcy's law ($v = ki$).

Consolidation laboratory tests, including a measurement of the permeability carried out on Saint Herblain clay, indicate a nonlinear permeability law. To be more precise, the flow velocity of the pore water is not proportionate to the pressure gradient until a threshold gradient called the "critical gradient" is produced. This phenomenon has already been observed on certain clays by other researchers like Hansbo [1], Abelev [2], and Law and Lee [3].

[1] Elève du C.U.S.T. de Clermont-Ferrand, 1983, Stagiaire Département Génie Civil, Ecole Nationale Supérieure de Mécanique (E.N.S.M.), 1, rue de la Noë, 44072 Nantes CEDEX, France.

[2] Assistant Département Génie Civil, Ecole Nationale Supérieure de Mécanique (E.N.S.M.), 1 rue de la Noë, 44072 Nantes CEDEX, France.

In order to test its incidence, a consolidation model has been worked out. In regard to Terzaghi's theory and so as to measure the phenomenon's consequences, only the hypothesis according to which the drainage obeys Darcy's law has been modified.

Laboratory Tests

The oedometer test is the most common laboratory method for defining the settlement and the consolidation velocity of a compressible soil. However, the conventional oedometer does not allow a direct measurement of the permeability and, moreover, the specimen being laterally confined, the normal and tangential stresses applied to the lateral face are unknown. In order to control the stress field applied to the specimen, the present one was set up in a triaxial cell.

Equipment

We have used a standard triaxial equipment fitted with a Bishop load pressure system; the cell is a 76.2 mm (3 in.) with a lengthened piston, and the specimen initially has a height of 12 mm and a diameter of 76 mm. A field of anisotropical stresses is applied to the specimen by loading a stirrup which is mounted on the axis of the cell. The general layout of the device is shown in Fig. 1. Throughout the test, the vertical deformation of the specimen is measured with a gage (1 unit = $\frac{1}{100}$ mm). Since the K_0-value is known from previous tests, we make sure that the specimen diameter deduced from the volume measurements is varied slightly during the test ($\Delta d < 0.1$ mm).

During the long-duration tests in the triaxial cell, the problem of the water-tightness of the rubber membrane which separates the specimen from the water contained in the cell occurs. In order to solve this problem, the membrane is doubled and covered in three layers of silicone grease. It has been verified that the interface thus obtained (silicone grease, membrane, silicone grease, membrane, silicone grease) is watertight, the leakage volume through interface being less than 0.1 cm³ (6.10×10^{-3} in.³) per month. Volume changes are defined thanks to the burettes used for the measurement of the outflow volume, and are 0.1 cm³ (6.10×10^{-3} in.³) when the burette is used during the consolidation phase or 0.01 cm³ (6.10×10^{-4} in.³) when the burette is used during the phase of permeability measurement. Leaks by evaporation in the burettes are measured by means of reference burettes placed next to them.

Test Principle

The loading process is similar to the one performed for an oedometer test, in which the stress couple (vertical σ_v, radial σ_r) is applied for 24 h. The chosen stress ratio σ_r/σ_v is close to the earth pressure ratio at rest, K_0, measured previously. The height variations of the specimen are measured continuously during the test. A very small secondary settlement ($\cong 3 \times 10^{-2}$ mm) is noticed during

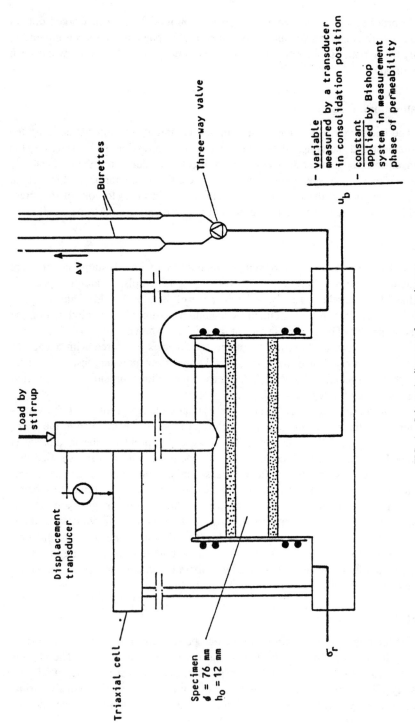

FIG. 1—*Schematic diagram of test equipment.*

the permeability tests. The diameter of the specimen, given by the volume measurement, remains practically unchanged through the test (initial diameter 76.10 mm, final diameter 76.16 mm).

During the consolidation phase, the specimen is drained only on its upper side, and the volume of the outflow Δv measured with a burette. The evolution of the pore pressure u_b is measured at the bottom of the specimen. The permeability tests are performed at the end of a loading. The specimen is always drained on its upper side, and a constant pressure u_b is applied to the bottom of the specimen to produce a constant pore pressure gradient. During the test, the burette which measures the outflow Δv is moved vertically for the level to remain the same (tests with a falling head are possible, too). The test procedure is specified in Fig. 2, the sequence of permeability tests performed through a random variation of the gradients. For instance, in the first test (σ_v = 47.7 kPa), the given gradients are respectively i = 29.1, 16.5, 10.2, 20.1, 34.9, 15.4, and 24.6.

Experimental Tests on Saint Herblain Clay

Characteristics of Saint Herblain Clay—The profiles of the main geotechnical and physical-chemical characteristics of the materials, which are rather recent alluvial deposits, are shown in Figs. 3 and 4. These clays are characterized by a high plasticity and are slightly or moderately organic and overconsolidated at the surface. The variation of the void ratio from a depth is practically linear [e = 2.0 to 3.5 m (11.5 ft); e = 2.8 to 20.0 m (65.6 ft)].

Characteristics of Compression and Permeability Tests Performed on Specimen Sampled at a Height of 4.15 m (13.6 ft)—The specimen had been anisotropically consolidated ($\sigma'_v/\sigma'_r \cong 0.63$) through loading stages applied during 24 h. The evolution of the void ratio e versus the effective stress σ'_v is indicated in Fig. 5, Curve 1, describing a lightly overconsolidated clay (σ'_{v0} = 32 kPa, σ'_p = 70 kPa) (σ'_{v0} = 0.32 tons/ft^2, σ'_p = 0.70 tons/ft^2). By comparison, Curve 2 represents the results of an oedometer test performed on a specimen sampled at a height of 4.30 m (14.1 ft) by the Laboratoire Régional des Ponts et Chaussées in Angers.

Three permeability tests after vertical loadings at 47.7, 125.1, and 231.0 kPa (6.9, 18.1, and 33.5 psi) have been performed; the outflow volume – time relationship is shown in Fig. 6. Thanks to the slopes of the obtained straight lines which define the flow velocities of the pore water for various hydraulical gradients, we are able to draw the graphs shown in Fig. 7. For the small critical gradients, the flow law is not clearly defined. In this area a stopping of the flow, in duration, has been observed, too, and the variation of the flow velocity according to the hydraulic gradient is strictly linear. However, the obtained straight lines, unlike Darcy's law, do not go through the origin; rather, they cut the axis of abscissas i for a value called i_0 or critical gradient. In fact, the i_0-values are rather important as they shift from approximately 14 to 20.

Remarks—The observed phenomena coincide with those already investigated

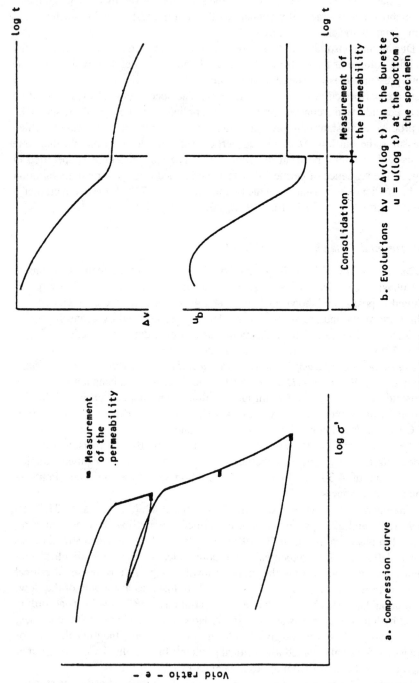

b. Evolutions $\Delta v = \Delta v(\log t)$ in the burette
$u = u(\log t)$ at the bottom of
the specimen

a. Compression curve

FIG. 2—*Test procedure.*

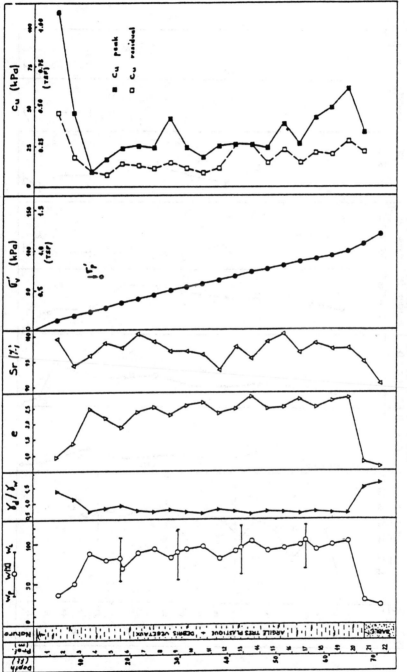

FIG. 3—*Saint Herblain clay: geotechnical characteristics.*

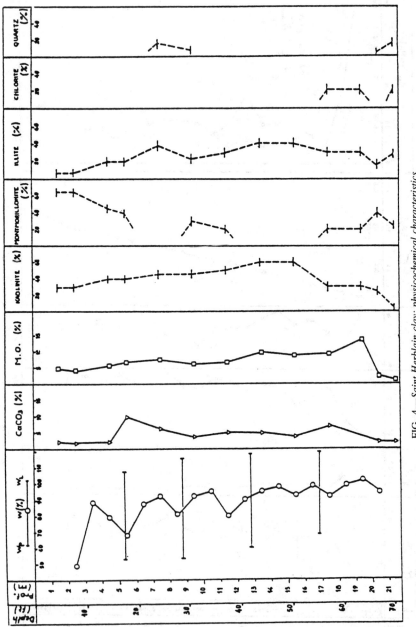

FIG. 4—*Saint Herblain clay: physicochemical characteristics.*

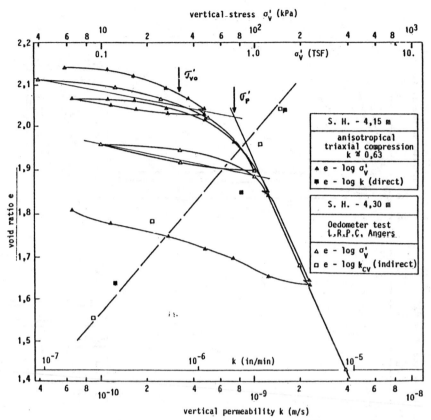

FIG. 5—*Saint Herblain clay: compression and permeability curve.*

on the Cordemais clay, a clay from the Loire estuary (Fig. 8). However, the i_0-values achieved here seem to be important—it is as though the flow of the pore water were ruled by two types of laws, either a fundamental macropermeability for major gradients or a micropermeability ruling the flow for small gradients. This had already been observed on peats, especially by Vautrain [4], who noticed a stabilization of the pore pressures during the consolidation. The role of the percentage of organic matter (i.e., 5.5% in the case of the tested specimen) is thus equally important to the permeability law.

Since the tests were carried out on a single specimen, we are not able to define precisely the permeability law which has to be adopted. However, if a major critical gradient (i_0) does exist, the latter is bound to have an important influence on the consolidation phenomenon. Next, we shall study the influence of this parameter.

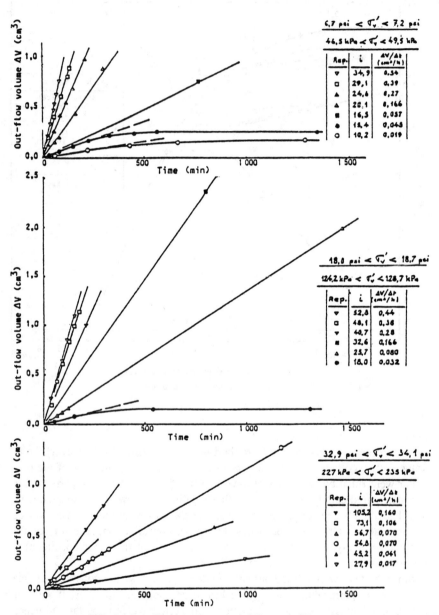

FIG. 6—*Saint Herblain clay: permeability tests,* $\Delta V = \Delta V(t)$.

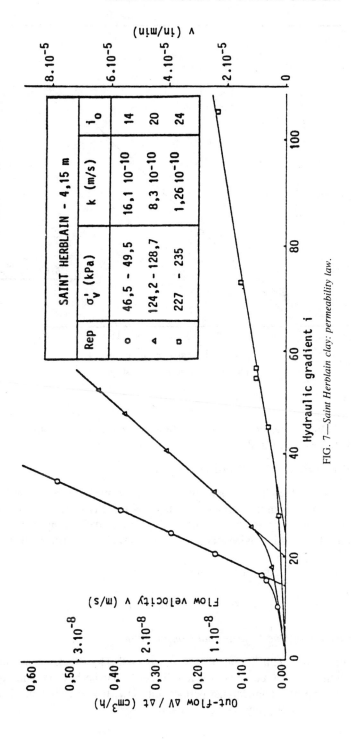

FIG. 7—*Saint Herblain clay: permeability law.*

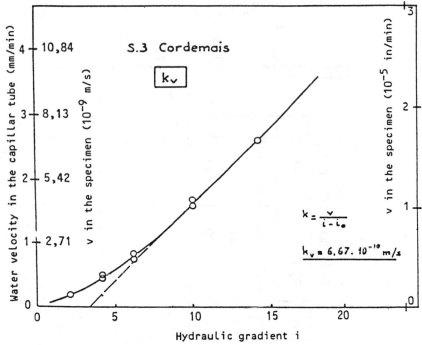

FIG. 8—*Cordemais clay: test "k_v".*

Study of One-Dimensional Consolidation in the Case of a Nonlinear Flow Law of Interstitial Fluid

Definition of Parameters Characterizing Flow Law

As a result of the phenomenon noticed in Fig. 9, we have defined a threshold gradient i_t from which the variations of the flow velocity v are linear and related to the gradient i. Up to the threshold, the flow is defined by a general law $v = \alpha k i^n$. The real law is best defined by the α and n parameters; that is,

$$v = \alpha k i^n \quad \text{for} \quad i_0 < i < i_t$$

$$v = k(i - i_0) \quad \text{for} \quad i_t \leq i$$

New Equation of One-Dimensional Consolidation

In comparison with Terzaghi's theory, only the hypothesis concerning the permeability law has been modified. In order to handle best the consequences due to the use of a non-Darcian permeability law, the latter is looked upon as being constant throughout the consolidation.

There are two equations of consolidation. For $i \geq i_c$, Terzaghi's equation is recognizable:

$$\frac{\partial u}{\partial t} = \frac{1 + e}{\gamma_w \, a_v} \cdot k \cdot \frac{\partial^2 u}{\partial z^2}$$

For $0 \leq i \leq i_c$:

$$\frac{\partial u}{\partial t} = \frac{1 + e}{\gamma_w \, a_v} \cdot k \cdot \frac{\partial^2 u}{\partial z^2} \cdot \frac{\alpha \, n}{(\gamma_w)^{n-1}} \cdot \left(\frac{\partial u}{\partial z}\right)^{n-1}$$

In other respects, it is possible to define α and n in functions of i_0 and i_c by using the solution of continuity for $i = i_c$; that is,

$$n = \frac{i_c}{i_c - i_0}$$

and

$$\alpha = \frac{i_c - i_0}{i_c^n}$$

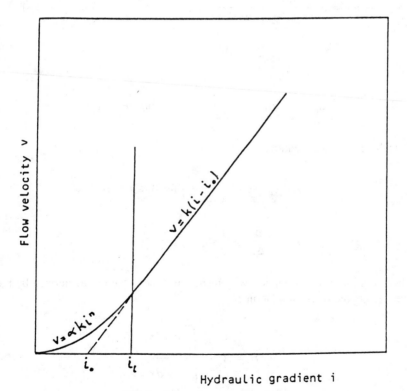

FIG. 9—*Schematic permeability law.*

This can be simplified by choosing a linear relationship between v and i in the $[0, i_t]$ interval, that is, $n = 1$ and, in this case, $v = \alpha k i$ for $0 \leq i \leq i_t$ and

$$\alpha = \frac{i_t - i_0}{i_t}$$

Resolution of the New Equation

In order to have this double equation solved, we have resorted to the standard finite-difference method. If we consider horizontal slices whose thickness Δz is assigned the factor i and time intervals Δt assigned the factor j, it is possible to calculate the pore pressure $u_{i,j}$ at any point in space and in duration. By considering

$$c_v = \frac{1 + e}{\gamma_w \, a_n} \cdot k \text{ for } 0 \leq i \leq i_t$$

$$u_{i,j+1} - u_{i,j} = c_v \frac{\alpha \, n}{(\gamma_w)n - 1} \frac{\Delta t}{(\Delta z)n + 1} (u_{i-1,j} - 2u_{i,j} + u_{i+1,j}) (u_{i+1,j} - u_{i,j})^{n-1}$$

$$\text{for } i \geq i_t$$

and

$$u_{i,j+1} - u_{i,j} = c_v \frac{\Delta t}{\Delta z^2} (u_{i-1,j} - 2u_{i,j} + u_{i+1,j})$$

Calculation stability requires

$$0 \leq c_v \frac{\Delta t}{\Delta z^2} \leq \frac{1}{2 \, \alpha n i_t^{n-1}} \text{ for } 0 \leq i \leq i_t$$

$$0 \leq c_v \frac{\Delta t}{\Delta z^2} \leq \frac{1}{2} \qquad \text{for } i \geq i_t$$

The evolution of consolidation with respect to time is then characterized by the average degree of consolidation U:

$$U = 1 - \frac{\int_0^H u_{i,j} \, dz}{\int_0^H u_{i,0} \, dx}$$

where $u_{i,j}$ is the pore pressure in the layer i at time j, and $u_{i,0}$ is the initial pore pressure in the layer i.

Choice of Parameters i_0, *a and* n

In order to test the influence of the parameters i_0, α, and n, a comparative study of the evolution of the consolidation has been worked out by resorting to four different permeability laws. The test has been performed from laboratory tests results achieved on Cordemais clay by Moulin [5] and for which the following can be written:

(1) $\left| \begin{array}{l} v = \alpha\, k_1\, i^n \quad , \; i\epsilon\ [0, 9] \quad \text{with } k_1 = 6.67\ 10^{-10} \text{m/s} \ (1.58\ 10^{-6}\ \text{in./min}) \\[4pt] i_0 = 3.25 \\[4pt] v = k_1(i - i_0)\, , \; i\epsilon\ [9, \infty[\quad \text{with } n = 1.56 \\[4pt] \phantom{v = k_1(i - i_0)\, , \; i\epsilon\ [9, \infty[\quad \text{with }} \alpha = 0.185 \end{array} \right.$

(2) $\left| \begin{array}{l} v = \alpha\, k_1\, i \quad , \; i\epsilon\ [0, 4] \\[4pt] v = k_1(i - i_0)\, , \; i\epsilon\ [4, \infty[\end{array} \right.$

(3) $\left| \begin{array}{l} v = 0 \quad , \; i\epsilon\ [0, 3.25] \\[4pt] v = k_1(i - i_0)\, , \; i\epsilon\ [3.25, \infty[\end{array} \right.$

(4) $v = k_2\, i$ – Darcy's law – with $k_2 = 6.2\ 10^{-10} \text{m/s}$ $(1.46\ 10^{-6} \text{in./min})$

The value chosen for the time t_{100} at the end of the consolidation should be the same for all four laws.

The evolution of the consolidation according to the four permeability laws is shown in Fig. 10. It may be noted that, on the one hand, Curves 1, 2, and 3 are very close and indicate a faster consolidation velocity than Curve 4. On the other hand, the influence of the critical gradient i_0 is fundamental and implies a stopping of the pore pressure for a while, although the consolidation is not over.

The parameters need to be studied separately for their influence to be better defined.

Influence of Critical Gradient i_0 *on Evolution of Pore Pressure*

A very simplified flow law is chosen

$$v = 0 \qquad\qquad \text{for} \quad i < i_0$$

$$v = k(i - i_0) \qquad \text{for} \quad i \geq i_0$$

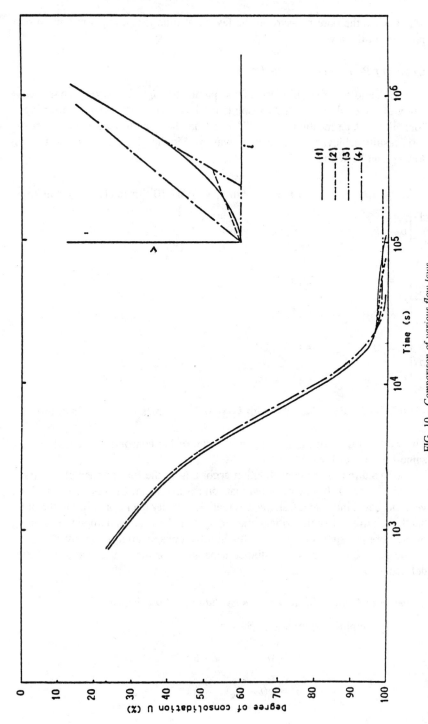

FIG. 10—*Comparison of various flow laws.*

enabling the phenomenon to be schematized so as to state the role of the critical gradient. With Δu = pore pressure increment, Δz = length increment of drainage path, H = length of drainage path, and u_0 = increase of initial pore pressure after loading, at the end of the consolidation, according to an element Δz, a threshold over-pressure Δu_f appears for which a stopping of the drainage occurs (i.e., $\Delta u_f = i_0 \gamma_w \Delta z$). By using the coordinate system z/H, u/u_0, the threshold pore pressure isochrone at the end of the consolidation is represented by a straight line such as (Fig. 11a)

$$\frac{\Delta u_f/u_0}{\Delta z/H} = \frac{i_0 \gamma_w H}{u_0}$$

The definition of the final pore pressure isochrone is thus related to the length of drainage path H. This phenomenon is schematized in Figs. 11b and 11c.

Evolution of Consolidation in Function of Parameters i_0, a, and H

A bilinear permeability law leads to a simplification

$$v = \alpha\, k\, i \quad , i \in [0, i_f[$$

$$v = k(i - i_0)\, , i \in [i_f, \infty[$$

The influence of each parameter is indicated in Fig. 12.

The consolidation stage due to the existence of a critical gradient i_0 occurs all the sooner since the i_0 value is important, its duration being related to the parameter α. The latter, which defines the permeability law for small gradients, rules the end of the consolidation, the decrease of α increasing the duration of the consolidation.

The higher the compressible layer, the more pronounced the phenomena which define a scale effect, since the consolidation of a specimen is ruled mainly by a permeability k, one of a compressible layer by a permeability αk. This phenomenon is illustrated by the comparison of the isochrones in Fig. 13.

Interpretations of Laboratory Tests

Given the simplifications in the hypothesis in the mathematical model, in particular that $a_v = de/d\sigma'$ is constant during loading, the evolution of the pore pressure alone has been tested, Fig. 14.

It must be noted that the proposed laws to determine correctly the level of pore pressure ($u/u_0 \simeq 0, 13$) of the consolidation stage only surround the real curve, since k is a constant. The model still needs to be improved by resorting to a permeability law which depends on the loading duration or, to be more precise, which depends on the void ratio e that evolves during the loading.

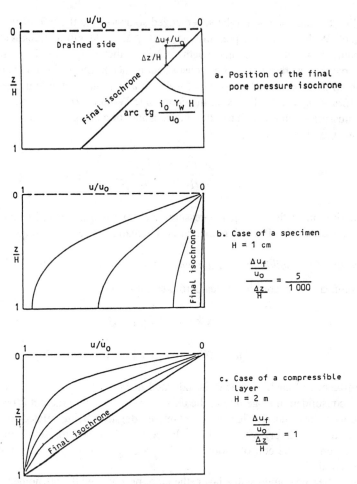

FIG. 11—*Influence of length of drainage path* H *on pore pressure isochrones. Case of a simplified permeability law with* $i_0 = 5$; $u_0/\gamma_w = 10$ m.

A law of the type $e = A + B \log k$ proposed by various authors such as Magnan [6,7], Leroueil et al [8], and Tavenas et al [9,10], in the case of a Darcian flow, will be adjusted to the case of a bilinear permeability law, as indicated in Fig. 15.

Conclusions

The few tests presented here are merely an approach to the problem; the precise knowledge of the permeability law of a clay is absolutely necessary because it is one of the main parameters of the clay consolidation. In particular, the consolidation velocities of *in situ* clays are not correctly defined by means of the coefficient of consolidation c_v, which is usually determined in the laboratory and

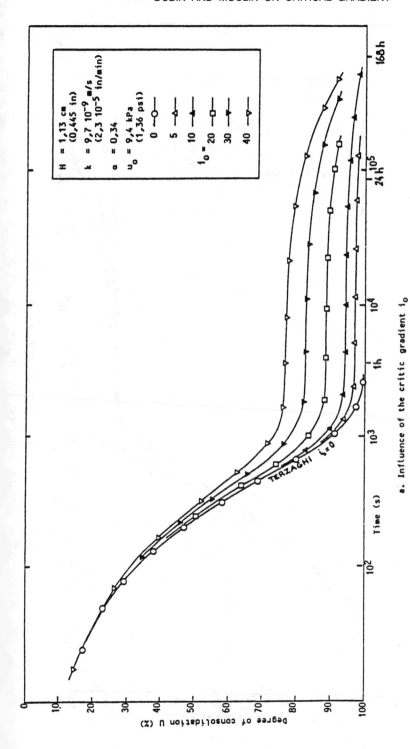

a. Influence of the critic gradient i_o

FIG. 12a—Curves $U = U(t)$: influence of critical gradient i_o.

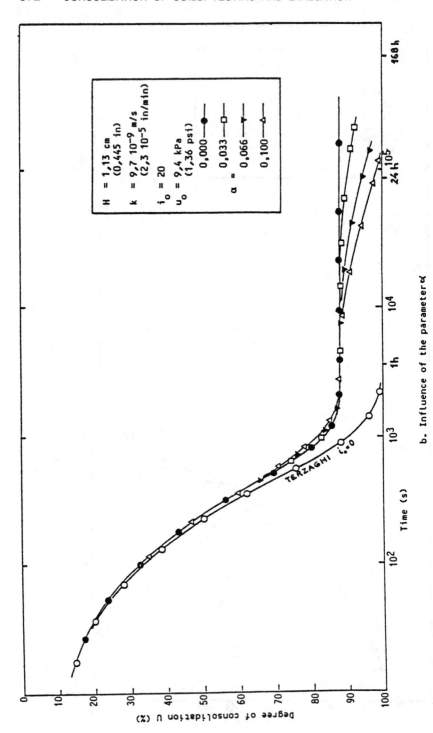

FIG. 12b—*Curves U = U(t): influence of parameter α.*

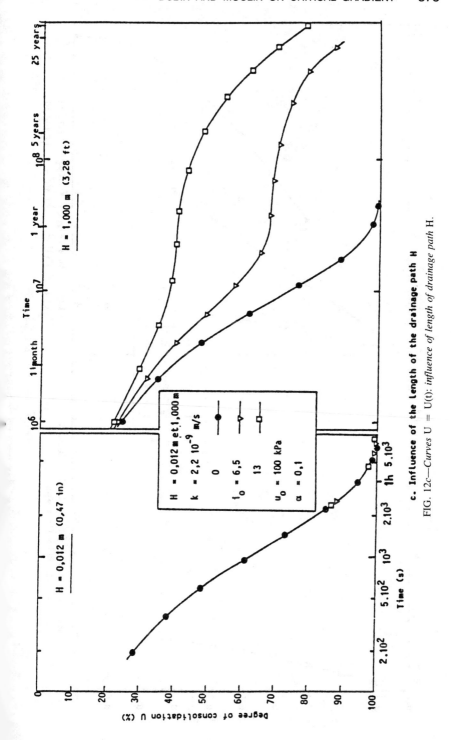

c. Influence of the length of the drainage path H

FIG. 12c—*Curves U = U(t): influence of length of drainage path H.*

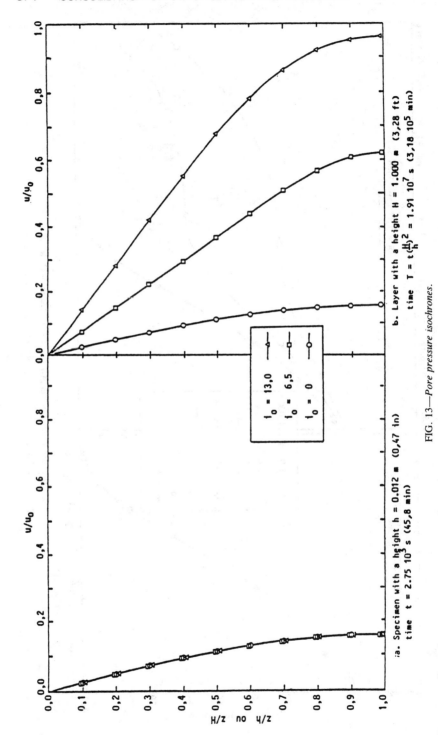

FIG. 13—*Pore pressure isochrones.*

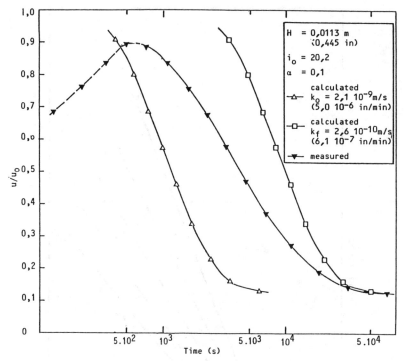

FIG. 14—*Evolution of the pore pressure.*

which is a function of k, e, σ', and c_c, these parameters being themselves changing in the course of the consolidation.

These consolidation velocities should be better anticipated by using the precise definition of the permeability law and by knowing whether a critical gradient exists or not. It is important to note, indeed, that the pore pressure gradients applied to a layer of *in situ* clay are from 100 to 1000 times smaller than those applied in the laboratory during an oedometer test. In the case of Saint Herblain clay, the presence of 5% to 10% organic matter might account for this nonlinear permeability law.

In order to specify the validity of the laws established in the case of Saint Herblain clays, we intend to continue with this research, in the laboratory, partly by performing tests on specimens of various heights and thus checking the influence of the length of drainage path, and partly *in situ*, by watching the variation of the pore pressures in duration while setting up an embankment.

In other respects, we intend to test the permeability on other types of natural clays so as to define the range of possible values concerning the critical gradient.

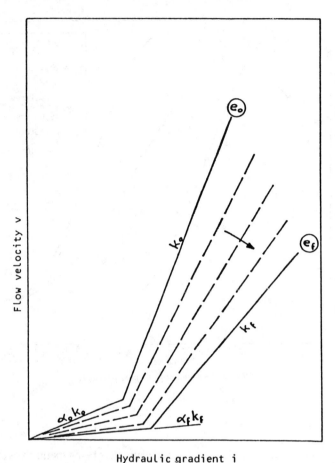

FIG. 15—*Evolution of permeability during a loading.*

References

[1] Hansbo, S., "Consolidation of Clay with Special Reference to Influence of Vertical Sand Drains," in *Proceedings,* Swedish Geotechnical Institute, No. 18, 1960.
[2] Abelev, M. Y., "Construction d'ouvrages sur les sols argileux mous saturés," *Technique et Documentation,* Paris, 1973 (Translation by J. P. Magnan, 1977).
[3] Law, K. T. and Lee, C. F., "Initial Gradient in a Dense Glacial Till," Stockholm, Vol. 1, 1981, pp. 441–446.
[4] Vautrain, J., "Etude d'un remblai sur tourbe à Caen," Rapport de Recherche Laboratoire des Ponts et Chausseés (L.P.C.), No. 41, 1975.
[5] Moulin, G., "Contribution à l'étude de la consolidation des argiles. Etude en laboratoire de la perméabilité," Ecole Nationale Supérieure de Mécanique (E.N.S.M.), Nantes, Internal Report, 1968.
[6] Magnan, J. P., "Progrès récents dans l'étude des remblais sur sols compressibles," *Bulletin Liaison,* No. 116, Nov.-Dec. 1981, pp. 45–56.

[7] Magnan, J. P., Baghery, S., Brucy, M., and Tavenas, F., "Etude numérique de la consolidation unidimensionnelle en tenant compte des variations de la perméabilité et de la compressibilité du sol, du fluage et de la non-saturation," *Bulletin Liaison Laboratoire des Ponts et Chausseés (L.P.C.)*, No. 103, Sept.-Oct. 1979, pp. 83–94.

[8] Leroueil, S, Le Bihan, J. P., and Tavenas, F., "An Approach for the Determination of the Preconsolidation Pressure in Sensitive Clays," *Canadian Geotechnical Journal*, Vol. 17, No. 3, 1980, pp. 446–453.

[9] Tavenas, F., Brucy, M., Magnan, J. P., La Rochelle, P., and Roy, M., "Analyse critique de la théorie de consolidation unidimensionnelle de Terzaghi," *Revue Francaise de Géotechnique*, No. 7, May, 1979, pp. 29–43.

[10] Tavenas, F., Leblond, P., Jean, P., and Leroueil, S., "The Permeability of Natural Clays," *Canadian Geotechnical Journal*, Vol. 20, 1983, pp. 629–659.

Mouslim Kabbaj,[1] Fusao Oka,[2] Serge Leroueil,[1] and François Tavenas[1]

Consolidation of Natural Clays and Laboratory Testing

REFERENCE: Kabbaj, M., Oka, F., Leroueil, S., and Tavenas, F., "**Consolidation of Natural Clays and Laboratory Testing,**" *Consolidation of Soils: Testing and Evaluation, ASTM STP 892,* R. N. Yong and F. C. Townsend, Eds., American Society for Testing and Materials, Philadelphia, 1986, pp. 378–404.

ABSTRACT: A numerical model has been developed, making use of a strain-rate-dependent compressibility law and a permeability-void ratio law, both experimentally established. The numerical program is used to simulate the behavior of clay specimens submitted to oedometric tests (MSL, CGT, creep, etc.). The following conclusions are drawn:

1. The effective stress-strain clay behavior depends on the type of test carried out. In particular, the preconsolidation pressure value varies from one test to another.
2. Each element throughout the specimen follows a specific stress-strain relation depending on its strain rate history. Moreover, the oedometric curve, as usually interpreted from a multiple-stage loading (MSL) test, does not agree with any of these relations, not even with the average curve followed by the whole specimen.
3. The pore pressure isochrones maintain the same shape during controlled-gradient (CGT) or creep tests, in particular when the effective stresses correspond to the preconsolidation pressure.

KEY WORDS: consolidation, natural clays, oedometric tests, strain-rate, normalized behavior, isochrone

Since the 1930s, the engineering practice for predicting the magnitude and rate of settlements of clay foundations has been based on the Terzaghi one-dimensional consolidation theory and on parameters measured in "conventional" oedometer tests in which the load is applied in steps with a stress increment ratio of 0.5 or 1.0 and a duration of 24 h. Schematically, the clay behavior is characterized by a stress-strain curve showing a preconsolidation pressure and by a coefficient of consolidation. While relatively simple in terms of equipment, procedure, and interpretation, the conventional oedometer test presents certain weaknesses. Its duration exceeds one week; it does not allow a precise determination of the stress-strain curve, particularly in the vicinity of the preconso-

[1] Department of Civil Engineering, Laval University, Quebec, Canada.
[2] Department of Civil Engineering, Gifu University, Yanagido, Gifu, Japan.

lidation pressure, and the determination of the coefficient of consolidation is sometimes difficult.

In the past 15 years, numerous new testing techniques have been proposed to eliminate these weaknesses: the constant rate of strain (CRS) test [1–4]; the controlled gradient (CGT) test [5]; the constant rate of loading (CRL) test [6,7]; the single-stage loading (SSL) test [8]; and the continuous loading (CL) test [9]. However, it has become evident that, depending on the testing technique used and on the duration of the test, the results could be very different [10,11].

Indeed, factors affecting clay behavior, in particular the viscosity of the clay skeleton, were evidenced a long time ago [12,13], and considerable work has been done to incorporate viscosity into the analysis of the consolidation process. Numerous rheological models have been proposed, but, in most cases, they were not sufficiently supported by experimental results, and finally, they did not significantly advance the understanding of clay behavior.

In recent years an experimental study of consolidation using different testing techniques [14] has provided a better understanding of clay behavior in laboratory conditions and has allowed the development of representative rheological equations for clays. The aim of the present study is to incorporate these results in a consolidation program and to re-evaluate the process of one-dimensional consolidation, as it happens in the laboratory when different testing techniques are used.

Consolidation and Clay Behavior

The process of consolidation of saturated soil is a combination of two phenomena:

1. The flow of water, and thus the strain rate, controlled by Darcy's law and the variation of the coefficient of permeability with void ratio. The governing equation can be written [15]

$$\frac{\partial \epsilon_v}{\partial t} = - \frac{1 + e_0}{\gamma_w} \frac{\partial}{\partial z} \left(\frac{k}{1 + e} \frac{\partial u}{\partial z} \right) \tag{1}$$

where

ϵ_v = strain,
z = depth related to initial thickness,
t = time,
e_0 and e = initial and present void ratios,
k = permeability,
γ_w = unit weight of pore water, and
u = excess pore pressure.

2. The constitutive law of the soil skeleton relating effective stresses to strains.

Different permeability-void ratio relations have been suggested for clays. However, Tavenas et al [16] showed that the one proposed by Taylor [17]

$$e_0 - e = C_k (\log k_0 - \log k) \tag{2}$$

where C_k is the permeability change index, and k_0 the permeability at the initial void ratio e_0, is well-representative of natural clays, at least for strains less than 20%.

Concerning the rheological law, the experimental study carried out by Leroueil et al [14] shows that, for strain rates usually observed in laboratory, the clay behavior is well-represented by a unique effective stress-strain-strain rate relationship (σ'_v, ϵ_v, $\dot{\epsilon}_v$). This relationship, established on the basis of constant rate of strain tests, controlled gradient tests, multiple-stage loading tests, and long-term creep tests, appears to be general, at least in the normally consolidated range, and is in agreement with the "isotache" model advocated by Šuklje [18,19]. More specifically, Leroueil et al [14] showed that the rheological behavior of the clay can be described by two curves:

$$\sigma'_p = f(\dot{\epsilon}_v) \tag{3}$$

$$\sigma'_v/\sigma'_p(\dot{\epsilon}_v) = g(\epsilon_v) \tag{4}$$

Figure 1 presents a typical example for a sensitive clay from Batiscan, Quebec. These two equations can be combined to give the general equation

$$\dot{\epsilon}_v = f^{-1}\left(\frac{\sigma'_v}{g(\epsilon_v)}\right) \tag{5}$$

The experimental results can be expressed in various forms. In particular, it has been found that Eqs 3 and 4 could be written for given ranges of strain rate and strain, respectively:

$$\ell n\ \dot{\epsilon}_v = A' + B'\ \ell n\ \sigma'_p \tag{6}$$

$$\ell n\ (\sigma'_v/\sigma'_p) = C' + D'\ \epsilon_v \tag{7}$$

where σ'_p is the preconsolidation pressure and A', B' and C', D' are constants for the given ranges. The last two equations can be combined to give

$$\dot{\epsilon}_v = \exp A'\ \exp [B'\ (\ell n\ \sigma'_v - C' - D'\ \epsilon_v)] \tag{8}$$

This equation is in fact a particular case of a more general equation, expressed in terms of void ratio, suggested by Battelino [20].

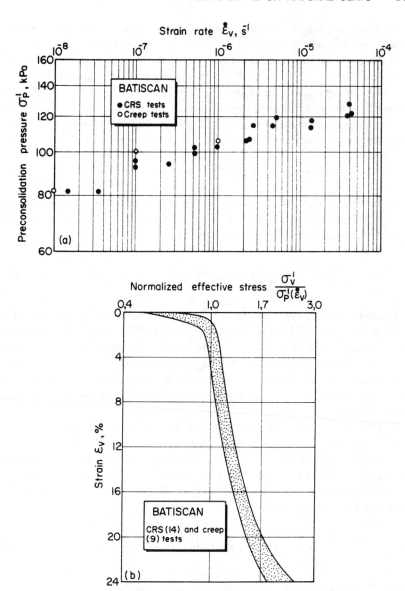

FIG. 1—σ'_p–ϵ_v and σ'_v/σ'_p – ϵ_v relations for the Batiscan clay (from Leroueil et al [14]).

While this model well represents the behavior of clay in oedometer tests in which the strain is always increasing, it should be noted that, defined in terms of total strains, it presents some shortcomings. In particular, this model cannot represent correctly the behavior observed in relaxation tests in which there is a decrease of the effective stress under constant strain. Such a decrease is most likely associated with the combined effects of elastic and plastic strains. In order to be generally applicable, the model should be modified accordingly.

In the experimental study carried out by Leroueil et al [14], it would have been very difficult, if not impossible, to separate elastic and plastic components. However, the model described by Eqs 3 to 5 and 6 to 8 has been developed essentially from experimental data in the normally consolidated region where plastic strains are strongly predominant. It follows that the form of Eq 8 is certainly representative of plastic strains and could be rewritten as

$$\dot{\epsilon}_v^p = \exp A'' \exp [B'' (\ell n \; \sigma'_v - C'' - D'' \; \epsilon_v^p)] \tag{9}$$

As previously mentioned, the elastic component of strain is not easy to determine and, as a first approximation, it is suggested to express it with the classical equation, derived from Cam-clay models [21]:

$$\epsilon_v^e = \kappa' \; \ell n \; (\sigma'_v/\sigma'_{v0}) = \frac{\kappa}{1 + e_0} \; \ell n \; (\sigma'_v/\sigma'_{v0}) \tag{10}$$

Equations 9 and 10 are thought to be realistic constitutive equations for clays in the normally consolidated region. In the overconsolidated range, viscous effects were also evidenced, but their real nature is more difficult to establish since the strains involved remain small. There is still a need for detailed studies of the clay behavior in this domain, but it appears to the authors that Eqs 9 and 10 could also be used as a first approximation of the real behavior of the overconsolidated clay.

In 1981, Oka established general constitutive equations for normally consolidated clays based on the principles of Cam-clay models [21] and on Perzyna's [22] elastic-viscoplastic theory [23]. For volumetric strains, he obtained equations having the same form as Eqs 9 and 10. If the stress ratio σ'_3/σ'_1 is assumed constant, the plastic volumetric strain rate (equal to the vertical strain rate in this particular case) can be written as

$$\dot{\epsilon}_v^p = C \exp [m' (\ell n \frac{\sigma'_v}{\sigma'_{vc}} - \frac{1 + e_0}{\lambda - \kappa} \epsilon_v^p)] \tag{11}$$

where C, m', and σ'_{vc} are constants and λ is defined by

$$\lambda = \frac{\delta\epsilon_v (1 + e_0)}{\delta \ell n \; \sigma'_v} \tag{12}$$

Equations 9 and 11 are exactly the same, indicating that the numerical model proposed by Oka is, in its principles, in agreement with observed behavior. However, Oka's model was used with unique m' and λ values and has to be modified to take into account the real behavior of the clay in which m' and λ vary, depending on the range of plastic strain rates and plastic strains considered.

For the Batiscan clay, the experimental results of which were presented in Fig. 1, the rheological model corresponding to Eq 9 or 11 is shown in Fig. 2 and the corresponding parameters presented in Table 1. As shown in Fig. 2, we chose to describe the $\sigma'_v/\sigma'_p - \epsilon_v^p$ curves with four "segments". A more detailed discretization could have been used. As for the variation of σ'_v with $\dot{\epsilon}_v^p$, two values for the parameter m' were taken. It must be noted that the elastic parameter κ has been chosen arbitrarily constant and equal to one fourth of the total strain parameter λ in the overconsolidated region.

Numerical Treatment

For the numerical treatment of the consolidation process, the finite-difference method has been used. The viscoplastic constitutive equation has been rewritten in the incremental form, using the implicit time integration form [24]

$$\Delta\epsilon_v^p = \frac{f_{\sigma'_v}\,\Delta t}{1 - f_{\epsilon_v}^p\,\Delta t}\,\Delta\sigma'_v + \frac{f_0\,\Delta t}{1 - f_{\epsilon_v}^p\,\Delta t} \tag{13}$$

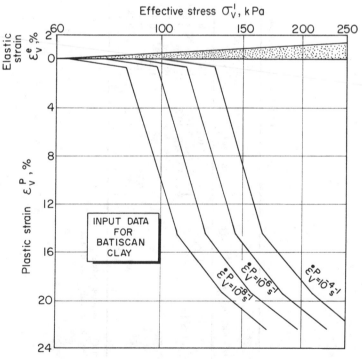

FIG. 2—Rheological model of the Batiscan clay used in the present analysis.

TABLE 1—Soil parameters for Batiscan clay.

Strain Rate ($\dot{\epsilon}_v^p$), s^{-1}	Soil Region	Strain (ϵ_v^p), %	B'' m'	exp A'' C, s^{-1}	exp C'' σ'_{vc}, kPa	λ	κ	$D'' = \dfrac{1 + e_0}{\lambda - \kappa}$
$\dot{\epsilon}_v^p \geq 10^{-6}$	OC^a	$\epsilon_v^p < 0.75$	16.0	0.749×10^{-4}	98.3	0.118	0.0296	36.0
	NC^b	$0.75 \leq \epsilon_v^p < 14.5$	16.0	0.123×10^{-5}	98.3	1.860	0.0296	1.74
		$14.5 \leq \epsilon_v^p < 19.3$	16.0	0.711×10^{-1}	124.9	0.690	0.0296	4.82
		$\epsilon_v^p \geq 19.3$	16.0	$0.506 \times 10^{+5}$	157.0	0.428	0.0296	7.98
$\dot{\epsilon}_v^p < 10^{-6}$	OC^a	$\epsilon_v^p < 0.75$	31.6	0.504×10^{-4}	85.0	0.118	0.0296	36.0
	NC^b	$0.75 \leq \epsilon_v^p < 14.5$	31.6	0.151×10^{-7}	85.0	1.860	0.0296	1.74
		$14.5 \leq \epsilon_v^p < 19.3$	31.6	$0.382 \times 10^{+4}$	124.9	0.690	0.0296	4.82
		$\epsilon_v^p \geq 19.3$	31.6	$0.139 \times 10^{+16}$	157.0	0.428	0.0296	7.98

^{a}OC = overconsolidated.
^{b}NC = normally consolidated.

where

$$f(\epsilon_v^p, \sigma'_v) = \dot{\epsilon}_v^p = \text{(Eq 11) viscoplastic strain rate,}$$

$$f_0 = \text{value of } f \text{ at } t = 0,$$

$$f_{\sigma'_v} = \partial f/\partial \sigma'_v,$$

$$f_{\epsilon_v^p} = \partial f/\partial \epsilon_v^p,$$

$$\Delta t = \text{time increment, and}$$

$$\Delta \sigma'_v = \text{effective stress increment.}$$

Using Eqs 10 and 13, the total strain increment is

$$\Delta \epsilon_v = \Delta \epsilon_v^e + \Delta \epsilon_v^p \tag{14}$$

$$\Delta \epsilon_v = \frac{\kappa}{1 + e_0} \frac{1}{\sigma'_v} \Delta \sigma'_v + \frac{f_{\sigma'_v} \Delta t}{1 - f_{\epsilon_v^p} \Delta t} \Delta \sigma'_v + \frac{f_0 \Delta t}{1 - f_{\epsilon_v^p} \Delta t} \tag{15}$$

or

$$\Delta \epsilon_v = \frac{1}{M'} \Delta \sigma'_v + g(\sigma'_v, \epsilon_v^p) \tag{16}$$

where

$$\frac{1}{M'} = \frac{\kappa}{(1 + e_0)\sigma'_v} + \frac{f_{\sigma'_v} \Delta t}{1 - f_{\epsilon_v^p} \Delta t} \tag{17}$$

and

$$g = \frac{f_0 \Delta t}{1 - f_{\epsilon_v^p} \Delta t} \tag{18}$$

The governing equation of one-dimensional consolidation (Eq 1) can be combined with Eq 16 and written, when assuming a constant σ_v during the time increment, in the finite-difference form

$$u_{t+\Delta t} = u_t + g M' + \frac{k}{\gamma_w} \frac{1 + e_0}{1 + e} \frac{\partial^2 u}{\partial z^2} M' \Delta t$$

$$+ \frac{1 + e_0}{\gamma_w} \frac{\partial}{\partial z} \left(\frac{k}{1 + e} \right) \frac{\partial u}{\partial z} M' \Delta t \tag{19}$$

where

$$\frac{\partial^2 u}{\partial z^2} = \frac{u_{i-1} - 2u_i + u_{i+1}}{\Delta z^2} \text{ for } i = 2 \text{ to } n_p - 1 \tag{20}$$

$$\frac{\partial^2 u}{\partial z^2} = \frac{u_{i+1} - 3u_i}{\Delta z^2} \text{ for } i = 1 \tag{21}$$

$$\frac{\partial^2 u}{\partial z^2} = \frac{u_{i-1} - u_i}{\Delta z^2} \text{ for } l = n_p \tag{22}$$

Here i denotes the nodal points (Fig. 3).

Analysis of Clay Behavior During Testing

The program previously described has been used to simulate various laboratory oedometer testing procedures (long-term creep tests, MSL tests with reloading at the end of primary consolidation or after 24 h, CGT tests) and to analyze the detailed behavior of elements located at different levels in the clay specimen during testing.

For reasons of clarity, only the set of parameters representative of the behavior of Batiscan clay is used in this paper. These parameters describing the constitutive law (Eqs 9 to 11) were developed mainly from CRS tests at various strain rates (Fig. 1). They are presented in Table 1 and the corresponding model is shown

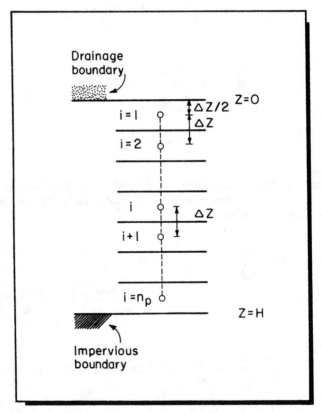

FIG. 3—*Grid system for the finite-difference model.*

in Fig. 2. The permeability parameters $k_0 = 9 \times 10^{-10}$ m/s and $C_k = 1.15$ were obtained from direct measurement in laboratory. The void ratio under the initial load ($\sigma'_{v0} = 65$ kPa) is $e_0 = 2.18$. Finally, all tests were simulated assuming a specimen height of 0.95 cm and drainage at the top only. The clay specimen was divided into ten elementary sublayers.

Direct comparisons between observed and computed results are first made with creep tests on the Batiscan clay in order to evaluate the performance of the model. For other types of tests, which were not carried out on the Batiscan clay, the computed results obtained from input data corresponding to this clay are compared on a qualitative basis with experimental data from tests on other similar clays.

Creep Test

In recent years, at least five series of long-term creep tests were performed at Laval University. The behavior observed in all series was systematically the same. The results obtained from tests in which the Batiscan clay was initially loaded to its *in situ* stress $\sigma'_{v0} = 65$ kPa and then reloaded to a final stress varying between 67 and 151 kPa (Fig. 4) also are typical of what has been previously observed [25–27].

The computed results, also presented in Fig. 4, show excellent agreement with the observations. For high stresses in excess of about 100 kPa, the curves relating strain and logarithm of time have the classical "S" shape and show $C_{ae} = \Delta e / \Delta \log t$ values continuously decreasing with time. At lower stresses, the $\epsilon_v - \log$

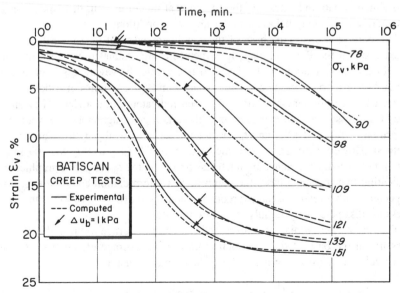

FIG. 4—*Creep oedometer tests on the Batiscan clay: experimental and computed strain-time relations (experimental data from Ref 14).*

t curves present a slope which is continuously increasing during the observation period, which is in excess of 140 days.

The same results are drawn in a log $\dot{\epsilon}_v$ − log t diagram (Fig. 5a). In such a diagram the Terzaghi solution is represented by an initial straight line with a downward slope $m = 0.5$, later curving down toward a vertical assymptote [28]; for soils developing secondary compression, the creep deformations are represented by continuous straight lines [29]. The two tests at stresses well beyond the preconsolidation pressure (121 and 151 kPa) show a similar behavior. Initially the strain rates are high and follow a relation with an m value of about 0.5. After times in the order of 100 to 1000 min, creep phenomena govern with values of m in excess of 1.0. The test at very low stress (78 kPa) remains in the overconsolidated range where it is governed by a creep law with $m = 0.85$.

The two tests at intermediate stresses (90 and 109 kPa) show a different behavior both experimentally and in the analysis. Initially the log $\dot{\epsilon}_v$ − log t relation follows that of the overconsolidated specimen, but after a certain time which varies depending on the magnitude of the applied stress, the strain rate remains momentarily constant until the log $\dot{\epsilon}_v$ − log t relation reaches the range of relations observed on normally consolidated specimens. Such S-shaped log $\dot{\epsilon}_v$ − log t curves were observed by Bishop and Lovenbury [30] and by Tavenas et al [29], but no explanation was provided. The present study demonstrates that the S-shape corresponds to the creep-delayed passing of the clay to a normally consolidated state. The arrows in Fig. 5a indicate this situation, which corresponds to a strain of 1% for the Batiscan clay.

The computed results are shown in Fig. 5b. The excellent agreement with the experimental data indicates that the rheological model used in this study is representative of the clay behavior under any test condition.

The pore pressure at the base of the specimens was not measured during the creep tests on the Batiscan clay. It is therefore impossible to establish the end of primary consolidation on this basis. It is possible, however, to get an idea of the end of primary from the computed results. In Fig. 4, points corresponding to an excess pore pressure of 1 kPa have been indicated by an arrow. They show that, for stresses well in excess of σ'_p, there is a good agreement between end of primary ($\Delta u_b = 1$ kPa) and t_{100} as determined graphically by the Casagrande method. However, for stresses in the vicinity of σ'_p ($\sigma'_v = 90, 98, 109$ kPa), the end of primary is reached in much less time than the t_{100} obtained graphically, indicating that in such a case, the common test interpretation could be false. As expected, with stresses in the overconsolidated region ($\sigma'_v = 78$ kPa), the pore pressure dissipates very rapidly, in a few minutes.

The strains and strain rates discussed above are average values. When pore pressure at the base of the specimen is known, an average effective stress can be calculated by assuming a parabolic isochrone and using

$$\sigma'_v = \sigma_v - u_0 - 0.66 (u_b - u_0) \tag{23}$$

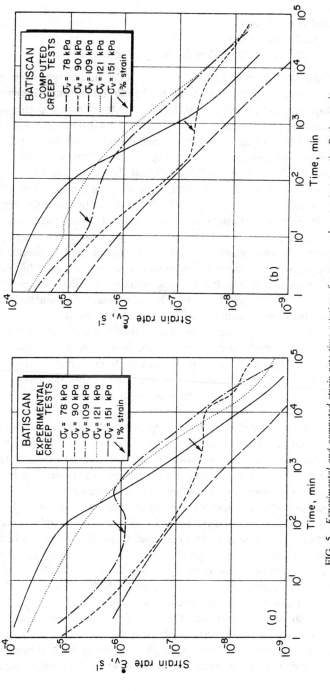

FIG. 5—*Experimental and computed strain rate-time relations for creep oedometer tests on the Batiscan clay.*

where

σ_v = total applied stress,
u_0 = applied back-pressure, and
u_b = pore pressure measured at the base of the specimen.

For example, for the simulated case σ_v = 151 kPa, the average effective stress-strain curve would have been as shown in Fig. 6b (dotted line). It can be seen that the shape of this curve is very different from the $\dot{\epsilon}_v = cst$ curves and thus different from the curves obtained in CRS tests. This difference simply reflects the reduction of the average strain rate during consolidation shown in Fig. 5.

The strain-strain rate relation followed by different elements in the clay specimen loaded to 151 kPa shows (Fig. 6a) that the strain rate history is quite different from one element to the other, depending on their position. Near the drainage boundary (first element) the strain rate is higher than near the impervious boundary (tenth element) during primary consolidation. When the effective stress becomes essentially constant, all the elements have the same strain rate at a given strain, as expected from the model used. Depending on their strain-rate history, the effective stress-strain curves followed by the elements are also different (Fig. 6b). The element located near the drainage boundary mobilizes a higher preconsolidation pressure than others and has, in the normally consolidated region, a smaller modulus of deformation (higher compression index).

Figure 7 shows pore pressure-logarithm of time curves followed in Elements 1, 2, 4, and 10. In each case, except for the first element for which it is not clear, there is a rapid decrease of pore pressure during the first minutes of the test, then a more or less pronounced step, and finally a slow decrease of Δu. On these curves, the pore pressure corresponding to the preconsolidation pressure mobilized in each element (Fig. 6b) is indicated by an arrow. It is shown that the "steps" are associated with the passing of the preconsolidation pressure of the corresponding element. Since the preconsolidation pressure varies from one element to the other, the pore pressure at the step also varies: the smaller the distance to the drainage boundary, the higher the strain rate, the higher the preconsolidation pressure, and the smaller the pore pressure.

Isochrones obtained at various times for the same test (σ_v = 151 kPa) are shown in Fig. 8. During the first 10 s, the pore pressure-height curve moves down and then keeps a more or less parabolic shape during the rest of the test. This seems to contradict results presented by Tavenas et al [31], which show that the shape of the isochrone is strongly modified when the clay is passing its preconsolidation pressure (Fig. 9). In fact, the CONMULT model used by Tavenas et al [31] did not take into account the viscous nature of the soil. The preconsolidation pressure was thus the same for all elements, leading to the behavior shown in Fig. 9. When a viscoplastic model is used, the preconsolidation pressure in the specimen decreases when the distance to the drainage boundary increases, moderating the influence of the passage of σ'_p on the shape of the

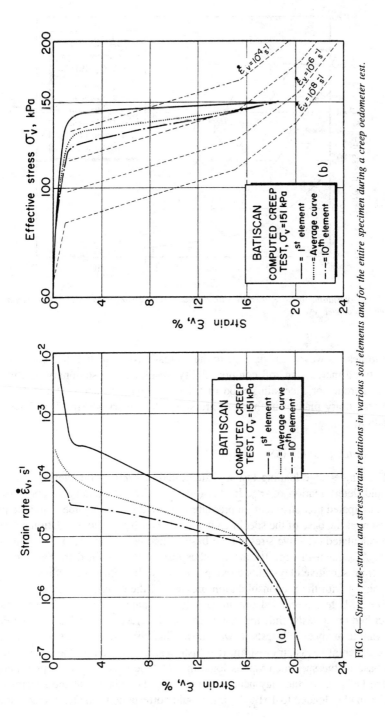

FIG. 6—*Strain rate-strain and stress-strain relations in various soil elements and for the entire specimen during a creep oedometer test.*

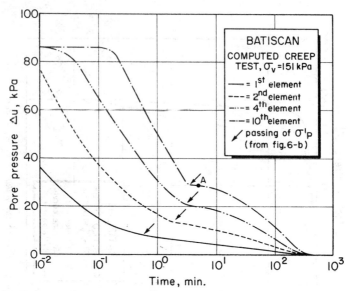

FIG. 7—*Variations of pore pressure with time at various depths within a clay specimen during a creep oedometer test.*

isochrones. However, the passage to the normally consolidated state corresponds to a strong increase in soil compressibility, resulting in a stationary position of the isochrone for a certain period of time (between 150 and 600 s in Fig. 8). The temporary plateau in the pore pressure – log t curves (Fig. 7) reflects this situation.

Single-Stage Loading (SSL) Test

The single-stage loading test was suggested in 1980 by Leroueil et al [8] for a rapid determination of σ'_p. In this test, a load corresponding to about twice the anticipated preconsolidation pressure is applied to the specimen, and the pore pressure at the base of the specimen is recorded during the consolidation process. The calculated effective stress at the base is then plotted versus the logarithm of time. As shown in Fig. 10 and in other test results obtained by Samson et al [32] from sensitive clays, this curve presents a well-defined step which has been associated with the preconsolidation pressure of the clay.

The SSL test is carried out in exactly the same manner as the previously described creep tests; thus the discussion on the creep test loaded to 151 kPa is relevant to analyze SSL tests. It was shown that different elements in a specimen have different stress-strain-strain rate histories. Consequently, the σ'_v (or u_b) at the base of the specimen versus logarithm of time curve is only representative of the behavior of the clay near the impervious boundary. For the specimen of Batiscan clay loaded to 151 kPa, this would correspond to the tenth element and the associated curve in Fig. 7. The preconsolidation pressure which could be estimated from this test by using the SSL method would be 122 kPa (total stress = 151 kPa minus the pore pressure at the step = 29 kPa at Point A in Fig. 7).

Multiple-Stage Loading (MSL) Test

In the multiple-stage loading test, the soil is loaded in steps. The conventional oedometer test with a load increment ratio of 0.5 or 1.0 and a reloading schedule of 24 h is a particular MSL test. If the pore pressure is measured at the base of the specimen during consolidation, it is possible to estimate an average effective stress by using Eq 23 and to associate it with the average strain in a stress-strain diagram. This has been done for two tests, carried out on the Joliette clay [32], in which the loading durations were, respectively, the time required for primary

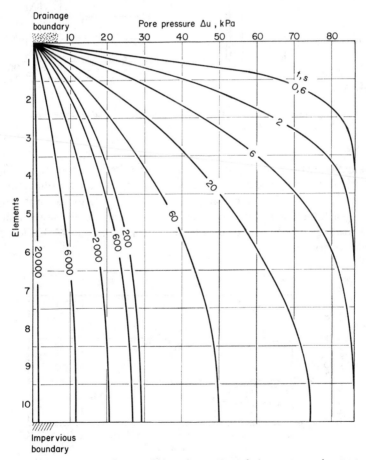

FIG. 8—*Pore pressure isochrones within a clay specimen during a creep oedometer test.*

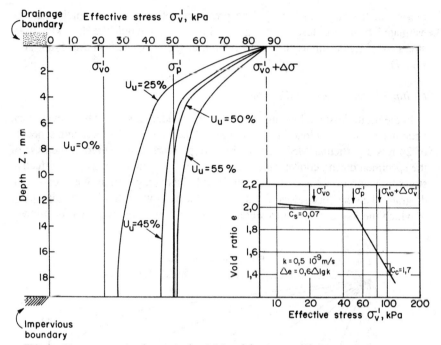

FIG. 9—*Pore pressure isochrones in the vicinity of the preconsolidation pressure, as computed from a consolidation model without creep* [31].

consolidation $(MSL)_p$ and 24 h $(MSL)_{24}$. The test results presented in Fig. 11 show that the effective stress increases very rapidly just after loading, and later much more slowly while strains are accumulating. It is also shown that for the $(MSL)_{24}$, in which there are some secondary deformations, the strain at the end of the loading period is larger than the one obtained in the $(MSL)_p$ test at the end of primary.

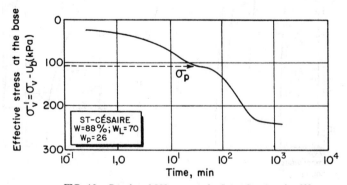

FIG. 10—*Results of SSL test on the Saint-Cesaire clay* [8].

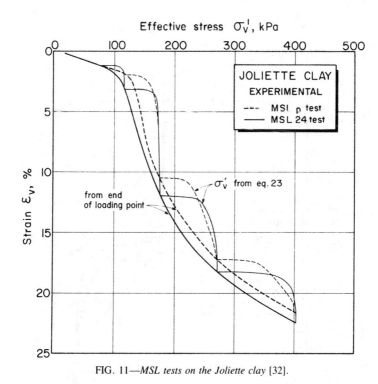

FIG. 11—*MSL tests on the Joliette clay [32].*

One $(MSL)_{24}$ test was simulated on the Batiscan clay. The stress-strain curves followed by the first and tenth elements, as well as the average stress-average strain curve of the overall specimen, are plotted in Fig. 12. This last curve is qualitatively very similar to the one observed on the Joliette clay. As for the stress-strain curves of the different elements, the behavior is very similar to the one observed in creep tests; each element follows its own curve, depending on its strain rate history. Just after loading, the pore pressures and thus the strain rates are high, and the clay goes to high stresses. Later, the strain rate decreases and the stress-strain curve crosses lines of equal strain rates. Such behavior was noted by Berre and Iversen [*33*].

Pore pressures are seldom measured or used when interpreting MSL tests. An "interpreted" stress-strain curve is rather drawn through the stress-strain points obtained at the end of the loading periods. Such a curve is shown in Fig. 12 for the Batiscan clay and it can be seen that it does not correspond at all to the stress-strain curves followed by various elements in the specimen. In particular, the preconsolidation pressure estimated from this "interpreted" curve is in the order of 90 kPa, while the preconsolidation pressures mobilized in the various elements range between 113 and 119 kPa.

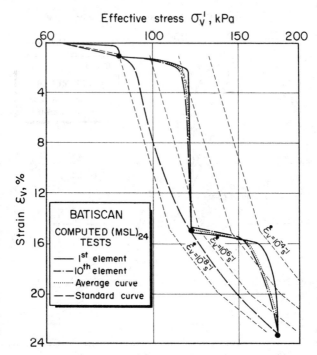

FIG. 12—*Simulated MSL test on the Batiscan clay.*

Controlled Gradient (CGT) Test

In a controlled gradient test, the applied load is continuously increased in such a way that the pore pressure measured at the base of the specimen remains constant. The average effective stress is calculated by Eq 23.

Two tests with pore pressure differences, Δu_b = 10 and 30 kPa, were simulated for the Batiscan clay. The average strain rate-average strain and the average stress-average strain curves are presented, respectively, in Figs. 13a and 13b.

As for the strain rate (Fig. 13a), the higher the pore pressure difference, the higher the strain rate. This is logical considering that the strain rate directly reflects the rate of outflow of pore water, which, in turn, is related by Darcy's law to the gradient (i.e., to the pore pressure difference). The shape of the $\dot{\epsilon}_v - \epsilon_v$ curve is the second concern. At the beginning of the test, the average strain rate decreases and then increases to reach a maximum at a strain corresponding approximately to the passage of the preconsolidation pressure. At larger strains, the strain rate continuously decreases. This behavior corresponds exactly to that experimentally observed by Leroueil et al [10,14] and by Tavenas et al [34]. The corresponding stress-strain curves are shown in Fig. 13b. According to the decrease of strain rate in the normally consolidated region, the stress-strain curves cross the $\dot{\epsilon}_v = cst$ curves and are consequently steeper. The stress-strain

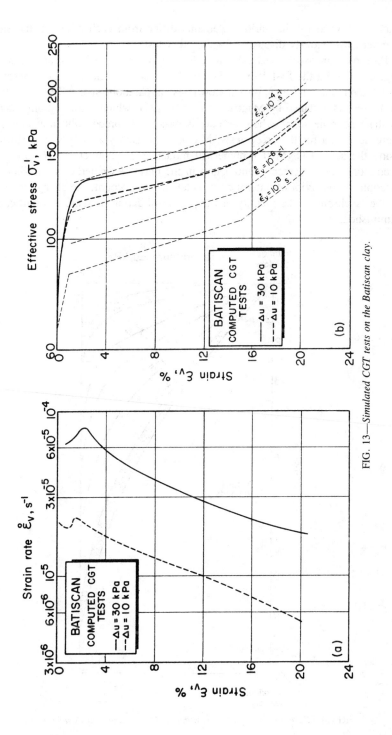

FIG. 13—*Simulated CGT tests on the Batiscan clay.*

curves followed by the various elements differ from each other, in the same manner as previously discussed for the other tests.

Pore pressure isochrones calculated for the $\Delta u_b = 10$ kPa test are shown in Fig. 14. During the first 10 s of the test, there is a flattening of the isochrone, which then maintains an approximately constant parabolic shape. This is contrary to the results obtained by Magnan and Deroy [35], which show a strong change of the isochrone shape when the clay is passing its preconsolidation pressure. Here again, as for the creep tests, Magnan and Deroy's study was performed using the same CONMULT numerical model as Tavenas et al [31] (i.e., not accounting for the creep phenomenon in the clay). As noted earlier, when the viscoplastic behavior of natural clays is taken into account, the change in shape of the isochrone in the vicinity of the preconsolidation pressure is significantly diminished.

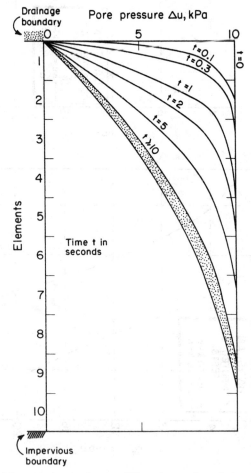

FIG. 14—*Pore pressure isochrones within a clay specimen during a CGT test.*

Discussion

The constitutive model used in the present study has been established mainly on the basis of CRS tests on small specimens using a range of strain rates usually encountered in such tests. The results presented above show that the numerical consolidation program using this model allows an excellent simulation of the clay behavior within specimens submitted to any kind of consolidation tests. In this sense, the numerical analysis fully confirms the conclusion of the experimental study presented by Leroueil et al [14].

The main result of this study is that, during a consolidation test, each part of the tested specimen follows its own effective stress-strain relation, depending on its strain-rate history. For example, in an MSL test (Fig. 12), an element near the drainage boundary deforms more rapidly than an element in the center of the specimen, therefore mobilizing higher effective stresses. The overall stress-strain relation observed in any test is thus an average of different local relations. This average relation also reflects the effect of strain rate. For example, in CGT tests the strain rate decreases with increasing strain and the effective stress-strain curves followed are steeper than those corresponding to CRS tests. This is shown in Fig. 13b and experimentally confirmed by the test results obtained from the Saint-Césaire clay (Fig. 15). In MSL tests (Figs. 11 and 12), the decrease in strain rate is also very rapid and the $\sigma'_v - \epsilon_v$ curve followed throughout each consolidation step is very steep. In controlled rate of loading and in continuous loading tests, the stress-strain behavior would probably be different, depending on the strain rate history and thus on the characteristics of the test and clay.

More generally it can be stated that the average stress-strain relation observed in any consolidation test reflects not only the deformability of the tested clay, but also the particular strain rate history imposed by the conditions specific to the test. It follows that the usual compressibility parameters such as the oedometer modulus $E_{oed} = \Delta \sigma'_v / \Delta \epsilon_v$ or the compression index C_c are not only characteristics of the soils, but also of the testing technique.

The strain-rate effect is particularly significant when measuring the preconsolidation pressure. For a given clay there is a unique $\sigma'_p - \dot{\epsilon}_v$ relation [11], and each test gives its own preconsolidation pressure value. For a CRS test, the value is associated with the imposed strain rate; in CGT tests, σ'_p is associated with the strain rate at which the specimen becomes normally consolidated. For a given u_b value, this depends on the permeability and compressibility of the clay. In MSL tests (Fig. 12) the σ'_p value, as usually determined, is dependent mainly on the stress-strain data points immediately following the preconsolidation pressure and it is associated, as suggested by Leroueil et al [10], with the strain rate obtained at the end of the loading stages corresponding to these data points.

Carrying out conventional and CRS tests on a wide variety of Champlain clays, Leroueil et al [11] have observed that for these clays, the preconsolidation pressure obtained in conventional oedometer test (24 h) corresponds to a strain rate in the order of 10^{-7} s^{-1}; a similar observation may be made for the test shown in Fig. 12. It follows that the difference between preconsolidation pressures obtained in

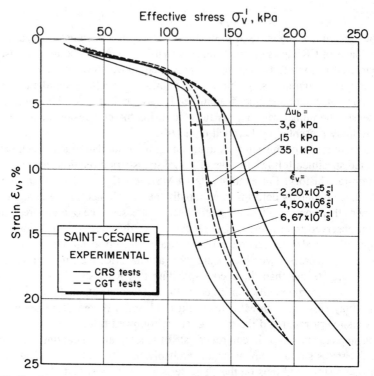

FIG. 15—*Experimental stress-strain curves from CRS and CGT tests on the Saint Cesaire clay* [14].

conventional tests and other more rapid tests may be important. In the case of CRS tests carried out at a strain rate of 4×10^{-6} s^{-1}, σ'_p values are about 28% higher than the conventional preconsolidation pressure (Fig. 16). A similar difference between the value of σ'_p computed in the MSL test (Fig. 12) and the value corresponding to the input data for a strain rate of 4×10^{-6} s^{-1} (Fig. 1) is obtained, as shown by the star in Fig. 16.

In SSL tests in which a load equal to 1.5 to 3 times the expected preconsolidation pressure is applied in one step, the strain rates are very high until the passage of the preconsolidation pressure. A few tests by Leroueil et al [11] indicate that the σ'_p values obtained from such tests are about 60% higher than those from conventional tests (Fig. 17). Here again, a similar difference is obtained with the preconsolidation pressures computed for the MSL and creep tests with an applied load of 151 kPa (star in Fig. 17).

The study presented in this paper applies strictly to consolidation processes in the oedometer equipment and at strain rates usually encountered in the laboratory. It is quite probable that the same strain-rate effects are also present in the field and that the numerical model used herein could serve for the consolidation analysis

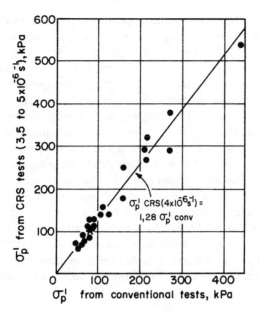

FIG. 16—*Correlation between the preconsolidation pressures obtained from CRS and conventional oedometer tests* [11].

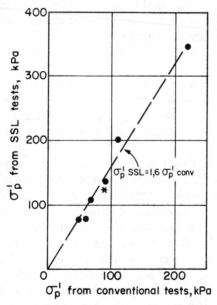

FIG. 17—*Correlation between the preconsolidation pressures obtained from SSL and conventional oedometer tests* [11].

of thick deposits. However, before such applications can be considered, it is necessary to investigate, by means of long-term field observations, the stress-strain-strain rate behavior of natural clay foundations. In particular, it is necessary to determine the magnitude of strain-rate effects in the range of very low strain rates usually prevailing in thick deposits. Such investigations are currently in progress at Laval University and will be reported in the near future.

In the model discussed in this paper, the numerical treatment of the basic constitutive Eqs 2, 10, and 9 or 11 necessitates the determination of permeability and rheological parameters. For the permeability parameters, Tavenas et al [34] recommend direct measurements by falling head tests in the oedometer, on specimens consolidated to various void ratios. For the stress-strain-strain rate behavior, CRS tests are recommended that give directly the shape of the normalized effective stress-strain relationship of the considered clay. The variation of σ'_p with strain rate could be obtained from tests carried out at different strain rates. For Champlain clays, it is worth mentioning that Leroueil et al [11] have produced a normalized $\sigma'_p - \dot{\epsilon}_v$ relation. Consequently, for these clays, only one CRS test would be required to completely define the constitutive equations.

Conclusion

In applying the findings of an experimental study reported by Leroueil et al [14], a numerical model has been developed for the analysis of the consolidation of natural clays in the usual laboratory test conditions. The deformation behavior of the clay skeleton is expressed in the form of a normalized effective stress-strain relation and of a preconsolidation pressure-strain rate relation. Darcy's law and the relevant permeability changes with void ratio are combined with Eqs 9 and 10 to obtain a complete description of the consolidation process in a viscoplastic material.

The finite-difference numerical method incorporating this model has been used for a detailed analysis of the behavior of natural clays in different types of consolidation tests. The behavior of creep tests, MSL tests, and CGT tests has been successfully reproduced, thus demonstrating the general validity of the model.

The results show that each element in a consolidating clay specimen follows a specific stress-strain relationship, related to the strain rate history of that element. The overall stress-strain relationship obtained from the usual interpretation of consolidation tests is an average of different, individual relations followed at different locations in the specimen. Consequently, that overall stress-strain relation reflects the testing conditions as much as the clay behavior itself. Differences in stress-strain parameters such as the preconsolidation pressure or the compression index obtained from different types of tests should be expected. Such differences are illustrated in the paper. In particular, the variations of the preconsolidation pressure measured in the various types of common consolidation tests are shown.

The present study was limited to the investigation of laboratory tests. The described model may also be applicable to the field situation. However, such application first requires that the magnitude of strain rate effects be checked for very low strain rates as normally encountered in thick clay deposits.

Acknowledgments

The present study was supported by National Sciences and Engineering Research Council of Canada (NSERC) Strategic Grant G-0851 and by NSERC Operating Grants A7724 and A7379. The major contribution of the second author to the development of the numerical model was made during his sabbatical stay at Laval University. The support of the Japanese Society for the Promotion of Science, of the National Research Council of Canada, and of Laval University in this connection is acknowledged.

The support of the technical staff and of R. Bouchard, former graduate student at Laval University, in developing the necessary laboratory data is also acknowledged.

References

[1] Hamilton, J. J. and Crawford, C. B., "Improved Determination of Preconsolidation Pressure of a Sensitive Clay," *ASTM STP 254*, American Society for Testing and Materials, Philadelphia, 1959, pp. 254–271.
[2] Smith, R. E. and Wahls, H. E., "Consolidation Under Constant Rate of Strain," *Journal of the Soil Mechanics and Foundation Division*, American Society of Civil Engineers, Vol. 95, No. SM2, 1969, pp. 519–539.
[3] Wissa, A. E. Z., Christian, J. T., Davis, E. H., and Heiberg, S., "Consolidation at Constant Rate of Strain," *Journal of the Soil Mechanics and Foundation Division*, American Society of Civil Engineers, Vol. 97, No. SM10, 1971, pp. 1397–1413.
[4] Sällfors, G., "Preconsolidation Pressure of Soft High Plastic Clays," Ph.D. thesis, Chalmers University of Technology, Gothenburg, Sweden, 1975.
[5] Lowe, J., III, Jonas, E., and Obrician, V., "Controlled Gradient Consolidation Test," *Journal of the Soil Mechanics and Foundation Division*, American Society of Civil Engineers, Vol. 95, No. SM1, 1969, pp. 77–97.
[6] Jarrett, P. M., "Time-Dependent Consolidation of a Sensitive Clay," *Materials Research and Standards*, American Society for Testing and Materials, Vol. 7, No. 7, 1967, pp. 300–304.
[7] Burghignoli, A., "An Experimental Study of the Structural Viscosity of Soft Clays by Means of Continuous Consolidation Tests," in *Proceedings, 7th European Conference on Soils Mechanics and Foundation Engineering*, Brighton, U.K., Vol. 2, 1979, pp. 23–28.
[8] Leroueil, S., Le Bihan, J. P., and Tavenas, F., "An Approach for the Determination of the Preconsolidation Pressure in Sensitive Clays," *Canadian Geotechnical Journal*, Vol. 17, No. 3, 1980, pp. 446–453.
[9] Janbu, N., Tokheim, O., and Senneset, K., "Consolidation Tests with Continuous Loading" in *Proceedings, 10th International Conference on Soil Mechanics and Foundation Engineering*, Stockholm, Vol. 1, 1981, pp. 645–654.
[10] Leroueil, S, Samson, L., and Bozuzuk, M., "Laboratory and Field Determination of Preconsolidation Pressure at Gloucester," *Canadian Geotechnical Journal*, Vol. 20, No. 3, 1983, pp. 477–490.
[11] Leroueil, S., Tavenas, F., Samson, L., and Morin, P., "Preconsolidation Pressure of Champlain Clays: Part II—Laboratory Determination," *Canadian Geotechnical Journal*, Vol. 20, No. 4, 1983, pp. 803–816.

[12] Buisman, A. S., "Results of Long Duration Settlement Tests," in *Proceedings*, 1st International Conference on Soil Mechanics and Foundation Engineering, Cambridge, MA, Vol. 1, 1936, pp. 103–107.

[13] Taylor, D. W., "Research on Consolidation of Clays," Massachussets Institute of Technology, Cambridge, MA, Serial 82, 1942.

[14] Leroueil, S., Kabbaj, M., Tavenas, F., and Bouchard, R., "Stress-Strain-Stain Rate Relation for the Compressibility of Sensitive Natural Clays," *Géotechnique*, June 1985.

[15] Berry, P. L. and Poskitt, T. J., "The Consolidation of Peat," *Géotechnique*, Vol. 22, No. 1, 1972, pp. 27–52.

[16] Tavenas, F., Leblond, P., Jean, P., and Leroueil, S., "The Permeability of Natural Soft Clays. Part I: Methods of Laboratory Measurement," *Canadian Geotechnical Journal*, Vol. 20, No. 4, 1983, pp. 629–644.

[17] Taylor, D. W., *Fundamentals of Soil Mechanics*, Wiley, New York, 1948.

[18] Šuklje, L., "The Analysis of the Consolidation Process by Means of the Isotache Method," in *Proceedings*, 4th International Conference on Soil Mechanics and Foundation Engineering, London, Vol. 1, 1957, pp. 200–206.

[19] Šuklje, L., *Rheological Aspects of Soil Mechanics*, Wiley-Interscience, London, 1969.

[20] Battelino, D., "Oedometer Testing of Viscous Soils," in *Proceedings*, Conference on Soil Mechanics and Foundation Engineering, Moscow, Vol. 1.1, pp. 25–30.

[21] Schofield, A. and Wroth, P., *Critical State Soil Mechanics*, McGraw-Hill, London, 1968.

[22] Perzyna, P., "The Constitutive Equations for Work-Hardening and Rate Sensitive Plastic Materials," in *Proceedings*, Vibrational Problems Symposium, Warsaw, Vol. 4, No. 3, 1963, pp. 212–240.

[23] Oka, F., "Prediction of Time-Dependent Behaviour of Clay," in *Proceedings*, 10th International Conference on Soil Mechanics and Foundation Engineering, Stockholm, Vol. 1, 1981, pp. 215–218.

[24] Zienkiewicz, O. C., *The Finite Element Method*, 3rd ed., McGraw-Hill, New York, 1977, pp. 450–499.

[25] Leonards, G. A. and Girault, P., "A Study of One-Dimensional Consolidation Test," in *Proceedings*, 5th International Conference on Soil Mechanics and Foundation Engineering, Paris, Vol. 1, 1961, pp. 213–218.

[26] Bjerrum, L., "Engineering Geology of Normally Consolidated Marine Clays as Related to the Settlements of Buildings, *Géotechnique*, Vol. 17, No. 2, 1967, pp. 83–119.

[27] Leroueil, S., "Quelques considérations sur le comportement des argiles sensibles," Ph.D. thesis, Department of Civil Engineering, Laval University, Québec, 1977.

[28] Parkin, A. K., "Coefficient of Consolidation by the Velocity Method," *Géotechnique*, Vol. 28, No. 4, 1978, pp. 472–474.

[29] Tavenas, F., Leroueil, S., La Rochelle, P., and Roy, M., "Creep Behaviour of an Undisturbed Lightly Overconsolidated Clay," *Canadian Geotechnical Journal*, Vol. 15, No. 3, 1978, pp. 402–423.

[30] Bishop, A. W. and Lovenbury, H. T., "Creep Characteristics of Two Undisturbed Clays," in *Proceedings*, 7th International Conference on Soil Mechanics and Foundation Engineering, Mexico City, Vol. 1, 1969, pp. 29–37.

[31] Tavenas, F., Brucy, M., Magnan, J. P., La Rochelle, P., and Roy, M., "Analyse critique de la théorie de consolidation unidimensionelle de Terzaghi," *Revue Française de Géotechnique*, No. 7, 1979, pp. 29–43.

[32] Samson, L., Leroueil, S., Morin, P., and Le Bihan, J. P., "Pression de préconsolidation des argiles sensibles," D.S.S. Contract 1SX79-00026, Report prepared for the Division of Building Research of the National Research Council of Canada, five volumes, 1981.

[33] Berre, T. and Iversen, K., "Oedometer Tests with Different Specimen Heights on a Clay Exhibiting Large Secondary Compression," *Géotechnique*, Vol. 22, No. 1, 1972, pp. 53–70.

[34] Tavenas, F., Jean, P., Leblond, P., and Leroueil, S., "The Permeability of Natural Soft Clays. Part II: Permeability Characteristics," *Canadian Geotechnical Journal*, Vol. 20, No. 4, 1983, pp. 645–660.

[35] Magnan, J. P. and Deroy, J. M., "Etude numérique de l'essai oedométrique à gradient contrôle," Laboratoire Central des Ponts et Chaussées, Internal Report, Paris, 1981.

Kouki Zen[1] and Yasufumi Umehara[1]

A New Consolidation Testing Procedure and Technique for Very Soft Soils

REFERENCE: Zen, K. and Umehara, Y., **"A New Consolidation Testing Procedure and Technique for Very Soft Soils,"** *Consolidation of Soils: Testing and Evaluation, ASTM STP 892*, R. N. Yong and F. C. Townsend, Eds., American Society for Testing and Materials, Philadelphia, 1986, pp. 405–432.

ABSTRACT: A new constant rate of strain consolidation testing technique for very soft soils with water content of 200 to 300% is proposed as an alternative of the conventional oedometer test. For the analysis of data from the constant rate of strain (CRS) consolidation test, three kinds of diagrams are constructed based on the finite-difference solutions for the basic equation of consolidation, taking into account the effect of large strain. It is shown that the consolidation constants are conveniently determined by using these diagrams. The validity of this method is examined by analyzing the consolidation phenomena of a model fill using the consolidation constants thus determined. It is concluded that the consolidation constants of very soft clays can be accurately determined from the CRS consolidation test.

KEY WORDS: cohesive soils, consolidation, laboratory tests, model tests, strain, pore pressure

Introduction

Very soft clayey sediments are likely to be produced as a result of the reclamation of dredged soils. With respect to both production and utilization of such a reclaimed land, it is of practical importance to predict the consolidation settlement, including the effect of self-weight of reclaimed soils. The method of settlement analysis for such very soft soils has been proposed by Mikasa [1], who derived the basic equation of consolidation taking into account the effect of self-weight and change of thickness in layer. Monte and Krizek [2] have also presented the one-dimensional mathematical model for large strain consolidation by introducing a unique reference state (stress-free state) characterized by the fluid limit into the basic equation of consolidation.

Any laboratory testing procedure for such very soft clays, however, has not been well-established so far. The conventional oedometer test based on Terzaghi's one-dimensional consolidation theory may not be applicable to very soft soils

[1] Chief, Soil Dynamics Laboratory, and Director, respectively, Soils Division, Port and Harbour Research Institute, Ministry of Transport, Yokosuka, Japan.

with high water contents because of limitations both in theory and in testing techniques. In the conventional oedometer test, the compression strain of the specimen is assumed so infinitesimal that the change of thickness has a negligible effect to the consolidation constants. When the specimen is consolidated from a very soft state, however, the change of thickness becomes too large and significant to be neglected. Furthermore, it is quite difficult to prevent the specimen from squeezing out along the inner surface of a consolidation ring, especially at the instant of loading, which tends to cause significant scatter in consolidation constants.

In the present paper, a new testing technique and procedure for determining the consolidation constants of very soft soils is proposed as an alternative method of the conventional oedometer test. A model fill is analyzed with the consolidation constants obtained from the proposed method in order to confirm the validity of this new testing technique.

Theoretical Basis

Mikasa [1] has proposed the extended consolidation theory including the effect of large strain and self-weight of the soil. In the case of laboratory consolidation tests on small specimens, the stresses arising from the self-weight of the specimen may be considered negligible compared with those applied. If the coefficient of consolidation C_v is assumed to be independent of depth at any time, the basic equation of consolidation is

$$\frac{\partial \zeta}{\partial t} = C_v \zeta^2 \frac{\partial^2 \zeta}{\partial z_0^2}$$

(1)

where

$$t = \text{time,}$$
$$z_0 = \text{original coordinate,}$$
$$\zeta = f_0/f = (1 + e_0)/(1 + e) = \text{consolidation ratio, and}$$
$$f \text{ and } f_0 = \text{volume ratio defined by } 1 + e \text{ and}$$
$$1 + e_0 \text{ at time } t \text{ and } 0, \text{ respectively.}$$

The consolidation ratio ζ can be related to the axial strain $\bar{\epsilon}$ by the relation $\zeta = 1/(1 - \bar{\epsilon})$.

In the constant rate of strain (CRS) consolidation test, the specimen is loaded at the constant rate of vertical deformation R, that is, at time t, the specimen of initial height H_0 is deformed by $\Delta H_0 = Rt$ at the surface of the specimen; hence the average vertical strain $\Delta H_0/H_0$ is

$$\frac{\Delta H_0}{H_0} = \frac{Rt}{H_0} = T'$$

(2)

in which T' is the dimensionless time.

Since Eq 1 is analytically unsolvable, it is replaced by the finite difference equation

$$\zeta(z_0, t + \Delta t) - \zeta(z_0, t) = \frac{C_v \Delta t}{\Delta z_0^2} \bar{\zeta}^2$$

$$\times \{\zeta(z_0 + \Delta z_0, t) - 2\zeta(z_0, t) + \zeta(z_0 - \Delta z_0, t)\} \quad (3)$$

Considering the following nondimensional variables:

$$Z_0 = \frac{z_0}{H_0}$$

$$\Delta Z_0 = \frac{1}{n} \quad (4)$$

$$\frac{C_v \Delta t}{\Delta z_0^2} = n^2 \left(\frac{C_v}{RH_0}\right) \frac{R \Delta t}{H_0} = n^2 \left(\frac{C_v}{RH_0}\right) \Delta T'$$

we can express Eq 3 by

$$\zeta(Z_0, T' + \Delta T') - \zeta(Z_0, T') = n^2 \left(\frac{C_v}{RH_0}\right) \Delta T' \cdot \bar{\zeta}^2$$

$$\times \{\zeta(Z_0 + \Delta Z_0, T') - 2\zeta(Z_0, T')$$

$$+ \zeta(Z_0 - \Delta Z_0, T')\} \quad (5)$$

in which $\bar{\zeta}$ is a certain value between $\zeta(Z_0, T')$ and $\zeta(Z_0, T' + \Delta T')$.

As the specimen has a freely drained layer at the top surface and the impervious boundary at the bottom in the CRS consolidation test, the initial and boundary conditions are given by

$$\zeta(Z_0, 0) = 1 \quad (6)$$

$$\zeta(0, T') = \phi(T') \quad (7)$$

$$\left(\frac{\partial \zeta(Z_0, T')}{\partial Z_0}\right)_{Z_0 = 1} = 0 \quad (8)$$

In addition, the following deformation condition under the constant rate of strain must be satisfied at any nondimensional time T':

$$\int_0^1 \bar{\epsilon}(Z_0, T')dZ_0 = \int_0^1 \left(1 - \frac{1}{\zeta(Z_0, T')}\right)dZ_0 = \frac{\Delta H_0}{H_0} \quad (9)$$

Since the $\phi(T')$ at the top boundary in Eq 7 is given in an implicit form, it is given as a first approximation so that for $T' = \Delta T'$:

$$\phi(T') = \frac{1}{1 - \bar{\epsilon}(0, \Delta T')} \doteq \frac{1}{1 - \Delta T'} \quad (10)$$

The value of $\phi(T')$ for the next step of calculation must be determined so that the displacement calculated by integrating strains in the specimen should be consistent with the deformation condition of Eq 9.

The boundary condition at the bottom of the specimen is introduced in the numerical calculation by establishing the imaginary point with the space mesh equal to $1/n$ below the impervious base, then assuming the same value of ζ as that for the space point on the impervious base.

The solution of Eq 5, namely value of consolidation ratio ζ satisfying initial condition of Eq 6, boundary conditions of Eqs 7 and 8, and the deformation condition of Eq 9 can be obtained by the finite-difference method. The solution is given as the function of two variables, that is, nondimensional space Z_0 and nondimensional time T', for a given nondimensional parameter C_v/RH_0. Figure 1 shows a typical example of the variations of strain distribution for 0.1, 1.0 and 10.0 of C_v/RH_0. It can be seen from Fig. 1 that the integration of strains is equal to the average strain of the specimen $\Delta H_0/H_0$. The compression strain $\bar{\epsilon}$ has the following relationship among ζ, f, and f_0.

$$\bar{\epsilon} = 1 - \frac{1}{\zeta} = 1 - \frac{f}{f_0} \quad (11)$$

Graphic Representation

Three kinds of diagrams, which are shown in Figs. 2 to 4, are constructed based on such computation results for various values of C_v/RH_0 as shown in Fig. 1 [3]. These diagrams indicate the variations of the consolidation ratio at the top $\zeta_1 = f_0/f_T$ with the average strain $\Delta H_0/H_0$; the variations of the consolidation ratio at the bottom $\zeta_B = f_0/f_B$ with the average strain $\Delta H_0/H_0$; and the variations of the ratio of bottom strain to top strain F for various values of C_v/RH_0. The ratio F is given by

$$F = \frac{\bar{\epsilon} \text{ at } z_n = 1}{\bar{\epsilon} \text{ at } z_0 = 0} = \frac{1 - \dfrac{f_B}{f_0}}{1 - \dfrac{f_T}{f_0}} = \frac{\left(\dfrac{f_0}{f_B} - 1\right) \times \dfrac{f_0}{f_T}}{\left(\dfrac{f_0}{f_T} - 1\right) \times \dfrac{f_0}{f_B}} \quad (12)$$

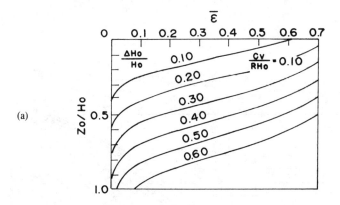

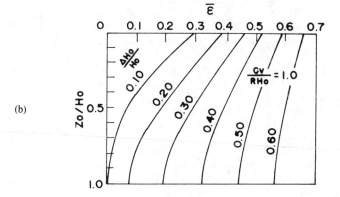

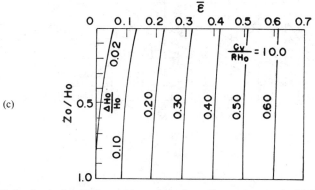

FIG. 1—*Strain distribution within specimen for various average strain.* (a) $C_v/RH_0 = 0.10$. (b) $C_v/RH_0 = 1.0$. (c) $C_v/RH_0 = 10.0$.

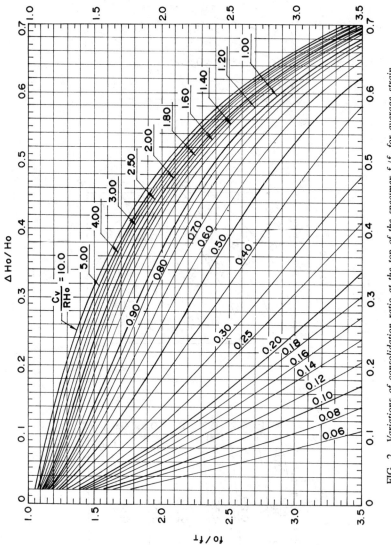

FIG. 2—*Variations of consolidation ratio at the top of the specimen f_0/f_T for average strain* $\Delta H_0/H_0$.

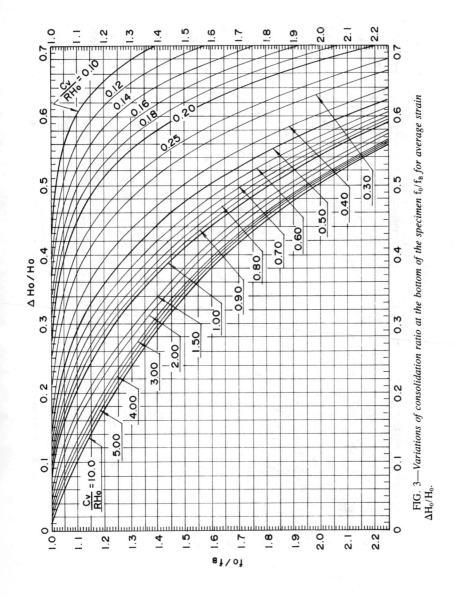

FIG. 3—*Variations of consolidation ratio at the bottom of the specimen f_0/f_B for average strain $\Delta H_0/H_0$.*

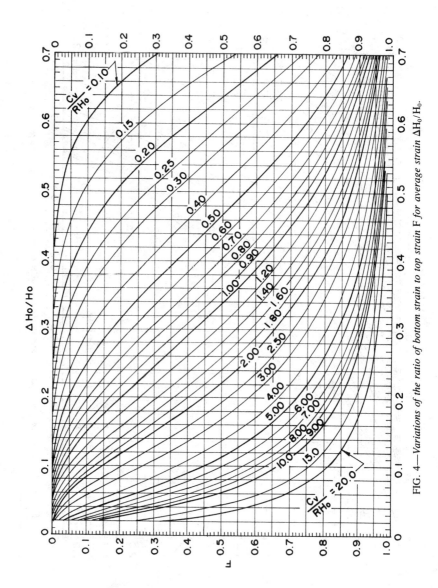

FIG. 4—Variations of the ratio of bottom strain to top strain F for average strain $\Delta H_0/H_0$.

where f_T and f_B indicate the volume ratio at the top and at the bottom of the specimen, respectively.

If the nonlinear stress-strain relation given by

$$f = f_0 - C_c \log\left(\frac{\sigma'}{\sigma'_0}\right) \tag{13}$$

holds, then the value of F given by Eq 12 can be represented by

$$F = \frac{\bar{\epsilon}_B}{\bar{\epsilon}_T} = \frac{\log(\sigma - u_h) - \log \sigma'_0}{\log \sigma - \log \sigma'_0} \tag{14}$$

where

σ = axial stress,
σ' = axial effective stress,
σ'_0 = effective stress at initial state,
C_c = compression index, and
u_h = excess pore water pressure at bottom of specimen.

Procedure

Equipment

A schematic drawing of the apparatus for the CRS consolidation test is shown in Fig. 5. The equipment consists of the loading apparatus, consolidation cell, and loading piston. The loading piston is used to apply a preload and a constant rate of strain after preloading. The inside surface of the consolidation ring is Teflon-coated and greased in order to minimize friction on the periphery of the specimen during loading. Two O-rings are used to prevent soils from squeezing out along the inner surface of the ring. As these two O-rings may inevitably increase some friction between the inner surface of the ring and the piston, two stress transducers are mounted in the pedestal and the loading piston to directly measure stresses at both ends of the specimen. The pore water pressure is measured through the porous stone mounted on the pedestal by a transducer installed at the base plate of the cell. Pore water is drained through the porous stone mounted on the loading piston.

Sample and Specimen Preparation

The clay samples employed in the present study are the marine clays taken from Honmoku in Yokohama Port and other ports in Japan. The physical properties of these samples are listed in Table 1.

In order to maintain the homogeneity of the specimen, the natural sample is completely remolded in a mixer, screened through the sieve, and finally mixed

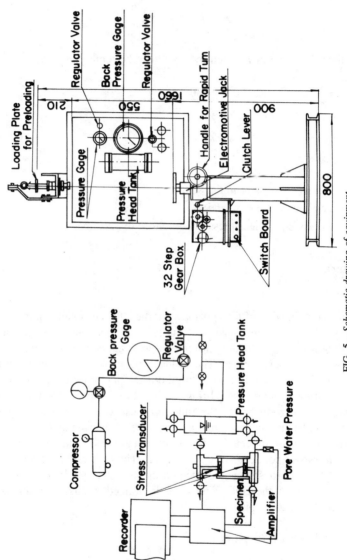

FIG. 5—Schematic drawing of equipment.

TABLE 1—*Physical properties of samples.*

Sample	G_s	w_L, %	w_p, %	I_p	Gradation, % Sand	Silt	Clay	Ignition Loss, %
Honmoku Clay	2.71	96.7	41.5	55.2	8.7	40.1	51.2	11.0
Tokyo Bay Mud	2.65	125.3	38.6	86.7	22.1	53.4	24.5	13.8
Yokohama Port Mud	2.71	73.4	39.2	34.2	24.1	40.9	35.0	10.6
Nagoya Port Mud	2.57	164.0	45.9	118.1	6.3	44.2	49.5	24.8
Osaka Bay Mud	2.59	102.8	45.8	57.0	20.0	43.0	37.0	18.1
Fushiki-Toyama Port Muds								
B-South	2.62	59.4	35.0	24.4	15.4	49.6	35.0	7.5
B-North	2.63	97.6	48.2	49.4	2.3	53.7	44.0	12.7
C-South	2.67	97.3	44.4	52.9	1.7	61.8	36.5	13.2
C-North	2.64	78.9	39.3	39.6	6.8	65.2	28.0	10.3
Minamata Bay Muds								
B-Site	2.72	92.2	39.2	53.0	24.9	39.3	35.8	22.7[a]
C-Site	2.70	88.0	35.7	52.3	13.5	50.5	36.0	14.6[a]
E-Site	2.71	66.3	32.3	34.0	8.0	61.0	31.0	11.8[a]
H-Site	2.70	70.4	34.9	35.5	12.0	53.0	35.0	12.6[a]
J-Site	2.71	96.0	38.5	57.5	10.1	45.9	44.0	18.4[a]

[a]These values were not obtained from the same samples as used for the laboratory tests, but from different samples at the site.

into a slurry with the water content of 200% to 230%. The slurry is placed under a vacuum and periodically agitated to remove entrapped air. It is then poured into the consolidation cell from 30 to 50 mm thick and allowed to undergo preloading to make a specimen whose initial state of stress is clearly specified.

Preloading

The initial stress in the specimen is required to be uniform for the interpretation of the test results using Figs. 2, 3, and 4, because those diagrams are constructed for the specimen with uniform initial volume ratio f_0. In the present study, preloading under a constant pressure is adopted to specify the uniform initial stress within the specimen prior to the constant rate of vertical deformation. A very light weight which can impose the initial stress of 1 to 4 kN/m² on the specimen is placed on the loading plate located at the top of the loading piston. Preloading is started by unlocking the loading piston and continued for about 24 h. During preloading the stress at the bottom of the specimen is measured by the stress transducer. Since the effective load decreases unavoidably because of the friction existing between the consolidation ring and O-rings, especially at the early stage of preloading, the additional small weight is loaded so as to keep the effective load constant according to the intensity of the stress monitored at the bottom of the specimen. After preloading is completed, the stresses at both ends of the specimen and the pore water pressure at the bottom are recorded. The height of the specimen after preloading is calculated by measuring the penetration of the loading piston.

Typical Record of Experiments

The CRS consolidation test is conducted by inducing a constant rate of vertical deformation to the upper surface of the specimen. During the test the axial deformation, the axial stresses both at the top and the bottom of the specimen, and the pore water pressure at the bottom are electrically measured. These data are recorded on a pen recorder or directly read by a digital voltmeter.

Figure 6 shows a typical example of the CRS consolidation test.

Determination of Consolidation Constants

On the basis of such data as shown in Fig. 6, the coefficient of consolidation C_v and the volume ratio versus log of effective stress f-log σ' relationship can be determined by the following procedure:

Step 1: Determination of σ'_0 in the initial state; σ'_0 is determined by measuring the stress just after the completion of consolidation due to preloading.

Step 2: Determination of the stress, pore water pressure, and displacement for each time lapse.

Step 3: Computation of the effective stresses both at the top and at the bottom of the specimen, and computation of F value of Eq 14.

Step 4: Computation of the average strain $\Delta H_0/H_0$ for each time.

Step 5: Determination of C_v—The values of F and $\Delta H_0/H_0$ are used to find C_v/RH_0 corresponding to $\Delta H_0/H_0$ at time t as shown in Fig. 7. R and H_0 being already known, C_v can be easily calculated.

Step 6: Computation of both f_T and f_B—The value of C_v/RH_0 obtained in Step 5 is used to find the volume ratio f_T and f_B as shown in Figs. 8 and 9.

Step 7: Determination of f-log σ'—The f-log σ' relationship is established by plotting f_T or f_B against each corresponding effective stress.

Step 8: Determination of σ'_{av}—The average effective stress σ'_{av} is determined

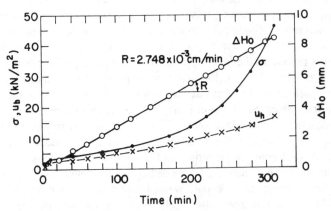

FIG. 6—*Typical record of tests.*

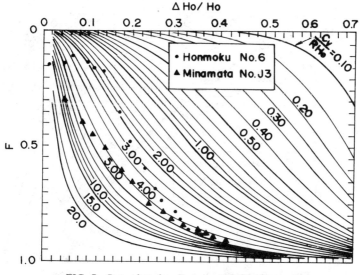

FIG. 7—*Data plotted on F versus* $\Delta H_0/H_0$ *diagram.*

by finding the stress corresponding to the average volume ratio $f_{av} = f_0 (1 - \Delta H_0/H_0)$ on the f-log σ' relationship established in Step 7.

Examples of the original data and the procedure of calculation are given in Tables 2 and 3.

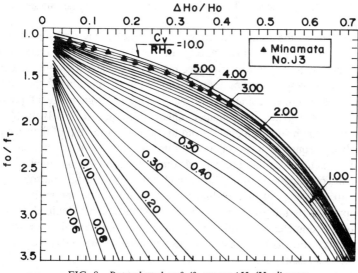

FIG. 8—*Data plotted on* f_0/f_T *versus* $\Delta H_0/H_0$ *diagram.*

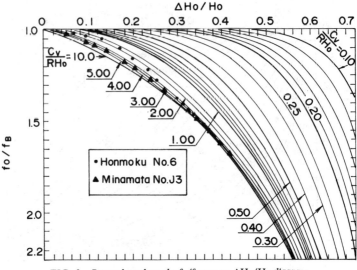

FIG. 9—*Data plotted on the* f_0/f_B *versus* $\Delta H_0/H_0$ *diagram.*

Presentation and Discussion of Test Results

The f-log σ' Relations

The volume ratio versus logarithm of effective stress relation is established both at the top and the bottom of the specimen. These two curves of the f-log σ' relations must coincide with each other, because the stress versus strain relation defined by Eq 13 should hold at the both ends of the specimen. In order to confirm this, the tendency of the f-log σ' relations is examined under various test conditions.

Figure 10 shows the f-log σ' relations for Honmoku clay determined by the constant rate of strain consolidation tests. In Fig. 10 the f-log σ' relations both at the top and at the bottom of the specimen are indicated by solid and open symbols, respectively. It can be seen that the f-log σ' relations at the top and bottom of the specimen coincide with each other for high range of pressure ($\sigma' = 50$ to 500 kN/m^2). For smaller range of pressure ($\sigma' = 10$ to 50 kN/m^2), however, the results calculated from the stresses at the top of the specimen are not always equal to those obtained from the stresses at the bottom. This disagreement is not essential, but depends on having selected too small an initial stress σ'_0 in comparison with the capacity (1000 kN/m^2) of the transducer.

The scattering range of nine oedometer test results is also shown in Fig. 10. The comparison of the f-log σ' relations with the results obtained from the conventional oedometer tests shows very good agreement, and this confirms the validity of this testing technique, though the comparison is limited to the relatively high effective stress range.

For the smallest range of pressure ($\sigma' = 1$ to 30 kN/m^2), the transducers with the capacity of 50 kN/m^2 are used to pick up more accurate axial stresses and pore water, and hence the f-log σ' relations are successfully obtained even at the very low range of effective stress where the conventional oedometer test is not applicable. The validity of this method in the low range of effective stress is examined in the later section by analyzing the consolidation phenomena of a model fill.

The log C_v-log σ'_{av} Relations

The average effective stress σ'_{av} is found as the stress corresponding to the average volume ratio $f_{av} = f_0(1 - \Delta H_0/H_0)$ on such a f-log σ' relation as shown in Fig. 10. The C_v versus σ'_{av} relation can be obtained by plotting the C_v and σ'_{av} at the identical time t on the log-log paper.

Figure 11 shows the log C_v-log σ'_{av} relations determined by the CRS consolidation test performed under various conditions. Solid symbols correspond to the results obtained from the tests using the transducer with the capacity of 50 kN/m^2, whereas open symbols represent the results obtained by employing the transducer with the capacity of 500 kN/m^2 for the pore water pressure and 1000 kN/m^2 for the axial stress. Scatters of nine conventional oedometer test results for normally consolidated states are also shown in Fig. 11 for comparison.

A solid line in Fig. 11 drawn by considering all test results indicates a representative line of the log C_v-log σ'_{av} relation for Honmoku clay. It is clear that the values of C_v are not constant over the wide range of the average effective stress, but increase monotonically as the average effective stress becomes larger. It is also found that the pressure dependency of C_v varies according to soil type, depends especially on the plasticity index, and becomes smaller for a high value of the plasticity index. This will be discussed in greater depth later.

In Fig. 11 the values of C_v are initially high but tend to fall onto the normal line as the average effective stress increases. This instability may be attributed to the poor reliability of transducers at the low level of pressures. This means that it is necessary to select appropriate transducers accurate enough to measure the pressure required for such practical use as an estimation of consolidation settlement of very soft reclaimed land.

Discussion on the Assumptions of Constant C_c and C_v

The analysis mentioned above has two major assumptions: the constant compression index and the constant coefficient of consolidation throughout the test. In the application of the analysis to obtain the material properties under the wide range of consolidation pressure, it is necessary to clarify the validity of these assumptions. For this purpose, sedimentation tests give a useful information on the compression index and the coefficient of consolidation of very soft soils in the very-low-pressure range [4].

The sedimentation test apparatus is shown in Fig. 12. After a specimen is poured into the inner tube, the cylinder is sufficiently shaken in the same manner

TABLE 2—*Data Sheet 1 (Original Data).*[a]

Port and Harbour Research Institute	CRS-Test (Data)		Date:
Sample Honmoku Clay		Test No. 6	

H_i^b	5.915 cm	G_s^f	2.712	C_f^j	kgf/cm²/mV
ΔH_i^c	3.545 cm	H_s^g	0.455 cm	C_B^k	1.549×10^{-4} kgf/cm²/mV
H_0^d	2.370 cm	f_0^h	5.205	C_u^l	2.307×10^{-4} kgf/cm²/mV
W_d^e	34.74 g	$\sigma_0'^i$	0.0081 kgf/cm²	R^m	2.748×10^{-3} cm/min

No.	t (min)	① Displacement ΔH_0 (cm)	② Axial Stress at Top Reading	σ_T (kgf/cm²)	③ Axial Stress at Bottom Reading	σ_B (kgf/cm²)	④ Total Pore Pressure reading	u (kgf/cm²)	⑤ Excess Pore Pressure u_h (kgf/cm²)
0	0				149	0.0230	65	0.015 $(=u_s)^n$	0
0	0	3.0×10^{-3}				(0.0081)			
1	1	28.5			216	0.0335	109	0.0251	0.0101
2	10	109.5			291	0.0451	153	0.0353	0.0203
3	40	164.7			338	0.0524	184	0.0424	0.0274
4	60	220.0			382	0.0592	214	0.0494	0.0344
5	80	274.0			435	0.0674	247	0.0570	0.0420
6	100				493	0.0764	283	0.0653	0.0503

No.						
8	120	330.0	565	0.0875	316	0.0729	0.0579
9	145	400.0	643	0.0996	331	0.0764	0.0614
10	170	468.0	773	0.1197	400	0.0923	0.0773
11	200	549.0	966	0.1496	459	0.1059	0.0909
12	220	602.0	1148	0.1771	502	0.1158	0.1008
13	240	658.0	1383	0.2142	554	0.1278	0.1128
14	260	714.5	1688	0.2615	608	0.1403	0.1253
15	270	742.0	1871	0.2898	625	0.1442	0.1292
16	280	769.0	2110	0.3268	669	0.1543	0.1393
17	290	796.0	2366	0.3665	699	0.1613	0.1463
18	300	824.0	2716	0.4207	757	0.1746	0.1596
19	305	838.0	2879	0.4459	767	0.1769	0.1619
20	310	852.0	3080	0.4771	792	0.1827	0.1677

a 1 kgf/cm² = 100 kN/m².
b Height of specimen before preloading.
c Settlement due to preloading.
d Initial height of specimen.
e Dry weight of specimen.
f Specific gravity.
g H_0/f_0.
h Void ratio.
i Initial effective stress in specimen.
j Calibration constant for stress at top.
k Calibration constant for stress at bottom.
l Calibration constant for pore pressure.
m Strain rate.
n Hydrostatic pressure.

TABLE 3—Data Sheet 2 (Procedure of Calculation).[a]

Port and Harbour Research Institute | CRS Test (Calculation) | Date:

Sample Honmoku Clay · Test No. 6

$\sigma'_0 = 0.0081$ kgf/cm² · $f_0 = 5.205$ · $H_0 = 2.37$ cm · $R = 2.748 \times 10^{-3}$ cm/min

No.	(6) $\Delta H_0/H_0$	(7) $= ③-④$ $\sigma - u_r$ $(\sigma_B - u)$	(8) σ $(\sigma_B - u_r)$	(9) $\log(\sigma - u_r)$ $-\log \sigma'_0$	(10) $\log \sigma - \log \sigma'_0$	(11) $= ⑨÷⑩$ $\dfrac{\log\left(\frac{\sigma - u_r}{\sigma'_0}\right)}{\log\left(\frac{\sigma}{\sigma'_0}\right)}$	(12) c_v/RH_0	(13) c_v (cm²/min)	(14) f_0/f_T	(15) f_T	(16) f_0/f_B	(17) f_B	(18) f_{av}	(19) σ'_{av} (kgf/cm²)
0	0											5.205		
1	0.001	0.0081	0.0081	0.0117	0.3598	0.032								
2	0.012	0.0083	0.0185	0.0838	0.5712	0.147								
3	0.046	0.0098	0.0301	0.0926	0.6655	0.139								
4	0.069	0.0100	0.0374	0.0838	0.7380	0.110	2.80	1.82×10^{-2}			1.01	5.153	4.845	0.017
5	0.093	0.0098	0.0442	0.1096	0.8119	0.135	2.00	1.30			1.03	5.053	4.721	0.021
6	0.116	0.0104	0.0524	0.1379	0.8808	0.157	1.80	1.17			1.04	5.005	4.601	0.025
7	0.136	0.0111	0.0614	0.2569	0.9529	0.270	1.60	1.01			1.05	4.957	4.481	0.032
8	0.169	0.0146	0.0725	0.3621	1.0200	0.355	1.95	1.27			1.11	4.732	4.325	0.040
9	0.197	0.0186	0.0846	0.5303	1.1125	0.477	2.40	1.56			1.16	4.487	4.180	0.048
10	0.232	0.0274	0.1017	0.7331	1.2216	0.600	2.50	1.63			1.23	4.232	3.997	0.066
11	0.254	0.0437	0.1346	0.8850	1.3024	0.678	3.00	1.95			1.27	4.098	3.883	0.100
12	0.278	0.0620	0.1621	1.0291	1.3919	0.739	3.50	2.28			1.33	3.914	3.758	0.135
13	0.301	0.0864	0.1992	1.1761	1.4844	0.792	3.30	2.15			1.37	3.799	3.638	0.180
14	0.313	0.1212	0.2465	1.2557	1.5316	0.820	4.00	2.60			1.40	3.718	3.576	0.190
15	0.324	0.1456	0.2748	1.3294	1.5865	0.838	3.900	2.54			1.43	3.640	3.519	0.230
16	0.335	0.1725	0.3118	1.405	1.6385	0.857	5.00	3.26			1.47	3.541	3.461	0.270
17	0.348	0.2052	0.3515	1.483	1.7008	0.872	5.00	3.26			1.49	3.493	3.394	0.330
18	0.354	0.2460	0.4057	1.522	1.727	0.881	5.00	3.26			1.51	3.447	3.362	0.360
19	0.359	0.2690	0.4309	1.561	1.757	0.889	5.00	3.26			1.52	3.424	3.336	0.390
20		0.2943	0.4621											

[a] 1 kgf/cm² = 100 kN/m².

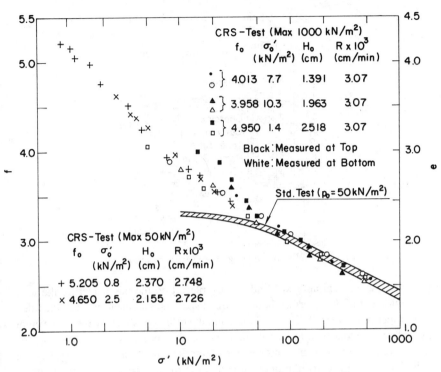

FIG. 10—*Volume ratio f versus log of effective stress σ′.*

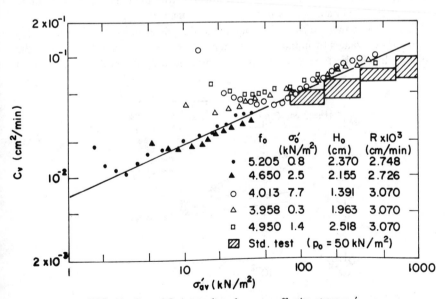

FIG. 11—*Log of C_v versus log of average effective stress σ_av′.*

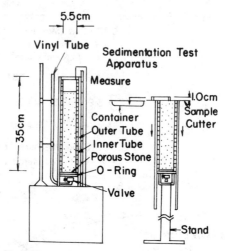

FIG. 12—*Schematic drawing of sedimentation test apparatus.*

as in the mechanical analysis, and quietly placed on the table. During the test, the settlement of the interface between the dispersion and clear water is recorded at a proper time interval. The void ratio and the effective stress by the self-weight of soil are calculated after the completion of the settlement by measuring the water content and the dry unit weight of the soil in each sliced ring. As the cylinder is composed of the outer tube and the inner tube which is made up of the rings of 1 cm high, this apparatus makes it possible to prevent each specimen from being compressed at the time of extrusion from the cylinder. Hence, more accurate water content and unit weight can be measured.

Figure 13 shows the *f*-log σ′ relationship thus determined on Honmoku clay with various values of initial water content. It can be seen from Fig. 13 that the relationship in the low-pressure range is influenced by the initial water content

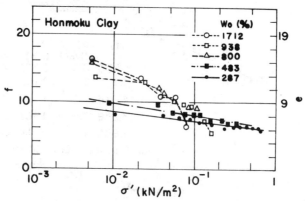

FIG. 13—*f-log σ′ relationship determined from sedimentation test (Honmoku clay).*

and becomes curved as it increases. The influence of the initial water content, however, becomes less significant as it decreases, and the f-log σ' relationship tends to be linear in such a very-low-pressure range.

Figure 14 shows the f-log σ' relationship for other marine sediments whose physical properties are presented in Table 1. These are obtained from the CRS consolidation test and the sedimentation test. It can be seen that the initial water content influences the compression index for each specimen the same as for Honmoku clay. The compression curve for the sample with a higher initial water content is subjected to a strong influence of the settling process and has a larger gradient compared with that of the compression curve obtained from the constant rate of strain consolidation test. As the initial water content decreases, the ratio of settlement due to sedimentation to the total settlement decreases. As a result, the total settlement is finally composed of only the settlement due to the selfweight of soil particles, and then the compression curves obtained from the CRS consolidation test and the sedimentation test for the specimens with approximately identical initial water content tend to fall onto the common straight line, as shown in Fig. 14.

From the discussion above, the approximately linear relation of the f-log σ' may hold in the wide range of pressure for very soft soils with an initial water content less than some value, within which the selfweight consolidation is dominant.

The selfweight consolidation process observed with the sedimentation test apparatus can reveal other useful information on the coefficient of consolidation in very low pressure range. By preparing the sample with relatively not so high water content whose settlement is caused mainly by the selfweight of soil particles, the C_v can be approximated by the curve fitting procedure using the computer program, CONSOLID [5], as typically shown in Fig. 15. Then it can be plotted against the average effective stress σ'_{av}. The σ'_{av} in this case is determined by assuming that $\sigma'_{av} = 1/2\Sigma\gamma'_i\Delta h_i$ in which γ'_i and Δh_i are the unit weight and the thickness of the ith sliced sample, respectively.

Figure 16 shows the variation of C_v with the average consolidation pressure obtained from the CRS consolidation test and the sedimentation test results by means of an approximating curve fitting procedure. It can be seen from Fig. 16 that the value of C_v is usually dependent on the pressure. In addition, its dependency seems to differ according to the type of sample. Figure 17 shows the correlation between the ratio of C_v to $\sigma'_{av} = 10$ kN/m^2 to C_v at $\sigma'_{av} = 1$ kN/m^2 and the plasticity index I_p. As shown in Fig. 17, the variation of C_v to the average effective stress tends to increase as the I_p becomes smaller.

Application to Settlement Analysis

The selfweight consolidation phenomenon of the model fill produced by Osaka Bay mud are observed by measuring the settlement and the excess pore water pressure at several points in the model fill. One of the reasons why Osaka Bay

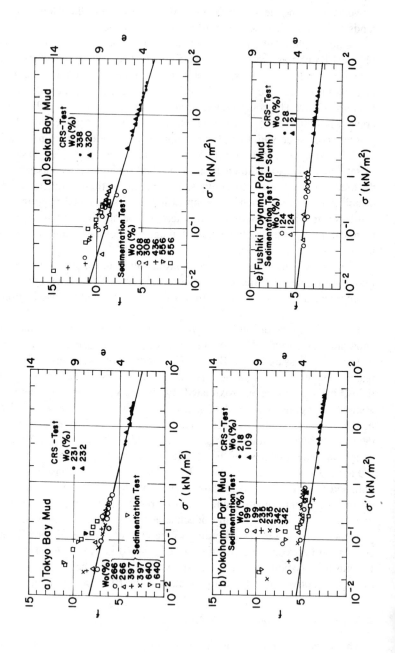

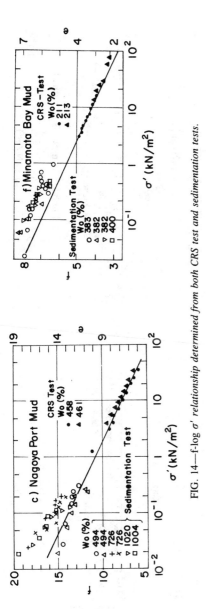

FIG. 14—f-log σ' relationship determined from both CRS test and sedimentation tests.

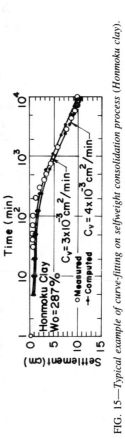

FIG. 15—Typical example of curve-fitting on selfweight consolidation process (Honmoku clay).

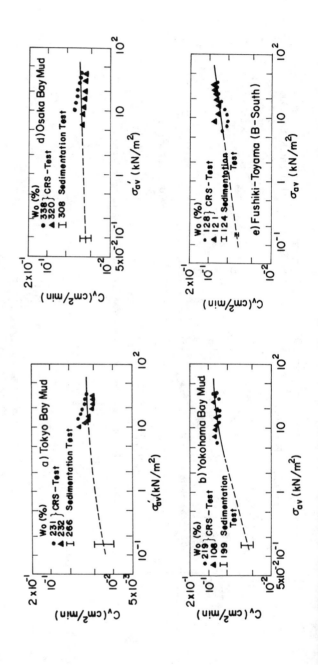

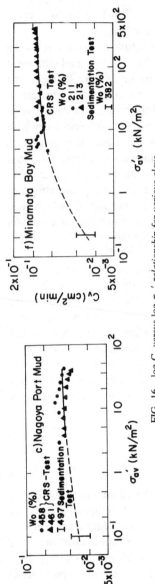

FIG. 16—log C_v versus log σ_{av}' relationship for various clays.

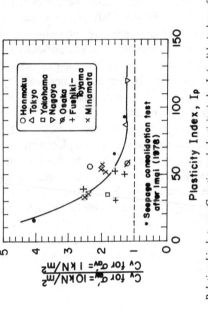

FIG. 17—Relationship between C_v ratio and plasticity index I_p (solid circles; after Ref 6).

mud is used is that the C_c and the C_v are constant over the wide range of effective stresses which will correspond to the stresses in the model fill. The initial state of the model fill is given in Table 4. The consolidation constants of Osaka Bay mud is given as 1.75 of the compression index C_c, 0.025 cm²/min of the coefficient of consolidation C_v and $f = f_1 - C_c \log \sigma' = 3.8-1.75 \log \sigma'$, where $f_1 = 3.8$ corresponds to the value of f for $\sigma' = 100$ kN/m².

The computed settlement at the location of 10 cm below the surface is shown together with the observed one in Fig. 18. It is found that the computed settlement shows a reasonably good agreement with the observed one. Figure 19 shows the comparison between the computed and observed excess pore water pressure at each depth for each different time. The computed distribution of excess pore water pressure with depth for ten days afterward shows a good coincidence with the observed one, while the computed distributions for either 50 days and 80 days after show a relatively poor coincidence with the observed one. This does not depend on the wrong determination of the consolidation constants, but could be attributed to the technical difficulty of the measurement in the long period of observation such as a time lag of pore water influx to manometers which may be caused by the decrease of permeability of porous stones installed in the model fill. In fact, in another model test no pore pressure change was observed in a few points of the fill after several months. Though some improvement of the pickup may be necessary to measure the very low excess pore pressure in the soil, the emphasis could be placed on the settlement analysis and the early stage of pore pressure analysis because the amount of pore pressure observed in the model fill after 50 days is too small to discuss strictly. It can be concluded from the discussion on Figs. 18 and 19 that the consolidation phenomenon of very soft clay fill is satisfactorily estimated by using the consolidation constants obtained from the proposed constant rate of strain consolidation test except the difference between the measured and the calculated pore pressure at very low range. The proposed method is especially advantageous for very soft soils, to which the conventional oedometer test is not applicable.

Conclusions

The constant rate of strain consolidation testing procedure and technique are proposed with a new interpretation of test results as an alternative of the conventional oedometer test for very soft clays. Three kinds of diagrams (Figs. 2 to 4) for the analysis of data from the constant rate of strain consolidation test are constructed based on the equation of consolidation, taking the effect of large strain into account. It is shown that these diagrams are very useful to conveniently determine the consolidation constants from the CRS consolidation test. The proposed method can be extended to the wide range of consolidation pressure by selecting appropriate transducers with accuracy enough to cover the pressure range required for practical use.

TABLE 4—*Initial state of model landfill.*

Clay thickness (H_0), cm	80
Water content (w_0), %	303.3
Void ratio (e_0)	7.849
Volume ratio (f_0)	8.849
Moist unit weight (γ_t), kN/m³	11.80
Dry unit weight (γ_d), kN/m³	2.93

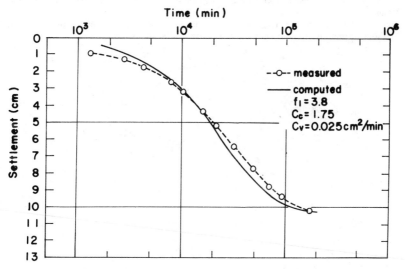

FIG. 18—*Settlement in the model fill due to selfweight of fill.*

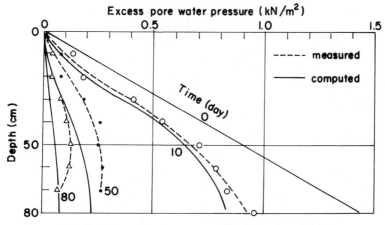

FIG. 19—*Excess pore water pressure distribution in model fill.*

References

[1] Mikasa, M., *The Consolidation of Soft Clay: A New Consolidation Theory and its Application*, Kajima Institution Publishing Co., Ltd., Tokyo, 1963 (in Japanese).
[2] Monte, J. L. and Krizek, R. J., "One Dimensional Mathematical Model for Large Strain Consolidation," *Geotechnique*, Vol. 26, No. 3, 1976, pp. 495–510.
[3] Umehara, Y. and Zen, K., "Constant Rate of Strain Consolidation for Very Soft Clayey Soils," *Soils and Foundations*, Vol. 20, No. 2, 1980, pp. 33–65.
[4] Umehara, Y. and Zen, K., "Consolidation Characteristics of Dredged Marine Bottom Sediments with High Water Content," *Soils and Foundations*, Vol. 22, No. 2, 1982, pp. 40–54.
[5] Umehara, Y. and Zen, K., "Determination of Consolidation Constants for Very Soft Soils," *Report of The Port and Harbour Research Institute*, Vol. 14, No. 4, 1976, pp. 44–55 (in Japanese).
[6] Imai, G., "Fundamental Studies on One-dimensional Consolidation Characteristics of Fluid Mud," Doctoral dissertation, University of Tokyo, 1978 (in Japanese).

V. Silvestri,[1] R. N. Yong,[2] M. Soulié,[1] and F. Gabriel[1]

Controlled-Strain, Controlled-Gradient, and Standard Consolidation Testing of Sensitive Clays

REFERENCE: Silvestri, V., Yong, R. N., Soulié, M., and Gabriel, F., "**Controlled-Strain, Controlled-Gradient, and Standard Consolidation Testing of Sensitive Clays,**" *Consolidation of Soils: Testing and Evaluation, ASTM STP 892*, R. N. Yong and F. C. Townsend, Eds., American Society for Testing and Materials, Philadelphia, 1986, pp. 433–450.

ABSTRACT: The objectives of this paper are threefold: (1) to compare the results obtained from constant rate of strain (CRS) and controlled-hydraulic gradient (CHG) consolidation tests with those obtained from incremental loading tests, (2) to compare the experimental data obtained in this study with the results of other investigators, and (3) to determine the possibility of using either CRS or CHG testing or both in geotechnical investigations dealing with soft clays of Eastern Canada.

The results of all the consolidation tests yield a unique relationship between preconsolidation pressure and strain rate. Both CRS and CHG tests performed following recommended procedures give values of the preconsolidation pressures which are higher than both those obtained in incremental load tests and those inferred from field studies. Both a simple creep law and a modified equation based on the rate process theory are found to adequately represent the observed soil response.

KEY WORDS: consolidation tests, incremental, controlled strain rate, controlled gradient, sensitive clays

Nomenclature

C_r Recompression index
C_c Compression index
C_v Coefficient of primary consolidation
C_α Coefficient of secondary consolidation
E_{oed} Constrained modulus of deformation
H Length of drainage path
e Void ratio
e_0 Initial void ratio

[1] Associate Professor, Professor, and Graduate Research Assistant, respectively, Department of Civil Engineering, École Polytechnique, Montreal, P.Q., Canada.

[2] William Scott Professor of Civil Engineering and Applied Mechanics, Director, Geotechnical Research Centre, McGill University, Montreal, P.Q., Canada.

n Creep parameter
t Time
u_b Excess pore pressure
Δ Increment
α Value of σ'_p at $\dot\epsilon_v = 0$
β Viscosity coefficient
ϵ_v Vertical strain
$\dot\epsilon_v$ Vertical strain rate
$\dot\epsilon_{v0}$ Arbitrary vertical strain rate
λ Time coefficient
σ'_p Preconsolidation pressure
σ'_{p0} Preconsolidation pressure at $\dot\epsilon_{v0}$
σ'_v Vertical effective stress
σ'_{v0} Overburden effective stress
R Average residual

The design of foundations on soft clays requires a prediction of the magnitude and rate of settlement. For the magnitude of one-dimensional consolidation settlement, only the knowledge of the vertical strain is required, without the need of using any particular theory [1]. It is only necessary to determine the change in void ratio corresponding to a particular applied stress increment by using an experimentally obtained stress-strain relationship. The rate of consolidation, on the other hand, involves the progressive development of effective stress as the interstitial fluid is expelled from the soil pores. In order to predict the rate at which such a phenomenon occurs, it is necessary to use a model which describes the flow of water through the pores, the relationship between permeability and void ratio, and the constitutive relationship in terms of effective stress.

For one-dimensional problems, the oedometer test and the Terzaghi theory have been the primary tools for many years to determine compressibility and consolidation parameters, and to predict the magnitude and rate of settlement in the field.

Among the various oedometer test techniques, the incremental load test is the most common one in present-day practice. In this test, the load is applied in several increments and each load increment acts for a certain amount of time. The load increment divided by the previous load is referred to as the load increment ratio (LIR) and the interval of time during which each load increment is acting is called the load increment duration (LID). In standard incremental consolidation testing (STD), the LIR is taken as unity and the LID is equal to one day.

The results of an STD test, which appear usually in the form of void ratio, e, as a function of the logarithm of the effective vertical stress, σ'_v, permit the determination of the recompression and compression indices, C_r and C_c, respectively, and of the effective vertical preconsolidation pressure, σ'_p. In addition, deformation-time curves allow the determination of the coefficients of primary and secondary compression, C_v and C_α, respectively.

Because of its particular procedure, the STD test is time-consuming since it requires an average of about ten days for its completion. In addition, for very soft plastic clays, it has been found that the STD test is unsatisfactory, since the clay might undergo sudden collapse and σ'_p will thereby be underestimated or totally obliterated [1,2]. Further, the spacing of the experimental points on the e-log σ'_v curve is such that it often gives a high degree of freedom in drawing the curve. Finally, as the specimen is loaded every 24 h, the amount of secondary compression will vary from load to load.

In order to have a better-defined stress-strain curve, several modifications have been proposed to the STD test. For example, in Eastern Canada it has become a standard practice to use an LIR of 0.5 for the testing of soft sensitive clays. However, as the LID is still equal to 24 h, the total test duration is longer than that of the STD test.

On the other hand, to speed up the test and at the same time to have a better defined stress-strain curve, Bjerrum [2] suggested a method where reduced load increments are used until the preconsolidation pressure is reached. These load increments are applied for a time interval corresponding to 100% primary consolidation. For pressures in excess of the consolidation pressure, the test is continued in the standard way. Compared with the STD test, Bjerrum's procedure gives higher values of both the preconsolidation pressure and the compression index [3]. It should be noted that the influence of both the LIR and LID on the oedometer stress-strain curve has been studied by several other investigators in the past 20 years [4–7].

In order to reduce the time involved in performing a consolidation test and to obtain a continuous stress-strain curve, other types of oedometer tests with continuous loading have been proposed and used in the past 15 years. Among these, the constant rate of strain (CRS) and the controlled hydraulic gradient (CHG) tests are the most widely used. Other types of tests, like the constant rate of loading (CRL) and the constant rate of consolidation (CRC) tests are also sometimes used but will not be discussed in this paper [3,8,9].

The CRS test was first mentioned by Hamilton and Crawford [10] as a rapid means of determining the preconsolidation pressure. Earlier references to this test were also made by Crawford [5] and Wahls and DeGodoy [11]. The theoretical development of the CRS test has been described by Smith and Wahls [12] and by Wissa et al [13]. In this test, the specimen is drained from the top surface only and is compressed at a constant speed. As such, a continuous recording of the deformation response of the clay is obtained. For the results obtained in the CRS test on plastic clays, it has been found that, in general, the higher the strain rates, the higher the preconsolidation pressures and the pore pressures generated [3,5,14–16].

The aims of developing the CHG test were [17] first to have uniform stress conditions and rates of compression, and second, to be able to extrapolate the results to field compression rates which are several order of magnitude lower than those in the test specimens [4,5]. Essentially, in the CHG test, the specimen is loaded at a rate such that the excess pore water pressure, u_b, generated at the bottom (undrained) end remains constant and equal to a predetermined value. Referring to the results of this type of test, Larsson [3] mentioned the fact that, for soft Swedish clays, higher values of the excess pore water pressure induce higher values of the preconsolidation pressures. Similar results have been found for sensitive clays of Eastern Canada [16,18].

The purpose of this paper is to compare the results of STD and modified incremental consolidation tests with those obtained by using both the CRS and CHG tests. The soil used in this study is a soft sensitive clay from Louiseville (P.Q.). It is shown that when the results of both the CRS and CHG tests are corrected for the effect of the strain rate, they compare very well with those of the incremental consolidation tests. In addition, the results found in this study are compared with those obtained for other clays, and it is shown that the correction for the effect of strain rate is unique, at least for clays of similar properties.

Limitations on space preclude a complete discussion of the test methods used and the observed results.

Soil Properties

The soil used in this study has been obtained from a test trench excavated on the outskirts of Louiseville (P.Q.), a town situated some 100 km northeast of Montreal. The site is located along Highway 40, on the north shore of the St. Lawrence River.

Undisturbed blocks of clay were recovered below the weathered crust, at a depth varying between 3 and 5 m. The general properties of the Louiseville clay are summarized in Table 1.

Testing Methods

Incremental Load Tests

For the incremental load tests, two values of the load increment ratio were used, namely 0.5 and 1.0. The load increment durations were: (1) the times to

TABLE 1—Geotechnical properties of the Louiseville clay.

Water content, %	82.0
Liquid limit, %	59.0
Plastic limit, %	26.0
Plasticity index, %	33.0
Liquidity index	1.7
Silt content, %	22.0
Clay content, %	78.0
Activity	0.4
Field vane strength, kPa	30.0
Sensitivity (by field vane)	20.0
Preconsolidation pressure, kPa	90.0

obtain 100% primary consolidation, (2) one day, and (3) one week. In all these tests, double drainage was used and the soil specimens measured 63.5 mm in diameter by 18.0 mm in height. In addition, the equipment used did not allow the application of a back-pressure to the soil specimens.

CRS Tests

For the CRS tests, two types of consolidometers were used. The first type was built of very heavy stainless steel tubing and was fitted onto a conventional triaxial press. The consolidation ring measured 101.6 mm in diameter by 20 mm in thickness.

A back-pressure of 140 kPa was applied at the top and bottom faces to ensure full saturation. The load, applied at the specimen by the loading ram, was measured at the top by a load cell and at the bottom by a pressure transducer. Pore water pressure was measured at the bottom (undrained) end of the specimen by means of a pressure transducer mounted in the base of the consolidometer. Height change was measured by a linear variable displacement transducer.

The second type of consolidometer used in this series of tests was obtained from Geotechnical Digital Systems of Walton-On-Thames, London. This equipment features a Rowe cell, is fully automatic, and is also able to perform STD, CHG, CRL, and cyclic consolidation tests. As such, it may be termed the "universal" consolidation apparatus. A back-pressure of 140 kPa was used to saturate the soil specimens, which measured 76.2 mm in diameter by 20.0 mm in height.

For all the CRS tests, the strain rates used varied between 1.0×10^{-7} and $7.0 \times 10^{-5} \text{s}^{-1}$. In addition, the experimental data were interpreted by means of the following equations that are based upon an assumed parabolic pore pressure distribution [14]:

$$\sigma'_v = \sigma_v - 2u_b/3 \qquad (1)$$

and

$$C_v = H^2 \, \Delta\sigma_v/2 \, \Delta t \, u_b \qquad (2)$$

where

σ'_v = average effective vertical stress,
σ_v = total vertical stress (minus back-pressure),
u_b = excess pore water pressure measured at the base of the specimen,
H = length of drainage path,
Δ = increment, and
t = time.

CHG Tests

For the CHG tests, two types of equipment were used. The first type, termed the "Anteus" consolidometer, was obtained from Anteus Laboratory Equipment

Inc. of Malden, Massachusetts. In this series of tests, the specimens measured
63.5 mm in diameter by 18.6 mm in height. A back-pressure of 140 kPa was
also used to saturate the specimens. The second type of equipment used was the
"universal" consolidometer referred to in the preceding section. The specimen
dimensions and back-pressures used were the same as those described previously.

In all the CHG consolidation tests, the excess pore pressure, u_b, varied between
5 and 75 kPa. The interpretation of this test is based also upon Eqs 1 and 2.

Test Results

Analysis and Discussion

Typical results obtained in the incremental load tests are shown in Fig. 1. In
order to compare these tests with one another, the ordinate in Fig. 1 (curves *b*)
refers to vertical strain, ϵ_v, and not to void ratio. In Fig. 1 (curves *b*) is shown
the variation of the coefficient of consolidation, C_v, as a function of log σ'_v.
Figure 1 (curves *a*) illustrates quite well the effect of both the LID and LIR on
the consolidation curve of the Louiseville clay. Extending the load increment
duration to one week reduces the preconsolidation pressure from approximately
90.0 to 70.0 kPa. The effect of the LIR on the stress-strain curve is also evident
in Fig. 1 (curves *a*), as is the dramatic drop in the value of C_v for pressures
straddling the preconsolidation pressure of the clay (curves *b*). The results shown
in Fig. 1 are similar to those reported by other investigators [3–5,16,18,19]. The

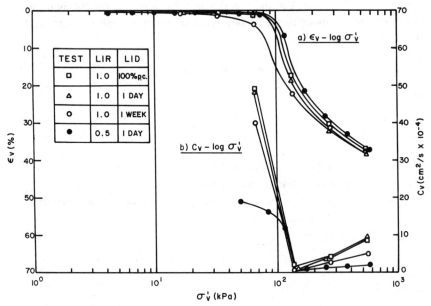

FIG. 1—*Incremental consolidation test results.*

variation of C_v shown in Fig. 1 (curves b) demonstrates that the results corresponding to $\sigma'_v \approx 135$ kPa do not yield a unique value of C_v; rather they should be viewed only as very small quantities.

Typical results obtained in the CRS tests are shown in Figs. 2 to 4. In Fig. 2 (curves a), the results are shown in terms of vertical strain, ϵ_v, and log σ'_v. Examination of the curves in this figure reveals that, increasing the strain rate, $\dot{\epsilon}_v$, from 1.0×10^{-7} to $7.0 \times 10^{-5}\mathrm{s}^{-1}$, increases the preconsolidation pressure from approximately 78.0 to 140.0 kPa. As evidenced in Fig. 2 (curves a), the parameters C_r and C_c are also affected by the strain rate. The figure also depicts (curves b) the variation of C_v as a function of σ'_v. Because of the large variation in the values of C_v, the data have been shown on a logarithmic scale. Once again,

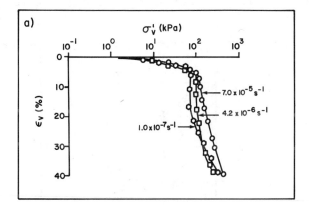

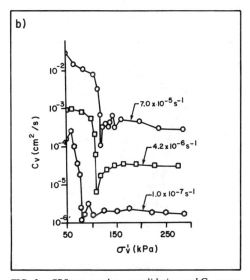

FIG. 2—*CRS test results: consolidation and C_v curves.*

for the range of strain rates used, Fig. 2 (curves b) illustrates the dramatic drop of C_v in the stress region that straddles the preconsolidation pressure of the clay.

Increasing the strain rate increases the value of C_v throughout the test, as shown in Fig. 2 (curves b) also. Following the work of Janbu [20], the results shown in Fig. 2 (curves a) have been plotted in Fig. 3 (curves a) on a simple arithmetic scale. Examination of the curves of this figure indicates, firstly, that the constrained modulus of deformation E_{oed} (i.e., the slope of the stress-strain relationships) is almost constant for each value of the strain rate and for pressures below the preconsolidation pressure of the clay; and second, that the constrained modulus of the clay increases for increasing values of the strain rate. Figure 3 (curves b) shows the values of the excess pore pressures, u_b, generated at the undrained end of the soil specimens. The curves in this figure indicate that by increasing the strain rate results in an increase of u_b, especially for pressures exceeding the

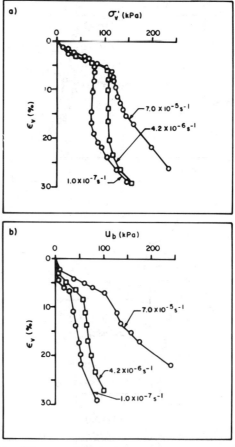

FIG. 3—*CRS test results: consolidation and pore pressure curves.*

preconsolidation pressure of the clay. For high values of u_b, it is possible that the parabolic pressure distribution [12,13] assumed in the analysis of the CRS test may no longer be valid, thus precluding the correct interpretation of the test, as mentioned also by Sällfors [13] and Gorman et al [14].

Results of CHG consolidation tests showed that the stress-strain curves were affected by the value of the excess pore water pressure, u_b. Some of the experimental results obtained in this series of tests are shown in Fig. 4 (curves a) and indicate that by increasing u_b from 5 to 25 kPa results in an approximate increase of 35 kPa in the value of the preconsolidation pressure. In order to make possible a comparison between CHG and CRS tests, the strains mobilized in the CHG tests were recorded as a function of time, as shown in Fig. 4 (curves b). In this

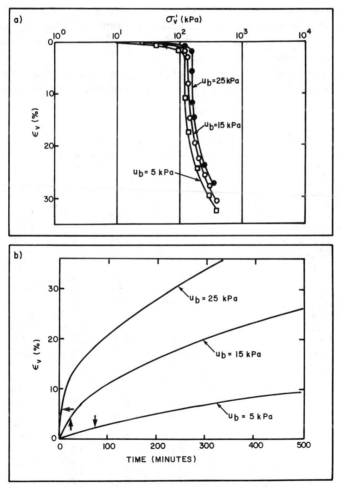

FIG. 4—*CHG test results.*

figure, the arrows indicate the points where the preconsolidation pressures are reached. Even though the maximum value of u_b used in this series of tests was equal to 75 kPa, the majority of tests was performed with u_b varying between 10 and 25 kPa.

Comparison Between Incremental, CRS, and CHG Tests

In order to compare the results obtained in the incremental load test program with those of the CRS and CHG consolidation tests, it was decided, in the light of experimental results, to use the strain rate operating in each test as a rational basis for comparison.

In the CRS tests the strain rate was obtained directly from the speed of the compression press and the height of the soil specimen. For the CHG tests, the slopes of the curves in Fig. 4 (curves b) were used to determine the strain rate at any particular time. For each incremental consolidation test, an equivalent strain rate was calculated by using the time elapsed from the start of the test to reach a certain strain (or stress) level.

All the results obtained in this study have been grouped in Table 2. Each entry is the average value of at least three individual experimental values. In this table,

TABLE 2—*Oedometer test results.*

Type of Test	$\dot{\epsilon}_v$, s^{-1}	e_0	σ'_p, kPa	C_r	C_c
Incremental					
LIR = 1.0, LID = 100% p.c.	. . .	2.10	92.0	0.040	2.9
LIR = 1.0, LID = 1 day	1.2×10^{-7}	2.29	89.0	0.040	3.3
LIR = 1.0, LID = 1 week	1.5×10^{-8}	2.30	70.0	0.060	2.4
LIR = 0.5, LID = 1 day	5.6×10^{-8}	2.36	86.0	0.068	3.2
Controlled Gradient					
$u_b = $ 5 kPa	1.0×10^{-5}	2.29	120.0	0.035	>5.0
$u_b = $ 10 kPa	2.5×10^{-5}	2.20	125.0	0.030	>5.0
$u_b = $ 15 kPa	3.0×10^{-5}	2.33	130.0	0.025	>5.0
$u_b = $ 20 kPa	4.0×10^{-5}	2.20	133.0	0.022	>5.0
$u_b = $ 25 kPa	7.0×10^{-5}	2.44	140.0	0.010	>5.0
$u_b = $ 75 kPa	3.3×10^{-4}	2.39	175.0	0.010	>5.0
Controlled Rate of	1.0×10^{-7}	2.26	78.0	0.087	>5.0
Strain	1.2×10^{-7}	2.28	78.0	0.070	4.8
	5.6×10^{-7}	2.33	92.0	0.069	>5.0
	1.1×10^{-6}	2.30	105.0	0.008	4.7
	2.1×10^{-6}	2.33	107.0	0.075	>5.0
	4.2×10^{-6}	2.28	110.0	0.080	>5.0
	1.0×10^{-5}	2.32	120.0	0.090	>5.0
	1.3×10^{-5}	2.31	122.0	0.014	>5.0
	3.8×10^{-5}	2.48	125.0	0.090	>5.0
	7.0×10^{-5}	2.08	140.0	0.029	4.8
	1.2×10^{-5}	2.36	173.0	0.018	1.9
	2.8×10^{-4}	2.26	143.0	0.010	>5.0

the strain rate reported for both the CHG and the incremental load tests refers to the average value calculated in the stress region below the preconsolidation pressure. Note also that the unusually large values of the compression index C_c reported in this table are due to the fact that they were calculated in the stress region located immediately above the preconsolidation pressure.

Because of the very large variation in the strain rate, as indicated in Table 2, it was at first thought that a plot of σ'_p versus log $\dot{\epsilon}_v$ would be useful. However, on the basis of experimental data on the creep of metals [21], it was deemed more appropriate to present the results on a log-log plot as shown in Fig. 5. The data shown in this figure indicate that the results are quite unique as long as the strain rate is taken into account. In addition, Fig. 5 indicates that the equivalent strain rate in an incremental consolidation test with an LIR of 0.5 and LID of one day, is on the order of $6.0 \times 10^{-8} s^{-1}$. As previous work [16,18] has indicated, for the sensitive clays of Eastern Canada this type of test appears to give the most representative value of σ'_p when compared with field behavior. It follows that CRS tests should be performed at such low values of strain rates for meaningful results to be obtained. It should be noted that such a strain rate is lower than that recommended in ASTM Test Method for One-Dimensional Consolidation Properties of Soils Using Controlled-Strain Loading (D 4186), which specifies a value of $1.0 \times 10^{-2}\%/min$ or $1.7 \times 10^{-6} s^{-1}$ for this type of clay. Had this value of strain rate been retained in a settlement analysis, for example, the preconsolidation pressure would have been overestimated by about 25% and, as a consequence, consolidation settlements would have been underestimated.

Because the preconsolidation pressure represents a basic parameter in geotechnical engineering, the analysis of the σ'_p values reported in Fig. 5 has been further extended. The relationship between log σ'_p and log $\dot{\epsilon}_v$ being approximately linear for the stress range used in this study, the experimental results may be analyzed by means of an equation based on the Norton-Hoff law, which is often used for the description of the high-temperature creep in metals [21,22]. This equation is

$$\sigma'_p = \sigma'_{p0} \left(\frac{\dot{\epsilon}_v}{\dot{\epsilon}_{v0}}\right)^{1/n} \qquad (3)$$

where

σ'_p = preconsolidation pressure at strain rate $\dot{\epsilon}_v$,

σ'_{p0} = normalizing preconsolidation pressure obtained at a strain rate equal to $\dot{\epsilon}_{v0}$,

$\dot{\epsilon}_v$ = strain rate,

$\dot{\epsilon}_{v0}$ = arbitrary or reference strain rate, and

n = creep parameter.

A nonlinear least-squares analysis [23] based on the data shown in Fig. 5

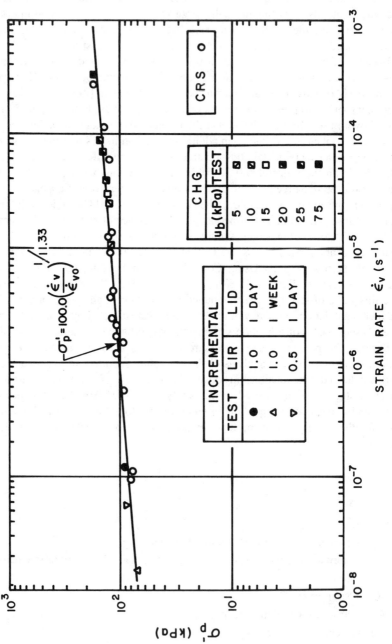

FIG. 5—Preconsolidation pressure – strain rate relationship: log-log plot.

yields $n = 11.33$. In Fig. 5 is also shown the curve given by Eq 3 and having $\sigma'_{p0} = 100.0$ kPa as the normalizing preconsolidation pressure, corresponding to an arbitrary strain rate, $\dot{\epsilon}_{v0}$, equal to 1.0×10^{-6} s^{-1}. The numerical approximation with $n = 11.33$ yields the best fit to the data, since it represents the curve having the minimum value of the average residual, R [23]. R is the standard deviation of the differences between the observed and estimated values of the variables. The minimum value of R is equal to 7.5 kPa in this case.

If one considers Eq 3 and the following equation obtained by Mesri and Choi [24]

$$\sigma'_p = \sigma'_{p0} \left(\frac{\dot{\epsilon}_v}{\dot{\epsilon}_{v0}} \right)^{C_\alpha / C_c} \tag{4}$$

where C_α is the coefficient of secondary compression, and C_c is the compression index, then it appears that the inverse of the creep parameter n is equal to the ratio C_α / C_c. Indeed, the value of $(1/n)$—that is, 0.088—obtained in this study fits quite well in the range of the values of C_α / C_c given by Mesri and co-workers [24,25].

When one considers Eq 3 or Eq 4, the problem of vanishing σ'_p appears; that is, if the results were extrapolated to very low strain rates, the values of σ'_p would tend to zero. Instead of viewing the data in such a perspective, it is more appropriate to consider the results as showing only an approximate linear relationship for the range of strain rates used in this study. Undoubtedly, had lower strain rates been used, the linear relationship in Fig. 5 would have sooner or later reached a finite horizontal asymptote corresponding to a certain minimum value of σ'_p. One would expect that for normally consolidated clays the minimum value of σ'_p would be equal to the existing effective overburden pressure in the field, σ'_{v0}. For overconsolidated clays, the minimum value of σ'_p would certainly be greater than σ'_{v0}.

The experimental data shown in Fig. 5 have been reinterpreted on the basis of the rate process theory [26] in order to take into account the effect of the nonvanishing σ'_p. For the analysis of the experimental results, the following modified equation has been used:

$$\sigma'_p = \alpha + \beta \sinh^{-1} (\lambda \, \dot{\epsilon}_v) \tag{5}$$

where

α = value of σ'_p at $\dot{\epsilon}_v = 0$,
β = viscosity coefficient, and
λ = time coefficient.

In order to obtain the best fit to the experimental data, non-linear least-squares regression analyses [23] were again performed and yielded the results shown in Fig. 6. Consistent with the usual conventional presentation of the data when

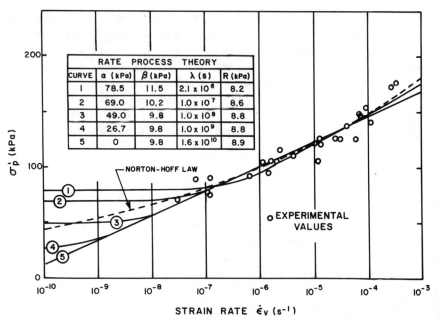

FIG. 6—*Preconsolidation pressure – strain rate relationship: semi-log plot.*

dealing with the rate process theory, the results in Fig. 6 are shown on a semilogarithmic plot. When considering the regression analyses performed, it should be noted that different numerical approximations correspond to different values of the parameter λ in Eq 5. The curves shown in Fig. 6 correspond, for each value of λ, to particular values of α and β which give the minimum value of the average residual R and which represent the best fits to the data. For example, curve 5, obtained for $\lambda = 1.6 \times 10^{10}$s, has $\alpha = 0$, and $\beta = 9.8$ kPa. This numerical approximation is the best fit to the data when $\lambda = 1.6 \times 10^{10}$s. In addition, it should be recalled from Eq 5 that the numerical value of α corresponds to the value of σ'_p at a strain rate of zero; that is, α represents the minimum horizontal asymptote for each curve.

For all tests performed on the Louiseville clay, the strain rates used were not sufficiently low to allow the determination of the parameter α. In fact, Fig. 6 shows that, for the range of strain rates used in this study, all the curves 1 to 5 adequately represent the observed response, with curve 1 being a slightly better approximation since the value of the average residual R is the minimum.

In order to determine whether Eq 3 or Eq 5 represents most adequately the observed response, the creep law has been also included in Fig. 6. Examination of all the curves in this figure indicate that both relationships give an adequate response in the range of strain rates used in this study. However, because of its simplicity, Eq 3 offers a distinct advantage over Eq 5.

Comparison with Related Work

In the technical literature are several papers on the dependence of the preconsolidation pressure of plastic clays on the strain rate. The results of pertinent studies [5,16,18,19,27] have been grouped in Fig. 7 together with the data obtained in this investigation. Examination of the curves in this figure shows quite well the effect of the strain rate on the value of the preconsolidation pressure. It was advantageous to normalize the curves by expressing the ratio (σ'_p/σ'_{p0}) as a function of $\dot{\epsilon}_v$ (Fig. 8). It should be noted that the reference strain $\dot{\epsilon}_{v0}$ was again arbitrarily chosen to be equal to 10^{-6}s^{-1}. Any other reference strain could have equally been chosen so long as one uses the corresponding σ'_{p0} as indicated in Eq 3.

Examination of Fig. 8 indicates that for these clays of marine origin, the experimental points fall in a very narrow band, implying quite similar values of the creep parameter n. It is quite possible that for these clays, the relationship shown in Fig. 8 is unique, as postulated also by Mesri and Choi [24] on the basis of the ratio C_α/C_c.

In addition, in order to assess the physical significance of the parameter α in Eq 5, the data reported by Leroueil et al [18] on the behavior of the Gloucester clay and shown in Fig. 7 have been examined in the light of the results shown in Fig. 6. These authors compare the preconsolidation pressure determined in the laboratory with that inferred in the field and indicate that for the Gloucester

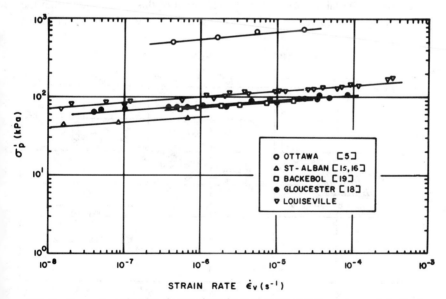

FIG. 7—*Comparison of studies showing dependence of preconsolidation pressure of plastic clays on strain rate.*

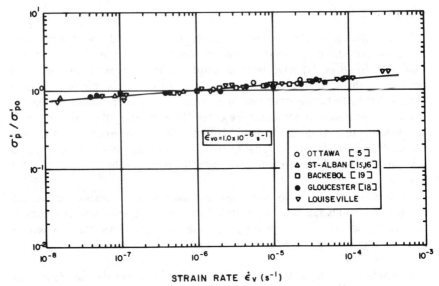

FIG. 8—*Normalized behavior.*

site the field values correspond to strain rates of the order of 10^{-10} to $10^{-9}s^{-1}$. The experimental results obtained by these authors appear in Fig. 9 together with the regression analysis based upon Eq 5. Note that the regression analysis was based only on the laboratory values. Examination of this figure shows that the curve having $\lambda = 10^7$ s, $\alpha = 64.2$ kPa, and $\beta = 4.9$ kPa yields the most

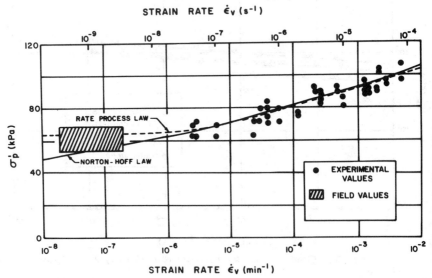

FIG. 9—*Gloucester clay (●, test points from Leroueil et al [18].*

adequate response. For comparison, the Norton-Hoff law is also shown in this figure.

In order to extend the application of either Eq 3 or Eq 5, further studies should be performed on clays of widely different properties. However, on the basis of the results presented in this section, it may be stated that while the predictive capability of this approach cannot be fully assessed in view of the limited data reported, a preliminary evaluation appears to indicate general consistency in the analysis of the results.

Conclusions

For the clays discussed in this paper, the following conclusions have been drawn:

1. The various consolidation tests used in this study give comparable results if the strain rate is duly taken into account.

2. In the CRS tests, the higher the strain rates used, the higher the preconsolidation pressures, the excess pore water pressures generated, and the constrained moduli of deformation.

3. In the CHG tests, higher excess pore pressures yield higher preconsolidation pressures.

4. A simple creep law is found to adequately represent the soil behavior in the range of strain rates used in this study. This creep law is related to a fundamental expression obtained by Mesri and Choi [24].

5. The experimental results were also interpreted by means of an equation based on the rate process theory. The first term of this equation was related to the value of the preconsolidation pressure corresponding to a strain rate of zero.

6. The results obtained in this study compare very well with those obtained by other investigators on similar marine clays. A unique relationship is obtained when the preconsolidation pressures are normalized with respect to an arbitrary or reference pressure.

Acknowledgments

The authors wish to express their gratitude to the National Research Council of Canada and the Ministry of Education of Quebec for financial support.

References

[1] Balasubranamiam, A. S. and Brenner, R. P. in *Soft Clay Engineering*, E. W. Brand and R. P. Brenner, Eds., Elsevier, New York, 1981, Chaper 7, pp. 481–486.
[2] Bjerrum, L. in *Proceedings*, 8th International Conference on Soil Mechanics and Foundations Engineering, Moscow, Vol. 3, 1973, pp. 111–159.
[3] Larsson, R., *Swedish Geotechnical Institute*, Report 12, 1981, pp. 1–157.
[4] Crawford, C. B., *Journal of the Soil Mechanics and Foundations Division*, American Society of Civil Engineers, Vol. 90, No. SM5, 1964, pp. 87–102.
[5] Crawford, C. B., *Canadian Geotechnical Journal*, Vol. 2, No. 2, 1965, pp. 90–97.

[6] Leonards, G. A. in *Proceedings*, American Society of Civil Engineers Specialty Conference on Performance of Earth and Earth-Supported Structures, Lafayette, IN, Vol. 3, 1972, pp. 169–173.

[7] Leonards, G. A. and Ramiah, B. K. in *Papers on Soils—1959 Meetings, ASTM STP 254*, American Society for Testing and Materials, Philadelphia, 1959, pp. 116–130.

[8] Jarret, P. M., *Materials Research and Standards*, Vol. 7, No. 7, 1967, pp. 300–304.

[9] Abashi, H., Yoshikuni, H., and Murayama, S., *Soil and Foundations*, Vol. 10, No. 1, 1970, pp. 43–56.

[10] Hamllton, J. J. and Crawford, C. B. in *Papers on Soils—1959 Meetings, ASTM STP 254*, American Society for Testing and Materials, Philadelphia, 1959, pp. 254–271.

[11] Walls, H. E. and De Godoy, N. S., *Journal of the Soil Mechanics and Foundations Division*, American Society of Civil Engineers, Vol. 91, No. SM3, 1965, pp. 147–152.

[12] Smith, R. E. and Wahls, H. E., *Journal of the Soil Mechanics and Foundations Division*, American Society of Civil Engineers, Vol. 95, No. SM2, 1969, pp. 519–539.

[13] Wissa, A. E. Z., Christian, J. T., Davis, E. H., and Heiberg, S., *Journal of the Soil Mechanics and Foundation Division*, American Society of Civil Engineers, Vol. 97, No. SM40, 1971, pp. 1393–1413.

[14] Gorman, C. T, Hopkins, T. C., Deen, R. C., and Drnevich, V. P., *Geotechnical Testing Journal*, Vol. 1, No. 1, 1978, pp. 3–15.

[15] Leroueil, S., Tavenas, F., Trak, B., LaRochelle, P., and Roy, M., *Canadian Geotechnical Journal*, Vol. 15, No. 1, 1978, pp. 54–65.

[16] Leroueil, S., Tavenas, F., Samson, L., and Morin, P., *Canadian Geotechnical Journal*, Vol. 20, No. 4, 1983, pp. 803–816.

[17] Lowe, J., Jonas, E., and Obrician, V., *Journal of the Soil Mechanics and Foundations Division*, American Society of Civil Engineers, Vol. 95, No. SM1, 1969, pp. 77–97.

[18] Leroueil, S., Samson, L., and Bozozuk, M., *Canadian Geotechnical Journal*, Vol. 20, No. 3, 1983, pp. 477–490.

[19] Sällfors, G., *Preconsolidation Pressure of Soft High Plastic Clays*, Ph.D. thesis, Chalmers University of Technology, Gothenburg, Sweden, 1975, p. 231.

[20] Janbu, N. in *Proceedings*, 7th International Conference on Soil Mechanics and Foundations Engineering, Mexico City, Vol. 1, 1969, pp. 191–196.

[21] Gittus, J., *Creep, Viscoelasticity and Creep Fracture in Solids*, Wiley, New York, 1975, p. 725.

[22] Ladanyi, B., *Canadian Geotechnical Journal*, Vol. 9, No. 1, 1972, pp. 63–80.

[23] Bard, Y., *Nonlinear Parameter Estimation*, Academic Press, New York, 1974, p. 341.

[24] Mesri, G. and Choi, Y. K., *Canadian Geotechnical Journal*, Vol. 16, No. 4, 1979, pp. 831–834.

[25] Mesri, G, and Godlewski, P. M., *Journal of the Geotechnical Engineering Division*, American Society of Civil Engineers, Vol. 103, No. GT5, 1977, pp. 417–430.

[26] Mitchell, J. K., *Fundamentals of Soil Behaviour*, Wiley, New York, 1976, p. 422.

[27] Vaid, Y. P., Robertson, P. K., and Campanella, R. G., *Canadian Geotechnical Journal*, Vol. 16, No. 1, 1979, pp. 34–42.

Sukomal Chakrabarti[1] and Robert G. Horvath[1]

Conventional Consolidation Tests on Two Soils

REFERENCE: Chakrabarti, S. and Horvath, R. G., **"Conventional Consolidation Tests on Two Soils,"** *Consolidation of Soils: Testing and Evaluation, ASTM STP 892,* R. N. Yong and F. C. Townsend, Eds., American Society for Testing and Materials, Philadelphia, 1986, pp. 451–464.

ABSTRACT: The consolidation behavior of laboratory prepared "young" specimens of two clay soils, kaolin and Dundas soil, was examined using conventional consolidation tests. The specimens were prepared under either isotropic or K_0 stress conditions. A fixed-ring type consolidometer was used for the tests. A 24-h loading time increment with a load increment ratio of 1.0 was used.

Test results indicate that the consolidation behavior of both soils was influenced by the past loading history and the test loading conditions. The soils were able to remember the past maximum stress.

Based on the test results, normalized e–σ'_v relationships are suggested for the two soils.

KEY WORDS: apparent preconsolidation pressure, past loading, loading history, quasi-static loading, normalized relations, consolidation testing, one-dimensional consolidation

Consolidation behavior of fine-grained soils is traditionally evaluated in the laboratory using the one-dimensional oedometer tests. The test is usually performed by incrementally loading thin right circular soil cylinders in the axial direction while the deformations in the lateral direction are restricted. In spite of the several shortcomings [1–7] the oedometer test remains popular, primarily because of its inherent simplicity and the quantity of performance data that has accumulated over the years with the method [8].

The important data obtained from the oedometer tests include the preconsolidation pressure [9,10], pore-water pressure behavior, and void ratio-effective stress behavior. However, there are several widely different views as to the interpretation of the test results [11–19].

The present study is concerned with the oedometer consolidation behavior of laboratory prepared "young" specimens of two soils, kaolin and Dundas soil. The major objectives of this investigation are to explore answers to the following questions:

[1] Graduate Student and Assistant Professor, respectively, Department of Civil Engineering and Engineering Mechanics, McMaster University, Hamilton, Ont., Canada L8S 4L7.

1. Do the "young" clays retain memory regarding the past loading history?
2. Is the pressure-void ratio relationship unique for the soil?

Experimental Investigation

Materials

Two materials were selected for the present study, Kaolin N.F. and a soil from the nearby city of Dundas. Kaolin had 81% finer than the 2-μm size and had a liquid limit of 74 and a plastic limit of 42. The Dundas soil was collected in a disturbed state, air-dried in the laboratory, carefully pulverized, and sieved. The portion of the soil passing U.S. Sieve #100 (150 μm) was stored in containers for use in the experimental work. The material so prepared had 16% finer than the 2-μm size and had a liquid limit of 23 and a plastic limit of 16. The physical properties of the soils investigated are summarized in Table 1.

Specimen Preparation

Isotropically Prepared Specimens—Triaxial cells were used for the preparation of specimens consolidated under isotropic stress conditions. The soil was kept moist for 24 h before the preparation of the specimen. A sufficient amount of distilled water was then added to the soil to make a slurry with a water content of 250 to 350%. The slurry was then sieved through U.S. Sieve No. 100 for a uniform mixing and to avoid lumps.

The soil slurry was poured inside a rubber membrane held by a flexible mold set up on the bottom plate of the triaxial cell. The cylindrical part of the mold

TABLE 1—*Physical properties of soils.*

Properties		Soils Used	
		Kaolin N.F.	Dundas Soil
Specific gravity		2.60	2.75
Liquid Limit (LL)		74	23
Plastic Limit (PL)		42	16
Plasticity Index (PI)		32	7
Activity		0.40	0.44
Grain Size Distribution			
% finer than	Size, μm	[a]	
	150.0	100	100
	75.0	100	91
	40.0	100	86
	10.0	100	60
	4.0	96	36
	3.0	92	28
(clay size)	2.0	81	16
	1.0	55	3
	0.5	28	...

[a]As per the manufacturer (Georgia Kaolin Co., N.J.).

consisted of two 200 by 300 mm flexible plastic sheets. The cell was assembled with the soil slurry inside the membrane. The drainage valves were closed and the soil specimen was subjected to vacuum (90.0 kPa average) for 24 h. The cell was disassembled. The top plate and the top porous stones were introduced, and then the cell re-assembled.

The assembled cell was filled with de-aired water. The applied isotropic pressure was maintained constant with the aid of a regulated air pressure supply. The drainage valves were opened and the soil slurry was allowed to consolidate by draining through both the top and the bottom porous stones.

The isotropic pressure was released after 72 h. The specimens so consolidated had an average size of 100 mm in diameter by 200 mm in height. The top and the bottom 50-mm portions were discarded and the middle 100-mm (average) portions were retained for testing. All specimens were carefully removed from the cell, wrapped in plastic foil papers, waxed, marked, and stored in a storage room maintained at a temperature of 21 ± 2°C and a relative humidity of 95 ± 2%. A period of 21 ± 1 day was allowed between the time of specimen preparation and testing.

Anisotropically Prepared Specimens—A few test specimens were prepared under anisotropic condition by consolidating the soil slurry one-dimensionally in a specially prepared mold. Distilled water was added to the soil to make a slurry. The slurry with a moisture content of about 150% was subjected to a vacuum of about 90.0 kPa for 12 h to expel air. The de-aired slurry was then carefully poured inside the mold. The mold was gently tapped from the outside by small laboratory vibrators during this operation. The slurry was then subjected to a vacuum of about 90.0 kPa for a further period of 24 h to expel entrapped air.

After the completion of the de-airing period, the mold was assembled and transferred onto the loading frame. The desired axial consolidation stress was then applied and maintained constant with the aid of a regulated air pressure supply. One-way drainage was applied and the soil slurry was allowed to consolidate for 24 h. The axial stress was then released and the mold disassembled.

The consolidated samples had an average size of 75 mm in diameter by 50 mm in height. The specimens were immediately wrapped in plastic foil papers after removal from the mold, waxed, and marked for identification and stored. A period of 21 ± 1 day was allowed between specimen preparation and testing.

Consolidation Testing

Consolidometer—A fixed-ring type consolidometer was used for all of the tests. The ring was made of highly polished stainless steel. The side friction between the specimen and the ring was minimized by applying a thin smear of silicon grease inside the ring. The dimensional details of the consolidometer are given in Fig. 1. The design of the consolidometer allowed the use of different sizes of rings up to a diameter of 75.00 mm. The present tests were conducted on specimens 50.70 mm in diameter and 18.29 mm in height. Only one-way

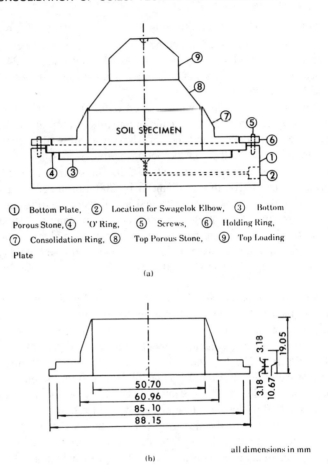

(1) Bottom Plate, (2) Location for Swagelok Elbow, (3) Bottom Porous Stone, (4) 'O' Ring, (5) Screws, (6) Holding Ring, (7) Consolidation Ring, (8) Top Porous Stone, (9) Top Loading Plate

(a)

all dimensions in mm

(b)

FIG. 1—*Consolidometer:* (a) *assembled;* (b) *consolidation ring.*

drainage through the top porous stone was allowed during the tests. The bottom porous stone was connected to a pressure transducer through 200-mm-long nondisplacement-type, 703-kPa-capacity tubing. Swagelok connectors and nondisplacement-type Whitey valves were used for all connections.

Instrumentation

Load cells and pressure transducers were used to measure the applied load and the induced pore pressures respectively. Dial gages as well as displacement transducers were used to measure the axial deformation.

Load cells were calibrated in the laboratory using dead weights. Pressure transducers were calibrated using the Budenberg Dead Weight Tester. A 24-channel data acquisition system was used to scan and record the data.

Testing

The soil specimen was taken out of the waxed plastic cover and trimmed with a soil trimmer to a size very slightly larger than the consolidation ring, which was then pushed into the specimen with sharp edge cutting. The top of the specimen was trimmed flush with the ring using a wire saw. The trimmed face of the specimen was then pressed against a ridged brass disk to create the recess at the top needed for the centering. The centering is very important because there is only about 0.05 mm clearance between the top porous stone and the consolidation ring. This tight fit was necessary to minimize the possible loss of soil during testing.

The consolidometer was assembled. Every possible care was taken so as to ensure that there was no air trapped in the system.

Loading frames with a lever arm ratio of 1:11 were used to load the specimens contained in the consolidometers. Two basic loading schemes were used for the tests. In the "A" loading scheme, the initial stress (σ'_{vi}) was 6.83 kPa. In Scheme "B" the initial stress (σ'_{vi}) was 27.31 kPa. A 24-h loading time increment with a load increment ratio of 1.0 was generally adopted for these tests.

Discussion

Effective Vertical Stress–Void Ratio Behavior

Isotropically Prepared Specimens—A comparison of Figs. 2 and 3 indicates that the slope of the virgin consolidation line is $C_c = -0.32$ for both loading schemes and thus appears to be independent of the initial stress used in the tests. This behavior is in agreement with the generally held conception that the value of C_c depends primarily on the type of material being tested. It can be seen from Figs. 2 and 3, however, that the location of the virgin consolidation line is essentially controlled by the magnitude of the initial stress used in the tests. The virgin consolidation line for the specimens tested under higher initial stress (Scheme "B": $\sigma'_{vi} = 27.31$ kPa) is located to the left of the line for the specimens tested under lower initial stress (Scheme "A": $\sigma'_{vi} = 6.83$ kPa.)

The slope of the recompression curve was found to be influenced by the magnitude of the initial stress used in the tests (Figs. 2 and 3). The values of the slopes were found to be $C_r = -0.037$ and -0.081 for the kaolin specimens tested under A and B loading schemes, respectively. The value of the slope of the recompression curve for Dundas soil was found to be $C_r = -0.032$ (Fig. 4). The slope of the virgin consolidation line for Dundas soil was found to be $C_c = -0.122$ (Fig. 4).

The slope of the virgin consolidation line for anisotropically prepared kaolin specimens was found to be $C_c = -0.31$ (Fig. 5). This value of C_c was almost identical to the value obtained for the isotropically prepared kaolin specimens.

The behavior regarding the slope of the recompression curves for the anisotropically prepared specimens is noticeably different from that for the isotropically

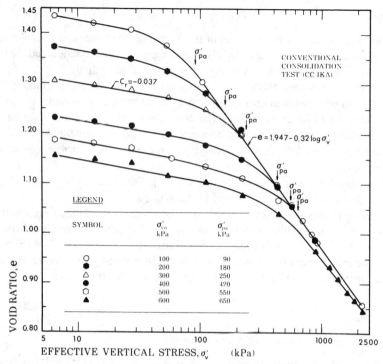

FIG. 2—*Effective vertical stress–void ratio relationship for isotropically prepared kaolin. Loading scheme A (σ'_{vi} = 6.83 kPa).*

prepared specimens. The recompression curves were found to be essentially parallel ($C_r = -0.097$ and -0.099 for loading schemes A and B, respectively), regardless of the loading scheme used (Fig. 5).

Comparison of Isotropically and Anistropically Prepared Specimens—A comparison of the effective stress versus void ratio relationships as presented in Figs. 2 to 5 indicates the following:

1. The recompression curves for the isotropically prepared specimens were initially linear (Figs. 2 to 4). The recompression curves of the isotropically prepared kaolin specimens showed more curvature than isotropically prepared specimens of Dundas soil (Figs. 2 to 4). The anisotropically prepared kaolin specimens, on the other hand, showed a somewhat different trend of behavior. The recompression curves for these specimens appear to consist of two straight line portions which do not show much of a curvature near the virgin consolidation line (Fig. 5).

2. The slopes of the virgin consolidation lines for the isotropically and the anisotropically prepared kaolin specimens were found to be nearly the same, $C_c = -0.32$ and -0.31, respectively, and were significantly smaller than the value of $C_c = -0.69$ obtained from one-dimensional consolidation of a kaolin

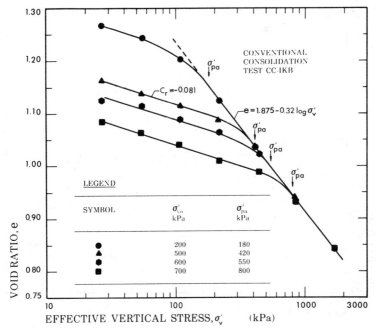

FIG. 3—*Effective vertical stress–void ratio relationship for isotropically prepared kaolin. Loading scheme B (σ'_{vi} = 27.31 kPa).*

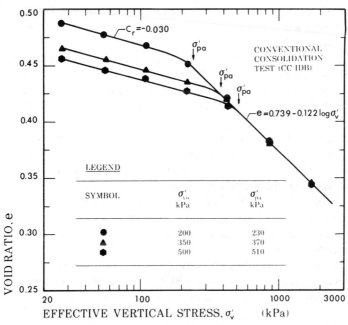

FIG. 4—*Effective vertical stress–void ratio relationship for isotropically prepared Dundas soil. Loading scheme B (σ'_{vi} = 27.31 kPa).*

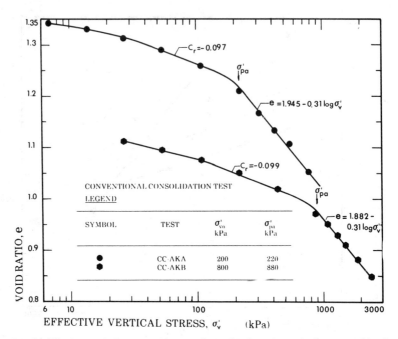

FIG. 5—*Effective vertical stress–void ratio relationship for anisotropically prepared kaolin.*

reported by Hambly [20] and $C_c = -0.58$ obtained from the Terzaghi-Peck empirical relationship [21]). The measured value of $C_c = -0.122$ for Dundas soil agrees well with the value of $C_c = -0.117$ estimated from the Terzaghi-Peck relationship [21].

3. The location of the virgin consolidation line was found to be dependent on the initial stress used in the tests. The equations of these lines for the different tests have been determined using the slope C_c and e_n, defined as the value of void ratio of the normally consolidated soil at an effective vertical stress of 1.0 kPa.

The foregoing observations regarding the recompression curves for the isotropically and the anisotropically prepared specimen of kaolin may be explained in terms of the fabric formed during the initial consolidation from the slurry. It has generally been recognized that the isotropic stress conditions during deposition tend to induce in clay an isotropic structure with a randomly oriented fabric, while anisotropic stress conditions tend to induce an anisotropic structure with the clay platelets oriented in the direction perpendicular to the direction of the major principal stress [22,23].

It can be construed that the isotropically prepared kaolin specimens may undergo some structural changes when tested under one-dimensional loading condition. The curvature of the recompression curves near the virgin consolidation line for the isotropically prepared specimens tested under the "A" loading scheme with a smaller initial stress ($\sigma'_{vi} = 6.83$ kPa) is believed to be due to the structural disturbances brought about by the anistropic loads. The fact that the isotropically prepared kaolin specimens did not exhibit the same degree of curvature of the recompression curves under the "B" loading scheme having a larger initial stress (σ'_{vi} = 27.31 kPa) may be due to the larger immediate structural disturbances caused by the larger initial stress.

The observation of significantly lower value of C_c for the anisotropically prepared kaolin specimens as compared with previously reported C_c values suggests that the anisotropic consolidation during the preparation of the specimens as used in the present study had been able to cause only minor differences in the structure of the specimens.

Pore Water Pressure

The pressure ratio (u_R), has been defined as the ratio between the pore water pressure measured at the base (u_b), and the load increment, $\Delta\sigma'_v$. The general trend observed in the pressure ratio versus time plots was that the excess pore-water pressures measured immediately after the load application became consistently lower as successive load increments were applied (Fig. 6). This is in agreement with the observation by other researchers [19], although pore-water pressure peaks were not always observed. The above behavior may be due to the changes in the specimen microstructure with increasing recompression stresses.

The excess pore-water pressure immediately after the load application was usually around 80% of the incremental load. However, smaller values were also observed for some specimens.

The system flexibility, air in the system, and the method of measurement are believed to have contributed towards the above behavior [24–26]. It may be

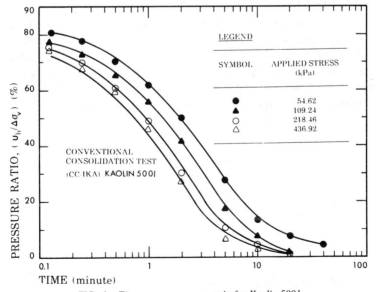

FIG. 6—*Time-versus-pressure ratio for Kaolin 5001.*

recalled that the pore-water pressure was measured at the base of the specimen and that the bottom porous stone was connected to the pressure transducer by a tube. It appears that better quality data could have been obtained had it been possible to measure the pore pressure at the midplane of the specimen.

The dissipation of excess pore-water pressure was found to be linearly related to the settlement. Similar trends have been observed by Crawford [2]. However, the time required for primary consolidation determined from the excess pore-water pressure dissipation was found to be significantly greater than that estimated by Square Root Fitting Method. This observation tend to support the suggestion that the division of the primary and the secondary consolidations may be arbitrary [2,27,28].

Preconsolidation Pressure

There are several definitions for determining the preconsolidation pressure from the e-log σ'_v data [9,10]. However, all these definitions give only the range of possible values. The preconsolidation pressure determined using laboratory data has, therefore, been designated as the apparent preconsolidation pressure (σ'_{pa}). The definition that σ'_{pa} is the axial stress corresponding to the junction between the recompression curve and the virgin consolidation line has been adopted for the present study. The value of σ'_{pa} can be rather easily determined using an arithmetic plot instead of the commonly used void ratio versus log σ'_v plots.

It can be seen from Figs. 2 to 5 that as per the above definition, σ'_{pa} values are always almost equal to the maximum past pressures (σ'_{v0}).

Normalized Pressure–Void Ratio Relationships

The well-defined behavior of specimens as presented in Figs. 2 to 5 suggests that the various points on the consolidation curve can be related to the maximum past pressure and the initial void ratio. The initial void ratio (e_0) is defined as the void ratio of the specimen at the start of the test (i.e., the void ratio of the specimen 21 $\pm$ 1 day after the initial consolidation from the slurry).

A normalized void ratio (e_N) similar to the strain measure [29,30] as defined by

$$e_N = \left| \frac{e_0}{e} - 1 \right| \tag{1}$$

and a normalized pressure (σ'_N) as defined by

$$\sigma'_N = \frac{\sigma'_v}{\sigma'_{pa}} \tag{2}$$

have been adopted for the present analysis.

The e-σ'_v data as presented in Figs. 2 to 5 have been normalized as above and are plotted in Fig. 7. The best-fit curve for all specimens tested under the ''A'' loading scheme is a straight line ''A''. The straight line ''B'' contains data points for all specimens tested under the ''B'' loading scheme. These two straight lines, ''A'' and ''B'', are parallel with a slope of 0.592.

Typical data for an undisturbed specimen of leda clay [2] have been similarly treated and plotted in Fig. 7. These data lie on a different straight line (C) having a slope of 1.910.

The steeper slope for the curve of best fit for leda clay is believed to be due to the sensitivity and the cemented nature of the clay. This material is highly sensitive and exhibits abrupt changes in the e-σ'_v behavior once a critical pressure is exceeded.

The normalization procedure was also employed for isotropic consolidation test data. The isotropic consolidation tests were performed on 38.00-mm-diameter and 76.00-mm-high (average) specimens of the two soils. The curves of best fit for kaolin and Dundas soil specimens have been found to be a set of two parallel straight lines having a slope of 0.94 (Fig. 8).

Several attempts have been made in the past to estimate the slope of the virgin consolidation line from the plasticity characteristics and other physical data [31–35].

The present approach suggests a simple method of predicting the complete

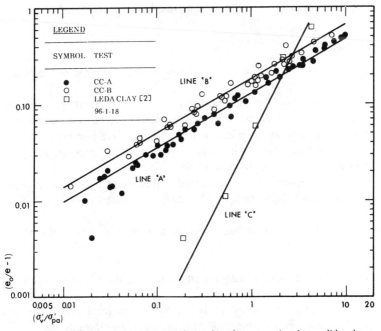

FIG. 7—*Normalized pressure-void ratio relationships for conventional consolidated tests.*

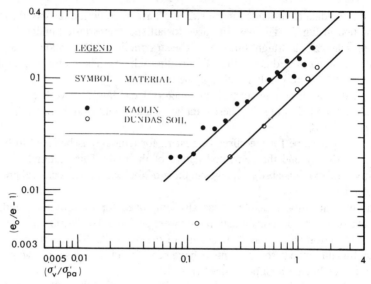

FIG. 8—*Normalized pressure-void ratio relationships for isotropic consolidation tests.*

consolidation path to be followed by a soil specimen. The form of relationship also provides for easy storage of the laboratory data. Further work is necessary to verify the approach for other materials and to investigate the various factors affecting the locations of the straight-line relationships presented herein.

Conclusions

The major conclusions of the study can be broadly summarized as follows:

1. The "young clays" of the study retained memory regarding the magnitude of past loading. The reload curves of the two soils were observed to be parallel. This observation supports Schmertmann's work on curve plotting [10].

2. An appropriate quasi-static application of the initial pressure (σ'_{vi}) is very important in case of conventional consolidation tests.

3. The normalization scheme presented in this report may be used to predict the complete consolidation path of laboratory sedimented soils.

The study indicates that further investigation is needed in the following major areas:

1. The various factors affecting the curvature near the historical consolidation pressure.

2. The findings of this study are based on the test results on two relatively low plasticity soils. Future studies should include high-plasticity soils.

3. Method of sample preparation.
4. Effects of the boundary conditions on the consolidation behavior of soils.

Acknowledgments

The study has been supported by the Dalley Fellowship, McMaster University, and a National Science and Engineering Council (NSERC) of Canada operating grant. The authors extend thanks to Mr. P. Koudys and Mr. M. Forget for their help during the testing program.

References

[1] Akai, K. and Adachi, T., "Study on the One-Dimensional Consolidation and the Shear Strength Characteristics of Fully Saturated Clay, in Terms of Effective Stresses," in *Proceedings*, 6th International Conference on Soil Mechanics and Foundation Engineering, Montreal, Vol. 1, 1965, pp. 146–150.

[2] Crawford, C. B., *Journal of the Soil Mechanics and Foundations Division*, American Society of Civil Engineers, Vol. 90, No. SM5, 1964, pp. 87–102.

[3] Lowe, J., Jonas, E., and Obrician, V., *Journal of the Soil Mechanics and Foundations Division*, American Society of Civil Engineers, Vol. 90, No. SM1, 1964, pp. 77–97.

[4] Lowe, J., Zaccheo, P. F., and Feldman, H. S., *Journal of the Soil Mechanics and Foundations Division*, American Society of Civil Engineers, Vol. 90, No. SM5, 1964, pp. 69–86.

[5] Newland, P. L. and Allely, B. H., *Geotechnique*, Vol. 10, 1960, pp. 62–70.

[6] Skempton, A. W., *Geotechnique*, Vol. 4, 1954, pp. 143–147.

[7] Skempton, A. W. and Bjerrum, L., *Geotechnique*, Vol. 7, 1957, pp. 168–172.

[8] Olson, R. E. and Ladd, C. C., *Journal of the Geotechnical Engineering Division*, American Society of Civil Engineers, Vol. 105, No. GT1, 1979, pp. 11–29.

[9] Casagrande, A., "The Determination of the Preconsolidation Pressure and its Practical Significance," in *Proceedings*, 1st International Conference on Soil Mechanics and Foundation Engineering, Harvard University, Cambridge, MA, Vol. 3, 1936, pp. 60–64.

[10] Schmertmann, J. H., *Transactions*, American Society of Civil Engineers, Vol. 120, 1955, pp. 1201–1227.

[11] Barden, L., *Geotechnique*, Vol. 15, No. 4, 1965, pp. 345–362.

[12] Barden, L., *Journal of the Soil Mechanics and Foundations Division*, American Society of Civil Engineers, Vol. 95, No. SM1, 1969, pp. 1–31.

[13] Barden, L. and Berry, P. J., *Journal of the Soil Mechanics and Foundations Division*, American Society of Civil Engineers, Vol. 91, No. SM5, 1965, pp. 15–35.

[14] Berre, T. and Iversion, K., *Geotechnique*, Vol. 22, No. 1, 1972, pp. 53–70.

[15] Leonards, G. A. and Altschaeffl, A. G., *Journal of the Soil Mechanics and Foundations Division*, American Society of Civil Engineers, No. SM5, 1964, pp. 133–155.

[16] Murakami, Y., *Soils and Foundations*, Vol. 17, No. 4, 1977, pp. 49–69.

[17] Murakami, Y., *Soils and Foundations*, Vol. 19, No. 4, 1979, pp. 17–29.

[18] Narain, J., Singh, B., Iyer, N. V., and Deoskar, S. R., "Quasi-preconsolidation Effects and Pore Pressure Dissipation During Consolidation," in *Proceedings*, 7th International Conference on Soil Mechanics and Foundation Engineering, Vol. 1, 1969, pp. 311–316.

[19] Sonpal, R. C. and Katti, R. K., "Consolidation—An Analysis with Pore Pressure Measurements," in *Proceedings*, 8th International Conference on Soil Mechanics and Foundation Engineering, Moscow, Vol. 1.2, 1973, pp. 385–388.

[20] Hambly, E. C., *Geotechnique*, Vol. 22, 1965, pp. 301–317.

[21] Terzaghi, K. and Peck, R. B., *Soil Mechanics in Engineering Practice*, Wiley, New York, 1967, p. 729.

[22] Terzaghi, K. and Peck, R. B., *Geotechnique*, Vol. 22, 1972, pp. 159–163.

[23] Kirkpatrick, W. M. and Rennie, I. A., *Geotechnique*, Vol. 22, 1972, pp. 166–169.

[24] Christie, I. F., "Secondary Compression Effects During One-Dimensional Consolidation Tests"

in *Proceedings*, 6th International Conference on Soil Mechanics and Foundation Engineering, Montreal, Vol. 1, 1965, pp. 198–202.

[25] Northey, R. D. and Thomas, R. F., "Consolidation Test Pore Pressures," in *Proceedings*, 6th International Conference on Soil Mechanics and Foundation Engineering, Montreal, Vol. 1, 1965, pp. 323–327.

[26] Perloff, W. H., Nair, K., and Smith, J. G., "Effect of Measuring System of Pore Pressures in the Consolidation Test," in *Proceedings*, 6th International Conference on Soil Mechanics and Foundation Engineering, Montreal, Vol. 1, 1965, pp. 338–341.

[27] Perloff, W. H., Nair, K., and Smith, J. G., *Canadian Geotechnical Journal*, Vol. 2, No. 2, 1965, pp. 90–97.

[28] Crawford, C. B. and Burn, K. N., "Long Term Settlements on Sensitive Clay," *Laurits Bjerrum Memorial Volume*, Norwegian Geotechnical Institute, Oslo, 1976, pp. 117–124.

[29] Purushotham, C. M., "On the Relationship Between Two Strain Measures in the Multiaxial Deformation," in *Proceedings*, CANCAM, Montcon, Canada, 1981, pp. 885–886.

[30] Seth, B. R., "Generalized Strain and Transition Concept for Elastic-Plastic Deformation, Creep and Relaxation," in *Proceedings*, 11th International Congress on Applied Mechanics, H. Gortler, Ed., Springer-Verlag, Munich, 1964, pp. 384–389.

[31] Azzouz, A. S., Krizek, R. J., and Corotis, R. B., *Soils and Foundations*, Vol. 16, No. 2, 1976, pp. 19–29.

[32] Butterfield, R., *Geotechnique*, Vol. 29, 1979, pp. 469–480.

[33] Cozzolino, V. M., "Statistical Forecasting of Compression Index," in *Proceedings*, 5th International Conference on Soil Mechanics and Foundation Engineering, Paris, Vol. 1, 1961, pp. 51–54.

[34] Herrero, R. O., *Journal of Geotechnical Engineering Division*, American Society of Civil Engineers, Vol. 106, GT11, 1980, pp. 1179–1200.

[35] Juárez-Badillo, E., "General Compressibility Equation for Soils" in *Proceedings*, 10th International Conference on Soil Mechanics and Foundation Engineering, Stockholm, Vol. 1, 1981, pp. 171–178.

John F. Peters[1] and Daniel A. Leavell[1]

A Biaxial Consolidation Test
for Anisotropic Soils

REFERENCE: Peters, J. F. and Leavell, D. A., **"A Biaxial Consolidation Test for Anisotropic Soils,"** *Consolidation of Soils: Testing and Evaluation, ASTM STP 892,* R. N. Yong and F. C. Townsend, Eds., American Society for Testing and Materials, Philadelphia, 1986, pp. 465–484.

ABSTRACT: A biaxial flow consolidation test was developed to determine the axial and radial permeabilities, k_a and k_r, respectively, of a transversely isotropic specimen. The test enables the additional information needed to determine the ratio of permeabilities, k_a/k_r, to be obtained by independently measuring the flow from the specimen ends and circumference during consolidation. By comparing the ratio of flows from the specimen ends and circumference to the theoretical relationship, the ratio of permeabilities can be determined. The radial permeability, k_r, is determined from the total time rate of consolidation.

Comparisons between various methods of permeability measurement and the proposed test are shown to be good. From the theoretical development, it was found that anisotropy in stiffness can have an important influence on test results. A method is presented for determining relevant stiffness parameters by combining the strain measured during consolidation and pore-pressure response.

KEY WORDS: permeability, anisotropy, consolidation, pore pressure, creep, elastic constants, clay shale

This paper describes a compression test developed to measure consolidation properties of transversely isotropic materials. The principal feature of the test is the ability to measure both horizontal and vertical permeabilities from a single load step. The device is useful for studying variations in the ratio of horizontal to vertical permeability for groups of specimens and for systematically studying variation in anisotropic permeabilities that result from loading history.

Measurement of Anisotropic Properties

The anisotropy of a material describes the directional dependence of its physical properties. Many natural and compacted soils are layered and their properties are directionally dependent, relative to the layering. For example, permeability may

---

[1] Research Civil Engineer and Civil Engineer, respectively, Geotechnical Laboratory, U.S. Army Engineer Waterways Experiment Station, Vicksburg, MS 39180.

be several times less when the flow is perpendicular rather than parallel to the layering. In contrast, permeabilities measured parallel to the layering are independent of the direction of flow. Compressibility is generally greatest for loading perpendicular to the layering and, as for the permeability, it is the same when loads are applied in any direction parallel to the layering direction. Materials displaying isotropy within one plane are referred to as transversely isotropic. Analysis of consolidation of an elastic, transversely isotropic material requires determination of eight constants: five elastic constants to define compressibility of the soil, one constant to describe the compressibility of the pore fluid, and two permeability constants. If the elastic constants are known, two independent measurements are needed from a consolidation or permeability test to determine the permeability constants.

The most common laboratory approach to measuring the permeabilities of a transversely isotropic soil is to perform one-dimensional tests on specimens having different orientations. One test series is performed with the flow perpendicular to the plane of transverse isotropy, and one series is performed with flow in the direction of isotropy. Either one-dimensional consolidation tests or falling head permeability tests, or both, may be used.

Although the method described above is relatively simple, it has one serious shortcoming: the tests are all performed on individual samples and differences in permeabilities must be attributed not only to anisotropy but also to testing error and statistical variations among the specimens. The histograms of the consolidation coefficient, c_v, computed from consolidation test data described by Rowe [1] are shown in Fig. 1. Clearly, the permeability measured perpendicular to the direction of layering is significantly less than the permeability measured in the direction of layering. However, there is also considerable variation for each test group. Although it is possible to estimate the variation in permeability for each direction, variations in the ratios of permeabilities cannot be determined. Also, systematic variations in the anisotropy caused by loading cannot be observed.

As an alternative to using independent specimens to obtain the two measurements needed to determine the anisotropic permeabilities, a scheme was developed to obtain two independent measurements from one load increment. The new test, shown schematically in Fig. 2, involves performing a consolidation test on a triaxial specimen with drainage on all boundaries. To obtain the two measurements needed for determination of the anisotropic properties, drainage from the specimen's ends and drainage from the specimen's circumference are collected in different measurement systems. The ratio of axial permeability, k_a, and radial permeability, k_r, can be determined from a plot of accumulated axial flow, V_a, versus accumulated radial flow, V_r. The radial permeability can be determined from a plot of total accumulated volume versus time. The test can be used to evaluate variations in anisotropic permeability for a variety of stress levels and stress paths.

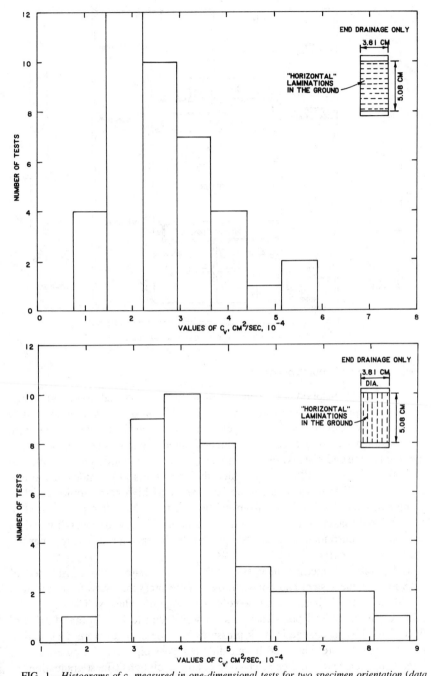

FIG. 1—*Histograms of* c_v *measured in one-dimensional tests for two specimen orientation (data from Rowe [1]).*

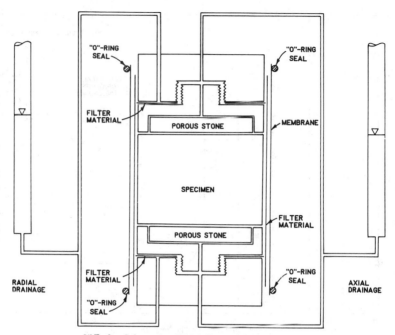

FIG. 2—*Schematic of biaxial consolidation test device.*

Test Equipment and Procedure

The biaxial consolidometer consists of a modified triaxial compression apparatus, the principal modification being the addition of a plumbing system to measure flow from the specimen's circumference and ends separately. Flow from the specimen's ends is collected through sintered stainless steel porous stones mounted in the end platens. Flow from the specimen's circumference is collected through a nonwoven filter fabric which drains to ports in the sides of the end platens. The filter fabric was chosen because of its high permeability and low compressibility. It was found from calibration tests that filter paper, used in conventional triaxial tests, is not suitable because of its low permeability and its tendency to consolidate at a rate comparable to the specimens having a permeability of 10^{-7} cm/s.

The pore-water measurement and back-pressure system was constructed of 3.18-mm stainless steel and copper tubing. Pore pressures were measured to a resolution of 0.7 kN/m² using low-volume change transducers. The compressibility of the combined pore-water system was found to be less than 0.01 cm³ over a 690 kN/m² change in back pressure. Volume changes were measured in burettes which were fabricated from calibrated glass pipettes. The burettes were interchangeable and could be selected to obtain resolutions (finest graduation) of 1.0, 0.1, and 0.01 cm³. The back pressure in the combined pore-water system was controlled by a common regulator which had a range from full vacuum to 690 kN/m²; the back pressure used for most testing was on the order of 350 kN/ m². The chamber pressure could be controlled by a differential regulator which could supply up to 210 kN/m² above back pressure. For pressures over 140 kN/ m² above back pressure, the chamber pressure control was switched to an absolute pressure regulator with a range of 0 to 3.5 MN/m². Generally, all pressures could be controlled to within 1.5 kN/m² over a period of several days.

The specimens tested were 10.16 cm in diameter and had a height-to-diameter ratio, h/d, of 0.60. This value of h/d was chosen, based on theoretical computations, so that an isotropic specimen would produce equal volumes of flow in axial, V_a, and radial, V_r, directions. Experimental data indicate that the compressibility of the specimen is significantly influenced by boundary constraints at the platens for smaller h/d ratios. Theoretical computations indicate that specimen shape, not anisotropy, controls V_a/V_r for larger values of h/d. Procedures for specimen preparation, including trimming, back-pressure saturation, and consolidation were the same as those used for triaxial testing.

The consolidation test is performed in a series of increments with each increment consisting of two steps. First, the specimen is loaded undrained and the pore pressure response is measured. Loading may consist of any combination of axial load and chamber pressure. Second, drainage valves are opened and consolidation monitored by measuring axial deformation, and volumes V_a and V_r versus time. The compressibility coefficients, used for analysis of the test, are computed from the measured deformations, volumes, and pore pressures using the procedure described below. Ratios of axial and radial permeabilities, k_a/k_r, are obtained by matching a plot of V_a versus V_r to theoretical curves. The radial permeability, k_r, is computed from the total volume versus time relationship, as for a conventional consolidation test.

Theoretical Behavior

The theoretical behavior of the specimen is based on the assumption that for an increment of loading the material is elastic with constant permeability and that the pore fluid is incompressible. Also, the material is assumed to be transversely isotropic with the plane of transverse isotropy perpendicular to the cylindrical axis of the specimen. The governing equations for consolidation are obtained from the continuity equation, Darcy's law, and equilibrium. They are

$$\frac{\partial \sigma'_r}{\partial r} + \frac{\sigma'_r - \sigma'_\theta}{r} = -\frac{\partial u}{\partial r} \tag{1}$$

$$\frac{\partial \sigma'_z}{\partial z} = -\frac{\partial u}{\partial z} \tag{2}$$

and

$$\frac{k_r}{\gamma_w}\left(\frac{\partial^2 u}{\partial r^2} + \frac{1}{2}\frac{\partial u}{\partial r}\right) + \frac{k_z}{\gamma_w}\frac{\partial^2 u}{\partial z^2} = -\frac{\partial e}{\partial t} \tag{3}$$

where

$$e = \epsilon_\theta + \epsilon_r + \epsilon_z,$$

ϵ_θ, ϵ_r, and ϵ_z = normal strains in θ, r, and z directions (assumed to be principal values),

σ'_θ, σ'_r, and σ'_z = normal effective stresses in θ, r, and z directions (assumed to be principal values),

u = excess pore pressure,

γ_w = unit weight of water, and

t = time.

These equations can be restated as a single equation in terms of volumetric strains e through application of the compatibility and stiffness relationship:

$$\sigma'_z = D_{11}\epsilon_z + D_{12}\epsilon_r + D_{12}\epsilon_\theta$$

$$\sigma'_r = D_{12}\epsilon_z + D_{22}\epsilon_r + D_{23}\epsilon_\theta \tag{4}$$

$$\sigma'_\theta = D_{12}\epsilon_z + D_{23}\epsilon_r + D_{22}\epsilon_\theta$$

The stiffness constants can be expressed in terms of the engineering constants as:

$$D_{11} = (1 - v_1^2)\,|D|$$

$$D_{12} = nv_2(1 + v_1)\,|D|$$

$$D_{22} = n(1 - nv_2^2)\,|D| \tag{5}$$

$$D_{23} = n(v_1 + nv_2^2)\,|D|$$

where

$$|D| = \frac{E_1}{(1 + v_1)(1 - v_1 - 2nv_2^2)}$$

and

E_1 = Young's modulus in axial direction,

v_2 = Poisson's ratio between axial and radial direction,

v_1 = Poisson's ratio in radial plane, and

n = ratio of radial to axial Young's modulus.

By combining Eqs 1 to 4 and compatibility equations, the governing differential equation, boundary conditions, and initial conditions are obtained as

$$\text{P.D.E.} = \phi^2 \left(\frac{\partial^2 E}{\partial R^2} + \frac{1}{R} \frac{\partial E}{\partial R} \right) + \alpha^2 \frac{\partial^2 E}{\partial Z^2} = \frac{\partial E}{\partial T}$$

$$\text{B.C.} = \begin{cases} E(R, 0, T) = E(R, 1, T) = 1; T > 0 \\ C_1 E(1, Z, T) + C_2 \int_0^1 R E dR = C_3; T > 0 \end{cases} \tag{6}$$

$$\text{I.C.} = E(R, Z, 0) = 0, 0 \le Z \le 1, 0 \le R \le 1$$

where

$$T = \frac{4 c_v}{d^2} t$$

$$c_v = \frac{D_{11} k_r}{\gamma_w}$$

$$E = \frac{e}{N P_a + 2 Q P_r}$$

$$R = 2 \frac{r}{d}$$

$$Z = \frac{z}{H}$$

$$\alpha^2 = \frac{1}{4} \frac{d^2}{H^2} \frac{k_z}{k_r}$$

$$\phi^2 = -\frac{\theta}{D_{11}} + \frac{A_{11}}{A_{12}} \frac{k_a}{k_r} \left(\frac{D_{22} + \theta}{D_{11}} \right)$$

$$A_{12} = D_{12} - D_{22}$$

$$A_{11} = D_{12} - D_{11}$$

$$A_{22} = D_{23} - D_{22}$$

$$C_1 = D_{12} + D_{11} \frac{A_{12}}{A_{11}} + \frac{1}{2} \frac{A_{22}}{A_{11}} \theta$$

$$C_2 = -\frac{A_{22}}{A_{11}} (\theta + D_{11})$$

$$C_3 = C_1 + \frac{1}{2} C_2$$

$$\theta = (MQ - N)^{-1}$$

$$N = \frac{1 - 2 v_2}{E_1}$$

$$Q = \frac{-v_2 + \dfrac{1 - v_1}{n}}{E_1}$$

$$M = (1 + v_1 + nv_2)$$

By a change of variable, $V = 1 - E$, the solution can be obtained by separation of variables as

$$E(R, Z, T) = 1 - \frac{8}{\pi} \sum_{m=1}^{\infty} \sum_{n=1}^{\infty} \frac{B_m}{\beta_m n} J_0 (\beta_m R) \sin n\pi z \exp (-\lambda^2 T) \qquad (7)$$

$$m = 1,2,3,...\infty; \qquad n = 1,3,5,...\infty$$

where

$$\lambda^2 = \alpha^2 \pi^2 n^2 + \phi^2 \beta_m^2$$

$$B_m = \frac{J_1(\beta_m)}{J_0^2(\beta_m) + J_1^2(\beta_m)}$$

where J_0 and J_1 are Bessel functions of zero and first order respectively and β_m is the mth root of

$$C_1 J_0(\beta_m) + C_2 \frac{1}{\beta_m} J_1(\beta_m) = 0$$

The average consolidation, U_{ave}, defined as the total volumetric strain up to T divided by the ultimate volumetric strain, is given by

$$U_{ave} = 1 - \frac{32}{\pi} \sum_{m=1}^{\infty} \sum_{n=1}^{\infty} \frac{B_m}{\beta_m^2 n^2} J_1(\beta_m) \exp (-\lambda^2 T) \qquad (8)$$

The flow from the specimen circumference divided by the ultimate volume change, V_a/V_T, is found by integrating over time the flow given by

$$Q_r = \int_0^1 \left(-\frac{k_r}{\gamma_w} \frac{\partial u}{\partial r} 2\pi a H \right) dZ$$

The resulting expression is

$$\frac{V_r}{V_T} = \frac{32}{\pi^2} \frac{\theta}{D_{11}} \sum_{m=1}^{\infty} \sum_{n=1}^{\infty} \frac{B_m}{n^2 \lambda^2} J_1(\beta_m)(1 - \exp (-\lambda^2 T)) \qquad (9)$$

The normalized flow through the specimen end is by definition

$$\frac{V_a}{V_T} = U_{ave} - \frac{V_r}{V_T} \tag{10}$$

For isotropic soil having $n = 1$ and $v_1 = v_2 = 0.5$, C_2 becomes zero and the solution is identical to the heat diffusion equations because there is no volume change of the soil matrix. For other values of v_1, v_2, and $n > 1$, the equations of flow are coupled to the equations of equilibrium. The uncoupled or heat diffusion case is often identified with the Terzaghi-Rendulic theory which assumes that total stress remains constant throughout consolidation. An important result of the uncoupled theory is that effects of elastic compressibility are completely accounted for by the consolidation coefficient c_v, whereas the coupled theory predicts a dependence between consolidation behavior and anisotropy (Figs. 3 and 4).

Determination of Elastic Constants

The elastic constants are determined from a combination of pore-pressure response and strain measurements [3]. For example, E_1 and v_2 can be determined from a drained uniaxial loading increment from

$$E_1 = \frac{\Delta\sigma'_a}{\Delta\epsilon_a} \tag{11}$$

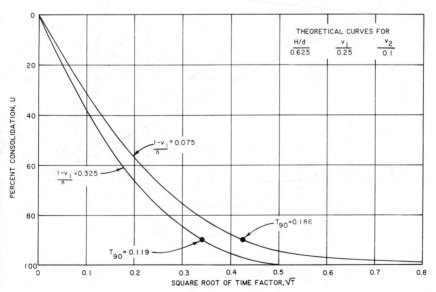

FIG. 3—*Theoretical consolidation curves.*

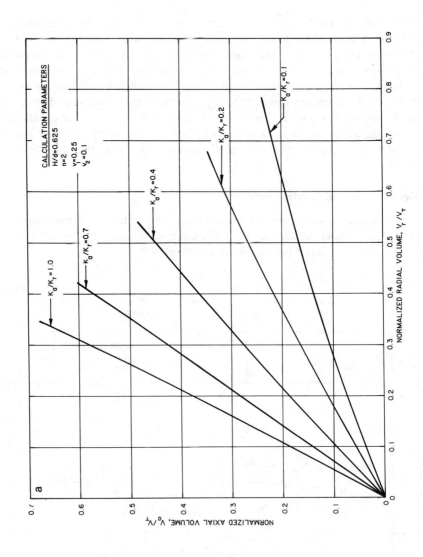

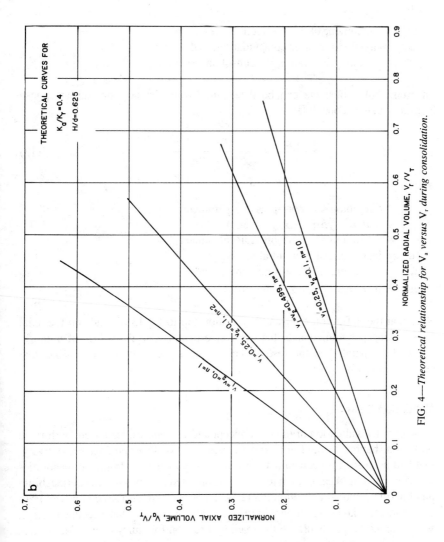

FIG. 4—*Theoretical relationship for V_a versus V_r during consolidation.*

and

$$v_2 = \frac{\Delta \epsilon_a - \Delta \epsilon_v}{2 \Delta \epsilon_a} \tag{12}$$

where

$\Delta \sigma'_a$ = increment of axial effective stress,
$\Delta \epsilon_a$ = axial strain after consolidation, and
$\Delta \epsilon_v$ = volumetric strain after consolidation.

The remaining constants can be determined from the pore-pressure response measured before consolidation:

$$\frac{1 - v_1}{n} = \frac{1 - 2v_2 + A(v_2 - 1)}{2A} \tag{13}$$

where

A = Skempton's A pore-pressure parameter,
  = $(\Delta u/B - \Delta \sigma_r)/(\Delta \sigma_a - \Delta \sigma_r)$,
B = Skempton's B pore-pressure parameter,
$\Delta \sigma_a$ = increment of axial total stress,
$\Delta \sigma_r$ = increment of radial total stress, and
Δu = increment of pore water pressure.

Note that the definition of A and its relationship to Eq 13 depends on the axial stress acting perpendicular to the transverse plane. Also the values of n and v_1 cannot be separated in the test. Generally, it is the value of $(1 - v_1)/n$ that determines test behavior.

Validation Tests

To obtain a validation of the test concept and procedures, tests were performed on kaolin, having a liquid limit and a plastic limit of 45% and 25% respectively, that had been initially consolidated from a slurry. The anisotropic permeabilities were obtained from calibration tests which consisted of constant head permeability and one-dimensional consolidation tests. The tests were performed on two 3.56-cm-diameter triaxial specimens with end-only drainage. One specimen, Kr-2, was trimmed perpendicular to the loading direction for slurry consolidation, and one, Ka-1, parallel to the loading direction. It had been determined from a procedure of drying and wet-slaking described by Hvorslev [4] that the clay had an anisotropic structure relative to these directions. The results of these tests (Figs. 5 and 6) indicated that the kaolin has a k_a/k_r ratio of about 0.77 and $k_r = 9.7 \times 10^{-8}$ cm/s.

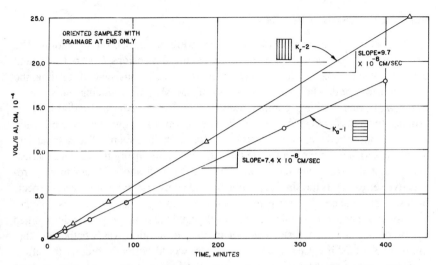

FIG. 5—*Comparison of permeability tests on oriented specimens of slurry-consolidated kaolin.*

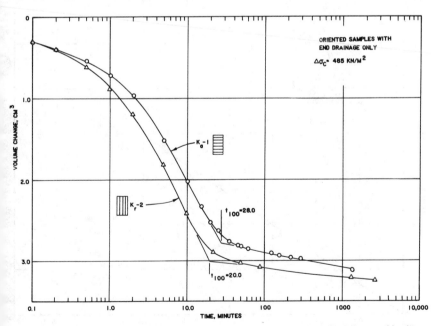

FIG. 6—*Comparison of one-dimensional consolidation tests on oriented specimens of kaolin.*

Biaxial Consolidation Results

Three series of biaxial consolidation tests were performed on the kaolin. The first series was performed on a specimen having dimensions $h = 6.10$ cm and $d = 10.19$ cm. The loading sequence in the series consisted of loading the specimen to a hydrostatic effective stress of 1.4 MN/m^2, reducing the stress to 625 kN/m^2, and applying combinations of incremental axial and radial loads to produce the stress path shown in Fig. 7. The undrained loading and consolidation sequence described previously was used for each increment. The second and third test series were performed on the same specimen using the first test series, except that the specimen height was reduced to 3.48 and 3.26 cm in each respective series. Thus the single specimen could be used to evaluate several aspects of test performance. The most important question to be answered was whether the test would be reproducible and if it would yield the same properties determined by the one-dimensional tests. Another important question was whether the theoretical relationships used to analyze the test would correctly predict the effect of sample shape on the V_r/V_a versus time relationship. Finally, it was important to determine the extent to which boundary effects might influence test results.

The results of the first test series (Fig. 7) show that the results are consistent for all loading increments. It is especially interesting to note that the first load

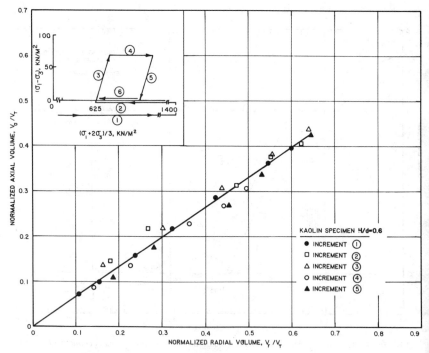

FIG. 7—*Measured* V_a *versus* V_r *for slurry-consolidated kaolin.*

increment, which produced relatively large plastic strain, displayed the same V_r versus V_a relationship as the smaller loading and unloading increments. The total volume versus time relationship, which depends on the stiffness D_{11}, as well as the ratios of stiffness, was found to be different for each load increment. However, the volume-versus-time plot varied in accordance with the stiffness measured during a particular load increment, as indicated on the plot of stiffness-versus-time for 90% consolidation shown in Fig. 8.

A comparison between the theoretical relationships and test data is shown on Fig. 9. It is seen that the difference in the plot of V_a versus V_r associated with differences in specimen height are consistent with the theoretical predictions. Thus the V_a versus V_r relationship appears to be little affected by end constraints. However, it was noted that the pore-pressure response to an axial load (A parameter) was greater for the short specimens; this implies a restraining effect of the end platens on lateral deformation. The theoretical computations shown in Fig. 8 were based on the properties determined from one-dimensional tests and agree best with data for long specimens. The lack of agreement with short

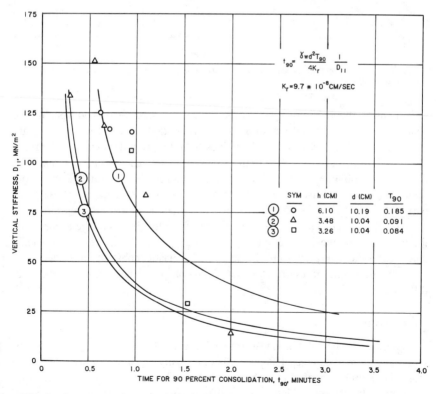

FIG. 8—*Comparison of measured stiffness, D_{11}, versus time for 90% consolidation for biaxial tests on kaolin.*

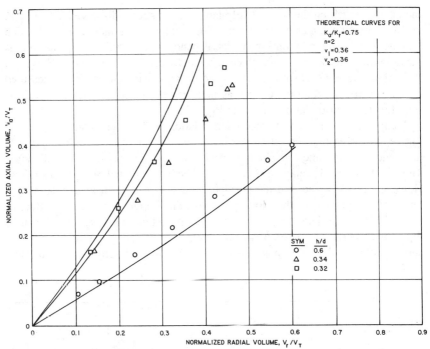

FIG. 9—*Comparison of measured and computed relationship for* V_a *versus* V_r *for kaolin specimens of different heights.*

specimens is the result of over estimating D_{11} as a result of end constraints. Values of k_r and k_a for all one-dimensional validation tests and biaxial tests are shown in Fig. 10. It is seen that the values obtained from the biaxial test having a h/d ratio of 0.6 agrees well with the validation tests. The differences in permeability values obtained from biaxial specimens with smaller h/d ratios are possibly the result of errors in measuring D_{11} from the shorter specimens. These comparisons highlight a shortcoming of the test shared by all compression tests: Because the incremental stiffness varies with each loading step, the accuracy of the computed permeability is closely tied to the ability to determine the stress-strain response for each step. Because three elastic constants are needed to interpret the biaxial test correctly, determination of permeabilities is especially difficult.

Other Tests

In addition to the validation tests on kaolin, a similar test series was performed on an undisturbed sample of Pepper Shale, a highly anisotropic clay shale taken from the foundation of Waco Dam in Texas. The shale displayed interesting coupling between consolidation and stiffness. Because of the limited scope of this paper, only a brief description can be given.

It was found that the V_a/V_r ratio was not a constant as predicted by the theory, but rather decreased steadily as the test proceeded. The variation in V_a/V_r is believed to be the result of creep. Unlike the kaolin, which also displayed creep, the shale is highly impermeable and the rate of consolidation is slow enough that viscoelastic behavior dominated consolidation. It was found from a series of undrained loading increments on triaxial specimens that the pore-pressure response (and as implied by Eq 13 the anisotropy) is time dependent. Upon initial loading the pore-pressure response is isotropic ($A = \frac{1}{3}$) but with time approached that of a transversely isotropic material ($A = 0.8$) (Figs. 11a to 11c). Importantly, in all cases it was found that the time-dependent response is transversely isotropic relative to the geologic layering; it is not related to the direction of loading (i.e., not stress induced). Preliminary computations using a simple linear viscoelastic model to describe the time-dependent anisotropy indicate that the observed trends in V_a/V_r are consistent with the interpretation given above. However, quantitative comparisons cannot yet be made using viscoelastic parameters measured in the undrained loading tests and the consolidation test.

Concluding Remarks

A new consolidation device for determining properties of anisotropic materials has been developed and shown to be capable of accurate property determination for "research" materials such as slurry-consolidated kaolin. It was also found

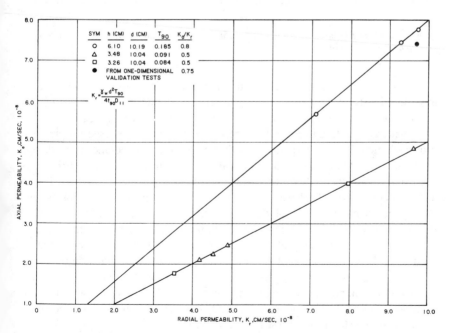

FIG. 10—*Comparison of permeabilities determined by various tests.*

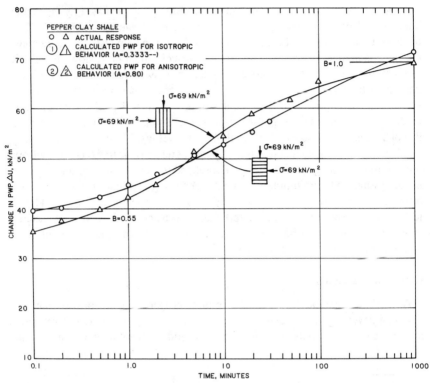

FIG. 11a—*Pore pressure versus time relationship for pepper shale showing the time-dependent development of anisotropy.*

that the effects of inelasticity (creep) produced interesting effects not observable in standard tests. It therefore seems appropriate to divide the conclusions of this paper into two categories: practical application and research application.

Considerable testing needs to be done before the test can be used for practical property determination. Experience with the device is limited and many details of the test procedure still need to be standardized. In particular, experience is needed in testing materials which are layered such as varved clay. Both the kaolin and the Pepper Shale displayed anisotropic ''fabric'' at microscopic dimensions. It may be found that the test does not perform as well in materials having macroscopic layering [1]. More importantly, if it is found that the inelastic stress-strain response has a strong influence on consolidation behavior, as indicated by the test on Pepper Shale, application of the test to routine engineering projects will be severely limited. This problem may be mitigated by a change in the test design. For example, a cube-shaped specimen should not display the same coupling as a cylinder. Also, a controlled gradient test may yield more consistent results than the step loading test used at present.

As a research tool, the test offers some interesting opportunities. First, the ability to determine both permeabilities and three elastic constants from one loading increment permits an evaluation of phenomena such as stress-induced anisotropy on consolidation properties. Second, because the V_a/V_r ratio depends on the pore-pressure distributions, the test offers an added dimension for studying the effects of anisotropy and inelasticity on consolidation. From the limited experience with the test on kaolins, it appears to us that the influence of anisotropic stiffness is not as dramatic as implied by the theory (Fig. 4b). In fact, the comparisons between experiment and theory (Fig. 9) can be improved if simple heat diffusion theory is used to analyze specimen performance rather than the fully coupled theory. On the other hand, experimental evidence for Pepper Shale suggests that anisotropic viscoelastic behavior does have a strong influence on results. Thus the device may provide data for a better understanding of coupling effects in elastic, viscoelastic, and elasto-plastic materials. For example, does the contractive-dilative response of granular material influence consolidation in a manner similar to anisotropy and creep?

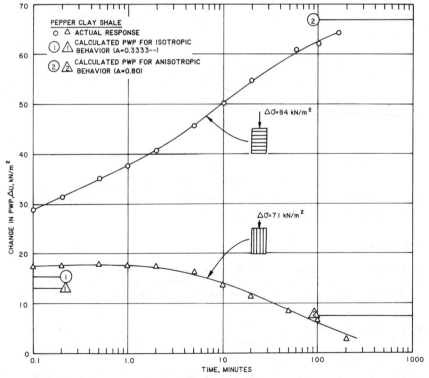

FIG. 11b—*Pore pressure versus time relationship for pepper shale showing the time-dependent development of anisotropy.*

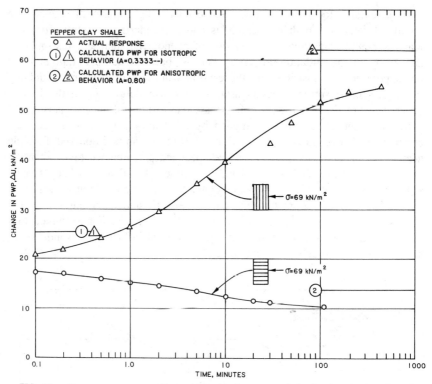

FIG. 11c—*Pore pressure versus time relationship for pepper shale showing the time-dependent development of anisotropy.*

Acknowledgments

This investigation was performed at the U.S. Army Engineer Waterways Experiment Station, Vicksburg, Mississippi, under the Civil Works R&D Program 31151 authorized by the U.S. Army Corps of Engineers.

References

[1] Rowe, P. W., *Geotechnique*, Vol. 9, No. 3, Sept. 1959, pp. 107–118.
[2] Leavell, D. A. and Peters, J. F., "Engineering Properties of Clay Shales; Laboratory and Theoretical Investigation of Consolidation Characteristics of Clay Shales," Technical Report S-71-6, Report 6, U.S. Army Engineer Waterways Experiment Station, Vicksburg, MS (in preparation).
[3] Leavell, D. A., Peters, J. F., and Townsend, F. C., "Engineering Properties of Clay Shales; Laboratory and Computational Procedures for Predictions of Pore Pressures in Clay Shale Foundations," Technical Report S-71-6, Report 4, U.S. Army Engineer Waterways Experiment Station, Vicksburg, MS, 1982.
[4] Hvorslev, M. J., "Physical Properties of Remolded Cohesive Soils," Translation 69-5, U.S. Army Engineer Waterways Experiment Station, Vicksburg, MS, 1969.

Robert W. Sarsby[1] and Brian Vickers[1]

Side Friction in Consolidation Tests on Fibrous Peat

REFERENCE: Sarsby, R. W. and Vickers, B., **"Side Friction in Consolidation Tests on Fibrous Peat,"***Consolidation of Soils: Testing and Evaluation, ASTM STP 892,* R. N. Yong and F. C. Townsend, Eds., American Society for Testing and Materials, Philadelphia, 1986, pp. 485–489.

ABSTRACT: Consolidation tests have been conducted on a fibrous peat using a 254-mm-diameter hydraulic oedometer which was modified to accommodate force and pressure transducers. These transducers were used to evaluate the side friction acting on the peat and to record the distribution of applied pressure across the width of a specimen. The peat specimens were subjected to consolidation pressure in the range of 20 to 120 kN/m². Significant side friction was recorded (up to 50% of the applied pressure) even when the insides of the oedometer were smeared with grease.

KEY WORDS: consolidation test, friction, peat

Peat deposits are highly compressible, and under the action of a load equivalent to only a metre of fill they may undergo strains of the order of 50%. Under these conditions the permeability may reduce by a factor of a hundred. A one-dimensional consolidation theory, which incorporates finite specimen strain, has been postulated by Berry [1]. The parameters used in this theory are determined from conventional one-dimensional consolidation tests. To evaluate accurately the permeability and consolidation characteristics of peat, however, a minimum specimen thickness of 100 mm is required. With the large strains associated with the compression of peat, significant friction may be mobilized between the specimen and the walls of the oedometer. This paper relates to an experimental investigation of the magnitude and effects of this side friction.

Apparatus

The basic apparatus was a 254-mm-diameter hydraulic consolidation cell, as described by Rowe and Barden [2]. The base of the cell was modified so that measurements could be made of the total stresses and pore water pressures on the base of the peat specimen. Drainage was from the upper face of the specimen

[1]Principal Lecturer and Head of Department, respectively, Department of Civil Engineering, Bolton Institute of Higher Education, Bolton, England.

through a flexible sintered bronze disk, 254 mm diameter, and the central spindle of the cell.

To measure the total pressure acting on the base of the specimen, a plate was manufactured which "floated" inside the oedometer. This plate was supported by three load transducers and the specimen was seated directly on this "floating base". The total vertical force on the base of the peat was thus recorded by the load transducers. The difference between the applied pressure (on the top of the specimen) and the total pressure calculated from the loads measured by the force transducers would then give some measure of the amount of side friction. To determine the actual distribution of vertical stress on the base of the peat, three pressure transducers were incorporated into the true baseplate of the oedometer such that their faces were mounted flush with the "floating" baseplate. They were located close to the periphery of the specimen, close to its center and at midradius. Two pore water pressure sensing elements were also located in the "floating base", one at the center of the specimen, the other at two-thirds radius.

The test procedure was essentially the same as that for conventional long-term vertical consolidation tests. In an attempt to eliminate side friction, the insides of the oedometer were smeared with high-vacuum silicon grease before installation of the specimen. All specimens were consolidated against a back-pressure of 36 kN/m^2, which represented the value of the pore pressure present in the peat specimens before they were removed from the ground. This back-pressure also ensured that any gas contained in the specimens would be dissolved so that the pore water pressure transducers would respond rapidly to pressure changes.

Test Data

The results of the side friction tests are presented in Figs. 1 to 4.

Figure 1 illustrates the typical variation of total pressures, pore pressures and settlement with time during the course of one load increment. Over the testing period, the total vertical stress at each measuring point remained essentially constant. In addition, the two pore pressure transducers recorded virtually the same values of pore water pressure throughout consolidation. We concluded that side friction does not significantly affect the pore pressure dissipation over the majority of the specimen area.

Figures 2 to 4 show the distribution of vertical effective stress on the base of the specimens for different applied total stresses. Side friction is clearly present to a considerable extent during consolidation testing of fibrous peat, and it greatly reduces the pressure at the base of a specimen. The effects of side friction are markedly less noticeable over the central area of the specimen, particularly at low pressure levels. Nevertheless, at the end of the primary consolidation stage the vertical effective stress at a distance of 38 mm from the center of the specimen was between 83 and 94% of the applied stress. As secondary compression proceeded, the magnitude of the side friction reduced so that the measured stress near to the center of a specimen approached the applied value. However, it should

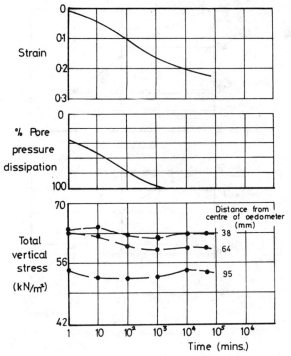

FIG. 1—*Variation of strain and stresses with time. Total pressure increment = 32 to 64 kN/m²; back-pressure = 36 kN/m².*

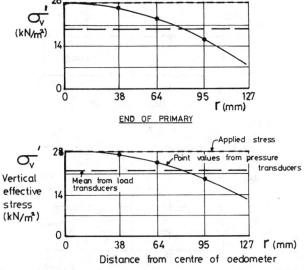

FIG. 2—*Effective stress distribution on base. Total pressure increment = 32 to 64 kN/m² and back-pressure = 36 kN/m².*

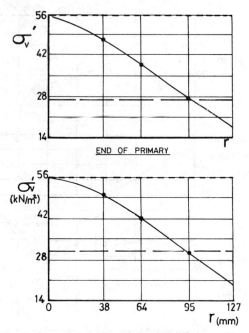

FIG. 3—*Effective stress on base. Total pressure increment = 23 to 92 kN/m².*

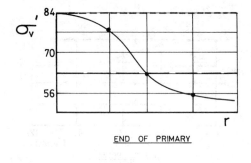

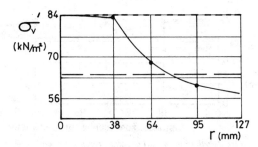

FIG. 4—*Effective stress on base. Total pressure increment = 20 to 120 kN/m².*

be noted that the secondary consolidation stage was much longer (i.e., 19 to 66 times) than the primary stage. The mean effective stress acting on the specimens (as deduced from the load transducer readings) was always considerably smaller than the applied pressure. Furthermore, the vertical effective stress at the periphery of a specimen (as estimated from the pressure distribution indicated by the pressure transducers) was between only 35% and 68% of the applied pressure even at the end of the test.

Smearing the insides of the consolidation cell with grease resulted in little, if any, reduction of side friction. This may have been due to the peat fibers penetrating the grease and rendering it useless (as has been observed in the testing of compacted granular fills) or it may have been due to a breakdown in the structure of the grease. Under the conditions of radial drainage that may be used in this form of oedometer, side friction effects will be more significant than for vertical drainage, since the specimen is then in direct contact with the "rough" plastic drain [3].

Conclusion

The magnitude of side friction when consolidating fibrous peat in an oedometer is considerable, and in the tests reported herein it has been observed to be as much as 50% of the applied stress.

References

[1] Berry, P. L. and Poskitt, T. J., "The Consolidation of Peat," *Geotechnique*, Vol. 22, No. 1, 1972, pp. 27–52.
[2] Rowe, P. W. and Barden, L., "A New Consolidation Cell," *Geotechnique*, Vol. 16, No. 2, 1966, pp. 162–170.
[3] Vickers, B., "A Study of the Consolidation Characteristics of Fibrous Peat," M.Sc. thesis, University of Salford, Salford, U.K., 1972.

Amar J. Sethi,[1] Donald E. Sheeran,[1] Nicolas Skiadas,[1] and S. Alammawi[1]

Determination of Clay Suspension Concentration Profile in a Settling Column Using the Fall-Drop Technique

REFERENCE: Sethi, A. J., Sheeran, D. E., Skiadas, N., and Alammawi, S., **"Determination of Clay Suspension Concentration Profile in a Settling Column Using the Fall-Drop Technique,"** *Consolidation of Soils: Testing and Evaluation, ASTM STP 892*, R. N. Yong and F. C. Townsend, Eds., American Society for Testing and Materials, Philadelphia, 1986, pp. 490–499.

ABSTRACT: The determination of clay suspension concentration can be done either by nondestructive or by destructive techniques. Non-destructive techniques such as transmission by radiation and total pressure measurement devices are either expensive or have a limited range of application, whereas destructive tests such as ovendrying require a large amount of sample, which can limit both the number of sampling ports and the frequency of sampling.

An apparatus has been developed which permits us to draw small-size specimens (less than 0.5 mL volume) from the settling columns, simultaneously from various depths, while at the same time causing minimum disturbance to the sample in the sedimentation column.

The determination of solids concentration in the small size specimen drawn from the sedimentation column is by "fall-drop" technique, whereby a drop of known volume (20 μL) is allowed to fall through a column of immiscible organic liquid. From the velocity of the falling drop the concentration of the suspension can be computed using a calibration curve. Various organic liquids with different densities at room temperature have been tested. The requirements for the selection of these liquids include: reasonable price, low melting point, high boiling point, low vapor pressure, nontoxicity, Newtonian, and immiscible with water, and no chemical reaction with salt solutions. The variations in density and viscosity of these liquids with temperature are presented in this paper. Based on this information, the proper choice of the organic liquid required for measuring the anticipated concentration of the suspension can be made so that the time measured for the fall of a drop between two points (400 mm apart) is sufficiently large (>30 s) and laminar flow prevails for the liquid through which the drop falls.

KEY WORDS: suspension density, clay suspension, sedimentation, fall drop, castor oil

The behavior of clay suspensions during sedimentation is profoundly affected by the nature and concentration of soluble salts and solid particles. At very dilute concentrations, the particles settle at velocities which are close to "Stokesian

[1] Division Chief, Division Chief, Division Chief, and Research Assistant, respectively, Geotechnical Research Centre, McGill University, Montreal, Quebec, Canada.

velocity''. At somewhat higher concentrations, the velocities of individual particles will be modified as a result of the presence of neighboring particles. With a further increase in solids concentration a sharp interface appears between the sediment and the supernatant liquid.

Coe and Clevenger [1] related the fall of the interface to the concentration of the suspension. Kynch [2] assumed that the settling velocity of the suspension is a function of the local particle concentration only. A number of empirical relations have been developed, but all of them include constants which had to be experimentally determined for each material type suspended in a given solution [3].

The determination of clay suspension concentration can be made either by non-destructive or by destructive techniques. Non-destructive techniques such as X-ray transmission [4], total pressure measurement devices [5], and radio-active solids [6] are either too expensive or have a limited range of application, whereas destructive tests such as sampling technique are based on the use of a modified Andreasen apparatus [7] which incorporates the usual pipette and sampling arrangement. After a large volume of specimen is withdrawn, it is allowed to dry either in the oven at 105°C or by evaporation using the Andreasen method for determining the solids concentration. This technique has been found to be very inflexible for the study of sedimentation in concentrated suspensions. The removal of each specimen causes some disturbance in the suspension and results in a change in the level of the suspension surface. Furthermore, the method is tedious and does not yield sufficient experimental results in a reasonable time. For these reasons, it was desirable to develop a technique which would alleviate the difficulties in sampling and solids concentration determination.

An apparatus has been developed which permits one to draw small size specimens (each less than 0.5 mL volume) from the settling column at various depths simultaneously and at the same time cause minimum disturbance to the sample in the settling column. The effectiveness of the apparatus was evaluated by sampling several columns of various known homogeneous concentrations and comparing the determined solids concentrations using the ''fall-drop'' technique described below.

The determination of solids concentration in the small size specimen drawn from the settling column is done by the fall drop technique whereby a drop of known volume (10 to 20 μL) is allowed to fall through a column of immiscible organic liquid. From the velocity of the falling drop, the concentration of the suspension can be computed using a calibration curve. The requirements for the selection of the liquid include: reasonable price, low melting point, high boiling point, low vapor pressure, nontoxicity, Newtonian, immiscible with water, and no chemical reaction with salt solutions. The variations in density and viscosity of the liquid with temperature are presented in this paper. Based on this information, the proper choice of organic liquid required for measuring the anticipated concentration of the suspension can be made so that laminar flow prevails as the drop falls through the liquid and the Reynolds number is less than 0.1.

Materials and Experimental Procedures

In this study Hydrite Flat D kaolinite, a product of the Georgia Kaolin Company, was used. The index properties are shown in Table 1.

Two different clay concentrations (41.0 and 106.6 g/L) dispersed in 15 meq/L soduim bicarbonate (NaHCO$_3$) were evaluated in the sedimentation column whereby the density profiles in the column were directly measured by using the falling drop method.

Transparent Plexiglas cylinders, 112.5 mm inside diameter and 320 mm height, with a bottom plate, were used as sedimentation columns.

An apparatus was designed to draw small size specimens (each less than 0.5 mL volume) simultaneously from 15 different levels in the column at any time. The time required to draw the specimens was less than 5 s. The specimens were withdrawn from heights of 4, 24, 49, 69, 85, 107, 133, 151, 170, 189, 210, 229, 249, 269, and 290 mm from the base of the column, by a stainless steel tubing (0.8 mm outside diameter) glued to the syringes. The apparatus consisted of three aluminum plates, each 205 mm square and 13 mm thick. Fifteen plastic syringes (10 mL) were arranged between the two plates and a suitable hole was made to fit the top and bottom diameters of the syringes. The two plates were fixed by four rods, which were fixed to the bottom plate of the sedimentation column.

The pistons of the syringes could be moved with the help of the third aluminum plate connected to the second plate by three threaded rods and whereby the total travel distance could be pre-fixed. The third plate had a handle and when it moved upward, all 15 pistons could draw an equal volume of specimen from the column at the same time. A schematic diagram of the apparatus is shown in Fig. 1.

After the specimens were withdrawn, the three plates along with the syringes were lifted up, and each piston was pushed down to collect the specimens in small sample vials which were used for measuring the suspension concentration.

The suspension concentration was measured by determining the rate of fall of the suspension drop through a water-immiscible fluid. Castor oil was used in this study as the water-immiscible fluid.

The viscosity of castor oil was measured at various temperatures after de-airing the specimen by using a Contraves Viscometer (Model Rheomat 15T). In addition, the reaction between the electrolytic solution such as sodium sulfate (Na$_2$SO$_4$), sodium chloride (NaCl), and NaHCO$_3$, and the castor oil was evaluated by measuring the viscosity of castor oil after mixing it with the electrolytic solution by using a magnetic stirrer, followed by separation through flotation.

TABLE 1—*Index properties of Kaolinite Flat D.*

Liquid Limit, %	Plastic Limit, %	Plastic Index	Specific Gravity (G_s)	D_{10}, μm	D_{50}, μm	D_{60}, μm	Cu
46	29.5	16.5	2.62	0.62	3.8	5.3	8.55

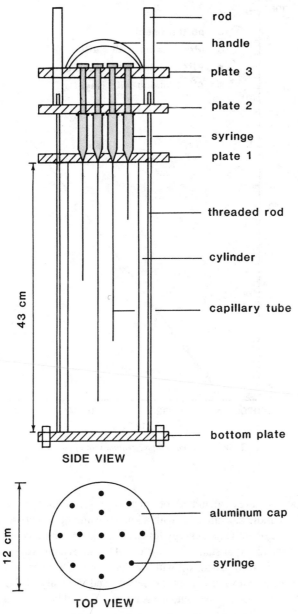

FIG. 1—*Schematic diagram of sampling apparatus.*

Figure 2 shows the results both with and without electrolytic solution mixing, plotted as log viscosity versus the reciprocal of the absolute temperature. These results indicate that the determination of the solids concentration will not be influenced, if the clay suspension contains any of Na_2SO_4, NaCL, or $NaHCO_3$ electrolytic solutions as dispersing fluids.

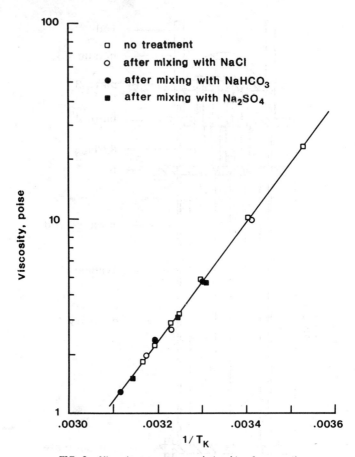

FIG. 2—*Viscosity-temperature relationship of castor oil.*

The activation energy calculated for the castor oil from these results was found to be 13.2 kcal/mole as compared with a known value of 4 kcal/mole for water. The higher value of activation energy for castor oil is corroborated with the higher volumetric thermal expansion of castor oil. This is evident when we compare the change in density of castor oil with that of water with temperature as shown in Fig. 3. One needs to know both the density and viscosity of the castor oil to choose the temperature at which the measurement should be made so that laminar flow prevails.

De-aired castor oil was poured into a jacketed glass column with a stopcock plug at the bottom which had a total height of 600 mm and inside diameter of 28.5 mm. A constant temperature of 30°C was maintained throughout the experiments by circulating water from a water bath in the jacket.

A known volume (20 μL) of a given clay dispersion specimen was dropped in the castor oil column as a droplet by using a micropipette (diameter of droplet is 3.36 mm if a perfect sphere is assumed). The terminal velocity of the droplet was measured by measuring the time required for the droplet to travel from the 200 to 500 mm mark from the top of the column.

The relationship thus obtained between the clay concentration (g/L) and the terminal velocity of the droplet is shown in Fig. 4. This relationship was used as a calibration chart for computing concentrations of unknown specimens. The Reynolds number R for the data shown in Fig. 4 may be calculated as

$$R = v\rho d/\mu$$

where

v = velocity of droplet, cm/s,
ρ = density of castor oil, 0.953 g/mL at 20°C (Fig. 3),
d = diameter of droplet, 0.336 cm, and
μ = viscosity of castor oil, 4.7 P at 20°C (Fig. 2).

For the two extreme values of velocity, 0.06 cm/s and 0.3 cm/s for corresponding concentrations of 20.0 g/L and 400.0 g/L, the Reynolds number is 0.041 and 0.0204, respectively, thus indicating laminar streamlines around the falling drop.

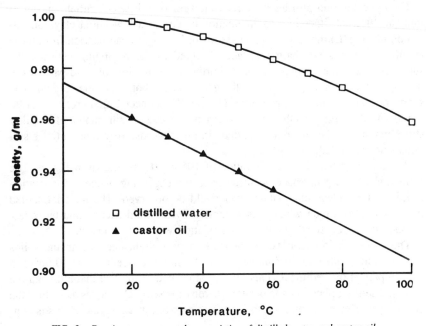

FIG. 3—*Density-temperature characteristics of distilled water and castor oil.*

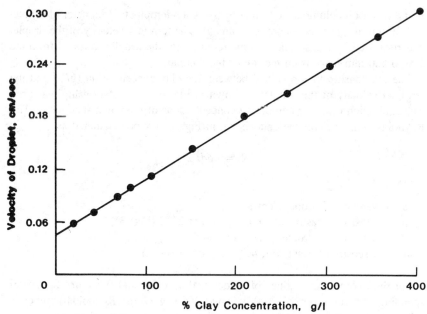

FIG. 4—*Relationship between velocity of droplet and clay concentration.*

Results and Discussion

The concentration profiles were measured for two different initial concentrations at different times in a sedimentation column. At low initial clay concentration (41.0 g/L) the results are as shown in Fig. 5. As can be seen from these results, there was no sharp interface observed in the beginning between the supernatant and the settling sediment. Furthermore, the loss of solid particles in the top of the column was equal to the gain near the bottom of the column. The sharp interface started to appear after 60 min. The concentration profile after 90 min shows that, although there was a sharp interface, a certain amount of turbidity still remained in the supernatant liquid. In other words, very fine particles still remained suspended.

Results for initial clay concentration of 106.6 g/L are shown in Fig. 6. The formation of sharp interface occurred from the beginning of the sedimentation process, and a clear supernatant layer could be observed. The material settled more or less as a plug which was supported by the fluid friction and buoyancy forces, although a small amount of segregation was also observed.

The concentration profiles so obtained show a slight decrease in the solids concentration with time in comparison to the initial concentration (10.66%) in the top part of the sediment. This may be attributed to the fact that although the hindered settling phenomenon occurs whereby the sediment moves as a plug, the coarser particles settle through the coherent network of aggregates. This is con-

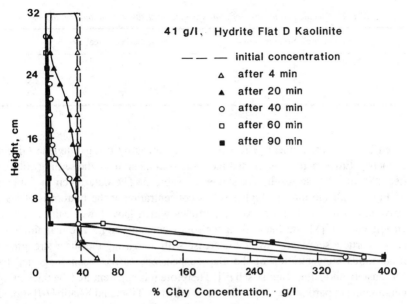

FIG. 5—*Concentration profiles with time (initial concentration = 41.0 g/L).*

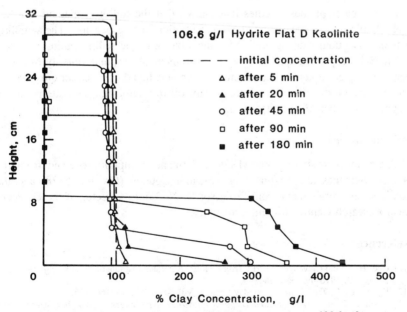

FIG. 6—*Concentration profiles with time (initial concentration = 106.6 g/L).*

TABLE 2—*Specific surface area of sediment specimens taken from different depths.*

Height from Bottom, mm	Specific Surface Area, m^2g^{-1}
4	5.6
24	18.2
85	20.5

firmed by the measurement of specific surface areas by using Ethylene Glycol Monoethyl Ether method on the sediment samples taken from the settling column after 200 min. These results are shown in Table 2. The data given in Table 2 and Fig. 6 indicate that the higher solids concentration at the bottom is at least partly due to the settling of coarser particles which have lower values of suspension volume [8] and thereby higher solids concentration. The magnitude of coarser particle segregation would decrease with increasing the shear strength or initial solids concentration of the clay suspension, as has been suggested by Morgenstern and Amir-Tahmasseb [9]. Therefore it is apparent that the thickening phenomenon in particulate suspensions as studied by Tiller and Khatib [10] should be evaluated in conjunction with the segregation phenomenon especially when the solids concentration in dispersion is low.

Conclusions

On the basis of these studies it is shown that the fall-drop method is a fast and reliable way of measuring clay concentration in a suspension. The sampling was done without causing any undue disturbance to the sedimentation column.

The fall-drop technique involves measurement of terminal velocity of a suspension droplet through a water-immiscible organic liquid (e.g., castor oil) whereby the suspension droplet falls through castor oil in a laminar flow pattern (i.e., a Reynolds number of less than 0.1).

Acknowledgments

The authors wish to acknowledge the financial support received from the Natural Sciences and Engineering Research Council (Grant A-5178) and the continual encouragement from Professor R. N. Yong, Director of the Geotechnical Research Centre, McGill University.

References

[1] Coe, H. S. and Clevenger, G. H., *Transactions*, American Institute of Mining Engineers, Vol. 55, No. 9, 1916, pp. 356–384.
[2] Kynch, C. J., *Transactions*, Faraday Society, Vol. 48, 1952, pp. 166–176.
[3] Vesilind, P. A., *Treatment and Disposal of Wastewater Sludges*, Ann Arbor Science, Ann Arbor, MI, 1979, p. 110.

[4] Gibson, R. E., Schiffman, R. L., and Cargill, K. W., *Canadian Geotechnical Journal,* Vol. 18, 1981, pp. 280–293.
[5] Michaels, A. S. and Bolger, J. C., *Industrial and Engineering Chemistry Fundamentals,* Vol. 1, No. 1, Feb. 1962, pp. 24–33.
[6] Richardson, J. F., *Transactions,* Institute of Chemical Engineers, Vol. 38, 1960, pp. 33–42.
[7] Andreasen, A. H. M., *Kolloid-Zeitschrift,* Vol. 49, 1929, pp. 253–265.
[8] Yong, R. N. and Sethi, A. J., *Journal of Canadian Petroleum Technology,* Vol. 17, 1978, pp. 76–83.
[9] Morgenstern, N. and Amir-Tahmasseb, I., *Geotechnique,* Vol. 15, 1965, pp. 387–395.
[10] Tiller, F. M. and Khatib, Z, *Journal of Colloid and Interface Science,* Vol. 100, 1984, pp. 55–67.

J. Don Scott,[1] Maurice B. Dusseault,[2] and W. David Carrier, III[3]

Large-Scale Self-Weight Consolidation Testing

REFERENCE: Scott, J. D., Dusseault, M. B., and Carrier, W. D., III, "**Large-Scale Self-Weight Consolidation Testing,**"*Consolidation of Soils: Testing and Evaluation, ASTM STP 892*, R. N. Yong and F. C. Townsend, Eds., American Society for Testing and Materials, Philadelphia, 1986, pp. 500–515.

ABSTRACT: Syncrude Canada Ltd. has developed self-weight consolidation tests with specimens 10 m high and 1 m in diameter. Used in a research and development program on oil sand waste slurry disposal, these large-scale tests determine whether or not the slurry material properties measured in large-strain slurry consolidation cells and in 2-m-high consolidation tests (when used in the finite-strain consolidation mathematical model) can predict the consolidation behavior of deep deposits of waste slurry such as those deposited in the field. The consolidation equipment and testing methods used in this research program are described, and typical test results are shown. Among the phenomena observed are the effective stress wave which travels up the cylinder during sedimentation, the development of thixotropic strength during the initial period of consolidation, the full pore pressure and total stress profile for the height of the cylinders at any time, and the consolidation with time and depth as shown by pore pressure dissipation and by the increase in solids content.

The rate of consolidation in the 10-m cylinders as shown by the settlement of the water-sludge interface is in good agreement with theory and indicates the validity of the test methods.

KEY WORDS: consolidation, laboratory testing, waste slurry, oil sands sludge, self-weight consolidation, finite-strain consolidation theory, sedimentation, pore pressure measurements, solids content measurements, oil sands

Self-weight consolidation tests with specimens 10 m high and 1 m in diameter have been developed as part of a research and development program on oil sand waste slurry disposal for Syncrude Canada Ltd. The purpose of these large scale tests is to determine whether the slurry material properties measured in large-strain slurry consolidation cells and in 2-m-high consolidation tests, when used in the finite-strain consolidation mathematical model, can predict the consolidation behavior of deep deposits of waste slurry such as those deposited in the field.

[1]AOSTRA Professor, Department of Civil Engineering, University of Alberta, Edmonton, AB, Canada T6G 2G7.

[2]Professor, Geological Engineering, University of Waterloo, Waterloo, ON, Canada N2L 3G1.

[3]Principal, Bromwell and Carrier, Inc., 202 Lake Miriam Drive, P. O. Box 5467, Lakeland, FL, 33807.

The Syncrude Canada operation is an open-pit oil sands mine where 90×10^6 metric tonnes of oil sand are mined and processed annually. Water is added to the oil sand in the bitumen extraction plant to separate the bitumen and to slurry the sand and fines to pump it to a tailings pond. The tailings stream amounts to 130×10^6 metric tonnes every year and the tailings pond required to store the waste material is approximately 12 km^2 in area. The tailings slurry, pumped at a solids content of 50 to 55%, segregates on deposition and the sand containing some fines is used to build the pond dykes and beaches while a fines stream composed of silt and clay and some lost bitumen flows into the pond. This fines stream settles out and consolidates to a solids content of 20% (where solids content = mass of dry solids/total mass of sample) fairly rapidly, but the large volume of mining results in approximately 14×10^6 m^3 of liquid slimes or sludge forming every year. Developing a method to more rapidly consolidate and dispose of this deposit of sludge is the purpose of the research program. The various disposal methods being considered by Syncrude have been outlined by Scott and Cymerman [1].

The consolidation equipment and testing methods used in this research program, especially the large-scale tests, are described in the following sections.

Determination of Sedimentation and Consolidation States

Sedimentation and consolidation tests on the fine grained sludge have been conducted during the past six years at the University of Alberta in cylinders of various heights [2]. The necessity to determine when sedimentation was complete and the progress of consolidation led to the development of instrumentation and sampling methods for tall self-weight consolidation cylinders.

Figure 1 shows a series of diagrams of the stages of a sedimentation test on a sludge sample which had an initial solids content of 10% and density of 1.05 g/cm^3. The initial pore pressure at the base of the column was equal to the total mass of the column of sludge, showing that no effective stresses existed in the slurry at the beginning of the test. During free sedimentation of the sludge it did not change density except at the bottom of the tube, where the density increased to about 1.07 g/cm^3. A small decrease in the pore pressure at this depth indicated that at this density, the sludge particles were touching and had developed effective stress between them. As the water-sludge interface settled, the effective stress front traveled upwards and after $2\frac{1}{2}$ days the full column had developed an initial interparticle stress. No significant consolidation had taken place at this time, however, and the sedimentation and consolidation stages were completely separate. Consolidation proceeded downward from the surface of the slurry and after 300 days was practically complete as shown by pore pressure and density measurements. The interface continued to slowly settle from secondary consolidation.

Such laboratory tests coupled with field measurements in the tailings pond showed that sedimentation was fairly rapid and could be considered to be complete

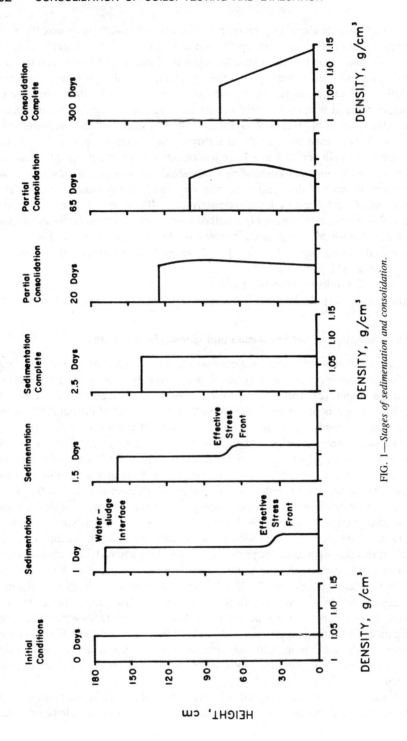

FIG. 1—Stages of sedimentation and consolidation.

when the solids content had increased to approximately 15%. Further increase in density of the sludge had to take place by consolidation; thus the research program has been directed towards a study of the consolidation mechanisms.

Imai [3] shows the results of similar sedimentation-consolidation small standpipe tests on kaolin and Osaka Bay mud. In these tests, however, the sedimentation and consolidation processes overlapped with self-weight consolidation occurring in the bottom part of the standpipe while sedimentation was still taking place in the upper part. This process may have occurred because of the small height of the slurries, 23 cm, which would have allowed the consolidation to take place more rapidly than in the 180-cm-high sample discussed here. Imai [3] points out that the solid content at which sedimentation is complete is not unique for a material but depends on the initial water content of the slurry. Similar results were found with the tailings sludge, but the range of solids content at the end of sedimentation was not large, between 10 and 15% for the fines stream entering the tailings pond with a solids content of approximately 8%.

Large-Strain Slurry Consolidation Cell

Slurry consolidometers have been developed by other investigators [4]. Their major difference from standard oedometers is that they are designed for samples 20 to 30 cm in diameter and 30 to 45 cm in height. During a test, strains up to 80% of the initial sample height may be measured. The maximum consolidation stresses used in the consolidometer testing may not be large, since the field stresses in slurries or dredged materials are often relatively low.

A typical experimental arrangement is shown in Fig. 2. Because finite-strain consolidation theory requires a knowledge of the variation of the coefficient of consolidation of the material during a test, the cell is designed so that stress-void ratio and void ratio-permeability relationships can be measured for each load increment. After consolidation under a load increment is complete, a permeability test is conducted on the sample.

Pane et al [5] discuss the need for measuring the permeability of very soft soils when a nonlinear finite-strain consolidation theory is employed to predict the rate of consolidation and they also review the difficulty in making such measurements. Very small hydraulic gradients usually exist in the field in self-weight consolidation of slurries, and the permeability tests in the laboratory should model these gradients and keep them sufficiently small so that seepage-induced consolidation is minimized.

Figure 2 shows a unique method of overcoming the tendency for seepage-induced consolidation. When consolidation under a load increment is complete, the loading piston is locked in place and a constant-head permeability test is performed. If seepage-induced consolidation starts, the effective stress applied by the loading piston is automatically reduced. With this technique, a hydraulic gradient can be used in the permeability test which applies a seepage stress up to the maximum past effective stress. If the slurry is a material which has high secondary consolidation characteristics, a smaller hydraulic gradient must be used.

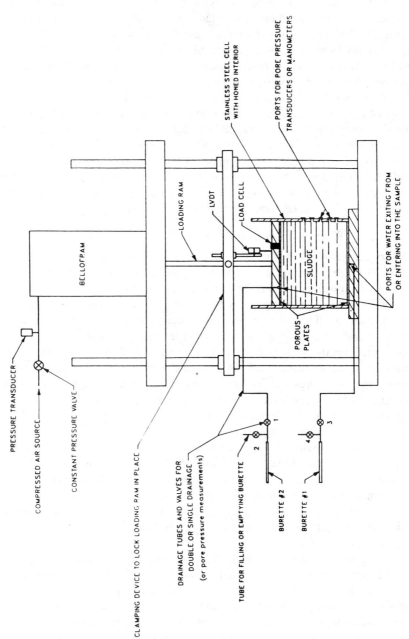

FIG. 2—*Large-strain consolidation and permeability cell.*

Two-Metre Slurry Consolidation Cylinders

Self-weight consolidation tests on the sludge and on sand-sludge and over-burden-sand-sludge mixes are first performed in 2-m-high test cylinders. Mixes that show promising consolidation characteristics can then be considered for testing in the 10-m-high cylinders. The 2-m cylinders (Fig. 3) are continuously monitored for settlement of the sludge-water interface, for pore pressure changes at 20-cm intervals down the column by small-diameter manometers, and also for density changes by sampling at 20-cm intervals. A 5-cm³ specimen is obtained from each sample port for density and solid content determination.

The advantages of the 2-m cylinders are that they cost relatively little, thereby allowing a large number of different trial mixes to be tested; their consolidation time is relatively short; and mixes of coarse and fine material can be tested for their segregation potential (i.e., whether the coarse material will settle through the matrix of fine-grained slurry).

A major disadvantage is that for low-solid-content slurries, the stresses are so small that small thixotropic gains in strength of the slurry affect the consolidation in the upper part of the cylinder. Initial consolidation rates and amounts are therefore smaller than predicted from incrementally loaded consolidation tests. A sludge with an initial solids content of 30% only has a maximum excess pore pressure at a depth of 2 m of 4.1 kPa. Tests indicate that for the 30% solids fine-

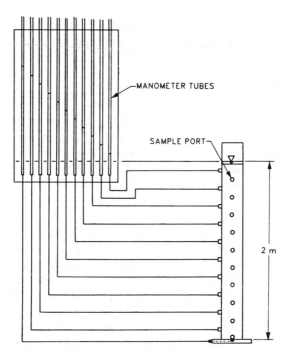

FIG. 3—*Two-metre self-weight consolidation cylinder.*

grained sludge, a thixotropic strength of 13 Pa develops in one or two days, and this bond strength appears to be large enough to prevent significant consolidation in the top metre of the column. A further indication of the effects of thixotropy is that negative pore pressures as great as 1 kPa have been measured near the top of the cylinder. Kolaian and Low [6] have shown that suctions of this magnitude develop in clays from thixotropy.

The experimental difficulty in measuring low excess pore pressure during long testing periods has also been noted by Umehara and Zen [7]. They show excess pore pressure distribution for an 80-cm-high standpipe containing Osaka Bay mud at an initial solids content of 25% over an 80-day period. Of interest is that the measured pore pressures in the upper part of the standpipe remained high, perhaps indicating that thixotropy was slowing the rate of consolidation.

Two-metre cylinder tests with sand-sludge mixes have more consistent results because these mixes have higher initial densities, and therefore stresses are not affected to the same degree by thixotropy.

The 2-m cylinder tests have a further disadvantage in that sampling of the sludge will affect the height of the slurry in the cylinder. Consideration was given to employing some indirect method of measuring density such as that used by Been and Sills [8], but the necessity for monitoring for segregation and chemical changes required samples. Sampling of the sludge is therefore done only when significant changes in the material are indicated by other measurements.

Following the foregoing procedures has allowed these tests to continue for well over a year when a high degree of consolidation is needed.

Ten-Metre Self-Weight Consolidation Tests

Two 10-m-high specimens, 1 m in diameter, have been consolidating under self-weight for over 1½ years. The test cylinders (Fig. 4) are similar to the 2-m-high cylinders except that pore pressure and sample ports are located at 1-m intervals. Additional ports at ½-m intervals are available to allow closer measurements, but these have not been needed except at the top and bottom of the cylinders where changes were taking place more rapidly.

One cylinder contains sludge which had an initial solids content of 31% and the other cylinder contains a sand-sludge mix at a considerably higher solids content and density. The 31% solids content sludge was chosen for the test because this degree of consolidation is reached fairly rapidly (probably within a year) in the tailings pond and because considerable depths of it exist.

The sampling mechanism is shown in Fig. 5. The sampling ports have an O-ring seal which allows the sample probe to be inserted to any distance across the diameter of the cylinder. Specimens are selected at different distances to reduce disturbance at any one location. The 12.7-mm ball valve remains closed until the probe tube passes the seal. The valve is then opened and the tube inserted to the desired location. A 55-cm³ specimen is taken which gives sufficient material for a number of tests. For high-solid-content slurries which do not flow easily,

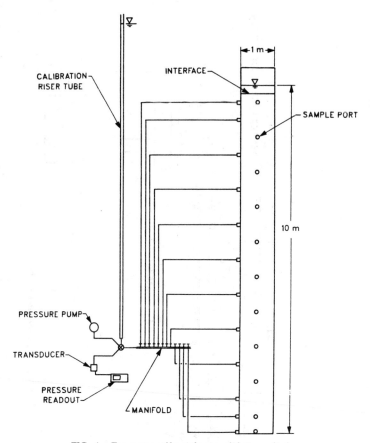

FIG. 4—*Ten-metre self-weight consolidation cylinder.*

a vacuum is put on the sample bottle. This sampling technique has worked very well and has been a key factor in allowing the changing characteristics of the sludge to be fully monitored.

The main measurements on the samples are density and solid content determinations; however, extraction and grain size distribution tests are also carried out to measure for segregation of bitumen or coarse sand grains in the mix. Chemical, biological and bacteria studies occasionally are made on specimens to check for long-term changes in these parameters. A flow chart (Fig. 6) shows the full suite of tests performed.

Temperature measurements are made of the sludge in the cylinder using a probe similar in diameter to the sampling probe, and of the sludge specimens when taken to ensure that temperature gradients are not affecting consolidation. The variation in temperature in the cylinder has been constant throughout the test, summer or winter, with an increase of about 3°C from the bottom to the top of the cylinder. This gradient is not considered sufficient to have any significant effect.

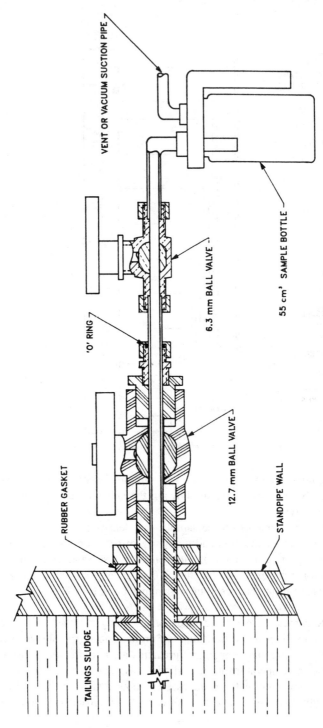

FIG. 5—*Sample port and sampling probe.*

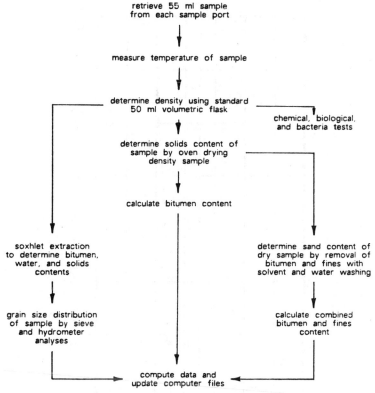

FIG. 6—*Flow chart for sludge sample analysis.*

Pore pressures are measured with a pressure transducer whose calibration is checked against a column of water, as shown in Fig. 4, before each set of readings. Figure 7 shows the pore pressures above the hydrostatic water pressure at depths of 1 m and 9 m for the sludge test. When the cylinder was first filled, the pore pressures were equal to the total stresses over the full height of the cylinder, indicating zero effective stress. There was a uniform drop of pore pressure of about 1.5 kPa during the first 30 days throughout the cylinder. As no measurable consolidation had taken place during this period, the drop in pore pressure is attributed to an establishment of an effective stress of this magnitude by thixotropy. Following this decrease the upper piezometers showed some increase between 75 and 125 days elapsed time, while the lower piezometers remained essentially steady. Following this period, the pore pressures in the upper 2 m have shown (Fig. 7) a slow, steady consolidation. Little further change in pore pressure has occurred below a depth of 3 m to date.

Although there has been some experimental variation in pore pressure readings during the test, the long-term trend indicates that the system can monitor excess pore pressures with an accuracy of about 0.2 kPa.

The dissipation of pore pressure with depth (Fig. 8) does not indicate at what

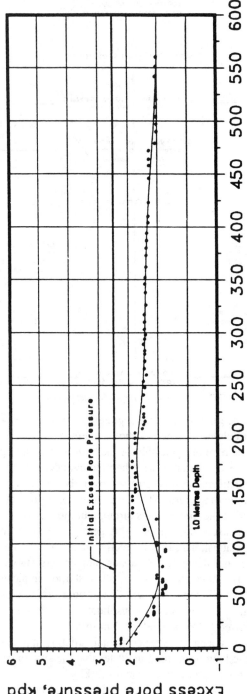

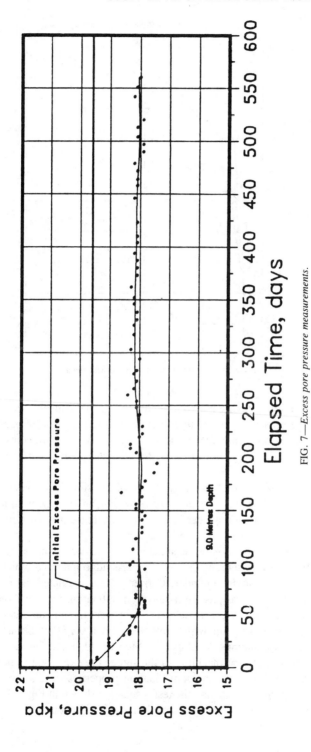

FIG. 7—*Excess pore pressure measurements.*

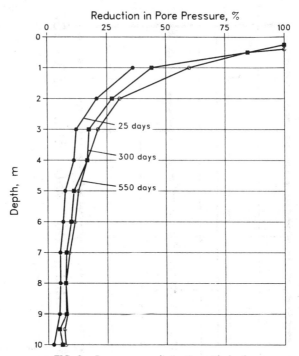

FIG. 8—*Pore pressure dissipation with depth.*

depth consolidation is taking place. It would appear that consolidation is mainly confined to the upper 3 m, since there has been little reduction in pore pressure below this depth following the initial drop. Density measurements (Fig. 9), however, indicate a fairly uniform increase in density and solids content with depth.

It is apparent that the high compressibility of the soil allows volume change to occur with little change in pore pressure. Although an increase in solids content from 31% to 34% does not appear large, it is a decrease in void ratio from 5.2 to 4.5.

The degree of consolidation of the sludge in the self-weight consolidation tests is being defined by the settlement of the water-sludge interface which has now reached 0.5 m in the sludge cylinder. Carrier [9] has predicted the rate of consolidation based on test results from slurry consolidation tests and on a finite-strain consolidation numerical model. This numerical model consists of a finite-difference computer program written by Somogyi [10,11]. It is based on the general, nonlinear equations developed by McNabb [12] and Gibson et al [13]. The compressibility and permeability of the sludge are expressed as power functions [14] in order to define the consolidation properties over a large range in void ratio. The predicted rate and the test results are shown in Fig. 10. The good agreement between theory and the large-scale test is encouraging. The rate is

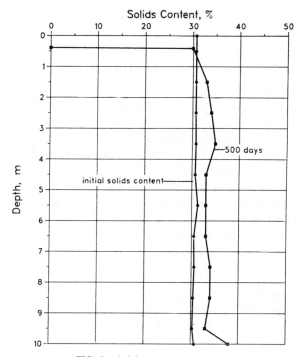

FIG. 9—*Solids content with depth.*

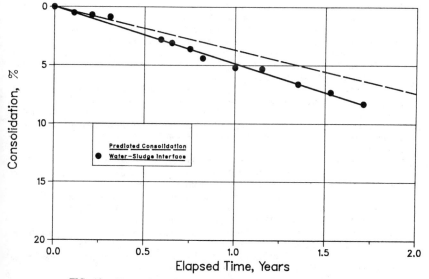

FIG. 10—*Comparison of predicted and experimental consolidation.*

essentially linear for the first two years because the tests are still at a very early stage in the consolidation process.

Discussion and Conclusions

Large-scale self-weight consolidation tests have been developed as part of a research and development program for oil sands sludge disposal. A combination of slurry consolidometer tests, 2-m-high self-weight consolidation tests, and 10-m-high self-weight consolidation tests has been found to be an economic and viable method of determining the consolidation properties of the tailings sludge for evaluating different disposal schemes.

Good agreement between the different test procedures has been found by employing the finite-strain consolidation mathematical model. Such agreement will allow the laboratory test results to be extrapolated with confidence to predict long term field behavior of the oil sand tailings sludge.

Acknowledgments

The authors would like to give credit to the Syncrude Canada Ltd. engineering and research staff with whom they have worked on this testing program, and acknowledge Syncrude for permission to give details of the test equipment and procedures.

References

[1] Scott, J. D. and Cymerman, G. J., "Prediction of Viable Tailings Disposal Methods" in *Proceedings,* Sedimentation Consolidation Models Symposium, American Society of Civil Engineers, San Francisco, 1 Oct. 1984, pp. 522–544.
[2] Scott, J. D. and Dusseault, M. B., "Behaviour of Oil Sands Tailings," in *Proceedings,* 33rd Canadian Geotechnical Conference, Calgary, Alberta, 24–26 Sept. 1980.
[3] Imai, G., "Experimental Studies on Sedimentation Mechanism and Sediment Formation of Clay Materials," *Soils and Foundations,* Japanese Society of Soil Mechanics and Foundation Engineering, Vol. 21, No. 1, March 1981, pp. 7–20.
[4] Carrier, W. D., III, and Keshian, B., Jr., "Measurement and Prediction of Consolidation of Dredged Material," presented at 12th Annual Dredging Seminar, Center for Dredging Studies, Department of Civil Engineering, Texas A&M University, College Station, TX, 1–2 Nov. 1979.
[5] Pane, V., Croce, P., Znidarcic, D., Ko, H-Y., Olsen, H. W., and Schiffman, R. L., "Effects of Consolidation on Permeability Measurements for Soft Clay," *Geotechnique,* Vol. 33, No. 1, March 1983, pp. 67–72.
[6] Kolaian, J. H. and Low, P. F., "Thermodynamic Properties of Water in Suspensions of Montmorillonite" in *Proceedings,* 9th National Conference on Clays and Clay Minerals, Pergamon Press, Monograph 11, Earth Science Series, Elmsford, NY, 1962, pp. 71–84.
[7] Umehara, Y. and Zen, K., "Consolidation Characteristics of Dredged Marine Bottom Sediments with High Water Content," *Soils and Foundations,* Japanese Society of Soil Mechanics and Foundation Engineering, Vol. 22, No. 2, June 1982, pp. 40–54.
[8] Been, K. and Sills, G. C., "Self-Weight Consolidation of Soft Soils: An Experimental and Theoretical Study," *Geotechnique,* Vol. 31, No. 4, Dec. 1981, pp. 519–535.
[9] Carrier, W. D. III, "Analysis of Ten Metre Cylinder Tests," Letter Report to Syncrude Canada Ltd., Bromwell and Carrier Inc., Lakeland, FL, Jan. 1984.

[*10*] Somogyi, F., "Analysis and Prediction of Phosphatic Clay Consolidation: Implementation Package," Bromwell Engineering, Inc., Final Report to Florida Phosphate Council, Lakeland, FL, 1979.
[*11*] Somogyi, F., "Large Strain Consolidation of Fine-Grained Slurries," presented at Canadian Society for Civil Engineering Annual Conference, Winnipeg, 1980.
[*12*] McNabb, A., "A Mathematical Treatment of One-Dimensional Soil Consolidation," *Quarterly of Applied Mathematics*, Vol. 17, 1960, pp. 337–347.
[*13*] Gibson, R. E., England, G. L., and Hussey, M. J. L., "The Theory of One-Dimensional Consolidation of Saturated Clays, I. Finite Non-Linear Consolidation of Thin Homogeneous Layers," *Geotechnique*, Vol. 17, 1967, pp. 261–273.
[*14*] Carrier, W. D., III, Bromwell, L. G., and Somogyi, F., "Design Capacity of Slurried Mineral Waste Ponds," *Journal of Geotechnical Engineering*, Vol. 109, May 1983, pp. 699–716.

Tadeusz Barański¹ and Wojciech Wolski²

Lateral Strain Measurement by an Ultrasonic Method

REFERENCE: Barański, T. and Wolski, W., "**Lateral Strain Measurement by an Ultrasonic Method,**" *Consolidation of Soils: Testing and Evaluation, ASTM STP 892*, R. N. Yong and F. C. Townsend, Eds., American Society for Testing and Materials, Philadelphia, 1986, pp. 516–525.

ABSTRACT: An ultrasonic method for the measurement of lateral strain of a soil specimen in triaxial tests has been developed. The measuring device combined with a triaxial cell consists of an ultrasonic detector and three ultrasonic heads which transform sound waves into electronic impulses. The ultrasonic wave, which penetrates into the triaxial cell, is reflected from the contour of the specimen, and comes back to the ultrasonic head. The measured values of lateral strains can be used not only for the evaluation of Poisson's ratio, but also for other parameters or functions required for different constitutive equations.

KEY WORDS: soil tests, lateral strain, strain characteristics, ultrasonic

New constitutive equations introduced in geotechnics require nonconventional soil parameters. For the evaluation of soil characteristics the strain-stress relationships are necessary. Triaxial tests, generally carried out to study stress-strain behavior, enable the measurement of vertical strain ϵ_1, but difficulties arise when horizontal strains ϵ_3 have to be measured.

It has been emphasized by many authors [1–6] that the horizontal strain test methods used so far have low accuracy due mainly to the fact that the measurements are performed in selected horizontal cross sections of the specimen. Therefore a new device based on ultrasonic measurements and providing more reliable values of horizontal strains has been developed.

Stress-Strain Behavior Measured by the Ultrasonic Method

Description of Ultrasonic Strain Indicator

The main part of the device is made of an ultrasonic detector adapted to nondestructive tests of soil specimens. The device consists of three 3.5 MHz ultrasonic heads, a voltage regulator and rectifier, and a control panel (Figs. 1 and 2). The heads spaced at 120-deg intervals are attached to the ring (Fig. 3)

¹ D.Sc., Department of Geotechnics, Warsaw Agricultural University, 02-766 Warsaw, Nowoursynowska 166, Poland.
² Professor, Head of Department of Geotechnics, Warsaw Agricultural University, 02-766 Warsaw, Nowoursynowska 166, Poland.

516

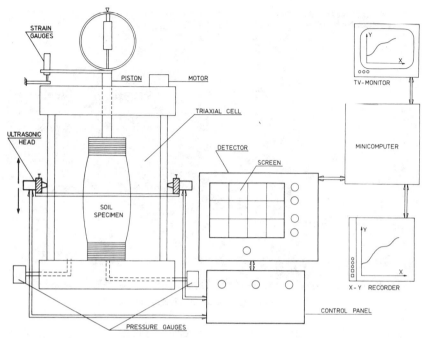

FIG. 1—*Schematic diagram of ultrasonic strain indicator.*

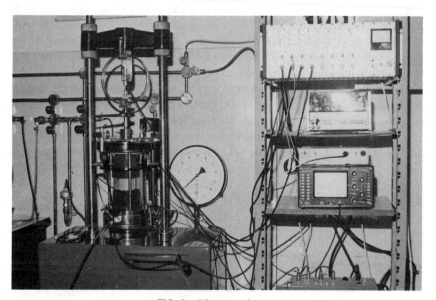

FIG. 2—*Ultrasonic device.*

FIG. 3—*Ultrasonic head attached to movable ring.*

which can be moved automatically in the vertical direction on the walls of the triaxial cell. The ring can also be turned in any horizontal position.

Principle of Operation

By means of three ultrasonic heads, the ultrasonic detector transforms electronic impulses into ultrasonic waves, which penetrate through the triaxial cell wall. The waves are reflected from the contour of the specimen and return to the ultrasonic heads where they are transformed back into electronic impulses (Fig. 4). After amplification the electronic impulses are transferred to an oscilloscope, where the results of the measurements appear on the screen. The light spot on the screen enables the evaluation of the distance between the ultrasonic head and the reflecting surface, as well as the shape of the deformed specimen. Measurement with any one of the ultrasonic heads can also be done separately (Fig. 5). All signals from the detector are sent to the minicomputer; thus it is easy to detect any changes in the diameter of the specimen and on the basis of these changes, the lateral strain can be determined. These results can be displayed directly on the minicomputer screen or can be plotted.

Determining Measurement Precision

Experiments were made to estimate the accuracy of the lateral strain measurements. For this purpose a plastic cylinder the same height as used in the triaxial tests (80-mm) was used. The cylinder was made of 16 uniform cylinders 5 mm in height, but each had a radius 0.1 mm smaller than the preceding one.

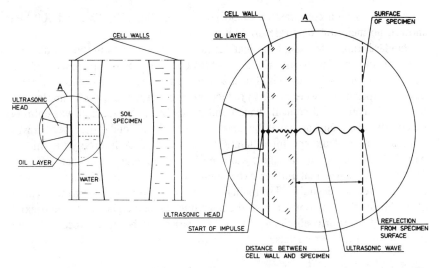

FIG. 4—*Principle of operation.*

Thus the diameter at the base was 36 mm and 33 mm on the top. Although the ultrasonic strain indicator (USI) provides the possibility of horizontal magnification up to ×30, a ×20 magnification was sufficiently adequate to determine the lateral strain within 0.2% accuracy.

There was also a risk that at high cell pressures the cell itself might deform. This could result in incorrect readings of the lateral deformation. However, a

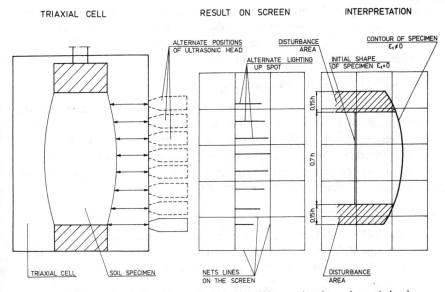

FIG. 5—*Interpretation of measurement of lateral strain; results of one ultrasonic head.*

series of tests with cell pressure of 50, 100, 150, 200, 250, 300, and 400 kPa showed no measurable cell deformability.

Standard triaxial tests indicate that the largest strain occurs in the middle part of the specimen. Due to the frictional forces between the soil specimen and the filter stones, the lateral strain is almost zero at the top and bottom of the specimen. Several experiments were made to reduce the end restraint by, for example, end platen lubrication and conical shaped end plates [7,8]. No method was entirely successful, and it appeared that the zone influenced by the end restraint effect was limited to 15% of the specimen height on each end. Therefore only the middle part, 70% of the specimen height, was used for the horizontal strain calculations. The authors realize that further experiments to eliminate or minimize the end restraint effect are desirable. The main measurement errors result from the position of the heads: the horizontal axis of the head must be perpendicular to the triaxial cell wall. This error is eliminated by the comparison of the signals received from the head with the test signals.

Capabilities of the USI

The USI and the lateral strain measurements enable various tests to be performed. On the basis of the lateral strain values ϵ_3, it is possible to determine elastic parameters, volume and deviator strain, dilatational characteristics, and strain characteristics in consolidation and creep tests. The proposed method of lateral strain measurements enables all the aforementioned characteristics of partly saturated soils to be determined.

Sample Test Results

The objective of the present experiments is to provide data for determining constitutive equations of soil behavior. Shear strength, consolidation, and creep tests were carried out on specimens of sand, clay, peat, and gyttja.[3] Examples of experimental results are presented below.

Strength and Dilatability Characteristics

The tests were carried out with medium sand, and various average stresses σ_m were applied. In the course of the test, stresses σ_1 and σ_3 as well as strains ϵ_1 and ϵ_3 were measured. On the basis of these measurements the characteristics of strength and dilatation were determined (Fig. 6). The tests showed that sand densification occurs in the initial stage of shearing, and only beyond a certain stress value does the sample volume increase.

Clay Testing

Elastic soil models or elastic ideally-plastic models applied in practice require an estimation of Poisson's ratio for the soil skeleton. As this parameter is not a constant value for soils, it is difficult to know what to use in practical analysis.

---

[3] ". . . dark pulpy freshwater anaerobic mud . . ." (*AGI Glossary of Geology*, 1974).

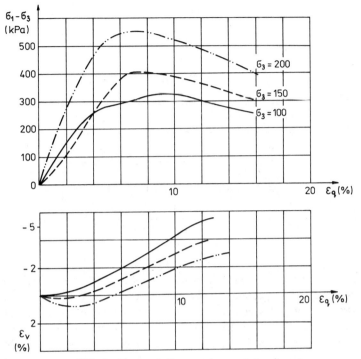

FIG. 6—*Shear and dilatancy characteristics of sand.*

Therefore it is advisable that a nonlinear stress-strain relationship be used for a description of the soil behavior [9]. In the nonlinear hyperbolic elastic model, a variable Poisson's ratio can be calculated according to the formula

$$\nu_t = \frac{G - F \log\left(\dfrac{\sigma_3}{Pa}\right)}{\left[1 - \dfrac{(\sigma_1 - \sigma_3)d}{E_t}\right]^2}$$

Parameters G, F, and d were estimated in accordance with Refs 4, 7, and 10. Test results of sandy clay and silty clay are given in Figs. 7 and 8.

Consolidation and Creep Tests

The tests were made on undisturbed specimens of peat and gyttja. Basic geotechnical characteristics of the tested soils are given in Table 1.

Two series of tests were performed: isotropic consolidation and creep.

The consolidation tests were performed in the triaxial cell and average values of the isotropic stress were 30, 45, and 85 kPa. The tests were continued until

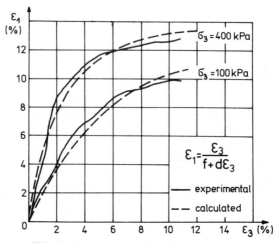

FIG. 7—*Deformation characteristics of silty clay.*

the deformation process ceased and the excess pore pressure dissipated. The measurements of the specimen height and diameter as well as the volume of the water drained from the specimen through the top and the bottom of the specimen were made. Then the volumetric strain ϵ_v of the specimen was calculated in two ways: (1) on the basis of the strains ϵ_1 and ϵ_3 which vary along the height of the specimen, and (2) on the basis of the drained water volume ΔV. In the first method the whole height of the specimen was taken into account. Experimental results are presented in Fig. 9.

The pattern of volume strain changes shows that the changes are dependent on the consolidation stress. Slightly greater volume changes were obtained from

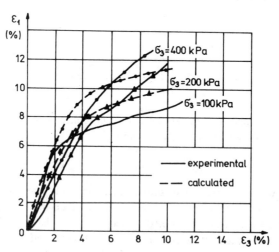

FIG. 8—*Deformation characteristics of sandy clay.*

TABLE 1—*Basic geotechnical characteristics of tested soils.*

Characteristics	Peat	Gyttja
Water content	310 to 340%	104 to 110%
Plasticity limit	185 to 198%	52 to 54%
Liquid limit	313 to 321%	101 to 104%
Density	1007 to 1120 kg/m³	1410 to 1430 kg/m³
Organic content	74 to 88%	32 to 38%

the lateral strain measurements than from the drained water volume. The discrepancies probably result from small differences in the compressibility of the soil skeleton. Thus, for more accurate data, the volume strain changes obtained from ϵ_1 and from the varying value ϵ_3 should be considered while determining soil behavior in the consolidation process.

Testing of the organic soil has shown that soil creep should not be disregarded, because it determines the character and size of secondary strains in the consolidation process. Experiments [11] have shown that the strain rate-shear stress relationship may be expressed by the equation based on the rate process theory:

$$\dot{\epsilon}_{q(\tau t)} = \beta \sinh (\alpha\tau)\left(\frac{t_0}{t}\right)^m$$

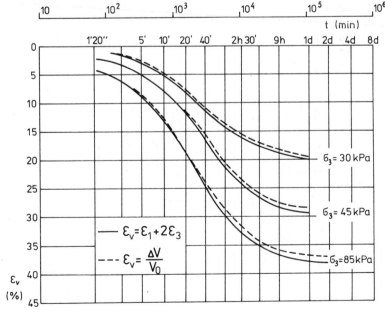

FIG. 9—*Strain-time relationship for peat during isotropic consolidation.*

Reological parameters α and β are calculated on the basis of the deviator strain rate-time relationship. The parameter m varies within small limits, depending on the deviator stress.

The tests have shown that in organic soil the deviatoric strain rate is dependent not only on shear strain but also on time (Fig. 10).

Conclusions

A device which permits the continuous monitoring of the lateral strain of the entire specimen in the triaxial apparatus has been described. The device is suitable for saturated as well as unsaturated soils. Lateral strain measurements provide data for determining the soil characteristics essential for identifying constitutive equation parameters. The tests have shown that the end restraint effect is applicable to 15% of the specimen height on both ends.

Tests on sand, clay, peat, and gyttja have shown that these kinds of soils exhibit a great variety of strain characteristics. Especially in organic soils, volumetric strains result not only from the drained water volume but also from skeleton compressibility.

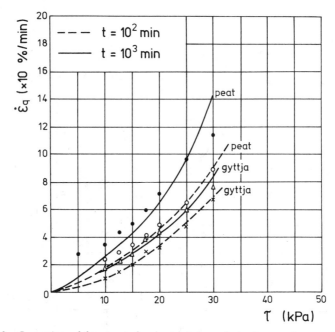

FIG. 10—*Comparison of the measured and computed correlation between the rate of* ϵ_q *and* τ.

References

[1] Cole, D. M., "A Technique for Measuring Radial Deformation During Repeated Load Triaxial Testing," *Canadian Geotechnical Journal*, Vol. 15, 1978.
[2] Félix, B., "Systeme de mesure des deformations radiales pour eprouvettes de sol," Bulletin 121, Laboratoire Central des Ponts et Chaussees, Paris, 1982.
[3] Khan, M. H. and Hoag, D. L., "A Noncontacting Transducer for Measurement of Lateral Strains," *Canadian Geotechnical Journal*, Vol. 16, 1979.
[4] N. V. Chuong, "Application of Finite Element Method to Earth Dam on Stability Analysis," Doctoral thesis, Warsaw Agricultural University, Poland, 1977.
[5] Paute, J. L., "Comportement des sols supports de Chaussees a l'appareil triaxial á chargements repeter," Bulletin 123, Laboratoire Central des Ponts et Chaussees, Paris, 1983.
[6] Wolski, W., Fürstenberg, A., and Barański, T., "Accommodation of Triaxial Apparatus to Poisson's Ratio Evaluation," in *Proceedings*, 5th Geotechnical Conference, Poland, 1978.
[7] Barański, T., "Stress-Strain Analysis of Clay Core of an Earth Dam During First Filling," Doctoral thesis, Warsaw Agricultural University, Poland, 1983.
[8] McDermott, R. J. W., "Discussion of 'Bedding Error in Triaxial Tests on Granular Media,'" *Geotechnique*, Vol. 32, 1982.
[9] Duncan, J. M. and Chang, C. Y., "Nonlinear Analysis of Stress and Strain in Soils," *Journal of the Soil Mechanics and Foundations Division*, American Society of Civil Engineers, Vol. 96, No. SM5.
[10] Desai, C. S. and Abel, J. F., *Introduction to the Finite Element Method*, Van Nostrand Reinhold, New York, 1972.
[11] Lechowicz, Z. and Szymański, A., "A Study of the Creep Behaviour of Organic Soils," in *Proceedings*, 7th Geotechnical Conference, Poland, 1984.

Masato Mikasa[1] and Naotoshi Takada[1]

Determination of Coefficient of Consolidation (c_v) for Large Strain and Variable c_v Values

REFERENCE: Mikasa, M. and Takada, N., **"Determination of Coefficient of Consolidation c_v for Large Strain and Variable c_v Values,"** *Consolidation of Soils: Testing and Evaluation, ASTM STP 892*, R. N. Yong and F. C. Townsend, Eds., American Society for Testing and Materials, Philadelphia, 1986, pp. 526–547.

ABSTRACT: Mikasa's advanced consolidation theory and Mikasa's method of determining c_v from the oedometer test are discussed. The latter consists of a curve rule method and correction of c_v by primary consolidation ratio, r. Theoretical time-consolidation relationships are shown for the cases of finite strain both with and without c_v change as the relation between time factor, T_v, and degree of consolidation in terms of settlement, U_s.

Examples on a clay and a peat show that the standard procedure of c_v determination using the curve rule method and correction by r gives practically the same f–c_v relationship as does the procedure based on the advanced consolidation theory, though a better curve fitting is always obtained by the latter.

Advanced consolidation theory, however, is necessary to interpret the physical meaning of consolidation test results and also to make a settlement prediction in the field.

KEY WORDS: consolidation, curve fitting, curve rule, oedometer test, finite strain, variable c_v, primary consolidation ratio, time-consolidation relation

Since the determination of the coefficient of consolidation, c_v, from an oedometer test is considered the reverse procedure of field settlement prediction with a given c_v value, the former, naturally, must be examined and refined as carefully and as critically as the latter on the basis of an advanced consolidation theory.

As is well-known, much progress recently was achieved in consolidation theory, extending its applicability to the cases of nonlinear stress-strain, variable permeability, finite strain, and selfweight consolidation [1–5]. "What are the effects of those factors on consolidation parameters?" is a question to be answered in dealing with consolidation in a more sophisticated way than the traditional Terzaghi theory.

This paper deals with the determination of c_v from oedometer tests on the basis of Mikasa's advanced consolidation theory. The error of the ordinary procedure

[1] Professor and Associate Professor, respectively, Civil Engineering Department, Osaka City University, Sugimoto Sumiyoshi-ku, 558, Osaka, Japan.

was examined and found not to be so serious in the standard oedometer test with load increment ratio not larger than unity, if the mean specimen height and mean c_v value in one load step of consolidation test is used instead of their initial values in that step. The treatment of secondary consolidation, an old problem, was found to be still the most important procedure in determining c_v.

This paper contains (1) Mikasa's consolidation theory as the basis of analysis of oedometer tests, and (2) two propositions on c_v determination in relation to secondary consolidation: curve rule method as an improved substitute for Casagrande's method and correction of c_v by primary consolidation ratio [2,6,7]. It also gives (3) the effect of finite strain and c_v change on time-consolidation relations calculated and illustrated in diagrams, and (4) two examples of determining c_v values for two soils with increasing and decreasing c_v values, respectively.

General Consolidation Equation for Oedometer Test

In the oedometer test the selfweight of specimen is negligibly small compared with the applied load. Therefore Mikasa's general consolidation equation [1] is reduced to

$$\frac{\partial \zeta}{\partial t} = \zeta^2 \left[c_v \frac{\partial \zeta^2}{\partial z_0^2} + \frac{dc_v}{d\zeta} \left(\frac{\partial \zeta}{\partial z_0} \right)^2 \right] \qquad (1)$$

where

t = elapsed time,
ζ = consolidation ratio $(= f_0/f$, where f denotes the volume ratio defined as (volume of soil)/(volume of soil particles) and is equal to $(1 + e))$,
f_0 = original volume ratio, and
z_0 = original coordinate or the coordinate in the "original state," in which the clay is assumed to have a certain volume ratio, f_0, throughout the layer.

Here, z_0 is measured positively in the downward direction, and f_0 is equal to the initial volume ratio that is uniformly distributed along the clay depth.

Equation 1 takes into account the changes of permeability k, volume compressibility m_v, coefficient of consolidation c_v, together with the effect of finite strain. The assumptions used in deriving Eq 1 are sevenfold: (1) clay is homogeneous, (2) clay is saturated, (3) one-dimensional consolidation occurs, (4) soil particles and water are incompressible, (5) Darcy's law is applicable, (6) f-log p and f-log k relationships are not time-dependent, and (7) the effect of selfweight of clay on consolidation is negligible.

When c_v is constant during consolidation, Eq 1 is simplified to

$$\frac{\partial \zeta}{\partial t} = c_v \zeta^2 \frac{\partial^2 \zeta}{\partial z_0^2} \qquad (2)$$

If the consolidation strain is small enough, Eq 2 is reduced to

$$\frac{\partial \epsilon}{\partial t} = c_v \frac{\partial^2 \epsilon}{\partial z^2} \qquad (3)$$

where ϵ is the compression strain and z is the real and the original coordinates at the same time measured positively downward. Rigorously, Equations 2 and 3 are equivalent and interchangeable; that is, Eq 3 is valid for any strain condition, but is not integrable for a finite strain, because depth z of a soil element changes during consolidation.

Equation 3 apparently gives the time-consolidation relation identical to that obtained from Terzaghi theory, although it allows for change in both permeability k and compressibility m_v (nonlinear stress-strain relationship) during consolidation under the condition that c_v is kept constant (k and m_v decreasing proportionally). This explains why the classical Terzaghi theory, which is based on the simple assumptions of linear stress-strain relationships and constant k values, has worked so well for the past 60 years. What we have compared with the time-settlement (not the pore pressure) curves of clays that have linear f-log p (not f–p) relations hitherto was not the Terzaghi equation but Eq 3. In this paper, therefore, clear distinction is made between the degree of consolidation in terms of strain, U_ϵ, and that in terms of stress, U_p. For a clay layer as a whole, degree of consolidation in terms of settlement, U_s, is used instead of U_ϵ defined for each element.

Standard Method of c_v Determination

Shown below are two propositions by Mikasa [1,2,6–8] in the procedure to determine the c_v value concerning the treatment of secondary consolidation in the oedometer test.

Curve Rule Method

Two curve-fitting methods have long been used to obtain c_v values from oedometer tests: the $\sqrt{t}$ method by Taylor and log t method by Casagrande. The former is reasonable enough, but the latter is not, and should be replaced by the curve rule method [2,8,9] that is based on the same principle, and gives almost the same results, as the $\sqrt{t}$ method. Casagrande's log t method is based on a different idea from, and generally gives a smaller c_v value than the other two methods (see Appendix).

Figure 1 shows a set of standard curve rules which are the theoretical time-

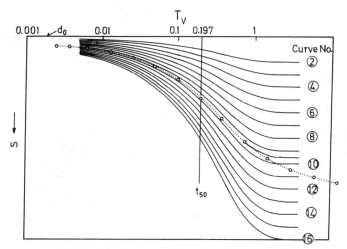

FIG. 1—*Example of curve-fitting (standard curve rule).*

settlement curves calculated from Eq 3 and plotted in different settlement scales against log T_v, where T_v ($= c_v t/(H/2)^2$) is the time factor. In practice it is drawn on a sheet of transparent paper to superimpose it on a test curve. An example of a fitted curve is shown as a dotted line in Fig. 1.

Figure 2 shows a test curve (the same as the dotted curve in Fig. 1) plotted against log t scale that has the same one log-cycle length as that in Fig. 1, the settlement scale arbitrarily taken.[2] Among the candidate curve rules drawn to fit the test curve, the longest fit curve rule numbered 9.5 defines the primary consolidation for this test curve.[3] All curves will fit the test curve at least on its earlier portion, so far as the test curve has a normal linear s–$\sqrt{t}$ relation in its earlier part as the theory indicates. It should be noted that only the longest fit curve rule can share the point of 90% degree of consolidation, the key point in $\sqrt{t}$ curve fitting method, with the test curve.[4]

Casagrande's fitting method that marks the beginning of secondary consolidation on the backward extension of the linear secondary consolidation trail always selects a curve rule of larger ordinate length versus the best fit one (10.5 or so in this example), thus yielding a larger t_{50} value and, consequently, a smaller c_v value than the other two methods, because it identifies the transient portion of consolidation curve between primary and secondary consolidation as its primary consolidation, contradicting the essential concept of primary consolidation that it should fit the theoretical consolidation curve (see Appendix and Fig. 15).

[2] If the test curve is drawn, for example, on the ½ settlement scale of Fig. 2, the best fit curve rule will be 4.75 (actually, 4.5 or 5.0 will work sufficiently well).
[3] In the fitting procedure, one must be careful not to tilt the curve rule against the test curve.
[4] In routine practice it is recommended to draw this best fit curve rule on the test plot to show how the test curve agrees with or deviates from the theoretical curve (Figs. 11 and 12).

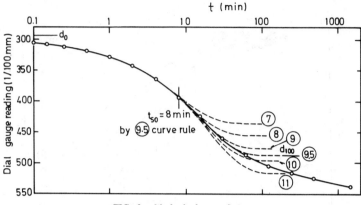

FIG. 2—*Method of curve-fitting.*

Once the best fit curve rule is found, we can read off the corrected initial reading d_0, final reading of primary consolidation d_{100} and 50% consolidation time t_{50} directly and simultaneously. Then c_v is calculated using this t_{50} from the equation

$$c_v = T_{50} \frac{(H_m/2)^2}{t_{50}} = 0.197 \frac{(H_m/2)^2}{t_{50}}$$

where the thickness of specimen H_m is the mean value during primary consolidation; and $H_m = (H_0 - \Delta H/2)$ in which H_0 is the initial clay thickness, and ΔH is the settlement of primary consolidation in that load step. The example in Fig. 2 gives $t_{50} = 8$ min, $H_m = 1.860 - (0.485 - 0.294)/2 = 1.765$ cm and $c_v = 0.0192$ cm^2/min.

Correction by Primary Consolidation Ratio

In the ordinary consolidation analysis the coefficient of consolidation c_v is taken as

$$c_v = \frac{k}{m_v \gamma_w} \tag{4}$$

in which m_v denotes the coefficient of volume compressibility for 24 hours' consolidation including secondary consolidation. The coefficient of consolidation obtained from the curve fitting shown in the previous section, however, is written as

$$c_{v1} = \frac{k}{m_{v1} \gamma_w} \tag{5}$$

where c_{v1} and m_{v1} denote the coefficient of consolidation and coefficient of volume compressibility, both corresponding to the primary consolidation, respectively, because any curve fitting procedure to define the primary consolidation in the test curve assumes an elastic consolidation model (Terzaghi model) that has the compressibility of m_{v1}. Since the c_v value obtained by a curve fitting is the $\bar{c}_{v1}$ of this elastic model, the value of k of the model is found as

$$k = c_{v1} m_{v1} \gamma_w$$

This k value is duly assumed as the k value of the clay at the same loading step whether in an oedometer or in the field; that is, the k value is assumed to be common in Eqs 4 and 5.

As a result of the above discussion, c_v values by Eqs 4 and 5 are found to be different by the ratio of $m_{v1}/m_v = r$, which is the primary consolidation ratio, generally less than unity.

Conclusively, the c_v value to be used in the field needs be corrected as follows:

$$c_v = \frac{k}{m_v \gamma_w} = \frac{k}{m_{v1} \gamma_w} \times \frac{m_{v1}}{m_v} = c_{v1} \times r$$

This correction, which has been utterly neglected heretofore, is a minimum requirement to secure the compatibility between the oedometer test and the field consolidation in dealing with the secondary consolidation, in so far as the curve fitting is considered valid at all.

The tentative code by the Japan Society of Soil Mechanics and Foundation Engineering (JSSMFE) (1969–1981) defined c_v and c_{v1} as in Eqs 4 and 5. The present Japan Industrial Standard (JIS) code, however, rejected the symbol c_{v1} and adopted the following symbols:

$$c_v = \frac{k}{m_{v1} \gamma_w}$$

$$c'_v = c_v \times r \left(= \frac{k}{m_v \gamma_w} \right) \tag{6}$$

This alteration was made to avoid confusion between the two c_v's with and without the correction by r. We admit that this is a necessary measure in the present situation without international consensus on this issue, though another confusion may occur between the original definition of c_v by Eq 4 and a new definition by Eq 6. In the following, the present paper adopts the symbols of Eq 6.

In the example of Fig. 2, $r = (0.485 - 0.294)/0.244 = 0.782$. Thus we obtain

$$c'_v = 0.0192 \text{ cm}^2/\text{min} \times 0.782 = 0.0150 \text{ cm}^2/\text{min}$$

Comments on c_v *Determination*

The procedure of getting c_v value described in the two preceding sections was proposed by the present senior author (Mikasa) in 1960 and was adopted as a tentative testing code of JSSMFE for eleven years beginning in 1969. The JIS code of consolidation test established in 1980, however, did not adopt the *r*-correction officially for some practical reasons, such as the field evidence to support this correction was not sufficient, and international consensus was not yet obtained on this point. However, we have consistently performed this correction in our case studies in field and laboratory for 25 years, and finding its validity in every case [1,2,10,11], are convinced of its necessity as a practical procedure. The curve-fitting, of course, must be performed by either the curve rule method or $\sqrt{t}$ method, both of which are already adopted in JIS.

It will be worthwhile to note here that the two recommended fitting methods combined with the correction by primary consolidation ratio *r* sometimes yield a c_v value similar to those by Casagrande's method without the *r*-correction. This, of course, does not justify the traditional Casagrande method, because it is merely a coincidence and cannot always be expected to occur.

A moral of the above story is that the curve rule method and $\sqrt{t}$ method, though better in their principle than Casagrande's, may yield poorer results than the latter if the *r*-correction is neglected.

The reader may request here some field evidences to support the newly proposed method. The aforementioned literature can be used as support, but clear-cut evidence of this single problem is difficult to seek in field case histories, which are always accompanied by a lot of difficulties both in ground and loading conditions. Therefore, direct verification of the validity of the newly proposed procedure should be sought in elaborately performed laboratory tests. Examples in the section entitled "The Case of Finite Strain and Variable c_v" will serve this purpose, we hope.

The Case of Finite Strain and Constant c_v

Figure 3 shows the normalized time-consolidation curves; that is, the relation between the degree of consolidation in terms of settlement, U_s, and the time factor, $T_{v0} (= c_v t/(H_0/2)^2)$, for the case of finite strain and constant c_v calculated from Eq 2, final strain $\bar{\epsilon}_f$ being 0.1, 0.2, 0.3, 0.4, and 0.5 under double drainage conditions [1],[5] together with the curve of $\bar{\epsilon}_f = 0$ obtained from the solution of

[5] $\bar{\epsilon}$ is the nominal (arithmetic) strain defined as

$$\bar{\epsilon} = \int d\bar{\epsilon} = \int_{f_0}^{f} \frac{-df}{f_0} = \frac{f_0 - f}{f_0} = 1 - \frac{1}{\zeta}$$

whereas ϵ in Eq 3 is the natural (logarithmic) strain defined as

$$\epsilon = \int d\epsilon = \int_{f_0}^{f} \frac{-df}{f} = \log_e \frac{f_0}{f} = \log_e \zeta$$

We must make a clear distinction between these two strains in finite strain problems.

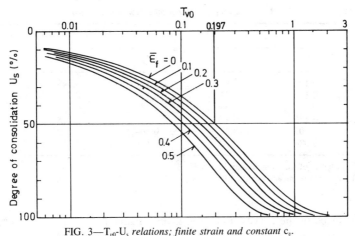

FIG. 3—T_{v0}-U_s relations; finite strain and constant c_v.

Eq 3. Equation 2 was numerically calculated by using a finite-difference method, in which the number of nondimensional depth mesh, n ($= H_0/\Delta z_0$), is taken as 40 and the nondimensional time mesh, ΔT_{v0}, is so chosen that the value of $(n^2 \Delta T_v \zeta_f^2)/4$ would not exceed ¼, where ζ_f is the final ζ value corresponding to $\bar{\epsilon}_f$, to secure the stability of calculation [11]. Table 1 gives the numerical T_{v0}-U_s relations.

Since the time factor T_{v0} in Fig. 3 is defined using the initial specimen height H_0, instead of the mean thickness H_m, which is common to all curves irrespective of the magnitude of $\bar{\epsilon}_f$, differences in T_{50}, for example, show the differences in real 50% consolidation time t_{50}. The values of T_{50}, T_{90}, and T_{90}/T_{50} of these curves shown in Table 2 will serve as the indices to compare the consolidation times

TABLE 1—Relation between T_{v0} and U_s (%) for finite strain.

Time Factor (T_{v0})	$\bar{\epsilon}_f = (f_0 - f_f)/f_0$					
	0^a	0.1^b	0.2	0.3	0.4	0.5
0.005	8.0	8.6	9.3	10.1	11.1	12.3
0.01	11.3	12.2	13.1	14.2	15.6	17.4
0.02	16.0	17.1	18.4	20.0	22.0	24.5
0.05	25.2	26.9	29.1	31.6	34.9	38.8
0.1	35.7	38.2	41.1	44.6	49.1	54.7
0.15	43.7	46.7	50.3	54.6	60.0	66.9
0.2	50.4	53.9	58.0	62.9	69.0	76.6
0.4	69.8	74.1	79.0	84.5	90.2	95.7
0.6	81.6	85.7	89.9	94.0	97.4	99.5
0.8	88.7	92.1	95.2	97.7	99.3	
1.0	93.1	95.7	97.8	99.2		
1.5	98.0	99.1	99.7			

$^a\bar{\epsilon}_f = 0$: analytical solution.
$^b\bar{\epsilon}_f = 0.1 \sim 0.5$: numerical solution.

TABLE 2—T_{50} and T_{90} for consolidation of finite strain.

$\bar{\epsilon}_f$	$T_{v0}{}^a$		$T_{vm}{}^b$		T_{90}/T_{50}
	T_{50}	T_{90}	T_{50}	T_{90}	
0	0.197	0.848	0.197	0.848	4.31
0.1	0.172	0.720	0.191	0.798	4.19
0.2	0.148	0.599	0.183	0.740	4.05
0.3	0.126	0.490	0.174	0.678	3.89
0.4	0.105	0.397	0.164	0.620	3.78
0.5	0.084	0.309	0.148	0.549	3.68

$^aT_{v0}$: T_v in terms of initial specimen height H_0.
$^bT_{vm}$: T_v in terms of mean specimen height H_m.

and the shape of the curves. Larger $\bar{\epsilon}_f$ gives smaller T_{50} and T_{90}; T_{90} decreasing more rapidly than T_{50}, the ratio T_{90}/T_{50} decreases with the increase of $\bar{\epsilon}_f$, changing the shape of the time-consolidation curves.

It is clear, therefore, that the effect of finite strain cannot be attributed solely to the decrease of specimen height,[6] which is usually taken into consideration by using the mean height H_m in the definition of time factor T_v as shown in the example in the Curve Rule Method section. Figure 4 shows the five curves in Fig. 3 redrawn against another time factor $T_{vm} = c_v t/(H_m/2)^2$ defined in terms of H_m. The curves change their positions laterally and come nearer to the standard curve of $\bar{\epsilon}_f = 0$, but do not coincide with it, still showing smaller T_{50} and T_{90} values than the standard curve as shown in the column headed T_{vm} in Table 2. From Fig. 4 and Table 2, we find that the error induced by the standard procedure with the mean specimen height is not serious, but cannot be neglected in precise analysis. If we shift these curves further to get common T_{50} value of 0.197, we will find that the curves in the time range $T_v < T_{50}$ virtually coincide with each other, but do not in the range $T_v > T_{50}$, in a somewhat similar way to Fig. 10.

The detailed procedure of c_v determination in the case of finite strain is the same as explained in the section on Standard Method of c_v Determination, except that curve rules should be drawn on the basis of the curves in Fig. 3 or Fig. 4 when H_0 or H_m is used, respectively (these are different only in T_{50} values). Curve-fitting examples are omitted in this section.

The Case of Finite Strain and Variable c_v

We shall discuss here the cases in which both finite strain and variable c_v are to be considered. The value of c_v increases during consolidation when m_v decreases at a rate higher than k, and decreases when k decreases at a rate higher than m_v. Generally speaking, in the normally consolidated region, ordinary inorganic clays show increasing, and peats decreasing, c_v values during consolidation [12].

---

[6] If it could, curves would not change their shape, but only shift along the time axis for different compression strains.

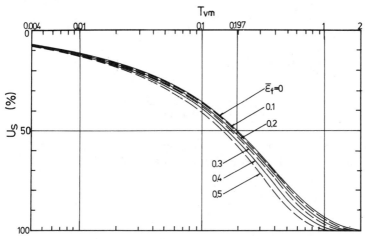

FIG. 4—T_{vm}-U_s *relations; finite strain and constant* c_v.

Theoretical Time-Consolidation Relation [13]

We shall take up the case in which the change of c_v in the normally consolidated region of a clay is linear on an f-log c_v plot as illustrated in Fig. 5. Then the relation between c_v and f is expressed as

$$f = f_0 - C_{cv}(\log c_v - \log c_{v0})$$

where C_{cv} is the index of c_v change defined as the slope of the f-log c_v line, and c_{v0} is the c_v value corresponding to f_0, the initial f value. Introducing a ratio $\delta = c_{vf}/c_{v0}$ where c_{vf} corresponds to the minimum final volume ratio, f_f, at the previous boundary, C_{cv} is expressed as

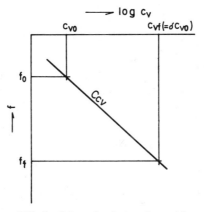

FIG. 5—*Schematic relation of* c_v *and* f.

$$C_{cv} = \frac{f_0 - f_f}{\log \delta}$$

Putting $c_v = c_{v0} \, \phi(\zeta)$, the function $\phi(\zeta)$ is expressed as

$$\phi(\zeta) = \exp \frac{(1 - 1/\zeta) \log \delta}{0.4343(1 - 1/\zeta_f)}$$

where ζ_f is the final consolidation ratio corresponding to the final volume ratio f_f. Taking the derivative with respect to ζ, we obtain

$$\frac{d\phi(\zeta)}{d\zeta} = \frac{\log \delta}{0.4343(1 - 1/\zeta_f)} \cdot \frac{\phi(\zeta)}{\zeta^2}$$

Equation 1 can be calculated numerically by transforming it into a finite difference equation and using the above functions for any value of $\bar{\epsilon}_f$ and δ. In the finite difference calculation the number of nondimensional depth mesh, n, was taken as 40 and the nondimensional time mesh ΔT_v was so chosen that the value of $(n^2 \Delta T_v \zeta_f^2 \delta)/4$ would not exceed $\frac{1}{4}$ [11].

Figures 6 to 8 show normalized time-consolidation curves (T_{v0}–U_s relations) calculated from Eq 1 for finite strain ($\bar{\epsilon}_f = 0 \sim 0.5$) and variable c_v ($\delta = 10 \sim 0.1$), where the time factor is defined as $T_{v0} = c_{v0}t/(H_0/2)^2$. Among them the curves of $\delta = 1$ are the same as those in Fig. 3. The values of T_{50} are shown in Figs. 6 to 8. Figure 9 shows T_{50} and T_{90} plotted against δ.

Figure 10 compares the shapes of U_s–log T_v curves with different δ in the case of $\bar{\epsilon}_f = 0$ by shifting the curves in Fig. 6 laterally so that they share the point of $T_{50} = 0.197$. In the time range of $T_v < T_{50}$, the curves almost coincide with each other, showing that all curves have almost a linear s–$\sqrt{t}$ relation in this

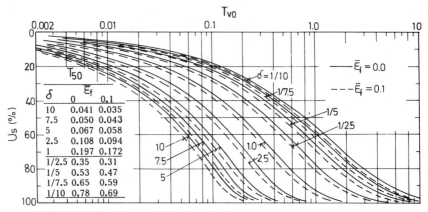

FIG. 6—T_{v0}-U_s relations and T_{50} values for finite strain and variable c_v; $\bar{\epsilon}_f = 0, 0.1$.

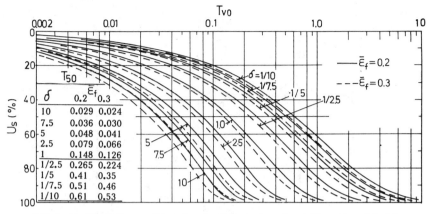

FIG. 7—T_{v0}-U_s relations and T_{50} values for finite strain and variable c_v; $\bar{\epsilon}_f$ = 0.2, 0.3.

part, while in the time range after T_{50}, large δ accelerates the consolidation conspicuously. Similar tendency is seen for all curves of $\bar{\epsilon}_f \neq 0$.

Table 3 gives the acceleration factors, or the ratios $T_{50}(\delta = 1)/T_{50}(\delta)$ for $\bar{\epsilon}_f$ = 0, 0.1, and 0.2 (Figs. 6 and 7), which are used to divide the t_{50} values obtained by the equation

$$t_{50} = \frac{T_{50}(\delta = 1)(H_0/2)^2}{c_{v0}}$$

without any consideration of the change in c_v value during consolidation in order to get the correct time of 50% consolidation in settlement prediction.

If we use the mean c_v value during the consolidation, c_v, instead of c_{v0}, then this correction may not be necessary. To check this point, this table also provides two sorts of mean values of 1 and δ: geometric (logarithmic) mean value $\sqrt{1 \times \delta}$ and arithmetic mean value $(1 + \delta)/2$.

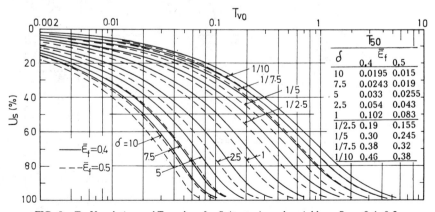

FIG. 8—T_{v0}-U_s relation and T_{50} values for finite strain and variable c_v; $\bar{\epsilon}_f$ = 0.4, 0.5.

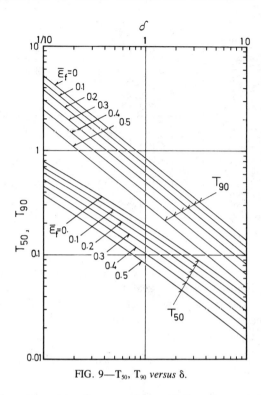

FIG. 9—T_{50}, T_{90} versus δ.

Comparing these two sorts of mean values with the above acceleration factors, we find that it may be practically admissible to use the arithmetic mean value of c_{v0} and c_{vf} (= δ c_{v0}) for the case of increasing c_v, and their logarithmic mean value for the case of decreasing c_v, to get the time of 50% consolidation, t_{50}, instead of applying the rigorous solution shown in Figs. 6 to 8. This approximation, however, is only valid for the prediction of the t_{50} value.

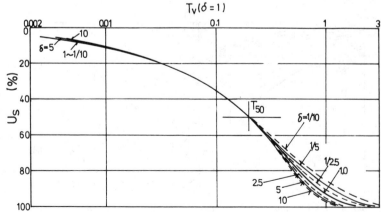

FIG. 10—*Comparison of shapes of time-consolidation curves with variable* c_v ($\bar{\epsilon}_f = 0$).

TABLE 3—*Acceleration factors for t_{50} values in consolidation with variable c_v.*

	Acceleration Factor $T_{50} (\delta = 1)/T_{50}(\delta)$				
	$\bar{\epsilon}_f$			Mean Value	
δ	0	0.1	0.2	$\sqrt{1 \times \delta}$	$\dfrac{1 + \delta}{2}$
10	4.80	4.91	5.10	3.16	5.50
7.5	3.94	4.00	4.11	2.74	4.25
5	2.94	2.97	3.08	2.24	3.00
2.5	1.82	1.83	1.87	1.58	1.75
1	1	1	1	1	1
1/2.5	0.563	0.555	0.558	0.632	0.700
1/5	0.372	0.366	0.361	0.447	0.600
1/7.5	0.303	0.292	0.290	0.375	0.567
1/10	0.253	0.249	0.243	0.316	0.550

Examples of c_v Determination

Two materials are chosen as the examples of c_v determination. One is a mixture of highly plastic alluvial clay ($w_L = 105.0\%$) from Osaka Bay and kaolinite powder ($w_L = 46.0\%$) at a dry weight ratio of 1:1, and the other is a peat taken from a swamp in Shizuoka prefecture, which consists mostly of decayed grass fiber (mainly reed), the ignition loss approximately 80%. The former was chosen as a representative of ordinary inorganic clay that has moderately increasing c_v, and the latter as an example in which c_v decreases conspicuously by consolidation.

The clay-kaolinite mixture was remolded with seawater to a slurry with an initial water content of about 100% and de-aired in a vacuum chamber. This slurry specimen, 10 cm diameter and 4 cm thick, was consolidated stepwise under double drainage conditions, starting from a low load of 0.44 tf/m² with an approximate load increment ratio of $\Delta p/p = 1$ ($p_f/p_0 = 2$) in the earlier five steps, and with $\Delta p/p = 2$ ($p_f/p_0 = 3$) in the last two steps. The t-s curves of the last two steps are shown in Fig. 11.

The peat was very fibrous and heterogeneous in its undisturbed state. Therefore to obtain a homogeneous and isotropic specimen it was moderately remolded in a home-use washing machine with an adequate amount of water to get sufficient workability. It was then poured into a consolidation ring of 20 cm diameter with an impermeable base to a certain depth so that the specimen would be about 5 cm thick after preconsolidated by the weight of a previous loading plate at a pressure of 0.089 tf/m². The specimen was then consolidated stepwise with a load increment ratio of $\Delta p/p = 3$ ($p_f/p_0 = 4$) under single drainage conditions. Figure 12 shows t-s curves at two loading steps.

For the t-s curves of clay-kaolinite mixture (Fig. 11), curve rules that take into account finite strain and variable c_v were applied together with the standard curve rules in Fig. 1. The curve rule of $\bar{\epsilon}_f = 0.1$ with $\delta = 1.5$ (interpolated from the curves in Fig. 6) showed the longest fit for both test curves of $p = 10 \rightarrow 30$ tf/

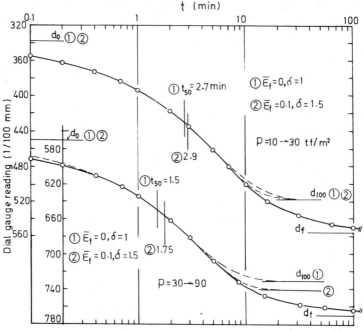

FIG. 11—*Examples of clay-kaolinite mixture curve-fitting.*

m² and $p = 30 \to 90$ tf/m². The values of d_0 were virtually identical with any curve rule (this is also the case for most soils). The values of d_{100} and t_{50} by both the best and the standard fittings happened to be similar for the curve of $p = 10 \to 30$ tf/m² and different for the curve of $p = 30 \to 90$ tf/m². Using the values of t_{50}, d_0, d_{100}, and the initial height H_0, the initial c'_v value, c'_{v0}, in that consolidation step is obtained by

$$c'_{v0} = T_{50} \frac{(H_0/2)^2}{t_{50}} \times r$$

where T_{50} takes different values for different curve rules as shown in Figs. 6 to 8, and r is the primary consolidation ratio.

The results are given in Table 4. The following is an example to obtain the values in Table 4 for the loading step $p = 10 \to 30$ tf/m², where H_0 is 2.299 cm. For the best-fit curve rule ($\bar{\epsilon}_f = 0.1$, $\delta = 1.5$, $T_{50} = 0.132$), the values of d_0, d_{100}, final reading d_f and t_{50} are read as 0.338 cm, 0.518 cm, 0.556 cm, and 2.9 min, respectively. The primary consolidation ratio, r, is $(0.518 - 0.338)/(0.556 - 0.338) = 0.826$. Therefore, the initial c'_v value, c'_{v0}, is calculated as

$$c'_{v0} = T_{50} \frac{(H_0/2)^2}{t_{50}} r = 0.132 \frac{(2.299/2)^2}{2.9} \times 0.826 = 0.0497 \text{ cm}^2/\text{min}$$

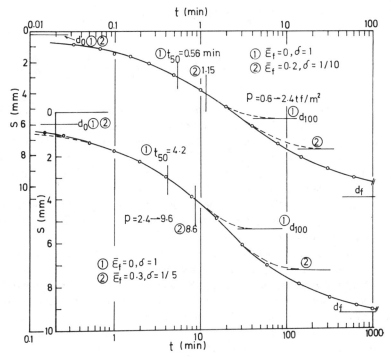

FIG. 12—*Examples of remolded peat curve-fitting.*

The mean c'_v value, $\bar{c}'_v$, is

$$\bar{c}'_v = c'_{v0} \frac{1 + \delta}{2} = 0.0497 \frac{1 + 1.5}{2} = 0.0621 \text{ cm}^2/\text{min}$$

For the standard curve rule ($T_{50} = 0.197$), t_{50} is 2.7 min. The mean specimen height for primary consolidation, H_m, is $H_0 - \Delta H/2 = 2.299 - (0.518 - 0.338)/$

TABLE 4—*Results of clay curve-fitting.*

		Mean c_v Value	
Consolidation Pressure (p), tf/m²	Initial Value (c'_{v0}), cm²/min	By Fitted Curve (c'_{vm}), cm²/min	By Standard Curve (c'_v), cm²/min
10 ↓ 30	0.0497	0.0621	0.0735
30 ↓ 90	0.0691	0.0863	0.105

$2 = 2.209$ cm, and r, accidentally, is the same as above. c'_v is then obtained as

$$c'_v = T_{50} \frac{(\bar{H}/2)^2}{t_{50}} = 0.197 \frac{(2.209/2)^2}{2.7} \times 0.826 = 0.0735 \text{ cm}^2/\text{min}$$

In this case, the value of c'_v by the standard curve rule method is 12% to 22% larger compared with $\bar{c}'_v$. This error can not be neglected in precise analyses.

For the two curves of remolded peat (Fig. 12), two curve fittings, the best fit and the standard, are shown. The best-fit curve rules are ($\bar{\epsilon}_f = 0.2$, $\delta = \frac{1}{10}$) for $p = 0.6 \to 2.4$ tf/m², and ($\bar{\epsilon}_f = 0.3$, $\delta = \frac{1}{5}$) for $p = 2.4 \to 9.6$ tf/m². Two fittings yielded conspicuously different d_{100} and t_{50} values. The part of the test curve which is considered the incipient part of the secondary consolidation by the standard curve rule fitting is found to be still a part of the primary consolidation by best-fit curve rules.

Figure 13 shows c'_v values of the peat in four load steps obtained in the following several ways and plotted against corresponding f values: the initial c'_v value, c'_{v0}, plotted against f_0; the final c'_v value in the primary consolidation, c'_{vf} $(= \delta c'_{v0})$, plotted against the volume ratio f_1 at the end of primary consolidation; the mean c'_v value obtained both by calculation of $\sqrt{1 \times \delta}\, c'_{v0}$, $\bar{c}'_v$, and by standard curve rule fitting, c'_v, plotted against the mean volume ratio both corresponding to respective 50% consolidation. The values of δ in the above explanation are those of curve rules. The mean c'_v values calculated by $\sqrt{1 \times \delta}\, c'_{v0}$ are at the midpoints between the two points, c'_{v0} and $\delta c'_{v0}$. The values $\delta c'_{v0}$ as well as $\sqrt{1 \times \delta}\, c'_{v0}$ are very near to the f-c'_{v0} line, showing the overall success in curve fitting.

It is also remarkable that the mean c'_v values obtained by the standard curve rule in the same consolidation step (though the t_{50} values are apparently incorrect

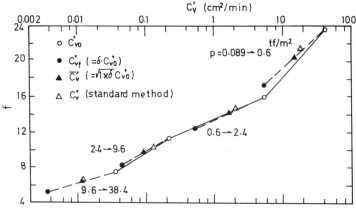

FIG. 13—c_v-f relation by different ways; remolded peat.

as shown in Fig. 12) fall on the f-log c'_v line obtained by the best fit curve rules that are based on more sophisticated consolidation theory with due consideration of finite strain and variable c_v value, provided that all c_v values are corrected by multiplying primary consolidation ratio r and are plotted against corresponding f values.

Procedure

Following are the recommended procedures in c_v determination and settlement analysis when the clay thickness and c_v value change considerably.

Research

For research purposes, the change in both specimen height and c_v value should be duly taken into account in dealing with the oedometer test results not only to obtain the correct c_v value, but also to know the physical meaning of the test data. For example, the peat shown in Fig. 12 has fairly large primary consolidation that follows the theoretical curve by the advanced elastic consolidation theory, whereas the standard curve rule gives an impression that almost half of the consolidation of this peat is the secondary compression. Thus a rigorous application of the advanced consolidation theory is necessary for research purposes, not only in the settlement prediction, but also in the curve fitting and c_v determination from oedometer tests.

Routine

In c_v determination in routine work, the examples in this paper give us an optimistic view on the ordinary oedometer test with load increment ratio of unity in that we can obtain satisfactorily correct c_v values by the standard curve rule method based on the simplest heat conduction type equation, provided that the mean specimen height H_m is used to obtain the mean c_v value during consolidation, (one of the two sorts of mean values of c_v in Table 4), and that the obtained c_v value is corrected by multiplying it by the primary consolidation ratio r. The corrected c_v value, c'_v, should then be plotted for the volume ratio at 50% consolidation. (In this case, ordinary plot of c_v value against the consolidation pressure p is not appropriate.)

In routine settlement analysis, a different approach is necessary to get tolerable results. In getting c_v from an oedometer test, we only deal with 50% consolidation time. But for the settlement prediction in the field, we must estimate the whole time-settlement relation, which is obtained only from the advanced consolidation theory that takes into account the change in both thickness of the clay layer and c_v value during consolidation and sometimes even the effect of selfweight of the clay. The curves shown in Figs. 6, 7, and 8 will serve as the solutions of time-settlement relation in the field when the effect of selfweight is not important. It should be noted that in using these figures the clay thickness H and c_v values should be taken as their initial values, H_0 and c_{v0}, the latter of which can be read off from the f-log c_v line such as in Fig. 13.

Conclusions

1. Mikasa's general consolidation theory and Mikasa's method of c_v determination are explained. The latter consists of curve fitting by curve rule method to be used in place of Casagrande's method, and correction of the c_v value by multiplying primary consolidation ratio r.

2. Time-consolidation curves for various finite strain are shown, and the effect of finite strain is discussed quantitatively. The use of the mean specimen height in calculating c_v value is shown to reduce the error to a tolerably small one.

3. Time-consolidation curves of various combinations of finite strain and change in c_v during consolidation are shown, and their characteristics are discussed. It was shown numerically that the change in c_v can be evaluated by using the mean c_v value in the calculation instead of its initial value, c_{v0}.

4. Examples are shown to deal with the time-settlement curves of oedometer test on two soils, a clay-kaolinite mixture and a remolded peat. The specimens are consolidated stepwise at a large load increment so that a large consolidation strain and a marked change in c_v are produced. Curve fitting is tried both by the standard curve rule and by the curve rule that considers finite strain and variable c_v. The results illustrate that the c_v values obtained by both methods satisfactorily agree with each other, provided that the mean specimen height be used to obtain mean c_v value in that consolidation step and that the correction by primary consolidation ratio be performed.

5. Recommended procedures in c_v determination and in settlement prediction are summarized for research purposes and for routine work. Generally speaking, there is no reason to evade the due application of the advanced consolidation theory. In obtaining f-log c_v relation from an ordinary oedometer test, however, the standard curve rule method can duly be applied as explained in the Examples in c_v Determination section. The settlement analysis, which was found not to be just the reverse operation of c_v determination from an oedometer test, should be calculated by the advanced theory when the clay generates large consolidation strain or substantial change in c_v or both.

APPENDIX

Why Casagrande's Method Should be Replaced by Curve Rule Method for Curve Fitting to Obtain c_v

Let us consider first the mechanism of secondary consolidation fundamentally. To simplify the discussion, we shall assume first a model of clay structure composed of two elastic springs and one dashpot that has high viscosity, combined as shown in Fig. 14. This model, a series connection of a spring and a Voigt model, was first proposed by Ishii [14], and can satisfactorily explain the phenomenon of secondary consolidation, at least qualitatively.

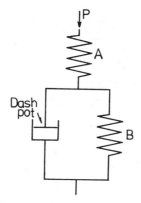

FIG. 14—*Ishii's consolidation model.*

In the early stage of consolidation, only Spring A will deform with the effective stress increase, causing "primary consolidation" that follows elastic consolidation theory. When the primary consolidation proceeds, Spring B will reluctantly be compressed to give "secondary consolidation." The later stage of primary consolidation and the early stage of secondary consolidation, therefore, will merge into each other, yielding a transient region between the pure primary and pure secondary consolidation, the latter of which is to proceed almost linearly against logarithm of time (though the simple model in Fig. 14 itself does not show a linear secondary consolidation). In this transient region, substantial excess pore water pressure should still remain to make the clay consolidate at a rate higher than the elastic consolidation model.

From the foregoing considerations we can conclude:

1. For curve fitting to obtain a c_v value, we assume an elastic model (Terzaghi model) that fits the primary consolidation of the test curve and has a compressibility of m_{v1}.

2. A test curve that fits a theoretical curve until about 90% degree of primary consolidation will deviate from and run below the theoretical curve in its later stage having a transient region to secondary consolidation, thus producing an apparent increase of compressibility from m_{v1} to m_v that corresponds to consolidation in 24 h.

3. Thus we can define the secondary consolidation as the process in which m_v value increases gradually under constant or somewhat increasing effective stress by the readjustment of soil grains or the "degradation of soil structure," which is expressed mathematically by Ishii's model or others.

4. The above conclusions are valid even if the Voigt model in Ishii's model is replaced by other types of mathematical models to represent the visco-elastic or visco-plastic behavior of the secondary consolidation, insofar as the model has an elastic Spring A representing the primary consolidation in series connection with that visco-elastic (plastic) model.

5. Conclusions 1, 2, and 3 provide a general guiding principle in dealing with the secondary consolidation in an oedometer test, which is summarized as "the test curve should always deviate from the primary consolidation curve in its lower side, never crossing it". The fitting technique of curve rule shown in Fig. 2 and well known $\sqrt{t}$ method is based on this principle, whereas Casagrande's method is not. It therefore gives a theoretical fitting curve that comes below the test curve and cross it as in the typical illustration of Fig. 15 [15], clearly showing a contradiction with the above guiding principle.

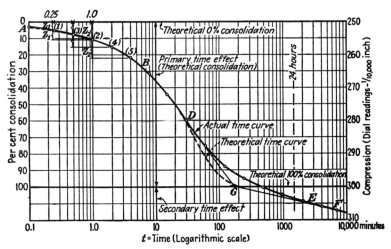

FIG. 15—*Graphic analysis of a time-compression curve obtained from the dial readings of one load stage during a consolidation test (after Casagrande and Fadum, 1964).*

References

[1] Mikasa, M., "The Consolidation of Soft Clay—A New Consolidation Theory and its Application," *Kajima Shuppan-Kai*, 1963 (in Japanese).

[2] Mikasa, M. and Ohnishi, H., "Soil Improvement by Dewatering in Osaka South Port, Geotechnical Aspects of Coastal Reclamation Project in Japan," in *Proceedings*, 9th International Society of Soil Mechanics and Foundation Engineering, Case History Volume, 1981, pp. 639–664.

[3] Gibson, R. E., England, G. L., and Hussey, M. J. L., "Theory of One-Dimensional Consolidation of Saturated Clays," *Geotechnique*, Vol. 17, 1967, pp. 261–273.

[4] Gibson, R. E. and Schiffman, R. L., "The Theory of One-dimensional Consolidation of Saturated Clays, II Finite Nonlinear Consolidation of Thick Homogeneous Layers," *Canadian Geotechnical Journal*, Vol. 18, 1981, pp. 280–293.

[5] Monte, J. L. and Kriezek, R. J., "One Dimensional Mathematical Model for Large Strain Consolidation," *Geotechnique*, Vol. 26, No. 3, 1976, pp. 495–510.

[6] Mikasa, M., "Testing of Mechanical Properties of Soils," Japan Society of Civil Engineers, Kansai Branch, 1960 (in Japanese).

[7] Mikasa, M., "On the Calculation of Parameters from Oedometer Test" in *Proceedings*, 19th Annual Convention of the Japan Society of Civil Engineers, 1964 (in Japanese), p. III-6.

[8] Committee of Shear Testing, "Tentative Method of Test for Consolidation of Soils," Japan Society of Soil Mechanics and Foundation Engineering, 1969 (in Japanese).

[9] Japanese Industrial Standard (JIS), "Test Method for Consolidation of Soils," 1980.

[10] Mikasa, M. and Takada, N., "Investigation of Settlement of Kobe Port Island," in *Proceedings*, Symposium on Sedimentation/Consolidation Models, American Society of Civil Engineers, 1984, pp. 481–500.

[11] Takada, N. and Mikasa, M., "Consolidation of Multi-layered Clay," in *Proceedings*, Symposium on Sedimentation/Consolidation Models, American Society of Civil Engineers, 1984, pp. 216–228.

[12] Taylor, D. W., *Fundamentals of Soil Mechanics*, Wiley, New York, 1948, p. 244.

[*13*] Takada, N., "Consolidation Process Associated with Variable Coefficients of Consolidation and Decreasing Thickness of Layers," *Tsuchi-to-Kiso*, Vol. 27, No. 11, 1979, pp. 61–64 (in Japanese).
[*14*] Ishii, Y., General Discussion, Symposium on Consolidation Testing of Soils, American Society of Civil Engineers, *Test Materials*, No. 123, 1951, p. 103.
[*15*] Tschebotarioff, G. P., *Soil Mechanics, Foundations, and Earth Structure*, McGraw-Hill, New York, 1952, p. 113.

Naotoshi Takada[1] and Masato Mikasa[1]

Determination of Consolidation Parameters by Selfweight Consolidation Test in Centrifuge

---

REFERENCE: Takada, N. and Mikasa, M., **"Determination of Consolidation Parameters by Selfweight Consolidation Test in Centrifuge,"** *Consolidation of Soils: Testing and Evaluation, ASTM STP 892*, R. N. Yong and F. C. Townsend, Eds., American Society for Testing and Materials, Philadelphia, 1986, pp. 548–566.

ABSTRACT: Consolidation parameters such as volume compressibility, m_v, and permeability, k, of very soft slurry clay are difficult to obtain from a conventional oedometer test. A selfweight consolidation test in a centrifuge is available as an alternative test. The m_v value is calculated from the f log p relationship which is obtained from a correlation between volume ratio f and effective overburden pressure along the specimen depth after self-weight consolidation, and k is obtained from the initial settlement rate of a singly drained selfweight consolidation test. Some examples of determination of these parameters are shown for four highly plastic clays. The effect of particle segregation on these parameters in a high centrifugal acceleration field is also discussed.

KEY WORDS: centrifuged model, consolidation parameters, particle segregation, self-weight consolidation, soft clay

Since the conventional oedometer test of a very soft clay is not easy to perform, several alternative methods have been proposed, such as the seepage consolidation test [1] and the constant-rate consolidation test [2]. The selfweight consolidation test, especially in a centrifuge, is basically a model test to simulate the field behavior of filled clay slurry. It can also be utilized as a method to obtain the consolidation parameters of this type of soft clay; that is, the f-log p relation from the water content distribution along the depth after selfweight consolidation, and permeability, k, from the initial settlement rate of selfweight consolidation under single drainage conditions. The coefficient of consolidation, c_v, is then synthesized from k and compressibility, m_v, the latter of which is calculated from the f-log p relation, both at a given volume ratio. This method is based on Mikasa's consolidation theory [3,4].

This report presents some examples of the determination of consolidation parameters of clay slurry by selfweight consolidation tests. They were conducted

---

[1] Associate Professor and Professor, respectively, Civil Engineering Department, Osaka City University, Sugimoto Sumiyoshi-ku, 558, Osaka, Japan.

either as a practical task to predict the settlement of a dredged clay fill [5], or as a research project to investigate the mechanism of selfweight consolidation [6]. Because they were done as three separate test series, from 1975 to 1978, the materials, test methods and test results are not necessarily consistent for all test series. They may, however, serve as the case studies to lead to the general conclusion on the availability and limitations of this method.

The selfweight consolidation in this report is somewhat analogous to, but different from, the sedimentation or the settling of soil suspension that has much higher water content [7] in that the former deals with the soil that has a definite soil skeleton and effective stress without, in principle, particle segregation. Actually, however, this phenomenon was observed in some tests in this report. There will be some discussion on this point.

Theoretical Consideration of Selfweight Consolidation and Similarity Law in Centrifugal Consolidation Test

General Consolidation Equation[2] and Similarity Law

The general equation of one-dimensional consolidation of saturated clay published by Mikasa in 1963 [3] is given as

$$\frac{\partial \zeta}{\partial t} = \zeta^2 \left[c_v \frac{\partial^2 \zeta}{\partial z_0^2} + \frac{dc_v}{d\zeta} \left(\frac{\partial \zeta}{\partial z_0} \right)^2 - \frac{d}{d\zeta} (c_v m_v \gamma') \frac{\partial \zeta}{\partial z_0} \right] \tag{1}$$

where z_0 is the original coordinate or the coordinate in the original state[3] in which the clay is assumed to have a certain uniform volume ratio f_0 and corresponding submerged unit weight γ'_0 throughout the depth, z_0 being measured positively in the downward direction, t is time, ζ is the consolidation ratio ($= f_0/f$, f denotes the volume ratio which is equal to $(1 + e)$) and γ' is the submerged unit weight.

This equation is free from the assumptions used in the Terzaghi consolidation theory that k, m_v, c_v, thickness of the clay layer, and consolidation pressure are all constant during consolidation, and that the selfweight of the clay does not affect the time-consolidation relationship.

The assumptions used in deriving Eq 1 are (1) clay is homogeneous, (2) clay is saturated, (3) one-dimensional consolidation applies, (4) soil particles and water are incompressible, (5) Darcy's law is applicable, (6) f-log p and f-log k relations are not time-dependent, and (7) soil water mixture should constitute a soil skeleton that carries an effective stress, though it may be feeble, and particle segregation is not allowed. This last assumption is a prerequisite to deal with a soil-water mixture as a "soil" and is usually not stated explicitly as an assumption.

---

[2] Gibson et al [8] derived another version of the consolidation equation for finite strain (1967), which is different in its expression, but equivalent in its function to Mikasa's equation.

[3] The original state is an imaginary state set for convenience of calculation of finite strain conditions. It may or may not be equal to the initial state (condition), from which the calculation starts.

By using the expressions

$$\phi(\zeta) = c_v/c_{v0} \tag{2}$$

$$T_v = c_{v0}t/(H_0/2)^2 \tag{3}$$

$$Z = z_0/H_0 \tag{4}$$

Eq 1 is transformed as

$$\frac{\partial\zeta}{\partial T_v} = \frac{\zeta^2}{4}\left[\phi(\zeta)\frac{\partial^2\zeta}{\partial Z^2} + \frac{d\phi(\zeta)}{d\zeta}\left(\frac{\partial\zeta}{\partial Z}\right)^2 - \frac{d}{d\zeta}(\phi(\zeta)m_v\gamma')H_0\frac{\partial\zeta}{\partial Z}\right] \tag{5}$$

where c_{v0} is the original c_v value, T_v is the nondimensional time, Z is the non-dimensional depth, and H_0 is the original thickness of clay. Equation 5 shows that the H^2-similarity rule of time-consolidation relationship is not applicable for the selfweight consolidation in a gravitational acceleration field of 1 g because of the existence of H_0 in the third term in the bracket. This term represents the effect of selfweight of clay and gives the consolidation process a significant effect when the clay is soft (m_v is large), or the clay layer is thick (H_0 is large), or both.

If a model clay specimen that has a thickness of H_0/n is subjected to a centrifugal acceleration of ng, its submerged unit weight becomes $n\gamma'$. By replacing γ' and H_0 in Eq 5 with $n\gamma'$ and H_0/n, respectively, the equation remains the same. This indicates that the consolidation of the prototype (H_0, 1 g) and the reduced and centrifuged model (H_0/n, ng) are to proceed similarly as a function of T_v given by Eq 3, and H^2-similarity rule is valid again between these two as follows:

$$\frac{t_m}{t_p} = \frac{H_m^2}{H_p^2} = \frac{1}{n^2} \tag{6}$$

$$\frac{s_m}{s_p} = \frac{H_m}{H_p} = \frac{1}{n} \tag{7}$$

where s denotes settlement, and subscripts m and p indicate the model and the prototype, respectively. We examined and confirmed these relationships, the similarity law, in the past [9,10] by comparing the behaviors of a series of models that have different specimen heights, different accelerations, and a same prototype height, thus making modelling of models.

It should be noted that the centrifuged model has a mode of consolidation, or a distribution of volume ratio and effective stress in the clay layer, similar to that of the prototype both during and after consolidation.

Initial Settlement Rate in Selfweight Consolidation

In the 1 g gravitational acceleration field, the saturated and submerged clay skeleton is subjected to two body forces: seepage force j and submerged unit weight γ'_0. Therefore

$$\frac{\partial p'}{\partial z_0} = j + \gamma'_0 \qquad (8)$$

where p' is the effective pressure. Here j is measured positive in the downward direction, and the subscript 0 denotes the original state. Using Darcy's law: $v = ki$ and the equation of seepage force: $j = i\gamma_w$ where i is the hydraulic gradient and γ_w is the unit weight of water in the 1 g gravitational acceleration field, the velocity of pore water flow relative to the soil skeleton is from the above relations:

$$v = \frac{k}{\gamma_w}\left(\frac{\partial p'}{\partial z_0} - \gamma'_0\right) \qquad (9)$$

where v is measured positively downwards.

Figure 1 shows a schematic illustration of isochrones of volume ratio during selfweight consolidation under single drainage condition as observed in our several analytical and experimental research projects [3,10,11]. This process of selfweight consolidation is simulated by the Terzaghi model (or simply by a coil spring) that settles in a viscous fluid under its selfweight. As shown in the figure, the upper part of the layer maintains its initial state for a certain period in the early stage of selfweight consolidation, where and when the value $\partial p'/\partial z_0$ in Eq 9 is zero.[4] Then

$$v = -k\frac{\gamma'_0}{\gamma_w} \qquad (10)$$

In selfweight consolidation under the single drainage condition, this velocity of pore water flow relative to the soil skeleton in the upper part of the layer in the early stage of selfweight consolidation is just the same in value as, but opposite in direction to, the settlement rate of the clay skeleton in the stationary water, because they are just the opposite views of the same phenomenon. Assuming that permeability is a function of volume ratio (Assumption 6 above), this settlement rate depends only on the volume ratio and is independent of the thickness

---

[4] Strictly speaking, a clay surface without any consolidation pressure may be soaked to some extent. But we can duly assume in the present problem that the surface of a clay slurry will not undergo any significant expansion and will maintain its initial state, f_0 and p'_0, throughout the test.

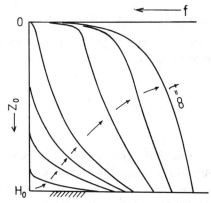

FIG. 1—*Isochrone of selfweight consolidation under single drainage condition.*

of clay layer. Then the value of k at the initial volume ratio f_0 is obtained from this initial surface settlement rate, $\dot{s}$, as

$$k = \dot{s}\,\frac{\gamma_w}{\gamma'_0} \tag{11}$$

where γ'_0 is the submerged unit weight of the clay at its initial state.

Let us consider next the case in which a clay specimen is put in a centrifugal acceleration field of ng as a $1/n$ scaled model of a prototype clay layer. Then Eq 8 is transformed as

$$\frac{\partial p'}{\partial z_0} = j + n\gamma'_0$$

while Darcy's law and the equation of seepage force remain the same as in 1 g gravitational acceleration field. Therefore, Eq 9 becomes

$$v = \frac{k}{\gamma_w}\left(\frac{\partial p'}{\partial z_0} - n\gamma'_0\right)$$

Thus in a centrifugal acceleration field of ng, Eqs 10 and 11 are transformed as

$$v = -k\,\frac{n\gamma'_0}{\gamma_w} \tag{12}$$

and

$$k = \dot{s}\,\frac{\gamma_w}{n\gamma'_0} \tag{13}$$

Comparing Eqs 11 and 13, we find that the value of $\dot{s}$ of a model centrifuged to ng is n times as large as that of the prototype in the 1 g gravitational acceleration field. Also, by combining Eqs 6 and 7, we get

$$\frac{\dot{s}_m}{\dot{s}_p} = \frac{s_m/t_m}{s_p/t_p} = \frac{s_m}{s_p} \frac{t_p}{t_m} = \frac{1}{n} \frac{n^2}{1} = n \tag{14}$$

This similarity between the model and the prototype is valid not only for the initial settlement rate, but also for the whole process of selfweight consolidation. Figure 2 illustrates t–s relations (t in arithmetic scale) of three models with the same initial water content: two models of respective thicknesses H and H/n in the 1 g gravitational acceleration field and a model of thickness H/n in a centrifugal acceleration field of ng. The initial settlement rates of the former two models are the same, while that of the last one is n times as high as the former two, its overall settlement behavior being similar to the model of thickness H in the 1 g gravitational acceleration field in accordance with the similarity laws of Eqs 6, 7 and 14.

Purpose, Materials, and Procedure of the Test

Four clays were used for the tests in this report: two clays taken from Tokuyama Bay in Yamaguchi prefecture, where the seabed had been contaminated by mercury-containing waste, and two clays from a site in the reclaimed land of Osaka

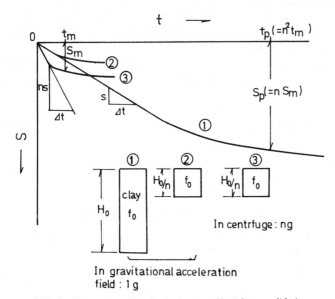

FIG. 2—*Illustration of similarity law in selfweight consolidation.*

South Port ("Osaka Nanko," in Japanese) and sieved through 0.3-mm mesh. Physical properties of these clays are shown in Table 1. Since the sampling date was different, Clays I and II from each site were not very similar to each other.

We shall explain here briefly the purpose of these test series. The tests on Tokuyama Clay I were conducted in 1974 [5] to predict the settlement rate of a presumed fill of soft clay dredged by a new pollution-free pneumatic suction pump for the purpose to determine the capacity of disposal pond for the contaminated clay. The clay was prepared at initial water contents of 400 to 500%, which were the estimated values in the field when the special pump was used. Since f-log p relation was not demanded, few such data were obtained for the clay, and the behavior of the selfweight consolidation of the presumed fill was directly observed in the centrifuge, especially on the initial rate of settlement after the clay was filled, because it would determine the necessary capacity of the disposal pond.

In the course of this test series, we realized the importance of particle segregation in selfweight consolidation of very soft clay under high centrifugal acceleration. The effects of this phenomenon on the observed settlement rate and f-log p relation were investigated in 1975 and 1976 with Tokuyama Clay II and Nanko Clay I, initial water content being chosen between 250 and 450%.

The test series on Nanko Clay II conducted in 1976 and 1977 [6] was a new research project to study the mechanism of selfweight consolidation in detail by the centrifugal model testing and the applicability of numerical analysis by Mikasa's consolidation theory to the centrifugal model test.[5] In this test series the initial water content was set at 120 to 200%, equal to or less than twice its liquid limit. These water contents were chosen so as to prevent particle segregation and to secure feasibility of numerical calculation, but not without the practical basis. According to our past experiences at several reclamation sites around Osaka Bay [3,12], a few weeks after the last filling the water content at the fill surface seldom exceeded twice the liquid limit of the clay. This is perhaps because of the rapid sedimentation or "hindered settling" of soil suspension with much

TABLE 1—*Physical properties of clays.*

Clay	Specific Gravity (G_s)	Liquid Limit (w_L), %	Plastic Limit (w_P), %
Tokuyama I	2.66	122.0	40.5
Tokuyama II	2.66	101.3	39.3
Nanko I	2.67	107.0	37.3
Nanko II	2.67	98.9	29.9

[5] The test series carried out from 1974 to 1976 were not analyzed at all. Though the guiding principle of selfweight consolidation test was based on Mikasa's general consolidation theory, it was found that the numerical calculation of selfweight consolidation of very soft clay with so feeble effective stress as, for example, 1×10^{-4} tf/m² is very difficult to perform, which was the case of Tokuyama Clay I.

higher water content discharged by an ordinary pump dredger, as exemplified by Imai [7].[6]

Figure 3 shows the Mark IV centrifuge (1975 to 1983) [13] in our laboratory that was used for the test series mentioned above; its maximum centrifugal acceleration is 200 g. It is, as all of our past centrifuges have been, a swing basket type without any device to fix the specimen container in flight. Therefore the resultant body force by the vertical 1 g gravitational and the horizontal centrifugal acceleration always acts in the direction of the specimen axis. For selfweight consolidation test, two steel vessels are attached to both ends of the rotor as shown in Fig. 3, and two transparent acrylic specimen cylinders with impermeable base are set in them for the single drainage test series reported in this paper.

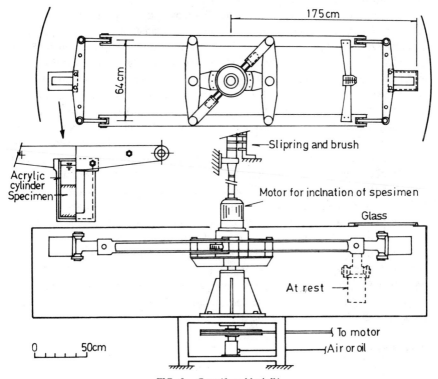

FIG. 3—*Centrifuge Mark IV.*

---

[6] It is very interesting, rather cynical, and practically important that a clay water mixture with a water content as high as $w_0 = 1000\%$ or more discharged from an ordinary pump dredger will settle sooner into a denser fill than a soft clay with water content about $w_0 = 450\%$ or so from a new sophisticated pneumatic suction pump dredger.

De-aired slurry clay was then poured into the cylinder to a predetermined depth, and seawater was added on it. In particular, some specimens were cured for 2 to 150 days before they were set on the centrifuge. Detailed conditions of the specimens including curing will be explained later in reference to their effects on test results.

For Tokuyama Clays I and II and Nanko Clay I, centrifugal acceleration was set at 150 g, and for Nanko Clay II at 100 g. It took 50 to 60 s to reach 150 g, and 30 to 40 s to reach 100 g. Time origin was taken at the midpoint of the accelerating period. Settlement was read visually to 0.2 mm at proper intervals by the help of a synchronized stroboscope. The test was terminated when the settlement rate in secondary consolidation versus logarithm of time was as small as 1 mm, or 1% in strain, per one log-cycle of time. The consolidation time was 1000 to 5000 min. The amount of secondary consolidation always occupied a very small percentage in the total settlement.

When the machine stopped, the water that covered the specimen was sucked out immediately to prevent swelling of the consolidated clay, and the specimen cylinder was detached from the machine. An undisturbed soil column of 5 cm diameter was then sampled for the whole specimen length using a thin-walled metal tube, and its water content was measured for every 2 or 3 mm thick slice continuously along the depth. The submerged unit weight γ' and volume ratio f at each depth were calculated from the water content distribution assuming full saturation, the specific gravity of soil grains and sea water being taken as $G_s = 2.67$ $(= 8/3)$ and $G_w = 1.03$, respectively.[7] The effective overburden pressure p' was obtained by integrating n γ' from the surface to that depth. The values of f and p' thus obtained were used to determine the f-log p relation of that clay.

Two series of selfweight consolidation tests in the 1 g gravitational acceleration field were also conducted for Tokuyama Clay I and Nanko Clay I for reference, with respective specimen size of 30 cm diameter and 100 cm thickness, and 10 cm diameter and 15 cm thickness. The test period of the former series was very long, amounting to 225 days, and yet two specimens with lower initial water content did not complete their primary consolidation in that period. For the fully consolidated specimen with higher initial water content, the water content distribution along the depth after the consolidation was measured for every 2.5 cm thick slice continuously along the depth.

Test Results

Permeability

Figures 4 to 6 show time-consolidation curves (in arithmetic time scale) of singly drained selfweight consolidation tests of Tokuyama Clay I under 150 and 1 g, and Nanko Clay II under 100 g, respectively.

---

[7] Using the relation $e = G_s(w/100)$, $w = 30\%$ will give $e = 0.8$ when $G_s = 8/3$ for a saturated clay. This relation is conveniently used as a rough calculation for ordinary saturated soils.

In the test series of Tokuyama Clay I shown in Fig. 4, the total height of solid H_s ($= H_0/f_0$) was kept approximately the same for a variety of initial water contents w_0; that is, the initial thickness H_0 of each specimen was changed proportionally to the f_0 value. This is because it was necessary to find the effect of the initial water content on the settlement behavior of the clay fill, total mass of contaminated clay being given. This resulted in almost the same final specimen height and the same final average volume ratio $\bar{f_f}$. To make this point clear, the ordinate of Fig. 4 is taken as the specimen height and the average volume ratio, $\bar{f}$, which are proportional to each other. All curves show steady linear settlement in the early 10 min (156 days in prototype) up to 60 to 70% consolidation, and then decreasing their slope converge to $\bar{f_f}$. The specimen with the highest water content shows a high rate of initial settlement, crosses the other curves, and remains below the others, though in the test series of double drainage condition (omitted in this paper) no such phenomena were seen. In some specimens of high water content, particle segregation may have occurred, but it was not ascertained in this test series.

The two test series shown in Figs. 5 and 6 were conducted with respective specimen heights of 100 and 10 cm. In these tests the specimens had different H_s values for different water contents, and thus the ordinate in the figure is taken as the settlement. Nanko Clay II (Fig. 6) with relatively low water content shows linear incipient time-settlement (t–s) relations under the centrifugal acceleration of 100 g. In this test series, particle segregation did not occur at all, as is evidenced in the next section. Two of the three tests on Tokuyama Clay I in the 1 g

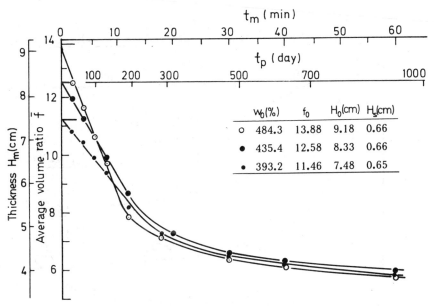

FIG. 4—*Time-consolidation curves: Tokuyama Clay I, 150 g in centrifuge.*

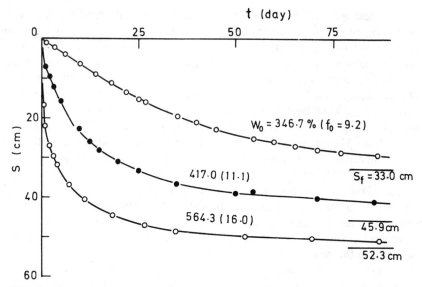

FIG. 5—*Time-consolidation curves: Tokuyama Clay I, in 1 g gravitational acceleration field, specimen height 100 cm.*

gravitational acceleration field (Fig. 5) did not show clear linearity of incipient *t–s* curve, the reason of which is not clear yet. Tokuyama Clay II and Nanko Clay I in centrifuge showed linear incipient *t–s* relations, which are omitted here, though their results are shown in Figs. 7, 8, and others.

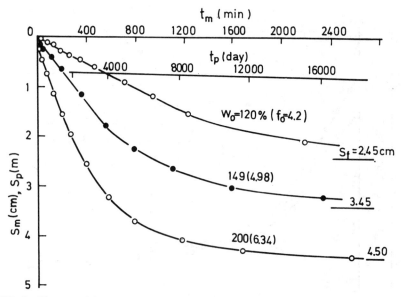

FIG. 6—*Time-consolidation curves: Nanko Clay II, 100 g in centrifuge, specimen height 10 cm.*

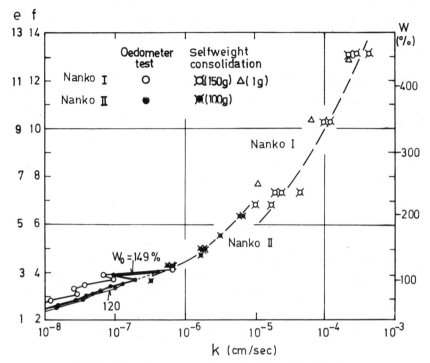

FIG. 7—*Permeability versus volume ratio: Nanko Clays I and II.*

From a set of incipient linear time-settlement relations with different f_0 values in a series of selfweight consolidation tests, we obtain a relationship between k and f. Figures 7 and 8 show the k–f relation thus obtained for the four clays both by selfweight consolidation and by oedometer test. The data of cured specimens are omitted here, because their f values after curing cannot be uniquely defined. Roughly speaking, the k value increases considerably with volume ratio, but decreases in the rate of increase.

Nanko Clay II provides a most persuasive result: centrifuged models yielded a definite f–k relation, which smoothly joins the result of oedometer test that is obtained by the procedure proposed by Mikasa [4,14–16].

The other soils give somewhat scattering k values for a f value, perhaps because of the initial unstable soil structure of the high water content specimens. Cured specimens will be saved from this uncertainty, but they cannot keep a uniform volume ratio owing to the gravitational acceleration of 1 g. (It would be too expensive, however, to cure them in a space shuttle!)

Compressibility

The f-log p relations of Nanko Clay I calculated from the final water content distribution along the depth of centrifuged models are shown in Fig. 9 in three

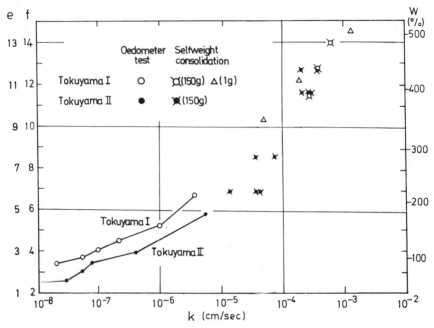

FIG. 8—*Permeability versus volume ratio: Tokuyama Clays I and II.*

groups of different nominal initial water contents of 450, 350, and 240%. To investigate the condition that would cause particle segregation, several different curing times t_a were assigned to specimens of the same water content: $t_a = 0$ (tested immediately after the specimen preparation), 2, 7, 30, and 90 days. Curing, just placing the specimen in the cylinder still in the laboratory, is inevitably accompanied by self-weight consolidation in the 1 g gravitational acceleration field. The change in specimen condition by "curing" is shown in Table 2. The initial height of specimens with different initial water content was chosen so that they have the total height of solid of $H_s = 1.15$ to 1.2 cm. The specimens of $t_a = 30$ and 90 days reached 100% consolidation by selfweight under the 1 g gravitational acceleration field, and those of $t_a = 7$ days reached 70 to 90%. One remarkable point is that the specimens of 450% nominal water content settled much faster and had denser state after curing than the specimens of 350% water content. This kind of phenomenon was also seen in the selfweight consolidation in centrifuge shown in Fig. 4. Since the cured specimens do not have a uniform f value along the depth, the values of k of the cured specimens were omitted in Figs. 7 and 8, though the deviations in f-log k plot were not always serious.

Among the eleven f-log p curves shown in Fig. 9, four specimens of high water content and short curing time show a remarkable drop in f value near the $p = 2$ tf/m² load, which corresponds to the bottom part of the specimens. This is due to the concentration of coarser particles in this part of the specimen, which

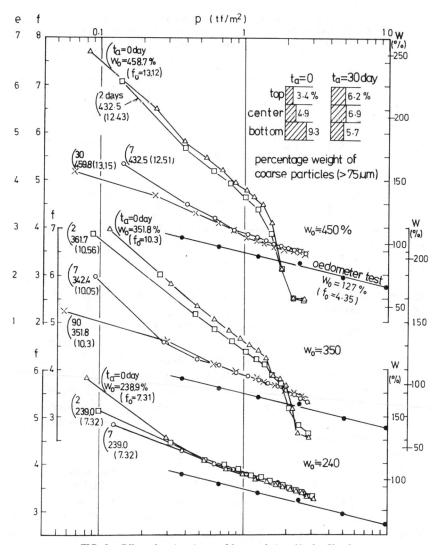

FIG. 9—*Effect of curing time on f-log p relation: Nanko Clay I.*

is clearly caused by the particle segregation in the early period of the test. In this figure two results of gradation tests for two specimens of nominal water content $w_0 = 450\%$ with different curing time are compared in the distribution of percentage weight of particles coarser than 0.075 mm, which supports the above interpretation.

The specimens either of longer curing time or of lower water content show linear f-log p relationships except the scattering in the very low stress range. They coincide with each other referring to the common f-log p line from an

TABLE 2—*Specimen conditions before and after curing.*

Initial Condition			Curing Time (t_a), day	After Curing[a]		
w_0, %	f_0	H_0, cm		H_a, cm	$\bar{w}$, %	$\bar{f}$
238.9	7.31		0	8.9	238.9	7.31
239.0	7.32	8.9	2	8.05	212.9	6.62
239.0	7.32		7	7.60	198.9	6.65
351.8	10.30		0	11.9	351.8	10.30
361.7	10.56		2	10.8	353.5	9.58
342.4	10.05	11.9	7	9.7	270.3	8.19
351.8	10.30		90	8.85	240.6	7.40
458.7	13.12		0	15.0	458.7	13.12
428.6	12.43		2	10.5	288.8	8.70
432.5	12.51	15.0	7	9.4	257.0	7.72
459.8	13.15		30	8.8	254.4	7.84

[a] H_a = thickness of specimen.
$\bar{w}$ = average water content.
$\bar{f}$ = average volume ratio.

oedometer test ($D = 10$ cm, $H_0 = 4$ cm) for the specimen remolded at a much lower initial water content of $w_0 = 127\%$.

Figure 10 shows the f-log p relations of Nanko Clay II with nominal initial water contents of 120, 150, and 200%, all maintaining linearity except in the very low stress range, showing that particle segregation did not take place at all. In this case the f-log p relation with higher initial water content remains above the others, and two f-log p lines from oedometer tests ($D = 10$ cm, $H_0 = 4$ cm) coincide well with the results of selfweight consolidation tests of the same initial water contents. An f-log p line of Nanko Clay I is also shown for reference, which is expressed in Eq 16.

Figure 11 shows representative f-log p relations of Tokuyama Clays I and II. The oedometer tests of the two soils were conducted in a special oedometer of 15 cm diameter and 6.35 cm specimen height, and with an initial water content as high as 250%, and yielded linear f-log p curves very similar to each other. The selfweight consolidation of Tokuyama Clay I under 1 g and Tokuyama Clay II under 150 g after 7 and 150 days' curing yielded f-log p relations close to those from oedometer tests, while Tokuyama Clay II without or with only two days' curing showed a marked evidence of particle segregation under the centrifugal acceleration of 150 g.

Coefficient of Consolidation c_v

The value of $c_v = k/(m_v \gamma_w)$ is calculated as a function of volume ratio from the obtained f-log k and f-log p relations using the relation

$$m_v = \frac{0.4343 \, C_c}{f} \cdot \frac{1}{p} \tag{15}$$

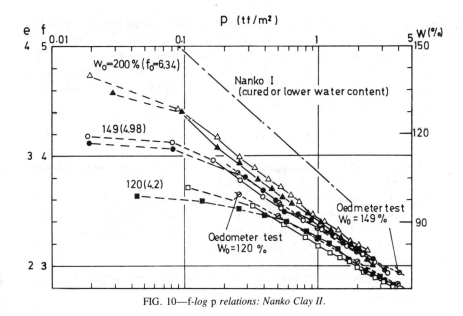

FIG. 10—f-*log* p *relations: Nanko Clay II.*

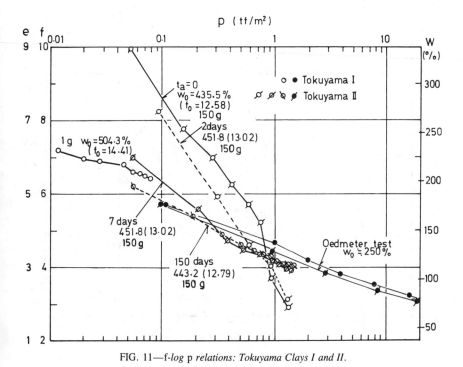

FIG. 11—f-*log* p *relations: Tokuyama Clays I and II.*

where C_c is the slope of f-log p curve at the point, which is constant (Compression Index) when the curve is linear. We shall determine here the f-log c_v relation of Nanko Clays I and II. The f-log k relations of both clays are assumed as the broken lines in Fig. 7. For Nanko Clay I with very high water content, f-log p relations from selfweight consolidation with a variety of initial water content in Fig. 9 can be expressed uniquely as

$$f = 3.85 - 1.1 \times \log p \ (p \text{ in tf/m}^2) \qquad (16)$$

For Nanko Clay II with moderately high water content, however, f-log p relations are dependent on the initial water content (Fig. 10).

Figure 12 shows the obtained f-log c_v relations of Nanko Clays I and II. Nanko Clay I has a unique f-log c_v relation, whereas Nanko Clay II has several different f-log c_v relations for different initial volume ratios. The dashed line is the locus of the initial c_v value, which will join smoothly to two f-log c_v relations from oedometer tests.

It is reasonably assumed that f-log k relation is determined uniquely for a clay

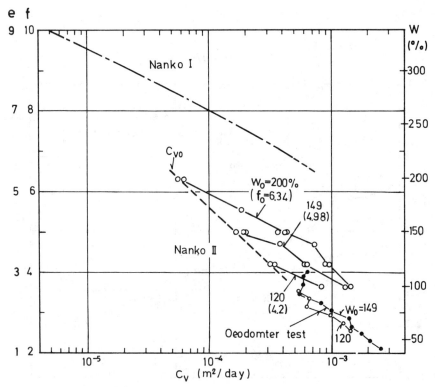

FIG. 12—*f-log c_v relations: Nanko Clays I and II.*

in principle. The f-log p relation, on the other hand, was found to be affected by some factors such as initial water content, curing time, or the value of gravitational acceleration. As expressed by Eq 15, m_v is as much a function of the consolidation pressure p as of f, and a difference in f-log p relation affects the m_v value considerably, because the change in p value for a given f value is often large for different f-log p relations.

If a c_v value obtained by any sort of consolidation test is likely to be affected by such miscellaneous factors as mentioned above, it may be preferable to determine k and m_v separately, considering the presumed field conditions, and to assemble them into a c_v value. Only in this way can we evaluate the difference between the consolidation test and the prototype behavior, either qualitatively or quantitatively, on the basis of physical meaning of the phenomenon.

Remarks

A geotechnical centrifuge can provide a small-scale model with similar stress patterns to the prototype; it enables us to predict both the time-settlement relationship (H^2 similarity law) and the ultimate state of a field selfweight consolidation directly. The determination of consolidation parameters discussed in this paper is a very useful by-product of the selfweight consolidation model test.

Selfweight consolidation test in the 1 g gravitational acceleration field also yields permeability and f-log p relation in the same manner as by a centrifugal test. However, it usually takes an impractically long period to complete the test, and acting effective stresses are restricted to a level too low to simulate field problems. If we pay attention only to the initial rate of settlement, however, this test will provide reliable data from a specimen of as small a thickness as 10 cm or so.

Besides the selfweight consolidation, there are some other test methods used to determine the consolidation parameters of very soft clays: the seepage consolidation test proposed by Imai [1] and the constant rate consolidation test proposed by Umehara [2]. Both methods require pore water pressure measurement, which necessitates elaborate techniques to get reliable results. The selfweight consolidation method is comparatively simpler in its principle, easier in its test procedure, and, above all, nearer to the prototype in its mode and mechanism of consolidation, though it needs a centrifuge.

One disadvantage of the centrifugal test is that particle segregation is more likely to occur than in the other methods when the water content is higher than a certain limit peculiar to a combination of material and centrifugal acceleration (e.g., 250% or so for highly plastic clays as reported in this paper under 100 g). Particle segregation makes the specimen behavior in centrifuge different from the prototype particularly in f-log p relation, and to a lesser degree in f-log k relation.

We have not come to any general conclusion yet about the condition to cause or prevent particle segregation. The type of soil, initial water content, centrifugal acceleration, curing time before the test, and possibly other factors may affect the condition. In spite of all such complexities, however, centrifugal model testing is valid for revealing the behavior of soft, but not too soft, soils under field stress conditions with an accuracy as tolerable as that of the oedometer test in a lower water content range.

References

[1] Imai, G., Yano, K., and Aoki, S., "Application of Hydraulic Consolidation Test for Very Soft Clayey Soils," *Soils and Foundations,* Vol. 24, No. 2, 1984, pp. 29–42.
[2] Umehara, Y. and Zen, K., "Constant Rate of Strain Consolidation for Very Soft Clayey Soils," *Soils and Foundations,* Vol. 20, No. 2, 1980, pp. 33–65.
[3] Mikasa, M., "The Consolidation of Soft Clay—A New Consolidation Theory and its Application," *Kajima Shuppan-Kai* (in Japanese), 1963.
[4] Mikasa, M. and Ohnishi, H., "Soil Improvement by Dewatering in Osaka South Port, Geotechnical Aspects of Coastal Reclamation Project in Japan," in *Proceedings,* 9th International Conference on Soil Mechanics and Foundation Engineering, Case History Volume, 1981, pp. 639–664.
[5] Mikasa, M., Takada, N., and Li, K., "Consolidation Characteristics of Very Soft Clay," in *Proceedings,* 11th Annual Meeting of Japan Society of Soil Mechanics and Foundation Engineering (in Japanese), 1976, pp. 185–186.
[6] Mikasa, M. and Takada, N., "Selfweight Consolidation of Very Soft Clay by Centrifuge," in *Proceedings,* Symposium on Sedimentation/Consolidation Models, American Society of Civil Engineers, 1984, pp. 121–140.
[7] Imai, G., "Settling Behaviour of Clay Suspension," *Soils and Foundations,* Vol. 20, No. 2, 1980, pp. 61–77.
[8] Gibson, R. E., England, G. L., and Hussy, M. J. L., "Theory of One-Dimensional Consolidation of Saturated Clays," *Geotechnique,* Vol. 17, 1967, pp. 261–273.
[9] Mikasa, M. and Takada, N., "Selfweight Consolidation Test in Centrifuge (2nd Report)," in *Proceedings,* 21st Annual Convention of Japan Society of Civil Engineers (in Japanese), 1966.
[10] Mikasa, M. and Takada, N., "Significance of Centrifugal Model Test in Soil Mechanics," in *Proceedings,* 8th International Conference on Soil Mechanics and Foundation Engineering, 1973, pp. 273–278.
[11] Mikasa, M., Takada, N., and Kishimoto, Y., "Selfweight Consolidation Test in Centrifuge (2nd Report)," in *Proceedings,* Annual Convention of Japan Society of Civil Engineers, Kansai Branch (in Japanese), 1965.
[12] Mikasa, M. and Maeda, K., "Settlement and Soil Volume in a Reclaimed Land," in *Proceedings,* 19th Annual Convention of Japan Society of Civil Engineers (in Japanese), 1964, III–56.
[13] Mikasa, M., Mochizuki, A., and Sumino, Y., "A Study on Stability of Clay Slopes by Centrifuge," in *Proceedings,* 9th International Conference on Soil Mechanics and Foundation Engineering, Vol. 2, 1977, pp. 121–124.
[14] Mikasa, M., "Testing of Mechanical Properties of Soils," Japan Society of Civil Engineers, Kansai Branch, 1960.
[15] Mikasa, M., "Determination of Consolidation Parameters from Oedometer Test," in *Proceedings,* 19th Annual Convention of Japan Society of Civil Engineers (in Japanese), 1964.
[16] Takada, N. and Mikasa, M., "Consolidation of Multi-layered Clay," in *Proceedings,* Symposium on Sedimentation/Consolidation Models, American Society of Civil Engineers, 1984, pp. 216–228.

James K. Mitchell[1] and Robert Y. K. Liang[1]

Centrifugal Evaluation of a Time-Dependent Numerical Model for Soft Clay Deformations

REFERENCE: Mitchell, J. K. and Liang, R. Y. K., **"Centrifugal Evaluation of a Time-Dependent Numerical Model for Soft Clay Deformations,"** *Consolidation of Soils: Testing and Evaluation, ASTM STP 892*, R. N. Yong and F. C. Townsend, Eds., American Society for Testing and Materials, Philadelphia, 1986, pp. 567–592.

ABSTRACT: A series of centrifuge tests on embankment-type soil structures founded on soft clay is being conducted to evaluate a time-dependent constitutive model that incorporates the combined effects of creep and hydrodynamic consolidation. Because final coding of this new constitutive model has not yet been completed, a two-dimensional consolidation program (CON2D, developed by Chang and Duncan) is being used for preliminary analysis of test results. This paper describes the building, testing, and monitoring of the centrifuge models. The results of the centrifuge tests are interpreted and shown to compare well with predictions made using the CON2D computer program.

Centrifuge testing is a valuable method for investigating the accuracy and usefulness of new soil behavior models, particularly in the absence of or as a supplement to detailed field records. The testing limitations imposed by centrifugal environment—the radial acceleration field and "curved" stress distribution—are examined. Stress histories and stress paths experienced by the centrifuged model are shown accountable in the numerical analyses. Data generated in this testing program can provide a reference for evaluating other constitutive relations or numerical models, thus assisting further development of predictive analytical techniques.

KEY WORDS: centrifuge, clays, numerical model, embankment, deformations, pore pressures

Predictions of the settlement and deformation behavior of embankments on soft clay foundations are often made using Terzaghi's consolidation theory. To better describe the geometrical conditions, numerical methods such as the finite-element method are often used. Recently there have been numerous new constitutive models proposed in an attempt to better represent the soil response to loading. Validation of the constitutive models is often difficult and inconclusive.

[1]Professor and Graduate Research Assistant, respectively, Department of Civil Engineering, University of California, Berkeley, CA 94720.

One approach, demonstrated by Wroth et al [1] and Kavazanjian et al [2], is to compare the predictions from numerical methods employing new constitutive models with the behavior of actual embankments. This approach is usually complicated because of the uncertainties such as nonhomogeneities, unknown boundary conditions, and stress histories, and the lack of sufficiently detailed field measurement data.

Fully instrumented centrifuge models offer an alternative method for investigating the accuracy and usefulness of new soil behavior models. Among the advantages are: (1) the same magnitude of stress levels are developed as in the full scale prototype, therefore no extrapolation is required for the constitutive laws to be applicable at prototype scale; (2) boundary conditions (both displacement and pore pressure) are well-defined; (3) spatial variations and uncertainties in material properties can be minimized by careful model preparation; (4) the effects of various stress histories, stress paths, and geometry conditions can be easily studied; (5) observations of the complete cross section of the model can be made, enabling predictions of displacements and strains throughout the entire foundation to be evaluated; and (6) time-dependent responses such as consolidation and creep can be studied in a short period of time.

The centrifuge tests described in this paper are being conducted as a part of a joint research program between Stanford University and the University of California at Berkeley for evaluating a time-dependent constitutive model that incorporates the combined effects of creep and hydrodynamic consolidation of soft clay. It was initially formulated by Kavazanjian and Mitchell [3,4]. Since then it has been validated by laboratory test programs [5] and improved through modifications by Bonaparte [6].

The essence of the constitutive model is that soil deformations can be divided into immediate and time-dependent deviatoric and volumetric deformations. The magnitude of each of these components is determined using simple constitutive functions and easily determined parameters to characterize the soil properties [4,6].

Currently, ongoing research includes development of a finite-element formulation of the constitutive relationships for analysis of embankments on soft clay foundations by the Stanford research group and centrifuge model testing by the Berkeley research group. Both groups are also analyzing well-documented case histories.

In this paper the centrifuge testing program is described, with particular reference to preparation of models, testing procedure, data acquisition, and interpretation of the results. Because the new computer code is not yet fully operational for the conditions of our centrifuge tests, the two-dimensional consolidation program CON2D developed by Chang and Duncan [7] is being used for preliminary analysis of the test results. CON2D uses aspects of critical state soil mechanics similar to those incorporated in our model, as modified by Bonaparte [6]. Creep deformations are not accounted for, however.

Centrifuge Tests

A series of model tests of embankments on soft clay foundations was conducted using the Schaevitz Centrifuge at the University of California, Davis. Clay foundations were prepared in the U.C. Berkeley laboratory, instrumented, and transferred to the centrifuge for testing. The embankment was constructed at the centrifuge site. A description of the materials, apparatus, and instrumentations used in the experiments and an outline of the testing procedures are given below.

Material

Two criteria are considered important in selecting materials for use in centrifuge tests. The first is test duration: a material of relatively high permeability is preferred because of the shorter time required for consolidation. The second consideration is sample uniformity: field samples may cause uncertainties in the analysis because of nonuniform material properties. It is therefore preferable to use a material prepared in the laboratory under controlled conditions for direct evaluation of predictive theories.

The clay chosen for the initial centrifuge tests was a commercial kaolin. Classification properties of the kaolin are: specific gravity, 2.61; liquid limit, 34%; plastic limit, 12%. It contains 25% sand size, 40% silt size, and 35% clay size particles by weight and is classified as (CL) in the unified system.

A comprehensive laboratory test program was conducted to determine the material properties required by the constitutive model. Triaxial tests were carried out on specimens trimmed from clay cakes that were prepared prior to the centrifuge testing program. The Kavazanjian et al [8] model requires a total of twelve soil parameters as listed in Table 1. These parameters can be divided into three categories according to the type of tests required to determine them:

1. *Consolidation and Secondary Compression Parameters*—The consolidation behavior is characterized by five soil parameters that can be obtained easily from a series of triaxial isotropic consolidation tests. They include slopes of the virgin isotropic consolidation and rebound curves (plotted as e versus $\ln \sigma'_{oct}$), λ and κ; the slope of secondary compression curve (plotted as e versus $\ln t$), ψ; the void ratio at $\sigma'_{oct} = 1$ along the virgin isotropic consolidation line, e_a; and permeabilities in the vertical and horizontal directions, K_x and K_y, respectively.

2. *Singh-Mitchell (S-M) Creep Parameters*—Three Singh-Mitchell [9] creep parameters, m, α, and A, were used in the constitutive model to describe the deviatoric creep deformations. A minimum of two undrained creep tests is required to establish the values of these parameters for any given soil. If two identical specimens are tested at different deviator stress levels, two plots may be prepared. The first, showing logarithm of strain rate versus logarithm of time, will yield a straight line with slope m. The second, showing logarithm of strain rate as a function of stress for different values of time, can be used to find A and $\bar{\alpha}$ from the intercept at unit time and the slope, respectively.

TABLE 1—*Material properties.*

Kaolinite			
Parameter	Symbol	Value	Type of Test
Virgin compression index	λ	0.075	triaxial
Recompression index	κ	0.0075	isotropic
Secondary compression index	ψ	0.0013	consolidation
Void ratio at $\sigma'_{oct} = 6.89$ kPa (1 psi)	e_a	0.877	test (IC)
Vertical permeability	K_y	2.1×10^{-6} cm/s	
Horizontal permeability	K_x	2.1×10^{-6} cm/s	
Singh-Mitchell creep parameters	A	1.2×10^{-5}/min	isotropically
	m	1.006	consolidated
	$\bar{\alpha}$	1.15	undrained creep test (ICU-C)
Hyperbolic parameters	a	0.004	isotropically
	b	1.20	consolidated
Slope of critical state line	M	1.37	undrained compression test (ICU)

Monterey #0 Sand	
Description	Value
Mean diameter	0.36 mm
Coefficient of uniformity	1.45
Specific gravity	2.65
Maximum void ratio	0.82
Minimum void ratio	0.55
Internal friction angle	40 deg for stress range of interest

3. *Undrained Compression Characteristics*—A series of isotropically consolidated undrained compression tests is carried out to determine the hyperbolic stress-strain parameters, a and b, and the slope of the critical state line, M.

Monterey #0 sand was used for constructing the embankment and the upper drainage layer of the model. The pertinent properties of the sand used in the analysis were obtained from Lade [10] and are also summarized in Table 1.

The University of California at Davis Schaevitz Centrifuge

The machine has a maximum capacity of 4500 g-kg at 1 m working radius. In the tests conducted by the authors, the packages weighed about 55 kg, and the maximum acceleration used was 70 g. The model was mounted in a swinging bucket. As the centrifuge accelerates, the bucket swings up toward a horizontal position. The exact position depends on the location of the center of gravity of the package. There are 20 available electrical slip rings and four hydraulic ports. Visual monitoring of the models during testing is achieved with a closed-circuit

TV camera. Still photographs of models during testing can be made using a Nikon F-1 camera mounted near the center of the centrifuge and directed radially toward an angled mirror in front of the transparent side of the model container. Figure 1 shows the internal arrangement of the centrifuge.

Apparatus

The major components of the test apparatus consist of the slurry mixer, the consolidation device, and the model container. The mixer is a large drum with motor-driven rotors. The consolidation device consists of a loading plate and an air-regulated loader that press the slurry from above. Drainage is permitted along the top through holes in the plate, and from a filter membrane along the bottom of the model container. The container is made of lightweight aluminum, lined with Teflon to minimize side friction and adhesion. The internal dimensions of the container are 0.42 m by 0.2 m in plan by 0.3 m in height. A transparent Plexiglas plate (2.54 cm thick) was used as one of the container's side walls to enable direct viewing of the specimen. An external water reservoir made of plastic tubing was attached to one of the end walls, and the passage of water between the box and reservoir were permitted through top and bottom inlets. Steady water flow into the reservoir during the centrifuge testing not only serves to maintain a constant water table elevation inside the box, but also helps minimize temperature rise inside the centrifuge chamber.

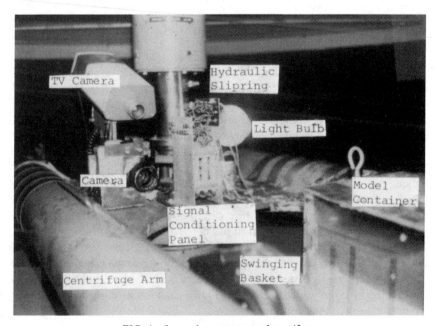

FIG. 1—*Internal arrangement of centrifuge.*

Instrumentation

Pore pressure measurements were made with five miniature semiconductor pressure transducers manufactured by Entran (EPB-125 Series). A stainless porous filter was secured to the transducer diaphragm by heat-shrinkable plastic tubing. The response time of the transducers is very fast because of the high diaphragm stiffness and the consequent small volume change associated with any pressure change. Calibration tests in the centrifuge showed that sensitivity was constant over a pressure range of up to 350 kPa. De-airing of the transducers was done in a glass chamber filled with de-aired water. The transducer was sealed in the chamber by a split rubber stopper which allowed the electrical leads to pass outside the chamber. Pressure in the chamber was reduced to about -0.9 atm using a vacuum pump.

Deformations in the foundation clay were determined by a photographic method. The rotating camera on board the centrifuge recorded positions of plastic marker beads installed along the Plexiglas side of the container in a 2.54 by 2.54 cm grid pattern. The coordinates of markers on each photo negative were digitized and stored using a comparator. The clay deformations that occurred between any two frames were calculated using a data reduction program [11]. The error associated with targeting the markers in the comparator is ± 10 μm. Since the grid had a 2.54 cm spacing, this measuring error corresponds to an error in linear strain of 0.04% on the film. The film is approximately twelve times smaller than the model; hence the maximum error in linear strain is about 0.5%.

Model Preparation

The kaolin clay powder was mixed with de-ionized water to form a slurry at 95% water content. After mixing for over 24 h, the slurry was carefully transferred into the model container so as to avoid trapping air bubbles. Once the slurry had been placed in the box, a filter membrane was laid on top, and the loading plate was inserted into the box. Consolidation to an effective pressure of 28 kPa required about a week. After consolidation the front and back face plates of the box were slid off the clay surface, and the clay bed was trimmed to the desired height. Next, the deaired transducers were inserted into pre-bored, horizontal holes in the clay bed. The holes were backfilled with a thin clay slurry injected using a long syringe. Each transducer was oriented with its pressure sensitive area parallel to the acceleration force to minimize centrifugal effects on its response. The plastic grid markers were then put in place.

With the clay fully instrumented, the front and back plates were sprayed with thin oil (WD-40), placed onto the specimen, and bolted to the container. A layer of sand for top drainage was constructed underwater. The sand embankment was constructed in layers using a template. Figure 2 shows a finished model ready for mounting onto the centrifuge bucket.

FIG. 2—*Centrifuge model ready for mounting.*

Testing Procedures

The centrifuge tests are done in two stages, as schematically shown in Fig. 3. The first stage, called "re-initialization," was intended to establish a prior consolidation stress in the clay bed. This was necessary because the clay foundation had undergone a partial release of consolidation pressure and disturbance during instrument installation and container transportation. The second stage was the "loading" stage, during which centrifugal accelerations were increased to a higher g-level.

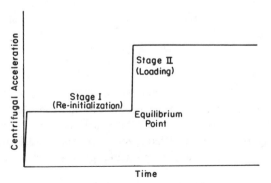

FIG. 3—*Centrifuge testing procedure.*

Test Program and Results

Two different series of centrifuge tests have been conducted: tests on models having the same dimensions and loading conditions (Series I), and a series of modeling of model tests (Series II). A listing of the tests is given in Table 2.

Series I

Four centrifuge tests were conducted with essentially the same model configuration and loading conditions to investigate the reproducibility of the testing procedure and results. During the process of specimen preparation and centrifuge testing, many factors (de-airing of pore pressure transducers, trimming the foundation clays to the desired dimension, constructing the embankments to the required density and geometry, etc.) may not be perfectly controlled and can thus cause scatter in the test results. A direct comparison of the measured pore pressures and deformations among the tests in Series I provides a means of evaluating data consistency.

The tests in Series I were conducted by the following procedure. The model was subjected to a 30 g acceleration level for 84 min. At the end of this stage, the acceleration was increased in about 1 min to 70 g and maintained constant for about 1 h.

After each test the foundation clays were carefully excavated to locate the exact positions of the pore pressure transducers. Figure 4 shows the measured final transducer positions for the Series I tests. Since the transducers were not located in the same positions in each test, direct evaluation of the repeatability of measured pore pressures was not possible. However, consistently good agreement was found between the measured and predicted pore pressures. The measurements of pore pressures are therefore considered reliable. Typical measured pore pressure responses are shown as dashed lines in Fig. 5.

A close examination of the measured clay foundation settlements indicated that each test in Series I had the same rate and amount of deformation. Figure 6 shows typical time-dependent movements of the plastic markers. The measured vertical settlements at four randomly selected locations of Test I-1 are plotted against time in Fig. 7.

Series II

Modeling of model tests was performed for two reasons: to obtain data for different model configurations to further evaluate the constitutive models; and to investigate the possible effects, if any, of model size. Three centrifuge models of similar configuration but of different size were subjected to different accelerations, N, in order to simulate one particular prototype, as given by

$$L_p = L_{m1} \times N_1 = L_{m2} \times N_2 = L_{m3} \times N_3 \tag{1}$$

TABLE 2—*Centrifuge tests.*

Test No.	Top Embankment Width, W_1, in.[a]	Bottom Embankment Width, W_2, in.	Embankment Height, H_1, in.	Sand Blanket Height, H_2, in.	Foundation Height, H_3, in.	Initial Acceleration, g	Final Acceleration, g	Initial Water Content of Clay, %	γ_{sat} of Clay, lbf/ft³	γ_{sat} of Sand Blanket, lbf/ft³	γ_m of Sand Embankment, lbf/ft³
I-1	3.25	10.0	2.75	2.0	6.1	30.0	70.0	39.5	110.0	130.0	113.7
I-2	3.25	10.0	2.75	2.0	6.1	30.0	70.0	42.0	110.9	129.6	114.0
I-3	3.25	10.0	2.75	2.0	6.1	30.0	70.0	40.0	109.5	129.6	113.8
I-4	3.25	10.0	2.75	2.0	6.1	30.0	70.0	39.0	109.5	126.8	111.0
II-1	2.8	8.70	2.4	1.8	5.3	32.0	75.0	40.5	110.5	127.6	113.5
II-2	3.28	10.18	2.81	2.1	6.2	27.5	64.0	41.0	111.9	126.4	113.5
II-3	2.17	6.73	1.90	1.4	4.1	41.4	97.0	40.0	111.5	127.2	111.0

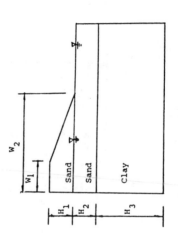

[a] 1 in = 2.54 cm.
[b] 1 lbf/ft³ = 16.018 kg/m³.

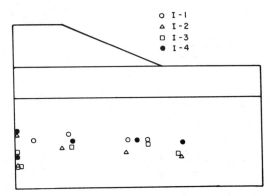

FIG. 4—*Transducer locations in Series I tests.*

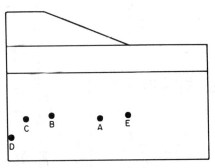

FIG. 5a—*Location of transducers.*

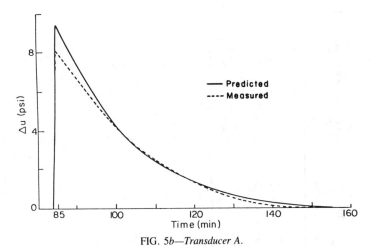

FIG. 5b—*Transducer A.*

FIG. 5—*Comparison of measured and predicted pore*

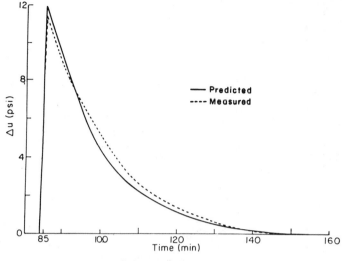

FIG. 5c—*Transducer B.*

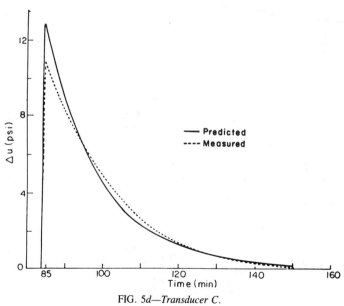

FIG. 5d—*Transducer C.*

pressure responses in Series I tests (1 psi = 6.895 kPa).   *Continues overleaf.*

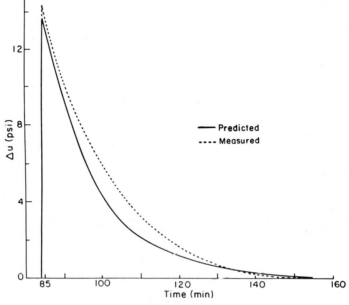

FIG. 5e—*Transducer D.*

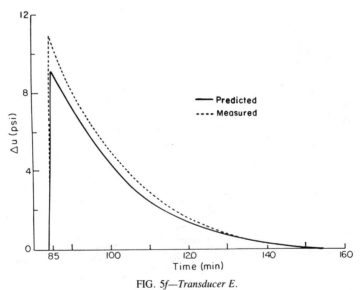

FIG. 5f—*Transducer E.*

FIG. 5 (continued).

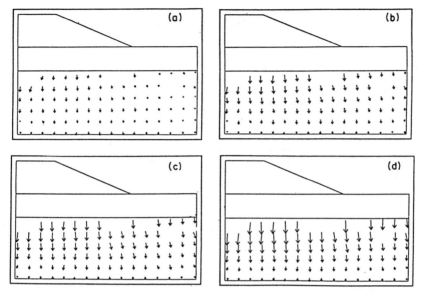

FIG. 6—*Measured deformations of Series I test at different time intervals after Stage II loading.* (a) t = *1 min.* (b) t = *9 min.* (c) t = *23 min.* (d) t = *65 min. Magnification factor = 4.0.*

where L_p is the dimension of the prototype, L_{m1}, L_{m2}, L_{m3} are the dimensions for centrifuge Models 1, 2, and 3, respectively, and N_1, N_2, N_3 are the required accelerations for centrifuge Models 1, 2, and 3. If there is no effect of model size, then the measurements of excess pore pressures and deformations at analogous positions in each model should be the same.

Figure 8 shows the positions of the pore pressure transducers and the measured pore pressure responses of several transducers. Comparisons of the measured values at similar transducer locations (for example, Fig. 8b) show good agreement. Therefore, model size in the range investigated does not significantly affect the generated pore pressures.

The measured deformations from the three tests in this series also agree closely, again indicating that model size is not significant. Figure 9 presents the vertical settlements of four randomly selected points in the clay foundation.

Analysis and Comparison

CON2D, originally developed by Chang and Duncan [7], is a plane-strain finite-element (FEM) computer program for analyzing the consolidation of saturated and partly saturated two-dimensional soil masses. The program combines a critical-state constitutive model (that is, modified Cam Clay [12] and Biot's [13] coupled consolidation theory) but neglects secondary compression and creep effects. Materials modelled as a modified Cam Clay are specified by five pa-

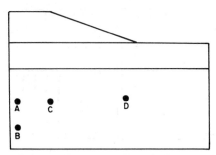

FIG. 7a—Location of settlement markers.

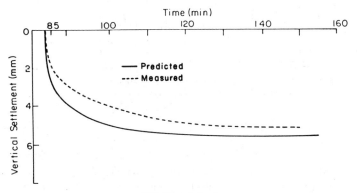

FIG. 7b—Point A.

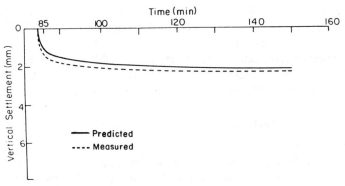

FIG. 7c—Point B.

FIG. 7—Comparison of measured and predicted vertical deformations at different nodal points in Series I tests.

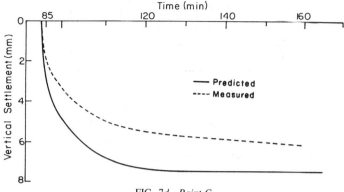

FIG. 7d—Point C.

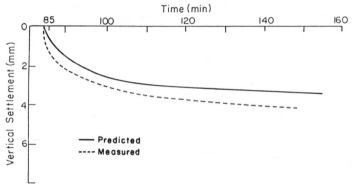

FIG. 7e—Point D.

FIG. 7 (continued).

rameters: (1) the slope of the critical state line M, (2) the slope of the virgin isotropic consolidation curve in the e versus ln σ'_{oct} plot λ, (3) the slope of the recompression curve κ, (4) Poisson's ratio ν, and (5) the initial void ratio e_0. In addition, vertical and horizontal permeabilities, K_x and K_y, are required for the consolidation analysis.

Isoparametric elements with four to eight nodes can be used with the computer program. The program is capable of analyzing both gravity turn-on and layered construction conditions. Details of the program capabilities are given by Duncan et al [14].

The general method used to analyze the centrifuge loading is depicted in Fig. 10. First, a stress analysis finite-element program FEADAM [15] and equivalent material unit weights are used to calculate the effective stresses in the clay foundation at the end of the re-initialization stage (Fig. 10a). These effective stresses can then be used in the CON2D analysis as the initial effective stresses.

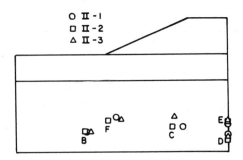

FIG. 8a—*Location of transducers.*

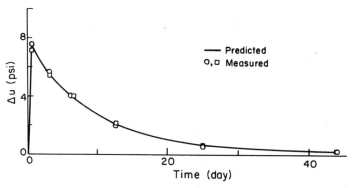

FIG. 8b—*Transducer B.*

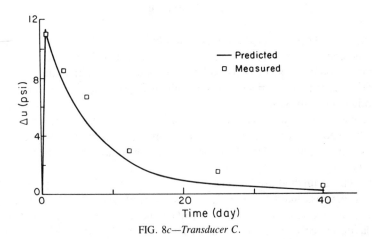

FIG. 8c—*Transducer C.*

FIG. 8—*Comparison of pore*

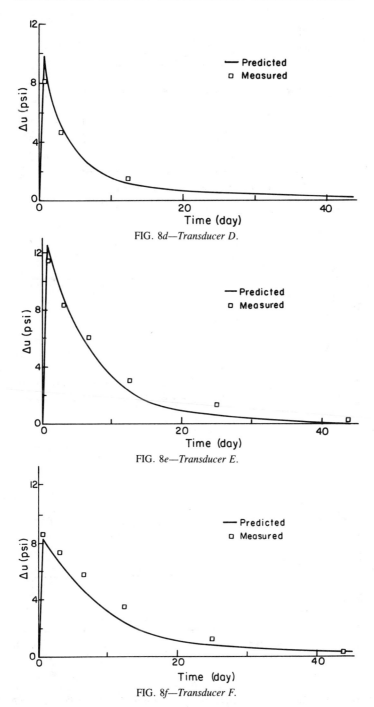

FIG. 8d—Transducer D.

FIG. 8e—Transducer E.

FIG. 8f—Transducer F.

pressure response in Series II tests.

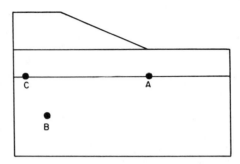

FIG. 9a—Location of settlement markers.

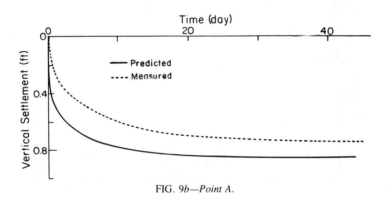

FIG. 9b—Point A.

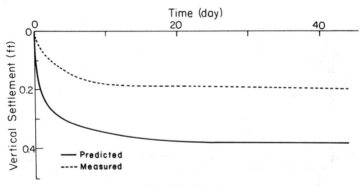

FIG. 9c—Point B.

FIG. 9—Composition of measured and predicted vertical settlements at different nodal points in Series II tests (1 ft = 0.305 m).

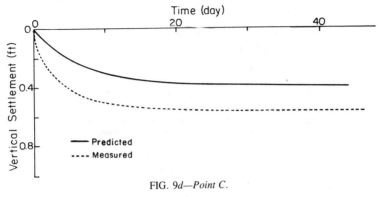

FIG. 9d—*Point C.*

FIG. 9 (continued).

Next, the self-weight induced excess pore pressures caused by the increase of acceleration during Stage II loading are calculated assuming the embankment was not present (Fig. 10b). This increment of excess pore pressure, Δu, corresponds to one-dimensional consolidation and is calculated according to

$$\Delta u = (\sigma_{vf} - U_{sf}) - \sigma'_{vi} \tag{2}$$

where

σ_{vf} = total vertical stress at the final g acceleration,
U_{sf} = hydrostatic pore pressure in the final g acceleration, and
σ'_{vi} = effective vertical stress in the initial g acceleration.

These calculated excess pore pressures were also used as initial conditions in the CON2D analysis. Finally, the sand embankment loading was simulated by using the equivalent unit weight γ_{equ} for the sand and zero unit weight for foundation materials (Fig. 10c). In essence, the Stage II centrifuge loading was simulated in the analysis by the combined effects of the foundation's self-weight-induced excess pore pressures and the additional embankment loading.

To predict the centrifuge model test results, the FEM program can be run using either the model dimensions or the prototype dimensions as a basis for discretization. The simplest approach is to treat the centrifuge model testing as a real event, using the model dimensions for FEM discretization and unit weights increased in proportion to the acceleration level. Using this method, no scaling is required in interpreting either the actual physical measurements or the numerical predictions. Since Series I only involves one geometry and loading condition, the test was analyzed following this fashion.

For analyzing Series II, which consisted of models having different dimensions but representing one prototype, the second approach is preferred because only one analysis is required. However, the times and deformations observed in the centrifuge test must be scaled up according to the proper scaling factors, N^1 and

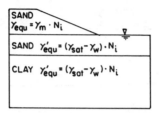

(a) Step I. FEADAM Analysis of Effective Stress
at End of Re-initialization of Acceleration N_i

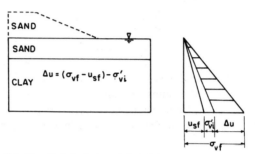

(b) Step 2. Calculation of Self-weight Induced Excess
Pore Pressure for Foundation Clay

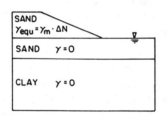

(c) Step 3. Embankment Loading - Use $\gamma_{equ} = \gamma \cdot \Delta N$ where
ΔN is the Increment of Centrifuge Acceleration
$\Delta N = N_f - N_i$

FIG. 10—*Schematic illustration of analysis procedure simulating Stage II centrifuge loading.*

N^2 for deformation and time, respectively, for direct comparison with the numerical prediction.

Series I

The finite-element mesh and boundary conditions used to represent the model in the Series I tests are shown in Fig. 11. The values of pore pressure predicted by the CON2D program are plotted as solid lines in Fig. 5. Comparisons of the predicted and measured pore pressures reveal the following information:

1. The initial response to "loading" was predicted reasonably well. The maximum deviations between measured and predicted values were in the range of 15 to 20%. One possible reason for this deviation is a computed pore pressure oscillation during undrained conditions that results from the inherently poor stability of the equations used to compute undrained pore pressures. With time, flow occurs and the oscillations disappear. For example, the pore pressure response predicted by the CON2D analysis for Transducer D (Fig. 5e) was very close to the experimental observations because Transducer D was located near the lower drainage boundary, allowing pore pressure to dissipate quickly. Another reason for the observed deviations between predictions and measurements is that for some transducers, the pore pressure nodal points in the finite-element mesh do not match the actual transducer locations. In these cases, interpolation between adjacent nodal pore pressures can introduce errors.

2. The rates of excess pore pressure dissipation were generally in good agreement with predictions. The rate of pore pressure dissipation is largely controlled by the permeability and compressibility of the soil skeleton, which, in turn, are stress-dependent. Further improvements in the predictions may be achieved by dividing the clay foundation into more layers and using representative permeability and compressibility values for each layer.

The vertical settlements predicted by the CON2D program are shown as solid lines in Fig. 7. A comparison of the predictions with the results of the centrifuge tests shows:

1. The CON2D analysis overpredicted the immediate deformation by 50% in areas beneath the central portion of the embankment, but underpredicted defor-

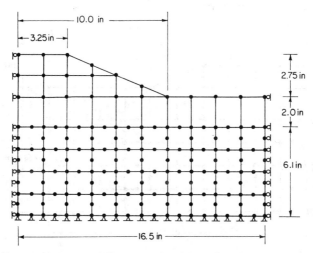

FIG. 11—*Finite element mesh and boundary conditions in Series I test (1 in. = 2.54 cm).*

mations in the regions near or beyond the toe of the embankment. The overprediction of initial settlements is probably due to the low Young's modulus assigned to the material. In the CON2D computer program the Young's modulus E was calculated as

$$E = 3B (1 - 2v) \tag{3}$$

where B is the bulk modulus and is related to the slope of the isotropic rebound curve κ according to

$$B = \frac{(1 + e) \, \sigma'_{oct}}{\kappa} \tag{4}$$

and where v is Poisson's ratio. Therefore both the values of the slope of rebound curve κ and Poisson's ratio v influence the computed immediate elastic undrained settlement in CON2D analysis. The underprediction of initial vertical settlement at areas close to the toe of the embankment is probably due to a deficiency of the numerical model in accounting for significant reorientation of principal stress directions.

2. The long-term settlements were well predicted.

Series II

Series II tests were analyzed as a single centrifuge model having the dimensions shown in Fig. 12 subjected to equivalent acceleration increases from 1.00 to 2.35 *g*. The dimensions shown in Fig. 12 were computed as a product of model

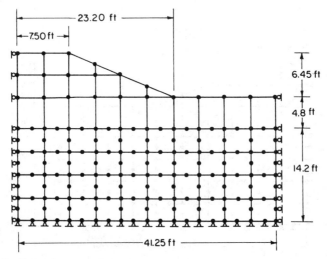

FIG. 12—*Finite element mesh and dimensions in Series II test analysis (1 ft = 0.305 m).*

dimensions and corresponding initial acceleration. An equivalent acceleration was calculated as the ratio of Stage II to Stage I accelerations—a single value of 2.35 for the three tests in Series II.

Some observations based on the measured data and numerical predictions can be made:

1. Shown in Fig. 8 are some of the pore pressure measurements and the corresponding predictions. Notice that a time-scaling factor N^2, where N is the acceleration during the re-initialization stage, was used in plotting the measured pore pressure responses in order to facilitate a direct comparison with the numerical predictions. The comparison shows that the response of pore pressure, both initial increment and subsequent rate of dissipation, is closely predicted by the analysis.

2. A comparison of the vertical settlements at several nodal points is shown in Fig. 9. The prediction of the foundation deformations exhibited the same trend as found in Series I tests. Basically, the immediate settlements were overpredicted in areas directly beneath the central portions of the embankment but were underpredicted in areas near the toe of the embankment. The long-term settlements were well predicted, however.

3. A typical comparison of pore pressures measured by transducers located in similiar positions in the Series II models is shown in Fig. 8b. The good agreement indicates that model-size effects are small.

Centrifuge Limitations

Centrifugal model testing, like other soil testing techniques, has some inherent limitations. It is important to take them into consideration when interpreting the test data. Among the more severe limitations are:

1. The centrifuge acceleration field is radial, as shown in Fig. 13. Soil elements that lie along a radial line normal to the soil surface (e.g., Element A in Fig. 13) are subjected to a "vertical" acceleration force. However, for other elements (e.g., Element B), the acceleration force is inclined at an angle $\Delta\phi = \Delta Y_0/R$. The error introduced by this effect depends upon the width of the box, w_0, relative to the effective radius, R_0, at mid-depth of the model. Hird [16] suggested that

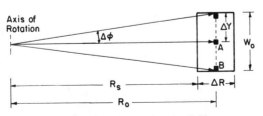

FIG. 13—*Radial acceleration field.*

radial divergence of the acceleration field was significant for w_0/R_0 ratios greater that 0.20. In our case the model width is 0.42 m and the effective radius is 1 m; therefore the w_0/R_0 ratio is about 0.42. Because of its significance, the radial acceleration field was taken into account by decomposing the acceleration forces into both normal and horizontal components during the analysis.

2. The vertical stress distribution in a centrifuge model is curved, as shown in Fig. 14. Schofield [17] examined the effect of variable acceleration with radius in the centrifuge and proposed the following expression for estimating the maximum error in vertical stress:

$$\frac{\Delta\sigma_v}{\sigma_v} = \frac{1}{2}\frac{\Delta R}{R_s}$$

where

$\Delta\sigma_v/\sigma_v$ = maximum error in vertical stress,
ΔR = $R - R_s$, in which R is the radius from the axis of rotation to the point of interest in the model, and
R_s = radius between the axis of rotation and the surface of the soil model.

Therefore, the maximum error in vertical stress is characteristic of the centrifuge facility and the model dimensions. In our case the clay foundation height (ΔR) is 0.15 m and the R_s is 1 m; thus the maximum stress error is about 8%.

3. The effect of friction and adhesion between the soil and model container can be significant in some cases. Rowe [18] observed that the effects of side friction can be minimized by increasing the width-to-height ratio of the model. According to Mair [19], there may be a 4% pressure deficiency at the base for

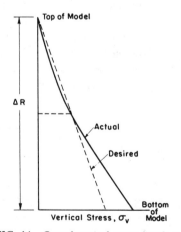

FIG. 14—*Curved vertical stress distribution.*

the model dimensions used in our tests. However, proper lubrication of the box sides can greatly reduce the side friction effect. Additionally, the side friction effect and curved vertical stress tend to cancel each other.

Summary and Conclusions

Two series of centrifuge tests of model embankments on soft clay foundations were performed to measure the time-dependent pore pressure and deformation behavior of soft clays. The initial results have been analyzed using the CON2D consolidation program, as secondary compression and creep effects are small for the kaolin clay being tested, and final coding of a new computer model is not yet complete. Based on these results, the following conclusions can be drawn:

1. The model generally predicted well the pore pressure responses—both the initial increment and the rate of subsequent dissipation. The predictive errors in the initial excess pore pressure increment are as high as 15% for some of the transducers. This is attributed to the poor numerical scheme used in calculating the undrained pore pressures as well as to the interpolation of analysis results necessitated by not being able to match nodal points in the analysis and transducer locations in the experiment. Further refinement of the prediction of pore pressure dissipation rate can be achieved by dividing the foundation into more layers and using more representative permeability and compressibility values for each layer.
2. The model failed to accurately predict the immediate settlement of the clay foundation but predicted the long-term settlement well. The model tends to overpredict the immediate deformation of the clay foundations beneath the central portion of the embankment but underpredicts them near the toe of the embankment.

Results from the modeling of model tests showed that model size in the range investigated did not affect the clay foundation behavior.

The limitations of centrifuge testing were examined. The effect of the radial acceleration field can be significant in situations where a large model is carried in a small centrifuge. Other limitations such as the curved vertical stress distribution and side friction are considered small in our case, and may compensate for each other to some extent.

Finally, it is concluded that centrifuge testing can be a useful tool for investigating the accuracy and usefulness of new soil behavior or numerical models. The data presented can also provide a valuable reference for evaluating other constitutive and numerical models.

Acknowledgments

This work forms a part of research under National Science Foundation Grant CEE-8203563. This support is gratefully acknowledged. The authors thank Professor James A. Cheney and Professor C. K. Shen of the University of California at Davis for their assistance during the experimental work. Edward Tse and Matthew Kuhn, Graduate Research Assistants at the University of California at Berkeley, assisted in the centrifuge testing and soil property determination.

References

[1] Wroth, C. P. et al, "Description of Method of Prediction," in *Proceedings*, Foundation Deformation Prediction Symposium, Massachusetts Institute of Technology, Cambridge, MA, 13–15 Nov. 1974.

[2] Kavazanjian, E., Jr., Borja, R. I., and Jong, H. -L., "Numerical Analysis of Time-Dependent Deformations in Clay Soils," in *Proceedings*, International Conference on Soil Mechanics and Foundation Engineering, San Francisco, 1985.

[3] Kavazanjian, E., Jr., and Mitchell, J. K., "A General Stress-Strain-Time Formulation for Soils," in *Proceedings*, Ninth International Conference on Soil Mechanics and Foundation Engineering, Tokyo, 1977.

[4] Kavazanjian, E., Jr., and Mitchell, J. K., "Time-Dependent Deformation Behavior of Clays," *Journal of the Geotechnical Engineering Division*, American Society of Civil Engineers, Vol. 106, No. GT6, June 1980, pp. 611–630.

[5] Kavazanjian, E., Jr., Mitchell, J. K., and Bonaparte, R., "Stress-Deformation Predictions Using a General Phenomenological Model," Position paper for the NSF/NSERC Workshop on Plasticity Theories and General Stress-Strain Modeling of Soils, McGill University, Montreal, May 1980.

[6] Bonaparte, R., "A Time-Dependent Constitutive Model for Cohesive Soils," Ph.D. Dissertation, University of California, Berkeley, 1981.

[7] Chang, C.-S. and Duncan, J. M., "Analysis of Consolidation of Earth and Rockfill Dams," Geotechnical Engineering Report, Department of Civil Engineering, University of California, Berkeley, 1977.

[8] Kavazanjian, E., Jr., Mitchell, J. K., and Bonaparte, R., "Three-Dimensional Time-Dependent Behavior of Clays," *Journal of Geotechnical Engineering*, American Society of Civil Engineers, to be published.

[9] Singh, A. and Mitchell, J. K., "General Stress-Strain-Time Function for Soils," *Journal of the Soil Mechanics and Foundation Division*, American Society of Civil Engineers, Vol. 94, No. SM1, 1968, pp. 21–46.

[10] Lade, P., "The Stress-Strain and Strength Characteristics of Cohesionless Soils," Ph.D. dissertation, University of California, Berkeley, 1971.

[11] Britto, A. M., "A User's Guide to the Soils Strain Calculating Program," Cambridge University, Cambridge, England, 1980.

[12] Roscoe, K. H. and Burland, J. B., "On the Generalized Stress-Strain Behavior of 'Wet' Clays," *Engineering Plasticity*, Cambridge, England, 1968, pp. 535–609.

[13] Biot, M. A., "General Solutions of the Equations of Elasticity and Consolidation for a Porous Material," *Journal of Applied Mechanics, Transactions of ASME*, Vol. 78, 1956, pp. 91–96.

[14] Duncan, J. M., "CON2D: A Finite Element Computer Program for Analysis of Consolidation," Report No. UCB/GT/81-01, Department of Civil Engineering, University of California, Berkeley, 1981.

[15] Duncan, J. M., Wong K. S., and Ozawa, Y., "FEADAM: A Computer Program for Finite Element Analysis of Dams," Report No. UCB/GT/80-02, Department of Civil Engineering, University of California, Berkeley, 1980.

[16] Hird, C. C., "Centrifugal Model Tests of Flood Embankments," Ph.D. Thesis, University of Manchester (UMIST), England, 1974.

[17] Schofield, A. N., "Cambridge Geotechnical Centrifuge Operations," *Geotechnique*, Vol. 30, No. 3, 1980, pp. 227–268.

[18] Rowe, P. W., "Large Scale Laboratory Model Retaining Wall Apparatus," *Stress Strain Behavior of Soils*, Publication Foulis, 1972, p. 441.

[19] Mair, R. J., "Centrifugal Modeling of Tunnel Construction in Soft Clay," Ph.D. Thesis, University of Cambridge, England, 1979.

*C. K. Shen,[1] J. Sohn,[1] K. Mish,[1] V. N. Kaliakin,[1] and
L. R. Herrmann[1]*

Centrifuge Consolidation Study for Purposes of Plasticity Theory Validation

REFERENCE: Shen, C. K., Sohn, J., Mish, K., Kaliakin, V. N., and Herrmann, L. R.,
"Centrifuge Consolidation Study for Purposes of Plasticity Theory Validation," *Consolidation of Soils: Testing and Evaluation, ASTM STP 892*, R. N. Yong and F. C. Townsend, Eds., American Society for Testing and Materials, Philadelphia, 1986, pp. 593–609.

ABSTRACT: The experimental phase of the research described in the paper involves a centrifuge model study of consolidation and surface settlement of a storage tank located on a soft soil foundation during a filling-storage-emptying cycle. The tank is placed on a thin layer of sandfill over a weak and compressible deposit of clay soil. The clay deposit is formed by consolidating a kaolin slurry in a 40.6-cm-diameter circular mold which is subsequently moved to the centrifuge for further consolidation and "field stress" initiation. The combined processes of consolidation cause the clay deposit to form an overconsolidated layer in the upper portion and a normally consolidated layer below. This situation simulates conditions often encountered in nature. Pore water pressure transducers are embedded in the clay deposit to monitor the rise and dissipation of water pressure as a result of loading and unloading. Furthermore, the surface settlement profile of the tank bottom is measured using a number of optical-electrical displacement transducers.

The second phase of the study involved a comparison of finite-element predictions with the experimentally measured quantities. These comparisons are part of a validation study for the recently developed bounding surface plasticity model for cohesive soils. The model is calibrated using the results of standard laboratory triaxial tests for the soil in question. A system of Fortran subroutines to numerically evaluate the model has been developed and incorporated into 2-D and 3-D finite-element consolidation programs. In this study the 2-D program is used to predict the soil response for comparison with experimental results. The predicted pore water pressures and surface displacements are presented and compared with the measured values. The agreement between the pore pressure values is very good while for the displacement it is adequate.

KEY WORDS: bounding surface, plasticity model validation, centrifuge model test, consolidation, finite-element analysis, geomechanics, plasticity theory, soil characterization

In recent years various plasticity models have been proposed to describe the behavior of soft clay soils. While advancements in geomechanical considerations give more precise physical interpretation of soil behavior and provide insight into the many significant factors governing the stress-strain-time relationships

[1] Professor, Graduate Student, Graduate Student, Graduate Student, and Professor, respectively, Department of Civil Engineering, University of California, Davis, CA 95616.

for soils, there remains an urgent need that these proposed soil models be properly calibrated and validated by laboratory or model tests or both under a variety of loading conditions. To date, well-documented soil data are scarce and in great demand.

During the past five years, the bounding surface plasticity concept proposed by Dafalias and Popov [1] has been successfully applied to describe the laboratory behavior of clay soils [2]. More recently, the numerical implementation of the bounding surface model in both 2-D and 3-D finite-element codes has been completed by Herrmann et al [3]. This paper summarizes a model study of consolidation and surface settlement during a filling-storage-emptying cycle of a storage tank placed on soft soil. The model study was carried out under elevated g-levels in a centrifuge. The purposes of this study are to further evaluate the predictive capabilities of the bounding surface model under complex loading and unloading conditions and to provide experimental data that can be used by other researchers working with alternative soil characterizations.

Centrifuge Model Study

Model tests were carried out in the Schaevitz Type B-8-D rotary accelerator. This centrifuge is capable of applying controlled centrifugal acceleration up to 175 g's or 10 000 g-lb at a nominal radius of 1 m.

The Model Package

The model package tested in the centrifuge consists of a model box, a model of a storage tank situated on a soft clay foundation, and a water reservoir. The model box, containing the soil, was made of 0.64-cm-thick aluminum alloy plate (6061-T6). The plate was rolled, and the two ends were welded together to form a 40.6-cm-inside-diameter tub. The tub is 33 cm high, closed at one end with a rectangular aluminum plate (43.2 by 45.7 cm). The model storage tank was made of 0.28-cm-thick aluminum pipe. The tank is 17.8 cm high and 16.3 cm inside diameter. The bottom of the tank and the foundation ring were fastened together with screws; a thin rubber sheet (1.5 mm thick) was made to fit between them as a flexible bottom plate. The 8200-cm^3 water reservoir was mounted on a steel frame which was fastened to the swing-up platform. A photograph of the overall package is shown in Fig. 1.

Specimen Preparation

The clay used in this study is a mixture of two commercial grades of kaolin, Snow-Cal 50 and Mono 90. All particles of Mono 90 are finer than the #325 standard U.S. sieve, while particles of Snow-Cal 50 are finer than the #250 sieve. Three parts of Snow-Cal 50 were mixed with one part of Mono 90. Laboratory determined physical properties of the mixture are given in Table 1 [4].

FIG. 1—*View of model package.*

The steps used in specimen preparation are as follows:

1. Forty-five kilograms (100 lb) of the kaolin mixture was mixed with distilled water for 12 h. The stirring was done with extreme care in order to minimize the trapping of air in the slurry. The slurry ($w/c = 85\%$) was then poured into the model box.

2. After placing the loading plate, an initial vertical pressure of 6.14 kN/m² was applied to the specimen. The load was doubled once a day until it reached the desired consolidation pressure of 49.2 kN/m².

3. After completion of primary consolidation under the desired pressure, the vertical load was reduced to zero. Two miniature pore pressure transducers made by Druck were inserted horizontally into the center of the clay specimen through holes drilled in the tub wall. Backfilling the holes was accomplished by injecting

TABLE 1—*Soil properties.*

Liquid limit (LL)	37%
Plastic limit (PL)	29%
Shrinkage limit (SL)	26%
[a]Coefficient of compressibility (a_v)	(3.96 to 7.92) $\times$ 10⁻³ m²/kN
[a]Coefficient of permeability (k)	(0.5 to 1.4) $\times$ 10⁻⁵ cm/s
[a]Coefficient of consolidation (c_v)	3.3 $\times$ 10⁻² cm²/s

[a] Determined at consolidation pressures between 27.6 and 193.2 kN/m².

kaolin slurry with a long syringe, and a watertight seal was applied around the transducer wires. The vertical pressure was then brought back to the desired consolidation pressure.

4. Twenty-four hours later, the vertical pressure was again reduced to zero, and the loading plate was removed from the model box. A 2.54-cm layer of saturated sand (Monterey No. 0) was placed on the clay surface and leveled by a steel edge.

5. An aluminum ring modeling the ring wall footing of the tank structure was placed on the sand layer. The whole system was again consolidated under the desired pressure for 24 h to reinitiate the consolidated state of the clay layer.

6. The system was transferred to the centrifuge and placed in one of the two swing-up platforms.

7. Prior to the centrifuge test, a thin, circular rubber sheet simulating a flexible tank bottom was fitted between the bottom of the storage tank and the ring wall. The storage tank was then fastened to the ring wall.

8. A steel frame was securely mounted to the swing-up platform. Six optical-electrical displacement (OED) transducers were positioned and clamped to a steel channel connected to the frame. The water reservoir was also securely fastened to the steel frame.

9. The required amount of water was placed in the reservoir for in-flight loading. Both static and dynamic balancing of the rotating arm were performed prior to testing, thus completing preparation of the model system for centrifuge testing.

Figure 2 shows the positions of the OED and pore pressure transducers, as well as the various dimensions of the model package.

Instrumentation and Data Acquisition

A microcomputer-based time-division data acquisition system [5] was used to monitor the model behavior. The model response was measured by two pore water pressure transducers imbedded in the clay layer, and six OED transducers in direct contact with the ground surface. Water level in the tank was controlled by two electrically operated solenoid switches under the command of the software program. Both surface displacement and pore water pressure readings were taken automatically at various time intervals throughout the entire testing period. Therefore, the surface displacement profile and the pore water pressure buildup and dissipation were conveniently recorded and plotted with time. In addition, a TV monitor permitted continuous viewing of the model during the test and a video tape recording of the events was made for later viewing.

Testing Program

The centrifuge model test performed for this study involved loading (filling) and unloading (drainage) of the storage tank. Table 2 presents the essential

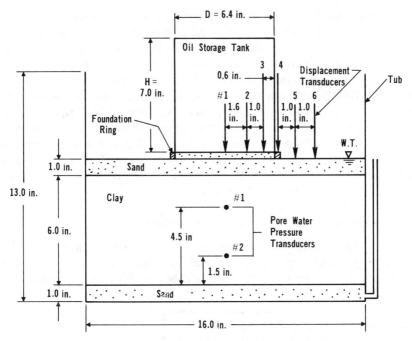

FIG. 2—*Dimensions of model package (1 in. = 2.54 cm).*

features of the test. As indicated, the model was tested under an equivalent gravitational field 60 times earth's gravity, and enough time was allowed for total pore pressure dissipation during each load increment (field stress initiation, loading, and unloading). There were four loading stages followed by two unloading stages. In the first stage the initial stress state is established at the 60-g level. This stress state varies from zero at the surface to a maximum at the bottom of the layer; the variation is due to the effects of overburden (and the effect of radius of rotation), and can be approximated by a linear variation with a maximum

TABLE 2—*Testing sequence.*

Test Segment	Function	Transient Loading (Unloading) Time, s	Total Loading (Unloading) Period, s	Load Increment (Decrement), psi	g-Level
1	initiation of field stress	. . .	6000	. . .	1 to 60
2	water loading	28	1800	2.6	60
3	water loading	28	1800	2.6	60
4	water loading	14	1800	1.3	60
5	water unloading	30	1800	−2.15	60
6	water unloading	30	1800	−2.15	60

NOTE: 1 psi = 6.9 kN/m².

error of less than 7%. In the precentrifuge consolidation of the specimen a uniform effective stress state of 49.2 kN/m² was developed throughout the clay layer due to 1-D consolidation (Fig. 3a); the solid line in Fig. 3b represents the effective vertical stress distribution in the clay layer at the 60-g level. The result was that the clay deposit developed an overconsolidated zone with variable overconsolidation ratio (OCR) values in the upper portion and a normally consolidated zone below. Since the excess pore water pressure was completely dissipated during the field stress initiation stage, the normally consolidated and overconsolidated zones are clearly defined. The remaining loading and unloading stages involved the adding or emptying of water from the tank. At no time did the intensity of the water loading approach the failure load of the foundation. The failure load for this system was studied by Kim [19].

Theoretical Model and Numerical Implementation

During the past five years there has been a great deal of research concerning the development of comprehensive constitutive equations for soils (for example, see Ref 6). The constitutive description for cohesive soils selected for evaluation

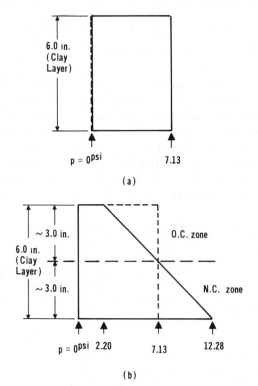

FIG. 3—*Effective stress distribution. (a) at 1 g. (b) at 60 g. (1 in. = 2.54 cm; 1 psi = 6.9 kN/m².)*

in this study is the bounding surface plasticity model developed by Dafalias and his co-workers [2,3,7–10]. To date, the evaluation of the model has been limited to numerous comparisons of predicted and experimental results for simple, homogeneous, laboratory tests (see cited references). The present study is the first evaluation for simulated field conditions. An overview of the plasticity model and its numerical implementation in finite-element codes is given in the following paragraphs; for a detailed description, the reader is referred to the cited references.

The bounding surface plasticity theory for cohesive soils provides a comprehensive description of the stress-strain response for arbitrary initial soil conditions and three-dimensional stress histories. One set of parameters (along with a knowledge of the soil's initial state) is sufficient to describe the soil behavior for all levels of overconsolidation for both monotonic and cyclic, three-dimensional stress histories. Unlike classical plasticity theory which assumes elastic behavior for all states inside the yield surface, the bounding surface theory, in accordance with experimental evidence for cohesive soils, allows for plastic deformation to occur within the bounding surface. Satisfactory comparisons of model predictions to experimental results for simple laboratory tests may be found in the cited references.

The fundamental concept of the theory is the existence in stress space of a bounding surface (Fig. 4) which controls the development of plastic deformation in a more flexible way than a yield surface (where I is the first stress invariant and J is the square root of the second deviatoric stress invariant; for clarity, the dependence upon the third invariant is not shown). At any point in the stress history the current size of the surface (I_0) is controlled by the values of appropriate plastic internal variables; for the isotropic behavior of cohesive soils the plastic void ratio change suffices for this purpose. For cohesive soils the actual form of the bounding surface was motivated by concepts from "critical state" soil mechanics [11]. For a given stress state within or on the bounding surface, an appropriate mapping rule [3] is used to establish an image point on the bounding surface. The actual and image stress points coincide when the former lies on the surface. The normal ∇F to the bounding surface at the "image" point determines the direction of plastic flow. Also associated with the image point is a bounding plastic modulus. The plastic modulus at the actual stress point is a function of this bounding plastic modulus and the distance δ (in stress space) between the two points. This expression contains one of the fundamental parameters of the theory, the "shape hardening parameter" which controls the plastic stiffness within the bounding surface for overconsolidated states.

In addition to the classical "critical state" soil parameters (note: $N = M_c/\sqrt{27}$), the model requires a number of additional material constants which define the shape of the bounding surface, the hardening parameter, the location of the projection point (I_c), and the size ($r - r/s$) of the elastic nucleus (if any). Several of these parameters are shown in Fig. 4. The model parameters are determined from standard consolidation tests, and triaxial tests performed in both extension and compression at two or three different overconsolidation ratios. Both manual-

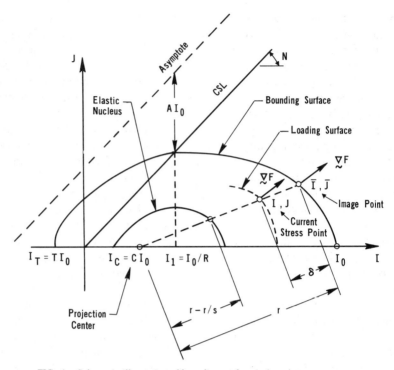

FIG. 4—*Schematic illustration of bounding surface in invariant stress space.*

and computer-evaluated calibration procedures have been developed for determining the model parameters from the experimental data. The calibration procedures are described in Ref *12* and properties for typical soils are given in Ref *13*.

For the present study, six isotropically consolidated undrained tests were performed using a triaxial testing device. Three tests were in extension and three in compression; the OCR's for each group were 1, 2, and 6. The results of the consolidation tests are given as solid and open points in Fig. 5; the measured undrained stress paths for the triaxial tests are shown in Fig. 6. From the results of the consolidation tests, initial values for the critical state parameters λ (0.0645) and κ (0.0073) were selected. During the subsequent calibration effort (selection of the remaining model parameters) it was found desirable to change λ and κ to 0.0745 and 0.0105, respectively. Both values fell within the experimental scatter, and significantly improved the numerical predictions for the undrained stress paths. The slopes M_c and M_e of the critical state lines in compression and extension were determined from Fig. 6 to be 1.35 and 0.90, respectively. For the elastic portion of the clay response the bounding surface model can be used either with a constant shear modulus or with a constant Poisson's ratio specified; in past work the latter has proven to be the most effective. In the absence of experimental evidence for this particular soil, a value for Poisson's ratio of 0.20 found in

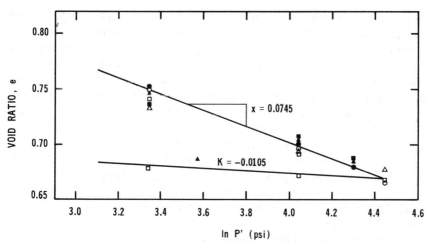

FIG. 5—*Isotropic consolidation results (1 psi = 6.9 kN/m²).*

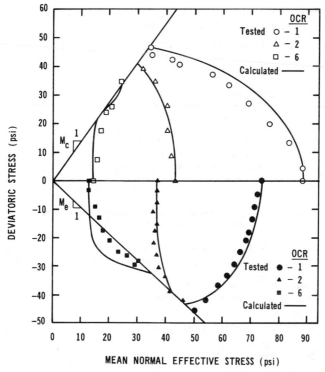

FIG. 6—*Undrained stress paths under triaxial conditions (1 psi = 6.9 kN/m²).*

previous work with a similar soil was selected. The remainder of the material and shape parameters which describe the bounding surface model were determined using the manual calibration procedure described in Ref 7; the values of these parameters are summarized in Table 3. Figure 6 compares the measured undrained stress paths with those predicted by the calibrated model; similar accuracy was found for the pore water pressure and deviatoric stress response. The calibration of the model was carried out independently of the centrifuge test, so the comparisons discussed in the following section are a true test of the predictive capabilities of the model. The calibration was based upon results from triaxial tests.

Consolidation parameters were determined from the one-dimensional consolidation test, and average values are tabulated in Table 1. Only the coefficient of permeability, however, is required for the analysis; a value of 1.4×10^{-5} was used. The saturated unit weights of the sand and clay are 2.02 g/cm^3 and 1.78 g/cm^3, respectively.

The incrementalization (by means of trapezoidal integration over the time step) of the rate equations of the bounding surface theory, for use in step-by-step solution algorithms, is described in Ref 3. In this form the model is easily incorporated into new and existing finite-element codes which use an approximate form (e.g., successive substitution or tangent stiffness [14] of the Newton-Raphson algorithm for determining the nonlinear response over a given time step. In Ref 15 detailed instructions are given for incorporating the subroutines which evaluate the model into new and existing codes. The modeling of water flow, and the development and dissipation of pore water pressure in saturated soils,

TABLE 3—*Parameters describing bounding surface model.*

Symbol	Description of Property	Value
λ	Slope of isotropic consolidation line for an e-ln p' plot	0.0745
κ	Slope of elastic rebound line for an e-ln p' plot	0.0105
M_c	Slope of critical state line in triaxial space (compression)	1.35
R_c		3.05
A_c	Parameters describing shape of bounding surface (compression)	0.175
T		0.010
P_t	Transitional value of confining pressure separating linear rebound curves on e-ln p' and e-p' plots	4.4
ν	Poisson's ratio	0.22
P_a	Atmospheric pressure (used for scaling and establishing units)	101.43 kN/m^2
Γ	Combined bulk modulus for soil particles and pore water	6.9×10^6 kN/m^3
m	Hardening parameter	0.02
h_c	Shape hardening parameter for compression	175.0
h_2	Shape hardening parameter on the I-axis	165.0
$n = M_e/M_c$		0.667
$\mu = h_e/h_c$	Ratio of extension-to-compression values	0.875
$r = R_e/R_c$		0.560
$a = A_e/A_c$		0.850
C	Projection center variable	0.485
S	Elastic zone variable	1.00

follows closely the development given in Ref *16*; details of the analysis can be found in Ref *17*. Descriptions of two of the finite-element codes in which the authors have implemented the model can be found in Refs *13* and *18*; the two-dimensional code [*13*] was used for the analysis reported herein. Since the finite-element analysis takes into account the actual boundary conditions and the elevated *g* force experienced in the model, a direct comparison of the model measurements and the theoretical predictions can be carried out without considering the scaling factors.

Results and Comparison

The entire test in the centrifuge lasted for 250 min. Information concerning the six segments of the test are tabulated in Table 2. Readings of the pore water pressure and OED transducers were taken by the microcomputer-based automated data acquisition system at 20 s intervals. The locations and the numbering of the sensors are given in Fig. 2. In order to judge experimental scatter, two "identical" tests were performed. Plots of pore pressure dissipation with time and the corresponding surface settlement at various locations for Test Segment 1 (initiation of field stress in Table 2) are shown in Figs. 7 and 8, respectively. It is interesting to note that some of the readings from the outside displacement transducers (Nos.

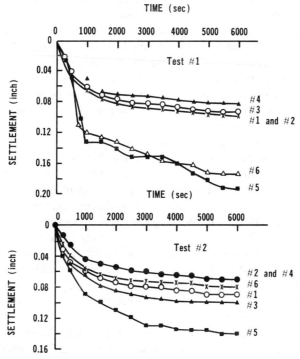

FIG. 7—*Surface settlement during initialization (1 in. = 2.54 cm).*

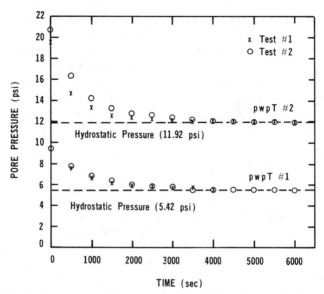

FIG. 8—*Pore pressure dissipation during initialization (1 psi = 6.9 kN/m²).*

5 and 6) are much larger than those registered by the inside transducers. Since surface settlement due to self-weight consolidation should be relatively uniform (a maximum of 4% error exists due to the effect of the radius of rotation), this difference cannot be satisfactorily explained. Based upon similar studies performed by Kim [19], it is believed that those settlement readings are in error, possibly as a result of local disturbance due to mounting of the tank structure. A total of 100 min was allowed for consolidation; both the pore water pressure and surface displacement readings indicate that a nearly completed consolidation process was achieved during this period.

The rise and fall of the pore water pressure with time due to each loading and unloading segment are shown as data points in Fig. 9 for both the No. 1 and No. 2 transducers. The No. 1 transducer is imbedded at a depth 3.8 cm below the surface of the clay layer, which is the midpoint of the overconsolidation zone, whereas, the No. 2 transducer is positioned at the midpoint of the normally consolidated zone, 11.4 cm below the surface of the clay layer. The load in each segment was maintained for a period of 30 min. It can be seen clearly from Fig. 9 that the loading (or unloading) induced pore water pressures are, by and large, dissipated at the end of the 30-min intervals. The surface settlement records are shown as data points in Fig. 10. As expected, since they are closely related for saturated clay soils, the shapes of the settlement curves seem to match well the pore water pressure dissipation curves.

The length of the time steps and the size of the finite-element mesh used in the finite-element analysis [13] were selected in order to give acceptable accuracy while keeping the computer cost within reason. The solution steps, and an indication of the number of iterations within each step, are described in Table 4. The finite-element grid used in the analysis is shown in Fig. 11. An analysis over part of the time history with a finer grid gave substantially the same results, thus demonstrating that the grid of Fig. 11 is sufficiently fine. The total central processor unit (CPU) time on a DEC VAX 780 computer was approximately 20 min.

Two uncertainties in the analysis related to the modeling of the sand and of the interface between the soil and the wall of the centrifuge bucket need to be mentioned. The only information available to the authors for the Monterey No. 0 sand is the "hyperbolic" elastic characterization given by Wong and Duncan [20]. Because the code used for the analysis does not have Duncan's nonlinear soil model as an option, it was necessary to use his equations to estimate an average modulus and treat the sand as linear elastic. It was estimated that E should be in the range 3.45×10^3 kN/m² to 34.5×10^3 kN/m²; Poisson's ratio was taken to be 0.3. A value of $E = 20.7 \times 10^3$ kN/m² was used in the first analysis. The pore water pressure predictions and the shape of the displacement curves were in good agreement with the experimental results; however, the magnitude of the displacements was underestimated. Because of the uncertainties in the sand modulus it was felt to be justifiable to modify this parameter; hence a second run with $E = 10.35 \times 10^3$ kN/m² was performed. The pore water

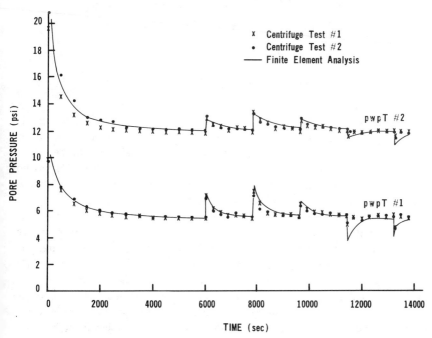

FIG. 9—*Pore pressure response to loading and unloading (1 psi = 6.9 kN/m²).*

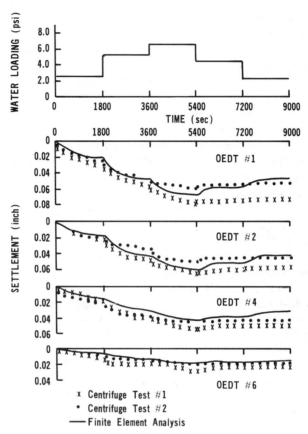

FIG. 10—*Surface settlement during loading and unloading (1 in. = 2.54 cm; 1 psi = 6.9 kN/m².*).

pressure predictions were virtually identical for the two analyses; however, the surface displacements were approximately 30% greater for the second. The results of this second run are used for comparison (in Figs. 9 and 10) with the experimental results. Finally, because a "frictional surface" capability [21] has not yet been incorporated into the code, it was necessary to approximate the interface between the bucket and the soil as one of rollers.

The predicted pore water pressures and surface displacements are presented and compared with experimental results as shown in Figs. 9 and 10, respectively. Considering the scatter in the experimental results, the agreement in the pore water pressure values is very good while that for displacement is adequate. The overprediction of the change in pore water pressure at the beginning of some loading/unloading segments is probably due to the inadequate response of the experimental equipment and/or oscillation in the solution due to the abrupt change in size of the time step. The slight underprediction in displacement is almost certainly due to the inadequate representation of the sand layer.

TABLE 4—*Solution steps, finite-element analysis.*

Solution Interval	Time Step Length, s	Number of Iterations
Self-Weight Consolidation	100	2
(Total duration: 6000 s)	100	3
	100	2
	100	2
	100	2
	500	4
	500	3
	500	2
	500	2
	500	2
	3000	3
Loading/Unloading Segments	40	5[a]
(Total duration: 1800 s/segment)	60	4
	100	4
	400	4[a]
	400	2
	400	2
	400	2

[a] Averaged over the five loading/unloading segments.

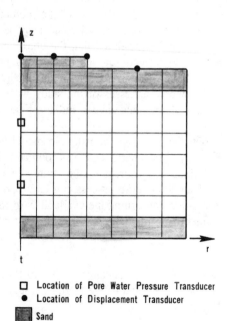

☐ Location of Pore Water Pressure Transducer
● Location of Displacement Transducer
▨ Sand
☐ Clay

FIG. 11—*Finite-element grid.*

Conclusions

Centrifuge modeling is an inexpensive means for providing the needed data, on realistic geotechnical soil and soil-structure problems, to verify the adequacy of proposed constitutive models for soils. For the given model test, a portion of the soil specimen was initially overconsolidated while the remaining was normally consolidated. During the course of the test both loading and unloading of the soil occurred, the stress state was decidedly nonhomogeneous, and at most points in the specimen the principal stress directions changed during the course of the tests. While the study reported herein, due to a limited budget and time, is not as complete as desired, it does provide another important step in the validation process of the bounding surface plasticity model for cohesive soils and will provide an important check case for investigations of other models.

Using the bounding surface model, the entire cohesive soil specimen was characterized by a single set of soil parameters. The nonhomogeneous initial state of the soil at the beginning of the storage tank test was accounted for by including the preconsolidation phase of the centrifuge test in the analysis. Alternatively, one could have input to the program as an "initial state description" the conditions at the end of the preconsolidation phase (Test Segment 1) and then have started the analysis at that point.

The independence of the calibration process for the cohesive soil and the finite-element analysis of the centrifuge test from the test itself needs to be emphasized. Thus, the study is a truly meaningful test of the predictive capabilities of the bounding surface model for cohesive soils and provides direct evidence concerning its validity. It should be noted that the question of whether or not the centrifuge can be used to correctly model field conditions is of little concern in this study as the analysis was carried out for the model structure, not for the field prototype. The two weaknesses of the study were the inadequacy of the constitutive modeling of the sand layers and the modeling of the soil-bucket interfaces.

Finally, the "robustness" of the constitutive model and its calibration, and the finite-element analysis in general, deserves comment. The calibration process followed a well-defined procedure and the finite-element analysis was quite routine in nature. Thus the overall process can be proposed as a working engineering tool. Information concerning the availability of the referenced computer codes can be obtained from the authors.

Acknowledgments

The authors wish to express their appreciation to Professor Yannis Dafalias, who reviewed the manuscript and offered valuable suggestions.

References

[1] Dafalias, Y. F. and Popov, E. P., "A Modeling of Nonlinearly Hardening Materials for Complex Loading," *Acta Mechanica*, Vol. 21, 1975.

[2] Dafalias, Y. F. and Herrmann, L. R., "A Bounding Surface Soil Plasticity Model," in *Proceedings,* International Symposium of Soils Under Cyclic and Transient Loadings, University of Swansea, U.K., Jan. 1980, pp. 335–345.

[3] Herrmann, L. R., Dafalias, Y. F., and DeNatale, J. S., "Numerical Implementation of a Bounding Surface Soil Plasticity Model," in *Proceedings,* International Symposium on Numerical Models in Geomechanics, Zurich, 1982.

[4] Fragaszy, R. J., "Drum Centrifuge Studies of Overconsolidated Clay Slopes," Ph.D. thesis, University of California at Davis, Sept. 1979.

[5] Shen, C. K., Li, X. S., and Kim, Y. S., "Microcomputer Based Data Acquisition Systems for Centrifuge Modeling," accepted for publication in *Geotechnical Testing Journal.*

[6] *Soils under Cyclic and Transient Loading,* O. C. Zienkiewiez and G. N. Pande, Eds., Wiley, New York, 1982.

[7] Dafalias, Y. F., Herrmann, L. R., and DeNatale, J. S., "Prediction of the Response of the Natural Clays X and Y Using the Bounding Surface Model," in *Proceedings,* American Society of Civil Engineers Workshop on Limit Equilibrium, Plasticity and Generalized Stress-Strain in Geotechnical Engineering, McGill University, Montreal, Part 1, May 1980, pp. 402–415.

[8] Dafalias, Y. F. and Herrmann, L. R., "Bounding Surface Formulation of Soil Plasticity" in *Soil Mechanics—Transient and Cyclic Loads,* O. C. Zienkiewiez and G. N. Pande, Eds., Wiley, New York, 1982, pp. 253–282.

[9] Dafalias, Y. F., "Bounding Surface Elastoplasticity-Viscoplasticity for Particulate Cohesive Media," in *Proceedings,* IUTAM Symposium on Deformation and Failure of Granular Materials, International Union of Theoretical and Applied Mechanics, Delft, The Netherlands, Aug. 1982, pp. 97–107.

[10] Anandarajah, A., Dafalias, Y. F., and Herrmann, L. R., "A Bounding Surface Plasticity Model for Anisotropic Clays," in *Proceedings,* 5th Engineering Mechanics Division Specialty Conference, American Society of Civil Engineers, University of Wyoming, Laramie, Aug. 1984.

[11] Schofield, A. N. and Wroth, C. P., *Critical State Soil Mechanics,* McGraw-Hill, London, 1968.

[12] DeNatale, J. S., Herrmann, L. R., and Dafalias, Y. F., "Calibration of the Bounding Surface Soil Plasticity Model by Multivariate Optimization," in *Proceedings,* International Conference on Constitutive Laws for Engineering Materials, University of Arizona, Tucson, Jan. 1983.

[13] Herrmann, L. R. and Mish, K. D., "User's Manual for SAC-2, A Two-Dimensional Nonlinear, Time Dependent Soil Analysis Code Using the Bounding Surface Plasticity Model," Report CR84.008, Naval Civil Engineering Laboratory, Port Hueneme, CA, Dec. 1983.

[14] Owen, D. R. J. and Hinton, E., *Finite Elements in Plasticity—Theory and Practice,* Pineridge Press, Swansea, U.K., 1980.

[15] Herrmann, L. R., Kaliakin, V. N., and Dafalias, Y. F., "Computer Implementation of the Bounding Surface Plasticity Model for Cohesive Soils," Report CR84.007, Naval Civil Engineering Laboratory, Port Hueneme, CA, Dec. 1983.

[16] Ghabousi, J. and Wilson, E. L., "Flow of Compressible Fluid in Porous Elastic Media," *International Journal for Numerical Methods in Engineering,* Vol. 5, 1973, pp. 419–442.

[17] Herrmann, L. R. and Mish, K. D., "Finite Element Analysis for Cohesive Soil, Stress and Consolidation Problems Using Bounding Surface Plasticity Theory," Report CR84.006, Naval Civil Engineering Laboratory, Port Hueneme, CA, Dec. 1983.

[18] Mish, K. D. and Herrmann, L. R., "User's Manual for SAC-3, A Three-Dimensional Nonlinear, Time Dependent Soil Analysis Code Using the Bounding Surface Plasticity Model," Report CR84.009, Naval Civil Engineering Laboratory, Port Hueneme, CA, Dec. 1983.

[19] Kim, Y. S., "Centrifuge Model Study of an Oil Storage Tank Foundation," Ph.D. thesis, University of California at Davis, June 1984.

[20] Wong, K. S. and Duncan, J. M., "Hyperbolic Stress-Strain Parameters for Nonlinear Finite Element Analyses of Stresses and Movements in Soil Masses," Department of Civil Engineering Report TE-74-3, University of California at Berkeley, July 1974.

[21] Herrmann, L. R., "Finite Element Analysis of Contact Problems," *Journal of the Engineering Mechanics Division,* American Society of Civil Engineers, Vol. EM5, Oct. 1978.

Stanley M. Bemben[1]

Brittle Behavior of a Varved Clay During Laboratory Consolidation Tests

REFERENCE: Bemben, S. M., ''**Brittle Behavior of a Varved Clay During Laboratory Consolidation Tests,**'' *Consolidation of Soils: Testing and Evaluation, ASTM STP 892,* R. N. Yong and F. C. Townsend, Eds., American Society for Testing and Materials, Philadelphia, 1986, pp. 610–626.

ABSTRACT: The brittle behavior aspects manifested during laboratory consolidation tests of a varved clay deposited in a glacial lake and thereafter having acquired natural cementation are presented. The effect of the quality of the test specimens on the shape of the compression curves from one-dimensional consolidation tests employing both incremental and constant rate of strain loading are presented. For good quality test specimens, the compression curves are arithmetically linear up to critical stress values at which breakage of the cement commences. A method is presented for estimating the maximum previous vertical effective stress remembered by the soil fabric of the intact natural state soil when one-dimensional test curves display brittle behavior. When the cement bonds of the natural-state soil are intact, the soil fabric can be underconsolidated or overconsolidated with respect to the existing vertical effective stress. The brittle behavior of the intact natural-state soil is attributed to the behavior of the cement bonds.

KEY WORDS: varved clay, soil structure, consolidation behavior

This paper is a continuance of previous work by the author [1]. The new laboratory work is on specimens obtained from the same sampling program. The new developments are based upon the new laboratory work as well as upon the previous presentations by Bemben [1] and Ladd and Wissa [2]. Except where noted, the reader should refer to these papers for details of the sampling and testing programs described here. The location of the site involved is Northampton, Massachusetts. The varved clay was deposited in glacial Lake Hitchcock.

Two typical laboratory stress-strain curves exhibiting brittle behavior and one not showing brittle behavior are shown in Fig. 1. So-called ''sensitive'' clays exhibit brittle behavior.

The portions of the conclusions by Bemben [1] which are pertinent starting points for the new developments are as follows:

---

[1] Retained Consultant, Goldberg–Zoino & Associates, Inc., Newton Upper Falls, MA 02164, and Associate Professor, Department of Civil Engineering, University of Massachusetts, Amherst, MA.

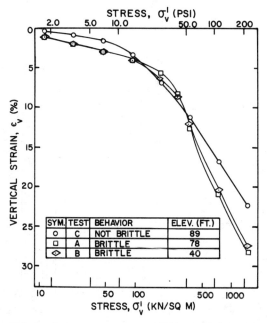

FIG. 1—*Oedometer test curves exhibiting brittle behavior.*

1. Post-depositional cementation throughout the deposit is the cause of brittle behavior in the natural state. The present *in situ* condition is a natural state.

2. In the natural state, the total undrained shear strength is the sum of the component due to natural-state inter-particle behavior plus the component due to cementation.

3. Breakage of the natural cement plus sufficient disturbance to the soil structure apparently cause the soil to change from its natural state toward a different state having a lower water content.

The purposes of this work are to show that specimens of intact soil exhibit brittle behavior during laboratory consolidation tests; to introduce a method for the determination of the previous vertical effective stress in existence at the onset of cementation; to show that the intact *in situ* soil fabric yet remembers the stress at onset of cementation; and to show the effects of specimen quality, compositional variability, and test procedures on the brittle behavior.

Geotechnical Aspects of the Deposit

The profiles of Atterberg Limits, natural water contents, vertical stress history, and undrained shear strength are shown in Fig. 2. Other sources of data presented by Ladd and Wissa [2] show the same general information has been obtained by others.

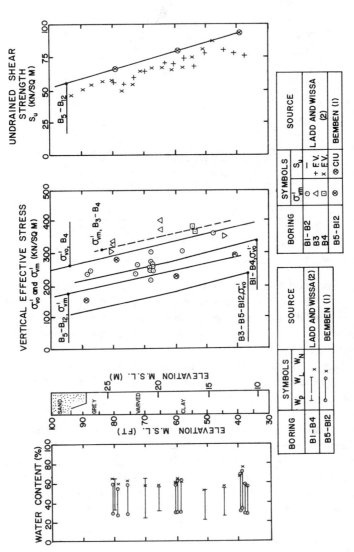

FIG. 2—*Profiles of Atterberg limits, stress history, and undrained shear strength.*

The profiles of natural water contents and Atterberg Limits are typical of brittle soils. The natural water content and liquid limit values of each bulk specimen and of each clay and silt portion of each seasonal layer are about equal in value irrespective of present vertical effective stress or of apparent maximum previous vertical effective stress. Houston and Mitchell [3] show that sensitive (brittle) behavior should always be expected for this situation and may even occur, in some instances, where the natural water contents are considerably less than the liquid limits.

The profiles of field vane strengths are essentially identical irrespective of the present vertical effective stress.

The profile of apparent maximum previous vertical effective stress values is very erratic. All values were obtained by the common Casagrande construction method which is performed on one-dimensional consolidation test curves. Nearly all of the one-dimensional test curves exhibited the brittle behavior aspect shown in Fig. 1. The method was developed for use with soils not having brittle behavior. One can readily envision that any sampling and testing procedures which affect the brittle shape of a one-dimensional test curve will similarly affect the apparent maximum previous vertical effective stress value obtained from the curve by using the common Casagrande method.

The apparent independence of the natural water contents and the field vane strength values from currently widely varying effective stress values and from apparently widely varying maximum previous vertical effective stress values lead to the following inquiry:

Is it possible that the onset of cementation in the glacial lake caused the existing water contents to be "locked in"? That is, could the cementation be preventing the subsequent water content changes in response to subsequent vertical effective stress changes?

An affirmative answer would be virtually certain if a method could be developed which, when applied to every one of the test curves, results in a single linear profile for maximum previous vertical effective stress "locked in" to the soil fabric. If an already accepted relationship between that line and the profile of undrained shear strength by both laboratory and field vane measurement could be shown to exist, the credibility of the "locked in" water contents and undrained shear strengths concept would be very high.

Balk [4] reports the elevations of the top of the main varved clay deposit at shoreline locations. The elevations vary from 49 m (160 ft) to 55 m (180 ft), with the higher values being associated with the location of deltas. Information within 3 km of the site involved in this work is provided. The surficial map by Jahns [5] is in agreement with the work of Balk.

Presentation of One-Dimensional Test Curves

Some pertinent data regarding four one-dimensional consolidation tests are presented in Table 1. The one-dimensional consolidation test data consisting of

TABLE 1—*One-dimensional test data.*

Test	Specimen Elevation, m (ft)	*In Situ* Vertical Efficiency Stress (σ'_{vo}), kN/m² (tons/ft²)	*In Situ* Water Content (W_N), %	Vertical Effective Stress Modulus (E_v), kN/m² (tons/ft²)	Initial Height, cm (in.)
A	23.8 (78)	134 (1.4)	63.3	4876 (51)	2.03 (0.80)
B	12.2 (40)	210 (2.2)	62.0	3824 (40)	2.03 (0.80)
C	27.1 (89)	115 (1.2)	56.3	2868 (30)	2.54 (1.00)
D	18.3 (60)	172 (1.8)	61.5	3824 (40)	2.54 (1.00)

NOTES:
1. Specimens for Tests A, B, and D are from 7.6-cm (3-in.)-diameter piston sample. Specimen C is from 2 7.6-cm (3-in.)-diameter Shelby tube without a piston.
2. All specimens are 6.35 cm (2.5 in.) diameter.
3. All tests are incremental, successive doubling of load type.
4. All tests are two-way drainage in fixed-ring Clockhouse oedometer.
5. All readings are manual.

vertical effective stress versus vertical strain for Tests A, B, and C are shown in Fig. 1.

The one-dimensional consolidation test data consisting of vertical effective stress versus vertical strain for Test D is shown in Fig. 3. It is a typical incremental successive doubling of load type loading test; the increments each lasted 70 to 81 min. Primary settlement portions were separated from secondary settlement portions, and the summation of primary portions alone comprises the uppermost curve. The summation of primary portions plus the secondary portions occurring between the end of primary and 60-min marks comprises the middle curve. Projections of the additional settlement which would have occurred if the loading increments each lasted 24 h were extrapolated and the summations of extrapolated primary plus secondary settlements for 24-h increments were calculated. These summations comprise the lower curve.

Plots of dial readings versus either square root of time or log of time are employed to separate the primary portions of settlements from the secondary portions. Plots of both functions are shown for two loading increments in Fig. 4. These plots are typical for a varved clay. Crawford [6] points out that for a clay exhibiting brittle behavior, the separation of primary and secondary consolidation is an arbitrary process, and the relative contributions of each depends on the method of loading. The present author postulates that the relatively large secondary settlements ongoing during primary phases cause a retardation effect on the primary portions so that the usual Terzaghi theoretical (t_{90}/t_{50}) ratio is not met when the usual constructions are used to determine those values; the actual (t_{90}/t_{50}) ratio is too large.

The postulated delaying action to the end of an ideal primary curve is discussed by Healy and Ramanjaneya [7] for incremental successive doubling of load-type loading tests on specimens from this same glacial lake deposit. Walker and Raymond [8] cite similar retardation of consolidation rates during consolidation

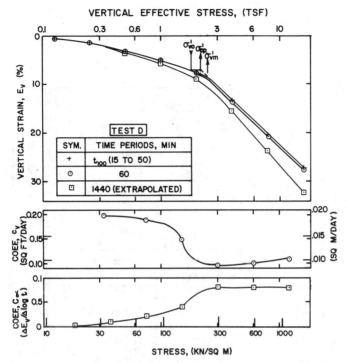

FIG. 3—*One-dimensional consolidation test values from an incremental loading test.*

tests with triaxial specimens on a similarly cemented clay. These authors, as well as the present author, attribute the retardation of consolidation rates for soils exhibiting brittle behavior primarily to the breaking of cement bonds and secondarily to the plastic lag of the soil fabric. Accordingly, neither Taylor's fitting technique on the square root of time-deflection plot nor Casagrande's fitting technique on the log of time-deflection plot should be applicable. Hamilton and Crawford [9] point out that structural breakdown is the important factor controlling the rate of compression for a clay exhibiting brittle behavior when near the apparent preconsolidation pressure; however, they do not mention the aspect of cementation.

The present author has developed an empirical technique to utilize with simplicity the work of Healy and Ramanjaneya. This technique recognizes that for virgin loading increments, the slope of the secondary on log time plot is always equal to about one fourth of the slope of the straight-line portion of the primary. Thus where slope *b* touches the curve is where 100% primary ends. The corresponding empirical geometry on a square root of time plot is to determine the slope of the straight-line portion and to construct a second straight line with a slope equal to the primary slope divided by 1.6. The intersection of the later line with the time curve marks when 100% primary ends. The constructions for 0% and 50% primary are, for either plot, in accordance with the usual procedures.

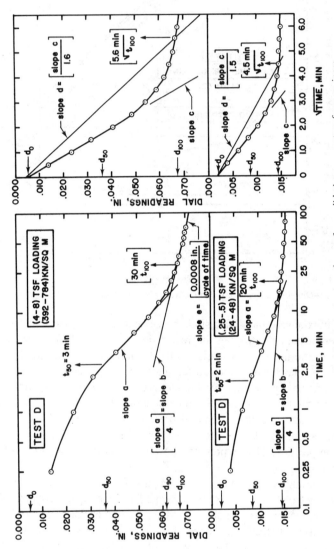

FIG. 4—Suggested method for determination of rate of consolidation parameters from an incremental loading test.

A great advantage of this log plot construction is that the point of the end of 100% primary can be recognized soon after it occurs during testing; the amounts of secondary settlements can thereby be minimized if such is desired. Likewise, a great advantage of this square root of time plot construction is that the point of 100% primary can be recognized as soon as it occurs.

The empirical slope ratios cited above for use with one-dimensional test will probably differ for other brittle clay deposits.

The postulated delaying effect of secondary compression upon the primary portion of settlement is less in the recompression region than in the virgin loading region. For log time plots, the cited empirical relationship between slope lines is yet reasonable, but for square root of time plots it is too drastic and a slope ratio of 1.5 or even 1.4 is more accurate. Fortunately, the amount of secondary settlement is essentially the same regardless of which slope correction factor is used. The important difference is in the calculation of the time to reach 100% primary.

Calculated values for the coefficients of consolidation, c_v, and the rate of secondary settlement, C_α, are shown for each load increment in Fig. 3. The values are typical for varved clays according to Ladd and Wissa [2] and Ladd [10]. The pattern of the changing values is such that one perceives that structural breakdown, as marked by increasing significant rates of secondary settlement is barely beginning with the 202-kN/m^2 (2 tons/ft^2) loading. The pattern of the values of c_v is similar to those reported by Smith and Wahls [11] and Walker and Raymond [8] for other varved clays exhibiting brittle behavior.

The shape of the downwardly displaced curve developed from the 24-h loading scenario is such that the Casagrande method for determination of the maximum previous vertical effective stress will give a lower value than for the uppermost curve.

A comparison of measured initial and final water content values with the measured stress-strain curve is a way to infer if elastic behavior exists while a specimen is subjected to recompression. If one assumes that the final water content and the stress-strain curve are in fact correct, one can calculate what the water content was at any point during the test. The point of interest to calculate is at a vertical effective stress equal to that felt by the specimen just prior to sampling. This calculated value can be compared to the actual measured value. If the two values are identical, the deductions are that the unloading and reloading portion of the curve (going from *in situ* vertical effective stress to zero stress and then back to the same stress) constitutes elastic behavior and that the laboratory curve is not lower than the field curve when they are at the same pressure. For Tests A and B, for example, the difference between computed water content and the measured natural water content was less than a 1% water content value in each case; one was too high and the other was too low. Other such tests by the author on specimens of this varved clay have produced similar results.

The one-dimensional test curves for 25 constant rate of strain consolidation (CRSC) tests are presented by Ladd and Wissa [2]. Of these, 19 are for a slow rate of straining and 6 are for a faster rate of straining. The slow rate prevents buildup of excess pore pressures throughout the specimens. The faster rate of straining allows calculations of the coefficient of consolidation by means of measurements of excess pore pressures at the base of the specimens for numerous instances during each test. The six tests at the faster rate of straining are with specimens carved from 12.7-cm (5-in.)-diameter samples and the slower rate of straining tests are with specimens carved from 7.6-cm (3-in.)-diameter samples. The actual rates of straining are not available.

The one-dimensional test curves for a specimen in each category are shown in Figs. 5 and 6. The pattern of behavior for the coefficient of consolidation for the fast rate of straining test is also shown in Fig. 6.

Maximum Past Pressure by the Casagrande Method

The determination of the maximum previous vertical effective stress by the Casagrande method for the four incremental loading tests and the 24 CRSC tests previously presented is plotted in Fig. 2. The values for the past pressure are designated as σ'_{vm}. The values for three of these tests are shown in Figs. 3 and 5. The values show wide variation at any one elevation. The values for the 12.7-cm (5-in.)-diameter sampling specimens which were all tested with the fast rate of straining are considerably higher than those for the 7.6-cm (3-in.)-diameter sampling specimens.

Maximum past pressure here includes the effects of cement bonding and is not indicative of the maximum previous vertical effective stress to which the soil fabric has been subjected in the field.

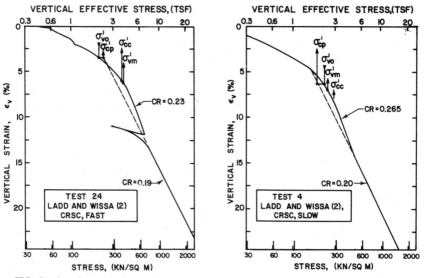

FIG. 5—*One-dimensional consolidation test values from a constant-rate-of-strain loading test.*

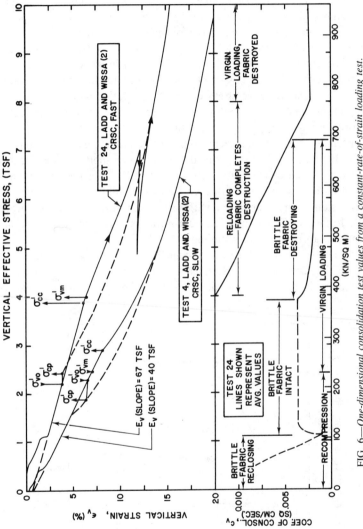

FIG. 6—*One-dimensional consolidation test values from a constant-rate-of-strain loading test.*

Maximum Past Pressure by the Breakpoint Method

The stress-strain curves of Fig. 6 each contain a straight-line portion. The slope of the line is an arithmetic volumetric strain modulus, E_v. According to Sangrey [12], this linear behavior is a manifestation of brittle behavior. The curves are not linear when approaching zero stress. The dotted lines, as projections of the straight-line portions, point to vertical strain values which comprise true origins for the elastic behavior portions of the stress-strain curves.

Values of the arithmetic modulus, E_v, for each test are shown in Fig. 7. The void ratio function, e_{00}, is the void ratio of the soil in the swelled state at zero pressure. As such, it is found by employing the combination of the natural water content and vertical effective stress information together with the linear stress-strain (swelling) line. The modulus values seem to be systematically related to void ratio for any one sampling condition. Sampling with a 7.6-cm (3-in.)-diameter tube without a piston seems to provide a stiffer and hence a more brittle specimen. Incremental doubling of loads together with short time loadings seems to develop the same stiffness in specimens with CRSC loading. Sampling with a 12.7-cm (5-in.)-diameter piston as opposed to sampling with a 7.6-cm (3-in.)-diameter piston seems to provide a stiffer specimen; however, all the specimens from the former samplings were tested with a faster rate of strain than any of the specimens from the latter samplings.

The highest stress for which linear behavior exists on the arithmetic plot is termed the "breakpoint". It is designated as σ'_{cc}. The method for determining the breakpoint is shown in Fig. 6. (The values are also shown in Fig. 5 for

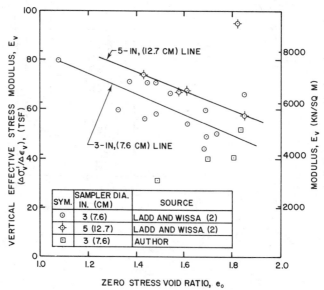

FIG. 7—*Relationship of elastic volume change compression modulus to zero stress void ratio.*

reference.) Values of the breakpoint, σ'_{cc}, for each test are shown in Fig. 8. They show a wide variation at any one elevation. The value of σ'_{cc} is generally very close to the values of σ'_{vm} for any one curve. This is seen in Fig. 6 for the two illustrated tests and by a comparison of σ'_{vm} values in Fig. 2 with the σ'_{cc} values in Fig. 8. It is seen that the controlling factor in the selection of the past pressure by both methods is actually the pressure region at which elastic behavior stops. The point at which elastic behavior ends does not have to coincide with maximum previous vertical effective stress to which the soil fabric has been subjected in the field.

The thesis by Masucci [13] written under the supervision of the author shows that the rate of testing during CRSC tests affects the values of the breakpoint, σ'_{cc}, and of the volumetric strain modulus, E_v. The tests were upon varved clay specimens from the same lake deposit. The sampling was in Amherst, Massachusetts, which is a few miles away from the Northampton, Massachusetts, site. All specimens were from a single block sample taken from a deep excavation. The effect of increasing the rate of strain from 8×10^{-7}/s to 1.3×10^{-4}/s was to systematically increase both the volumetric strain modulus, E_v, and the breakpoint, σ'_{cc} values. At the lowest rate, these aspects of brittle behavior are essentially removed, while at the highest rates these aspects of brittle behavior are

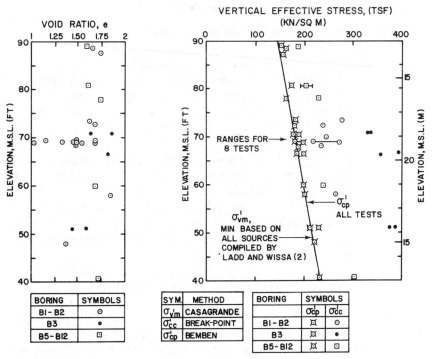

FIG. 8— *Comparison of apparent maximum previous vertical effective stresses by various methods.*

maximized. The moduli values increase from about 2400 to 5300 kN/m² (25 to 55 tons/ft²); the breakpoint values increase from about 145 to 290 kN/m² (1.5 to 3.0 tons/ft²) in response to the cited changes in rates of strain. Crawford [14] similarly reports increasingly larger values for moduli with increasing rates of strain.

Masucci also shows that CRSC tests with a rate of strain of about 10^{-5}/s produced stress-strain curves similar to incremental, successive doubling of load, type loading tests with time intervals equal to t_{90} values. The CRSC tests with a rate of strain of about 10^{-4}/s produced stress-strain curves similar to incremental, 48 kN/m² (0.5 ton/ft²) constant increments, type loading tests with time intervals equal to t_{90} values. Based upon this limited evidence, it seems that the higher E_v values in Fig. 5 and the higher σ'_{cc} values of Fig. 8 which are noted for specimens with a 12.7-cm (5-in.)-diameter sampling in comparison to a 7.6-cm (3-in.)-diameter sampling are attributed mostly, if not solely, to the rate of testing of specimens rather than to a difference in quality of the specimens. All specimens were of good quality.

It should also be remembered that variations in volumetric strain modulus values, E_v, and breakpoint values, σ'_{cc}, due to varying amounts of cementing agent in the ground are also to be expected.

Maximum Past Pressure by the Bemben Method

The method is based upon the following principles:

1. The present-day *in situ* water content is essentially identical to the water content at the onset of cementation. Changes in effective stress to *in situ* elements do not cause changes in water content as long as the cement remains intact.

2. The volume changes experienced by a specimen during relief of stresses and return of stresses during laboratory test loading represent elastic behavior as long as the cement remains intact.

3. The elastic behavior during loading of a laboratory specimen can extend beyond the maximum previous vertical effective stress experienced by the soil fabric prior to the onset of cementation.

4. After the breakpoint is reached, the stress-strain curve moves away from the cement controlled regime. With increasing stress, the stress-strain curve eventually becomes one of soil alone. That is, the formerly intact brittle fabric becomes destroyed.

The method requires the availability of the following information:

1. A one-dimensional stress-strain curve based upon CRSC or conventional incremental successive doubling of loads but with short time periods of loading. The loading must be to sufficiently high enough values that the straight-line portion on the log pressure plot is distinctive.

2. A knowledge of the vertical effective stress felt by the test specimen at the time of sampling, σ'_{v0}.

The method consists of the following steps:

1. On the log pressure plot, extend the straight-line portion of the curve from the high-pressure region backwards into the low-pressure region.
2. Determine the point on the stress-strain curve which corresponds to the *in situ* vertical stress, σ'_{v0}.
3. Draw a horizontal line through the point of the stress-strain curve which was determined in Step 2.
4. Determine the point of intersection of the lines from Steps 1 and 3. The stress coordinate of this point represents the maximum previous vertical effective stress at the onset of cementation; it is termed σ'_{cp}.

Illustrations of the method are shown in Figs. 5 and 6. The maximum previous vertical effective stress σ'_{cp} can be either greater or lesser than the present vertical stress.

The values of the maximum previous vertical effective stress for all of the tests are shown in Fig. 8. The average of the values is a straight-line profile; there is very little scatter to the data. The observed slope of the line is equal to the measured average buoyant weight of the soil. The upward projection of the line infers that the top of the clay deposit was at an elevation of about 49 m (160 ft). This is in excellent agreement with the previously mentioned geological evidence.

The underconsolidated stress state of all specimens from Borings 1, 2, and 4 (such as that of Test 4) is due to the fact that the soil fabric has been prevented from coming to the equilibrium water content of soil fabric alone. This is because of the resistance to compression provided by the intact cement.

The effect of the state of the brittle fabric on the measured values of the coefficient of consolidation during a test, as well as on the already mentioned stress-strain curve, is shown in Fig. 6. The important notion regarding the coefficient of consolidation is that the intact natural state soil prevents the occurrence of the ordinary primary settlements at the stress levels between the maximum previous vertical effective stress, σ'_{cp}, and the breakpoint, σ'_{cc}. Similar plots of the average effect of the state of the brittle intact natural state on the measured coefficient of consolidation were obtained by Masucci [13].

A check on the basis of the intact versus the destroyed states of the natural soil as they affect the stress-strain curve is readily made. The slope of the destroyed state of the natural soil is in essence also the slope of the soil fabric alone. On a plot of log pressure versus strain, this portion of the curve is the common compression ratio, *CR*. Measured value of *CR* versus the natural water content are shown in Fig. 9. These are the lower points of each paired set of points. They do correlate well with values from soils not having any cement

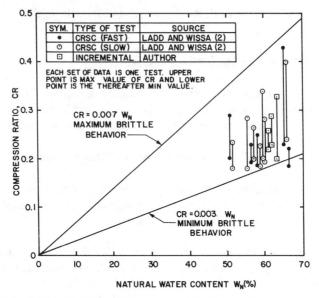

FIG. 9—*Relationship of virgin compression index values to natural water content.*

fabric. By looking at instantaneous tangents to the same curve at pressure values just beyond σ'_{cc}, one can determine an instantaneous *CR* value. The largest such instantaneous value is of interest as its value with respect to the true *CR* value is an indicator of the amount of brittle behavior being indicated by the test. The largest such instantaneous values are also shown in Fig. 9; they are the upper points of each paired set of points. The paired values agree with values cited by Ladd and Wissa [2] for this and other varved clay deposits.

Correlation to Field Undrained Shear Strengths

If the *in situ* soil is yet in the locked-in condition of the onset of cementation, the *in situ* measurements of shear strength should be about the same at any elevation irrespective of the subsequent stress history. The field vane results shown in Fig. 2 indicate that that is approximately the case.

The work of Sambhandharaksa [15] indicates that the soil fabric alone has an undrained strength to consolidation stress ratio of about 0.25. The work of Bemben [1] suggests that an additional component of strength due to the cement bonds amounts to about 25 kN/m² (4 psi). With the use of σ'_{cp} for consolidation stress, the predicted *in situ* shear strength values are in close agreement with the measured values. For example, at an elevation of 18.3 m (60 ft) the predicted shear strength is [(0.25) (200) + 25], which equals about 75 kN/m² (11 psi). This is in good agreement with the measured field vane values shown in Fig. 2.

Conclusions

1. Post-depositional cementation throughout the portion of the deposit investigated is the cause of brittle behavior during laboratory consolidation testing.

2. Test specimens made from good to excellent quality samples exhibit brittle elastic behavior during laboratory consolidation tests.

3. The shape of a stress-strain curve during consolidation in the brittle elastic stress range is significantly effected by the quality of the test specimen and by the loading conditions.

4. Constant rate of strain compression tests at rates of strain which produce only small pore pressures and incremental (successive doubling of loads) type tests with primary time periods of loading produce similar brittle-type stress-strain curves. For the two types of tests, systematic variations in either the rate of strain or the magnitude of the load increments produce systematic variations in the brittle behavior of the stress-strain curves.

5. Breakage of the cement causes each stress-strain curve to change from a state of brittle behavior to a state of nonbrittle behavior.

6. The Casagrande method for determining the past maximum vertical effective stress remembered by the soil fabric alone is not appropriate when brittle behavior exists. The method is appropriate for determining the apparent past pressure of the intact natural-state soil.

7. A method for determining the true past maximum vertical effective stress remembered by the soil fabric alone is presented.

8. The natural-state soil fabric frequently is in an underconsolidated state because of the intact cemented state.

Acknowledgments

The author extends his gratitude to several members of the firm of Goldberg–Zoino & Associates, Inc., Newton Upper Falls, Massachusetts. Mr. William Hover, Senior Geotechnical Engineer of the Vernon, Connecticut Branch Office, read the manuscript and offered constructive criticisms; Mr. Donald Schulze, Laboratory Director, conducted Tests A and B; and Mr. Gardner Hayward, Jr., Chief Draftsman, supervised preparation of the figures. The company provided a monetary award to the author.

References

[1] Bemben, S. M., "Brittle Behavior of a Varved Clay During Triaxial Undrained Shear Strength Tests," in *Geotechnical Properties, Behavior, and Performance of Calcareous Soils, ASTM STP 777*, K. R. Demars and R. C. Chaney, Eds., American Society for Testing and Materials, Philadelphia, 1982, pp. 252–276.

[2] Ladd, C. C. and Wissa, A. E. Z., "Geology and Engineering Properties of Connecticut Valley Varved Clays with Special Reference to Embankment Construction," Soils Publication 264, School of Engineering, Massachusetts Institute of Technology, Cambridge, MA, 1970.

[3] Houston, W. N. and Mitchell, J. K., "Property Interrelationships in Sensitive Clays," *Journal of the Soil Mechanics and Foundation Division,* American Society of Civil Engineers, Vol. 95, SM4, July 1969, p. 1059.

[4] Balk, R., "Geology of Mount Holyoke Quadrangle, Massachusetts," *Geological Society of America Bulletin,* Vol. 68, 1957, pp. 481–485.

[5] Jahns, R. H., "Surficial Geology of the Mount Toby Quadrangle, Mass.," USGS Quadrangle Map GQ-9, United States Geological Society, 1951.

[6] Crawford, C. B., "Interpretation of the Consolidation Test," *Journal of the Soil Mechanics and Foundation Division,* American Society of Civil Engineers, Vol 90, No. SM5, Sept. 1964, pp. 87–102.

[7] Healy, K. A. and Ramanjaneya, G. S., "Final Report, Consolidation Characteristics of a Varved Clay," Project 64-3, School of Engineering, University of Connecticut, Storrs, CT, 1970.

[8] Walker, L. K. and Raymond, G. P., "The Prediction of Consolidation Rates in a Cemented Clay," *Canadian Geotechnical Journal,* Vol. 5, No. 4, 1968, pp. 192–216.

[9] Hamilton, J. J. and Crawford, C. B., "Improved Determination of Preconsolidation Pressure of a Sensitive Clay," in *Papers on Soils—1959 Meetings, ASTM STP 254,* American Society for Testing and Materials, Philadelphia, 1960, pp. 254–269.

[10] Ladd, C. C., "Foundation Design of Embankments Constructed on Conn. Valley Varved Clays," *Geotechnical Publication 343,* School of Engineering, Massachusetts Institute of Technology, Cambridge, MA, 1965.

[11] Smith, R. E. and Wahls, H. E., "Consolidation Under Constant Rates of Strain," *Journal of the Soil Mechanics and Foundation Division,* American Society of Civil Engineers, Vol. 95, No. SM2, March 1969, pp. 519–539.

[12] Sangrey, D. A., "Naturally Cemented Sensitive Soils," *Geotechnique ,* Vol. 22, No. 1, March 1972, pp. 139–152.

[13] Masucci, R. E., "The Effect of Testing Procedures on the Consolidation Test Results of a Varved Clay," M.S. thesis, Department of Civil Engineering, University of Massachusetts, Amherst, MA, 1978.

[14] Crawford, C. B., "The Resistance of Soil Structure To Consolidation," *Canadian Geotechnical Journal,* Vol. 5, No. 4, May 1965, pp. 90–115.

[15] Sambarhandharaksa, S., "Stress-Strain Strength Anisotropy of Varved Clays," D.S. thesis, Massachusetts Institute of Technology, Cambridge, MA, 1977.

Masaharu Fukue,[1] Shigeyasu Okusa,[2] and Takaaki Nakamura[1]

Consolidation of Sand-Clay Mixtures

REFERENCE: Fukue, M., Okusa, S., and Nakamura, T., **"Consolidation of Sand-Clay Mixtures,"***Consolidation of Soils: Testing and Evaluation, ASTM STP 892,* R. N. Yong and F. C. Townsend, Eds., American Society for Testing and Materials, Philadelphia, 1986, pp. 627–641.

ABSTRACT: Consolidation properties of bentonite-sand mixtures, as reclamation materials, were investigated in the one-dimensional consolidation test. From the relationships between the void ratio of nonclay fraction and consolidation pressure was found the existence of a threshold void ratio at which the frictional resistance becomes dominant during consolidation. The threshold value is slightly greater than the maximum void ratio of sand used for the mixtures.

A new standard, which utilizes values of clay content, dry density, and threshold void ratio, is proposed to clarify the boundary between sandy and clayey soils.

The salt infiltration into the mixtures causes a rapid settlement but greatly reduces the swelling in unloading stage. As long as the void ratio of nonclay fraction is less than the threshold value, however, the influence of salt disappears.

KEY WORDS: consolidation, compressibility, clay-sand mixture, bentonite, classification, electrolyte solution

In coastal regions, clay sediments are usually very soft and have a low bearing capacity [1]. For reclamation in such regions, soft sediments are often needed to be displaced with good-quality materials to lessen settlement. This situation may cause a lack of good-quality materials and may produce another problem concerning the disposal of the soft sediments.

Generally, good-quality materials mean sand and gravel soils because clayey soils are often regarded as poor materials in engineering practice. If a proper mixture of sand and clay is found, however, it can be properly used for reclamation or for an embankment. Remaining materials should be of good quality or should be improved.

The advantages of using sand-clay mixtures may be summarized as follows:

1. Cohesion can be expected and may reduce the liquefaction potential.

[1]Associate Professor, Department of Ocean Civil Engineering, Tokai University 3-20-1, Orido, Shimizu, Japan 424.

[2]Professor, Department of Ocean Civil Engineering, Tokai University 3-20-1, Orido, Shimizu, Japan, 424.

2. Regarding settlement, the compressibility may be as low as that of sand or gravel, if a proper mixture is obtained.

3. Most soft sediments, which must be displaced, can be utilized for mixture.

4. Soils containing a certain amount of clay can also be used for reclamation if they satisfy the new standard for materials.

This study examines the consolidation properties of sand-clay mixtures, and a new standard for reclamation materials is proposed. The influence of salt infiltration on consolidation properties also is examined.

Fundamentals of the Study

It is well known that the interaction between soil particles and pore fluid is quite different for clay and sand. The interaction between clay particles is described in terms of both the attractive and repulsive electrical forces between clay surfaces [2-4].

However, the interaction between nonclay particles depends mainly on the intergranular friction at the contacts, and is distinguishable from that of clay particles. Therefore, if no skeleton of nonclay in clay fractions forms (Fig. 1a), the consolidation properties of mixture can be governed primarily by the interaction between clay particles.

As Skempton [5] and Seed et al [6] found, the liquid limit for various mixtures with the same clay minerals is proportional to clay content. This means that the liquid limit to clay fraction (WL/C) is constant, irrespective of clay content or nonclay content. Considering that the liquid limit is approximately the water content at a given shear strength, it is deduced that the shear strength for mixtures with the same minerals and different clay content is almost constant, so far as the water content of clay component is constant. In other words, so far as clay component is concerned, the effect of nonclay component on shear strength can be negligible for mixtures with relatively high clay content.

At a range of small amount of clay content, however, the plastic limit is independent of clay content or increases slightly with the increase of nonclay content. This independence of clay content or the increase in plastic limit is

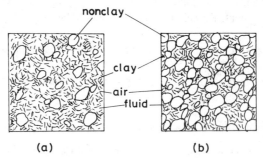

FIG. 1—*Simple structural model of sand-clay mixture.*

considered to be due to the frictional resistance between nonclay particles [6,7].

Thus the formation of skeleton of nonclay fractions (Fig. 1*b*) may reduce settlement because the frictional resistance is dominant. Note that the states of mixture illustrated in Figs. 1*a* and 1*b* are not only dependent on clay content, but also on void ratio or density of clay components. Accordingly, in this sense, it is not sufficient to classify soils using clay content only.

Hence it is necessary to distinguish clay fraction from nonclay fraction for comprehensive interpretation of the consolidation properties of mixture.

Four-Phase Soil Model

In this study, soil components are divided into four phases: gas, pore fluid, and clay and nonclay fractions (Fig. 2). Here v denotes the volume and m denotes the mass. The physical quantities of soil, which can be defined from the four-phase soil model, are presented in Table 1. The important relationships between the physical quantities in the four-phase and the conventional three-phase soil models are also shown in Table 1. It is noted that all the quantities can be determined from the conventional soil testing without complicated calculation. For example, the void ratio of nonclay fraction (e_s) can be evaluated from the lowest equation presented in Table 1, if the specific gravity of nonclay (G_{ss}) is obtained, where ρ_w ρ_d and M are usually known. Furthermore, if the G_{cs} is known, the void ratio of clay fraction (e_c) can also be evaluated using values of e_s, G_{ss} and M.

In this study, clay and nonclay components are separated significantly by a particle size of 2 μm, but the specific gravity of nonclay component is conveniently represented by that of particles greater than 74 μm.

Experimental Procedure and Materials

The consolidation tests on bentonite and bentonite-sand mixtures were performed using double-drained cylindrical specimens, 6 cm in diameter and 2 cm in height. The consolidometer used is illustrated in Fig. 3. The consolidation

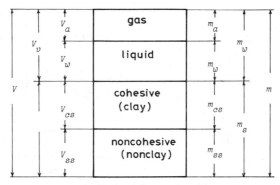

FIG. 2—*Four-phase soil model.*

TABLE 1—*Definitions of physical properties in four-phase model.*

Physical Properties	Definition
Dry density of nonclay component	$\rho_{ds} = m_{ss}/V$
Dry density of clay component	$\rho_{dc} = m_{cs}/V$
Specific gravity of nonclay component	$G_{ss} = m_{ss}/V_{ss}\,\rho_w$
Specific gravity of clay component	$G_{cs} = m_{cs}/V_{cs}\,\rho_w$
Void ratio of nonclay component	$e_s = (V - V_{ss})/V_{ss} = V/V_{ss} - 1$
Void ratio of clay component	$e_c = V_v/V_{cs}$
Water content of clay component	$w_c = m_w/m_{cs}$
Ratio of clay to nonclay in mass	$M = m_{cs}/m_{ss}$
Clay content	$C = m_{cs}/m_s\;(\times\;100)\;(\%)$

Important relationships between physical quantities:[a]
$\rho_d = \rho_{dc} + \rho_{ds},\ \rho_{ds} = \rho_w\,G_{ss}/(e_s + 1),\ M = C/(100 - C)$
$e_s = \rho_w\,G_{ss}\,(1 + M)/\rho_d - 1 = M\,(e_c + 1)\,G_{ss}/G_{cs}$

[a]ρ_d = dry density of mixture.
ρ_w = density of water.

ring was made of brass and well lubricated. The ring friction was possibly less than 5% of the applied load.

From X-ray analysis, the commercially available bentonite contains mainly montmorillonite, probably the most undesirable clay mineral for soil foundations. Toyoura sand, Japanese standard sand, was used for the mixtures. The soil properties and the specimens used are summarized in Tables 2 and 3, respectively.

The effects of nonclay content on consolidation properties are examined by the relationships between void ratio of nonclay components (e_s) and effective consolidation pressure ($\bar{p}$), while the influence of salt infiltration on consolidation properties is found from the relationships between the void ratio of clay components (e_c) and $\bar{p}$ for both specimens, with and without salt infiltration.

The soil samples were mixed with the desired amount of distilled water, sodium chloride (NaCl) solution, or calcium chloride (CaCl$_2$) solution. The specimens were compacted into a thin-wall ring, 6 cm in diameter and 2 cm in height, with a knife, and were then carefully pushed into the consolidation ring with a piston.

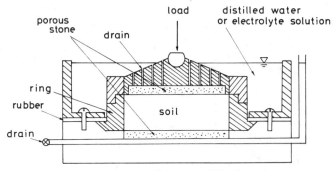

FIG. 3—*Consolidometer used in the study.*

TABLE 2—*Physical properties of specimens.*

Specimens	Bentonite A	Bentonite B	Toyoura Sand
Liquid limit, %	354.5	338.5	...
Plastic limit, %	34.1	47.9	...
Specific gravity	2.6	2.6	2.65
Clay content, %	72.0	70.0	0
Maximum size, mm	0.06	0.065	0.4
Minimum size, mm	...	...	0.07
e (maximum)	...	...	0.98
e (minimum)	...	...	0.58

In all the specimens, the saturation degree exceeded 90% and the initial water content of each specimen was slightly lower than the liquid limit.

In the specimens with sodium chloride or calcium chloride solution, the same type and the same concentration were used as fluid surrounding the consolidation ring. For the specimens mixed with distilled water, the initially distilled water surrounded the ring was exchanged with 0.6 M NaCl solution at a given consolidation pressure during test, except one specimen for the same type of mixtures, as shown in Table 3.

The consolidation pressures were applied in the normal manner, using an 11-to-1 lever-type loading system, and the load increment ratio of $\Delta p/p$ was one. For each consolidation pressure, the readings were taken from 6 s until the settlement rate dropped to 0.005 mm per day as measured by the dial gage.

Results and Discussion

Void Ratio and Effective Pressure Relationships

The typical relationships between final void ratio and effective consolidation pressure for the specimens with distilled water are shown in Fig. 4. The relationships, therefore, include the secondary compression. In the figure the higher void ratio in this conventional relationship does not mean the higher void ratio of clay fraction. (This is discussed later.)

The void ratio of clay or nonclay fractions can be easily calculated using the relationships shown in Table 1. The relationships between the e_s (larger than 2 μm) and $\bar{p}$ are shown in Fig. 5. The curves shown in the figure consist of three stages: initial low rate, secondary high rate, and final low rate. The first and second stages can be interpreted by the ordinary consolidation mechanisms of clay [8].

Considering that the clay content is less than 50%, the final low stage is due to the frictional resistance between nonclay particles. This stage should be distinguished from the reduction phenomenon of compressibility due to creep strain or the decrease in permeability which occurs in clay [8,9]. Therefore in this study the maximum void ratio of nonclay fraction in which their friction acts is

TABLE 3—Description of specimens.[a]

Specimen No.	Soil Type	Pore Fluid Initial	w, %	C, %	Sr, %	WL, %	WP, %
1	bentonite A	distilled water	230.2	72.0	>90	354.5	34.1
2	bentonite A	0.26 M NaCl	148.4	72.0	>90	169.2	23.4
3	bentonite A	0.43 M NaCl	124.1	72.0	>90	139.2	32.2
4	bentonite B	0.01 M CaCl$_2$	242.0	70.0	>90	325.0	40.0
5	bentonite B	0.05 M CaCl$_2$	167.4	70.0	>90	163.0	37.9
6	bentonite A + sand	distilled water	96.5	26.0	94.7	...	...
7	bentonite A + sand	distilled water[b]	111.0	26.0	94.0	...	...
8	bentonite A + sand	distilled water[b]	63.9	26.0	92.6	...	...
9	bentonite A + sand	distilled water	63.9	26.0	95.8	...	...
10	bentonite A + sand	distilled water[b]	63.9	26.0	93.8	...	...
11	bentonite A + sand	distilled water	73.3	20.3	97.4	...	...
12	bentonite A + sand	distilled water[b]	73.3	20.3	97.4	...	...
13	bentonite A + sand	distilled water	109.2	35.4	97.8	...	...
14	bentonite A + sand	distilled water[b]	109.2	35.4	99.0	...	...

[a] w = water content, C = clay content (<2 μm), Sr = saturation, WL = liquid limit, WP = plastic limit, 1 M = 1 mol/L.
[b] Salt diffusion.

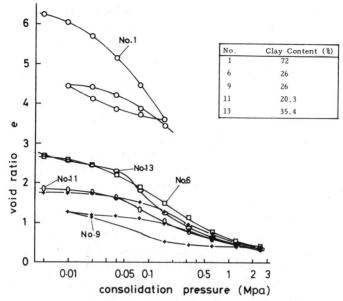

FIG. 4—*e–log p̄ relationships of bentonite and bentonite-sand mixtures with a distilled water as pore fluid.*

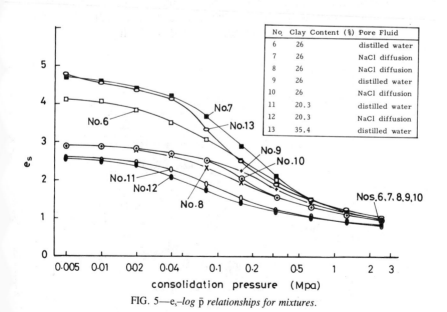

FIG. 5—*e_s–log p̄ relationships for mixtures.*

defined as "the threshold nonclay void ratio" and is evaluated from the intersection of the tangents of curve at a high pressure stage (Fig. 6).

From the curves shown in Fig. 5, the threshold nonclay void ratio of Specimen 13, with a clay content of 34%, is found to be 1.48. Similarly, for Specimens 6 to 10, the threshold nonclay void ratio, e_s (threshold), is 1.4, while 1.25 is obtained for Specimens 11 and 12, with a clay content of 20.3%.

Seed et al [6] assumed that a value of critical void ratio in which clay component fills up the void of nonclay skeleton was 0.8. This value is considerably smaller than the e_s (threshold) obtained in this study.

Since a void ratio of 0.8 in pure sands is not very loose, the angle of shearing resistance is as high as 30° [10,11]. Therefore it is natural that the e_s (threshold) is greater than 0.8, though the size distribution and shape of nonclay fraction definitely affect on the value.

From the experimental results, the value of e_s (threshold) is the same for the same clay content mixtures, although it increases slightly with increasing of clay content, as shown in Fig. 7. This probably means that the clay particles adhere on the surfaces of nonclay particles and that they exist between some of the nonclay contacts, but the friction between nonclay is dominant. Therefore, at the final stage of consolidation, the degree of orientation in clay particles may be different for the different mixtures. It is noted that the maximum void ratio of sand is also the threshold nonclay void ratio and that the threshold nonclay void ratio is not influenced by the salt infiltration.

Herein, the mixture can be classified by the e_s (threshold). If a mixture has a void ratio of nonclay less than the e_s (threshold), the soil is classified as sandy. It has low compressibility and high strength, because the frictional resistance and cohesion are expected during consolidation or shearing in this range of nonclay void ratio.

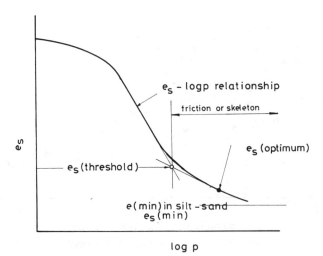

FIG. 6—*Definition of threshold nonclay void ratio.*

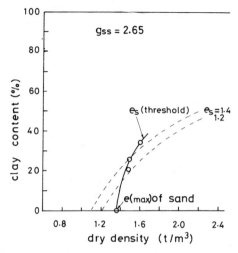

FIG. 7—*Threshold nonclay void ratio of bentonite-sand mixtures.*

Generally, in mixtures that have a low nonclay void ratio the angle of shearing resistance is comparable to sands in undrained condition, and the shear strength is much higher than that of sand because of cohesion [7]. Accordingly, in this range of nonclay void ratio, mixtures or soils behave like sand with properties of clay, though their properties are very complicated in terms of pore water pressure and volume change.

Figure 8 shows the relationships between clay content and dry density of materials for a given specific gravity of nonclay, in terms of nonclay and clay void ratios. The figure can be easily obtained for any specific gravities of each component by the calculation of e_s and e_c using the relationships between the physical quantities presented in Table 1. Accordingly, the compressibility is dependent on a change in void ratio from the state considered to the minimum nonclay void ratio, which is comparable to the minimum void ratio of sand. After soil reaches to the threshold value the maximum compressibility is therefore not very large in comparison with ordinary sands. The small change in G_{ss} will not greatly influence the lines in the figure.

If clay content and dry density are known, e_s is immediately obtained using Fig. 8, where the threshold nonclay void ratio may vary slightly with the size distribution and shape of nonclay particles. It is unlikely, however, that the maximum void ratio of ordinary sands is considerably higher than about one.

Therefore significant classification between clayey and sandy soils can be made by a diagram presented in Fig. 9. This classification method for mixtures utilizes clay content, dry density, and threshold nonclay void ratio, and is based on the mechanical properties of soils unlike the conventional method which is based on clay content or the plasticity chart.

In the procedure to determine the Atterberg limits, either water will be added

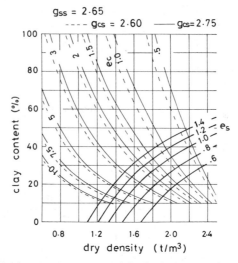

FIG. 8—*Relationships between clay content and dry density in terms of a given value of* e_s *or* e_c.

to the sample or the soil will be dried. Therefore it should be recognized that the e_s and e_c of the sample are not in the Atterberg limit test.

Effect of Salt Infiltration

Figure 10 shows the e_c–log $\bar{p}$ relationships in loading and unloading conditions for Specimens 8, 9, and 10. For Specimen 9, the test pore fluid was distilled water. With Specimens 8 and 10, the salt diffusion was allowed to be initiated at a load of 0.005 MPa and 0.04 MPa, respectively.

The e_c–log $\bar{p}$ relationships indicate that the salt infiltration causes a rapid settlement and reduces the apparent yield point. This is considered to be due to

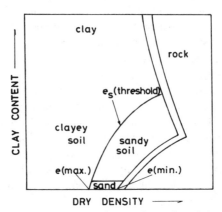

FIG. 9—*A new classification diagram for sandy and clayey soils.*

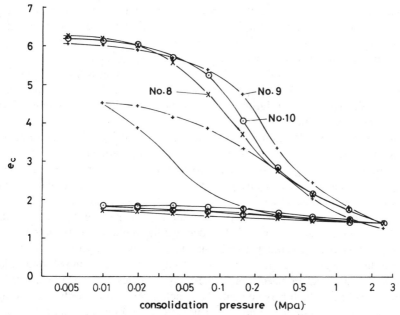

FIG. 10—e_c–log $\bar{p}$ relationships of mixtures with a clay content of 26%. Specimens 8 and 10: NaCl diffusion, Specimen 9: distilled water.

the reduction of repulsive forces and/or of shearing resistance between the clay particles [2,4]. Furthermore, the swelling of the mixtures is greatly decreased by the salt infiltration. This effect in some clay is also well known and has been studied by many researchers [12].

The details of the interaction between clay particles will not be discussed; this study will treat the consolidation properties only from the engineering point of view.

It is well known that the consolidation and strength properties in bentonite-electrolyte solution system are influenced by the type and concentration of solution, possibly by the dielectric constant of solution [13].

In specimens similar to Nos. 1 to 4, the Hvorslev's constants (i.e., true cohesion and true angle of friction), which were obtained by a similar manner followed in Gibson [14], depended on the activity of the specimens. For example, the plasticity index decreased from about 300% with distilled water to about 100% with 0.6 M NaCl solution (Table 3). In this case the reduction of activity increased the true angle of friction but considerably decreased the true cohesion. It is notable that the Hvorslev's constants can be presented as a function of activity for relatively high clay content soils [15].

Accordingly, during consolidation, the salt infiltration gradually would cause the reduction of plasticity, which consequently would reduce the cohesion of clay components during the diffusion of salt, though a reverse effect with volume change may also occur at the same time. It is important to note that in some clay minerals, such as illitic clay, the leaching of salt decreases the shear strength, as in Norwegian quick clay. Therefore the effect of salt infiltration on the consolidation properties may depend on the type of clay mineral.

The influence of salt infiltration is shown in a different manner in Fig. 11, where e_c is concerned in the specimens with distilled water and e'_c is concerned in the specimens with salt infiltration. Therefore the ratio e'_c/e_c is one in which no effect of salt infiltration is seen.

In Curves A and B, the ratio e'_c/e_c decreases gradually as the consolidation pressure increases at a relatively low pressure stage, and recovers up to one at a given pressure in high-pressure stage. As described earlier, in this high pressure stage the frictional resistance between nonclays becomes dominant. This means that an optimum nonclay void ratio where consolidation load is mostly supported by the nonclay skeleton may exist. From the experimental results, we see that the optimum nonclay void ratio is less than the threshold nonclay void ratio and must be between the maximum and minimum void ratios of sand-silt fractions.

Because Specimens 2 and 3 contain salt initially and their initial void ratios are lower than that of Specimen 1, the ratio is far less than one. Since these specimens have a high clay content, the nonclay void ratios hardly reach to the threshold value during consolidation. Curves C and D therefore do not show a large change in the ratio.

Thus, if active clays are used for mixtures, the salt infiltration influences the consolidation properties at a range of relatively high nonclay void ratio. If the nonclay void ratio reaches to the optimum value beyond the threshold value, however, the influence of salt disappears completely.

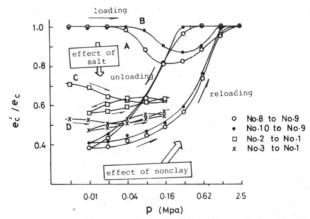

FIG. 11—*Influence of salt infiltration under loading, unloading, and reloading. Specimens 1 and 9: distilled water. Specimens 2 and 3: mixed with 0.26 and 0.43 M NaCl, respectively. Specimens 8 and 10: NaCl diffusion.*

Compression Index

Figure 12 shows the linear parts which show the highest rate of compressibility in e_c–log $\bar{p}$ relationships for most of the specimens subjected to the tests. As far as e_c is concerned, the linear parts may be identified with the normal consolidation of clay. The difference between the curves shown in Fig. 12 is not only dependent on the initial void ratio but on the electrolyte concentration of pore fluid. Generally, in cohesive soils the higher the initial void ratio, the faster the reduction rate in void ratio [16]. For the same initial void ratio, the addition or infiltration of salt reduces the reduction rate in void ratio.

The compression index of remolded clay has been regarded empirically as a function of liquid limit [17]. This trend is seen for the curves on Specimens 1 to 4 in Fig. 12, where the specimens have an electrolyte solution initially. The salt infiltration also makes the rate slow, which is clearly seen by comparing the curves of Specimens 8 to 10.

Therefore the compression index, defined as $\Delta e_c / \Delta$log $\bar{p}$, depends on the initial void ratio and electrolyte solution as pore fluid. Figure 13 shows the relationship between the compression index defined in this study and a value of e_c at a consolidation pressure of 0.1 MPa. This relationship is apparently linear, though one plot obtained for Specimen 1 is exceptionally low. This probably means that no effect of nonclay is involved within this range of void ratio.

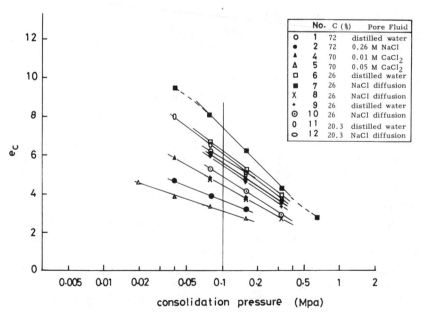

FIG. 12—e_c–log $\bar{p}$ *relationships for bentonite and mixtures.*

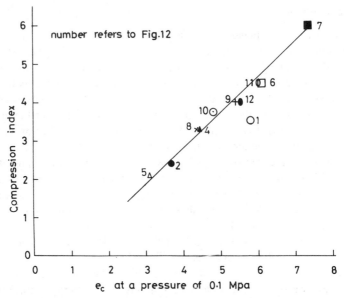

FIG. 13—*Relationships between compression index in* e_c*–log* $\bar{p}$ *relationship and* e_c *at a pressure of 0.1 MPa.*

Conclusion

The consolidation properties of clay-sand mixtures which involved the most undesirable clay minerals were examined in order to establish a new standard in relation to the quality of reclamation materials. The relationships between the void ratio of nonclay fraction and consolidation pressure of the mixtures show that the compressibility decreases greatly if the nonclay void ratio reaches a given value, defined as the threshold nonclay void ratio.

The threshold nonclay void ratio of mixtures is slightly greater than the maximum void ratio of sand used for mixtures, and increases slightly with clay content. If the clay content and dry density of a mixture are known, the soil is immediately classified using the new classification diagram given in this study.

Accordingly, if mixtures are compacted beyond this threshold nonclay void ratio, the soil has very low compressibility and can be properly available as reclamation materials.

The effect of nonclay and salt infiltration can be seen in the ratio e'_c/e_c. For the bentonite and mixtures, the ratio varied from about 1 to 0.4. In the case of no salt influence or in the case where the nonclay void ratio is less than the optimum nonclay void ratio, the ratio e'_c/e_c is unity. The addition or infiltration of salt lessens the ratio.

The compression index, defined as $\Delta e_c/\Delta \log \bar{p}$, is correlated to a value of e_c at a consolidation pressure of 0.1 MPa. The relation is linear for the various specimens.

Acknowledgments

The authors wish to acknowledge Dr. R. N. Yong, Professor of McGill University, for his keen interest and encouragement during this study. The part of this study was financially supported by the Ministry of Education, Japan.

References

[1] Umehara, Y. and Zen, K., "Consolidation Characteristics of Dredged Marine Bottom Sediments with High Water Content," *Soils and Foundation,* Japanese Society of Soil Mechanics and Foundation Engineering, Vol. 22, No. 2, 1982, pp. 40–54.

[2] MacEwan, D. M. C., "Short-Range Electrical Forces Between Charged Colloid Particles," *Nature,* Vol. 174, 1954, pp. 39–40.

[3] Mesri, G. and Olson, R. E., "Consolidation Characteristics of Montmorillonite," *Geotechnique,* Vol. 21, No. 4, 1971, pp. 341–352.

[4] Sridharan, A. and Jayadeva, M. S., "Double Layer Theory and Compressibility of Clays," *Geotechnique,* Vol. 32, No. 2, 1982, pp. 133–144.

[5] Skempton, A. W., "The Colloidal Activity of Clay," in *Proceedings,* 3rd International Conference on Soil Mechanics and Foundation Engineering, Vol. 1, 1953, pp. 57–61.

[6] Seed, H. B., Woodward, R. J., and Lundgren, R., "Fundamental Aspects of the Atterberg Limits," *Journal of Soil Mechanics and Foundation Engineering,* American Society of Civil Engineers, Vol. 90, No. SM6, 1964, pp. 75–105.

[7] Fukue, M. and Okusa, S., "A Consideration of Mechanical Properties of Sand-Clay Mixture," *Journal of the Faculty of Marine Science and Technology,* Tokai University, Shimizu, Japan, Vol. 14, 1981, pp. 247–261.

[8] Leonards, G. A. and Altschaeffl, A. G., "Compressibility of Clay," *Journal of Soil Mechanics and Foundation Engineering,* American Society of Civil Engineers, Vol. 90, No. SM5, 1964, pp. 133–155.

[9] Mesri, G. and Olson, R. E., "Mechanisms Controlling the Permeability of Clays," *Clays and Clay Minerals,* Vol. 19, No. 3, 1971, pp. 151–158.

[10] Bishop, A. W. and Eldin, A. K., "The Effect of Stress History on the Relation Between ϕ and Porosity in Sand" in *Proceedings,* 3rd International Conference on Soil Mechanics and Foundation Engineering, Vol. 1, 1953, pp. 100–105.

[11] Nash, K. L., "The Shearing Resistance of Fine Closely Graded Sand," in *Proceedings,* 3rd International Conference on Soil Mechanics and Foundation Engineering, Vol. 1, 1953, pp. 160–164.

[12] Kleijn, W. B. and Oster, J. D., "A Model of Clay Swelling and Tactoid Formation," *Clays and Clay Minerals,* Vol. 30, No. 5, 1982, pp. 383–390.

[13] Sridharan, A. and Rao, G. V., "Mechanisms Controlling Volume Change of Saturated Clays and Role of Effective Stress Concept, *Geotechnique,* Vol. 23, No. 3, 1973, pp. 359–382.

[14] Gibson, R. E., "Experimental Determination of the True Cohesion and True Angle of Internal Friction in Clays," in *Proceedings,* 3rd International Conference on Soil Mechanics and Foundation Engineering, Vol. 1, 1953, pp. 126–130.

[15] Fukue, M. and Okusa, S., "Activity Indices and Hvorslev's Constants in Bentonite-Sodium Chloride System," *Journal of the Faculty of Marine Science and Technology,* Tokai University, Shimizu, Japan, No. 15, 1982, pp. 169–179.

[16] Nishida, Y., "A Brief Note on Compression Index of Soils," *Journal of Soil Mechanics and Foundation Engineering,* American Society of Civil Engineers, Vol. 82, No. SM3, 1956, pp.1–14.

[17] Terzaghi, K. and Peck, R., *Soil Mechanics in Engineering Practice,* Wiley, New York, 1948.

Vito A. Guido[1] and Norbert M. Ludewig[2]

A Comparative Laboratory Evaluation of Band-Shaped Prefabricated Drains

REFERENCE: Guido, V. A. and Ludewig, N. M., "**A Comparative Laboratory Evaluation of Band-Shaped Prefabricated Drains,**" *Consolidation of Soils: Testing and Evaluation, ASTM STP 892*, R. N. Yong and F. C. Townsend, Eds., American Society for Testing and Materials, Philadelphia, 1986, pp. 642–662.

ABSTRACT: Theoretical studies pertaining to the design and application of vertical drainage systems, and recommendations for the selection of design spacing for band-shaped prefabricated (wick) drains, were examined. It was the intent of this study to examine the effects of band-shaped prefabricated drains on the consolidation characteristics of a clay soil, and in addition to evaluate and contrast the efficiency of the prefabricated drains in a laboratory experiment in a way that would guide the engineering application of prefabricated drains in the field. An apparatus referred to as a "wick drain consolidometer" was designed and constructed to fulfill the above-stated intent of study. Five different prefabricated drains were tested: Castleboard, Franki-Kjellman, Mebradrain, Geodrain, and Alidrain. All wick drains with a central core and loosely fitted filter jacket performed better than the Franki-Kjellman wick, which had the filter glued to the core. Based on compression versus time data, the Alidrain wick performed the best, with Castleboard, Mebradrain, and Geodrain all performing approximately the same.

KEY WORDS: vertical drainage system, comparative laboratory study, geocomposites, band-shaped prefabricated drains, wick drains, efficiency, consolidometer, kaolin, compression, water release

Nomenclature

A_w Cross sectional area of band-shaped prefabricated drain
a_v Coefficient of compressibility
C_c Compression index
c_h Coefficient of consolidation for horizontal flow
c_v Coefficient of consolidation for vertical flow
d_e Diameter of zone of influence
d_w Diameter of drain well
e Void ratio of soil

[1] Associate Professor, Civil Engineering Department, The Cooper Union for the Advancement of Science and Art, Cooper Square, New York, NY 10003.
[2] Geotechnical Engineer, Port Authority of New York and New Jersey, One World Trade Center, New York, NY 10048.

e_i Initial void ratio of soil prior to consolidation process

H Length of longest drainage path

k_h Coefficient of permeability in horizontal direction

k_s Coefficient of permeability in remolded zone

k_v Coefficient of permeability in vertical direction

k_w Coefficient of permeability of well backfill (sand drain) or permeability of band-shaped prefabricated drain

ℓ Length of drain when open at one end only (half-length when both ends are open)

p Intergranular pressure

q_w Specific discharge

r Radius, a coordinate in the cylindrical system

r_e Radius of zone of influence

r_s Radius of remolded zone

r_w Radius of drain well

S Spacing of drain wells

T_h Dimensionless time factor for radial flow

T_v Dimensionless time factor for vertical flow

t Time of consolidation

$\bar{U}_r$ Average degree of consolidation due to radial flow

$\bar{U}_{r,v}$ Average degree of consolidation due to combined radial and vertical flow

$\bar{U}_v$ Average degree of consolidation due to vertical flow

u Excess pore-water pressure

u_i Initial uniform excess pore-water pressure

$\bar{u}_r$ Average excess pore-water pressure due to radial flow

z Coordinate in cylindrical system, or distance from open end of drain in Eqs 16 and 17

γ_w Unit weight of water

With the increased use of marginal construction sites, the *in situ* treatment of soil has risen markedly. In many instances, these marginal construction sites are located in areas where underlying deposits of soft, highly compressible strata are found. These strata can yield large settlements over long periods of time.

Until recently, especially in the United States, vertical sand drains were used almost exclusively to speed up this settlement process, having been used successfully for over half a century. Techniques for installation of vertical sand drains are mandrel-driven, hollow stem-continuous flight auger, and jetted hollow or closed-end pipes. All three techniques yield large soil disturbance effects, such as a reduction in the permeability of the soil surrounding the sand drain. Other drawbacks of sand drains are: large quantities of water are needed if the "jetted" sand drain installation technique is used; cost of the large quantities of specifically graded sand used in the drains is high; and sand drain installation is labor-intensive.

In the 1930s, Kjellman [1] developed an alternative to vertical sand drains, a cardboard strip-type drain. However, the cardboard proved to be an inadequate material in terms of its physical strength and poor permeability. Over the last quarter of a century, considerable work has been done in Europe in the development of band-shaped prefabricated drains, commonly called wick drains because of their resemblance to old oil lamp wicks. These advancements in prefabricated drains have been made possible with the advent of plastics and filter fabrics (geotextiles). The basic band-shaped drain consists of a plastic core or base which is wrapped in a filter jacket, and so is sometimes referred to as a geocomposite (Fig. 1). Numerous channels in the core allow water conduction along the length of the wick.

In a few short years prefabricated drains have replaced sand drains in over 80% of all applications nationwide. Why have these drains become the preferred approach over sand drains? Comparable reliability to sand drains with less environmental impact, less soil disturbance, and ease of installation are the reasons.

At the present time, there is a lack of standard tests for wick drains, especially a comparative test between the different wicks available on the market. It was the intent of this study to examine the effects of band-shaped prefabricated drains on the consolidation behavior of a clay soil, and to evaluate and contrast the wicks' efficiency in a laboratory experiment in a way that would guide the engineering application of prefabricated drains in the field. The primary objectives of the testing program were to:

FIG. 1—*Prefabricated drains (left to right: Castleboard, Mebradrain, Alidrain, Franki-Kjellman, and Geodrain).*

1. Design and construct an apparatus to investigate the effects of band-shaped prefabricated drains on the consolidation process.

2. Determine whether prefabricated drains accelerate the consolidation process and to assess any potential benefits from their usage.

3. Test commercially available band-shaped prefabricated drains.

4. Determine whether laboratory testing of the prefabricated drains is a practical way of distinguishing one from another.

5. Contrast the respective performance of the prefabricated drains.

Theory of Consolidation

One-Dimensional Consolidation

Terzaghi [2] was the first to propose a theory for the time rate of one-dimensional consolidation. The basic differential equation for the one-dimensional case is

$$c_v \frac{\partial^2 u}{\partial z^2} = \frac{\partial u}{\partial t} \tag{1}$$

where

$$c_v = \frac{k_v(1 + e_i)}{a_v \gamma_w}, \quad a_v = \frac{-de}{dp}$$

and flow is in the vertical direction. For a uniform initial hydrostatic excess (u_i), the average degree of consolidation due to vertical flow is

$$\bar{U}_v = 1 - \sum_{m=0}^{m=\infty} \frac{2}{M^2} \exp(-M^2 T_v) \tag{2}$$

where $M = (2m + 1)\pi/2$, and $T_v = c_v t/H^2$.

Two- or Three-Dimensional Consolidation

Theories for the consolidation of fine-grained soils due to both vertical and radial flow to wells have been available since the 1940s. Barron [3] studied the effect of radial drainage as well as the effects of smear and well resistance on a drain well. Smear is a remolding of the soil adjacent to the vertical drain, whereas well resistance is a significant reduction in flow capacity due to finite drain well permeability. Barron considered two fundamental cases: free strain and equal strain. In the free strain case differential surface settlements occur but do not have an effect on the redistribution of stresses in the soil by the arching of the soil. In the equal strain case, however, horizontal sections through the soil remain horizontal during consolidation. Richart [4] made a comparison of the free and equal strain solutions and found that little difference exists between the two cases.

For symmetrical flow toward a central drain well, the differential equation for consolidation by three-dimensional flow is

$$c_h \left[\frac{1}{r} \frac{\partial u}{\partial r} + \frac{\partial^2 u}{\partial r^2} \right] + c_v \left[\frac{\partial^2 u}{\partial z^2} \right] = \frac{\partial u}{\partial t} \qquad (3)$$

where

$$c_h = \frac{k_h(1 + e_i)}{a_v \gamma_w}$$

and r and z are cylindrical coordinates as defined in Fig. 2. Carrillo [5] demonstrated that independent solutions can be obtained for the two terms on the left-hand side of Eq 3 and then superimposed to yield a complete solution. Therefore the average degree of consolidation due to combined radial and vertical flow is

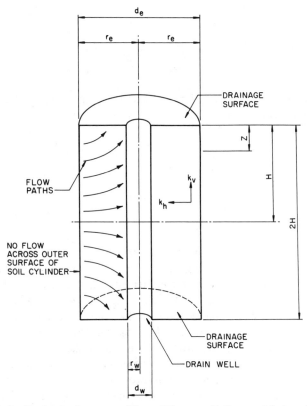

FIG. 2—Fundamental concepts of flow within zone of influence of drain well [3].

$$1 - \bar{U}_{r,v} = (1 - \bar{U}_r)(1 - \bar{U}_v) \tag{4}$$

where $\bar{U}_v$ can be obtained from Eq 2 and $\bar{U}_r = (\bar{u}_r - u_i)/u_i$. Barron [3] obtained expressions for $\bar{u}_r$ for an ideal well, and for wells taking into consideration the effect of peripheral smear and well resistance. The average degree of consolidation due to radial flow for an ideal well is

$$\bar{U}_r = 1 - \exp(\lambda) \tag{5}$$

where

$$\lambda = -8T_h/F(n)$$
$$T_h = c_h t/d_e^2$$
$$n = d_e/d_w$$

and

$$F(n) = \frac{n^2}{n^2 - 1} \log_e(n) - \frac{3n^2 - 1}{4n^2} \tag{6}$$

Considering only smear, $\bar{U}_r$ is

$$\bar{U}_r = 1 - \exp(\xi) \tag{7}$$

where

$$\xi = -8T_h/F(n,s,k_h,k_s),$$
$$s = r_s/r_w$$

and

$$F(n,s,k_h,k_s) = \left[\frac{n^2}{n^2 - s^2} \log_e\left(\frac{n}{s}\right) - \frac{3}{4} + \frac{s^2}{4n^2} + \frac{k_h}{k_s}\left(\frac{n^2 - s^2}{n^2}\right) \log_e(s) \right] \tag{8}$$

Considering smear and well resistance, $\bar{U}_r$ is

$$\bar{U}_r = 1 - \exp[\xi \cdot f(z)] \tag{9}$$

where

$$f(z) = \frac{\exp[\beta(z - 2H)] + \exp(-\beta z)}{1 + \exp(-2\beta H)} \tag{10}$$

and

$$\beta = \left[\frac{2k_h(n^2 - s^2)}{k_w r_e^2 F(n,s,k_h,k_s)} \right]^{1/2} \tag{11}$$

The expression for $f(z)$ is numerically equal to one when the effects of well resistance are not considered.

Consolidation by Band-Shaped Prefabricated Drains

The vertical drain well is assumed to have a specific cylindrical zone of influence governed by an external radius, r_e. Figure 3 illustrates typical drain well patterns. The diameter of the drain well influence zone is obtained by comparing equivalent cross-sectional areas for a square grid pattern and for a triangular grid pattern. For the latter,

$$d_e^2 = 2\sqrt{3}\, S^2/\pi \tag{12a}$$

or

$$d_e = 1.05S \tag{12b}$$

For a rectangular grid pattern,

$$d_e^2 = 4\, S^2/\pi \tag{13a}$$

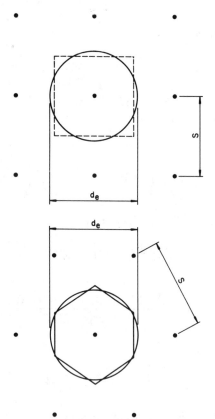

FIG. 3—*Equivalent drain diameters for square and triangular grid patterns* [6].

or

$$d_e = 1.13 \, S \tag{13b}$$

Because the rate of consolidation is proportional to the square of the drainage path, the overall effect of the equivalent cylinder concept is to overestimate the average consolidation at the end of a particular time interval. Atkinson and Eldred [6] maintain that an error of about 5% on the unsafe side will result for a square grid. For a triangular grid, the error will be less.

In estimating the degree of consolidation obtained using band-shaped prefabricated drains, additional errors result due to the shape of the drains. Barron's analytical solutions employ a circular cross section for the drain, whereas the band-shaped prefabricated drain introduces an error due to the altered flow of pore water near the drain. Normally the wick drains are modelled as an equivalent cylindrical drain according to the relationship

$$d_w = 2(t + b)/\pi \tag{14}$$

where

d_w = equivalent drain diameter,
t = thickness of drain, and
b = width of drain.

For a typical prefabricated drain 100 mm wide and 4 mm thick, the equivalent drain diameter is approximately 66 mm. The band-shaped drains available today are installed with a mandrel, which upon insertion and removal from the soil always creates a smeared zone. This smeared zone is approximately 10 mm thick for a 100-mm-wide band-shaped drain, creating a zone in which $r_s = 1.3 \, r_w$. It has been shown that migration of finer soil particles into the drain can cause clogging problems, but correct selection of filter material can avert much of this problem. According to Atkinson and Eldred [6], a bridging network of larger particles adjacent to the wick often forms a natural filter around the wick, removing a substantial amount of the smear caused by installation. Essentially, the drain fabric must not have too large or too small an effective pore size. Field and laboratory experience have shown that an optimum pore size of 10 to 20 μm is appropriate for most applications.

Hansbo [7,8] suggests a simple formula for the design of band-shaped prefabricated drain installations based on Barron's classical solution. In Hansbo's approach, the wick drain is considered as a vertical drain with a diameter equivalent to a cylindrical column of the same cross sectional area as the wick. Furthermore, he recommends that an equivalent diameter of approximately 50 mm be used rather than the equivalent diameter given by Eq 14. The average degree of consolidation due to radial flow for an ideal well can be obtained from Eq 5 with $F(n)$ approximated by

$$F(n) = \frac{n^2}{n^2 - 1} [\log_e(n) - 0.75 + n^{-2}] \qquad (15a)$$

Since n will usually be on the order of ten or greater, $F(n)$ can be further simplified to

$$F(n) = \log_e(n) - 0.75 \qquad (15b)$$

A modification of the expression for $F(n)$ in Eq 15b permits the well resistance parameter to enter into the drain spacing design:

$$F(n) = \log_e(n) - 0.75 + \pi z(2\ell - z)\frac{k_h}{q_w} \qquad (16)$$

where $q_w = A_w \cdot k_w$. An expression for $F(n)$ with both the smear and well resistance parameters accounted for is

$$F(n) = \log_e\left(\frac{n}{s}\right) + \frac{k_h}{k_s}\log_e(s) - 0.75 + \pi z(2\ell - z)\frac{k_h}{q_w} \qquad (17)$$

Hansbo suggests that since the effect of well resistance and smear is to delay the consolidation process, one may choose to use a reduced value of equivalent drain diameter rather than utilize Eqs 16 and 17, which are defined as functions of the depth z and specific discharge capacity q_w. Alternatively, a lower value for the horizontal coefficient of consolidation may be used.

The Experiment

The test program was devised to model ideally the behavior of a band-shaped prefabricated drain in a clay soil. The primary objectives of the testing program were given in the introductory section. These objectives were addressed by Ludewig [9], who designed and constructed an apparatus referred to as a "wick drain consolidometer"; see Fig. 4. The parameters considered included the type and amount of soil, its water content, all aspects of soil preparation and wick drain insertion into the consolidometer, the length of the wick drain, and the size of the consolidometer. In addition, it was necessary to determine if single or double drainage would be more practical, if this drainage layer should be at the top or bottom, how the water was to be drained from the apparatus, and what was a suitable drainage material and thickness.

Soil

The clay used throughout the testing program was kaolin at a water content equal to its liquid limit of 50% and having a plastic index of 17. At this water content it had a soft consistency and a permeability on the order of 10^{-8} cm/s.

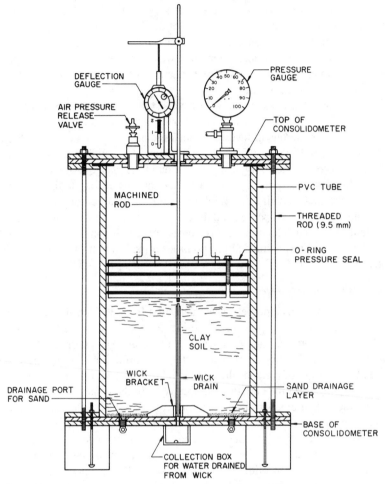

FIG. 4—*Wick drain consolidometer.*

From consolidation tests performed in accordance with ASTM Test for One-Dimensional Consolidation Properties of Soils (D 2435), it was ascertained that the consolidation of the kaolin in the wick drain consolidometer would be one of virgin compression. Nominally 0.45 kN of kaolinite, at a 50% water content, were placed into the wick drain consolidometer. This was prepared from powdered Florida EPK-kaolin and gradually mixed into a homogeneous mass. Care was taken to maintain the vertical position of the wick by uniformly placing and lightly compacting the clay in small handfuls until the consolidometer was full.

Band-Shaped Prefabricated Drains

In all, some 36 tests were undertaken in the testing program. Each wick tested was 38.1 cm long and placed vertically into the consolidometer. Five different band-shaped prefabricated drains were tested; see Table 1. The test results presented herein are based on seven tests. All of the wicks were tested once, except for the Alidrain wick, which was tested twice to check repeatability and consistency of the testing procedure. One test was performed with no wick placed in the consolidometer to contrast the performance of the wicks without the benefit of a wick embedded in the clay.

The Castleboard drain consists of a special chemical fiber non-woven cloth with a polyolefin base or interior. A hot-melt adhesive bonds the cloth to the base which has numerous grooves 2.5 mm wide and 2.5 mm deep. The Franki-Kjellman drain is fabricated from a white plastic base with a thick gray filter fabric glued to both sides but not wrapped around the plastic. Although the filter is thick, the plastic base itself is not, and there are fewer longitudinal channels. The Mebradrain, Geodrain, and Alidrain drains have a central core of plastic with a filter material wrapped around the core but not glued to it with an adhesive. The Mebradrain and Geodrain cores consist of longitudinal grooves, but the Alidrain wick features a studded plastic base, folded in half, and around which a filter sleeve is fitted.

Wick Drain Consolidometer

The wick drain consolidometer was fabricated from 30.2 cm inside diameter, 1 cm wall thickness, gray polyvinyl chloride (PVC) tubing with a height of 51.4 cm. The consolidometer was equipped with two drainage systems; one to determine the amount of water coming from the clay through a sand drainage layer placed at the bottom of the consolidometer and the other to determine the amount of water coming from the clay through the wick. In the center of the base of the consolidometer is a bracket made from clear plastic which holds the wick drain in place during a test. In the narrow slot of the bracket are a series of holes to drain water coming from the prefabricated drains. Four holes near the edge of the base act as drainage ports for the water coming from the sand drainage layer at the bottom of the consolidometer. This system allows separate collection of water from the sand drainage layer and from the wick drain (Fig. 5). The drainage layer at the bottom of the consolidometer consisted of saturated Whitehead's Port Elizabeth, N.J., sand No. 90P at a 1-cm thickness.

The clay was consolidated at 86.2 kPa for 72 to 96 h, or when measurable compressive strain or water release had ceased. The load was applied pneumatically through a piston seal at the top of the consolidometer, where deflection readings could also be obtained.

Testing Sequence

Three phases constitute the testing sequence:

1. Filling the consolidometer.
2. Testing.
3. Emptying the consolidometer.

For a complete description of the testing sequence see Ludewig [9].

TABLE 1—Band-shaped prefabricated drain technical specifications.

	Mebradrain	Geodrain	Castleboard	Franki-Kjellman	Alidrain
Filter jacket material	Typar (polypropylene)	polyester/cellulose	nonwoven cloth	cloth fabric	cloth fabric
Core material, type	100% polypropylene plastic with grooves	polyethylene LD plastic with grooves	polyolefin plastic with grooves	plastic with grooves	plastic studded
Weight (including filter), N/m	0.90	1.47	0.88	NA	NA
Width (including filter), mm	100	95	94	100	100
Thickness (including filter), mm	3	4	2.6	3.5	7
Discharge capacity, m^3/year	378	10	1134	NA	NA
Water permeability, m/s	4.5×10^{-4}	3×10^{-7}	1.5×10^{-4}	NA	3×10^{-6}
Number of grooves	38	27	38	10	no grooves
Free surface, mm^2/mm[a]	200	130	NA	NA	180
Free volume, mm^3/mm[b]	180	170	NA	NA	470
Equivalent diameter (d_w), mm	66	66	62	66	68

[a] Area of filter allowing water to enter directly into drain voids of channels without obstruction by core material.
[b] Volume of voids allowing flow of water within drain core.

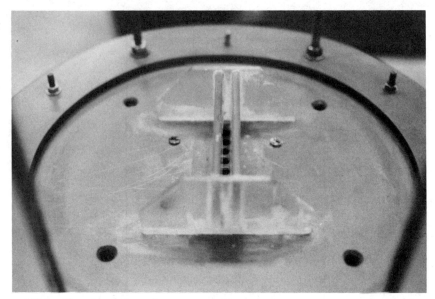

FIG. 5—*Base of wick drain consolidometer, with wick bracket and drainage ports for the wick and sand drainage layer.*

Results

The kaolin placed in the wick drain consolidometer had a specific gravity of 2.70. If the soil was initially saturated, a 0.45-kN specimen at a 50% water content would be at an initial void ratio of 1.35. From consolidation tests in accordance with ASTM D 2435, a compression index, C_c, of 0.26 was obtained for a water content of 50%. For a pressure increment from 0 to 82.6 kPa, the theoretical ultimate compression for a clay specimen 38.1 cm thick would be 3.7 cm. The theoretical time rate of consolidation for the no wick test can be obtained from

$$t = \frac{T_v H^2}{c_v} \tag{18}$$

where $H = 38.1$ cm, since only single drainage existed at the bottom of the consolidometer; for the 0 to 82.6 kPa pressure increment $c_v = 0.094$ mm^2/s (from consolidation tests in accordance with ASTM D 2435); and T_v values are from a source such as Taylor [10]. Figure 6 plots the results of the consolidation test with no wick drain in the consolidometer and the theoretical consolidation curve for this case. The laboratory-derived curve of compression versus time for the no-wick test was nearly identical to the theoretical curve, especially for time less than 250 min. For time greater than 250 min, the observed settlement rate was slightly greater than the theoretical curve.

Figure 7 illustrates the consolidation behavior of the Alidrain wick tests. This plot represents the results of two separate consolidation tests. It can be seen that the two tests yielded nearly identical consolidation curves. Comparing the two Alidrain tests, after $t = 2000$ min, the second Alidrain test indicated a slightly greater settlement rate. In Fig. 8 the theoretical consolidation curve for combined radial and vertical flow is plotted in addition to the results of the wick drain testing program. Values of $\bar{U}_{r,v}$ were obtained from Eq 4, and values of $\bar{U}_r$ were obtained from Eq 5, where $c_h = c_v = 0.094$ mm^2/s; $d_e = 30.2$ cm, the diameter of the consolidating soil cylinder (i.e., the diameter of the consolidometer); d_w = drain well diameter, for a wick drain determined from Eq 14, and is approximately 66 mm; and $n = d_e/d_w = 4.58$. Theoretically, at $t = 4000$ min, $\bar{U}_{r,v}$ was determined to be 95%, with a corresponding settlement of 3.5 cm.

Of the five prefabricated drains tested, no drain exhibited the expected theoretical behavior for the entire duration of the experiment. For $t < 40$ min, the Mebradrain curve was closely approximated by the $\bar{U}_{r,v}$ curve, after which they diverged notably. When t was between 40 and 1400 min, the two Alidrain tests exhibited the characteristics of the $\bar{U}_{r,v}$ curve, after which they diverged notably. At $t = 4000$ min, three wick drain tests attained settlements greater than the theoretical prediction of 3.5 cm, while three tests attained settlements which were less. All tests performed with a wick drain in the consolidometer attained set-

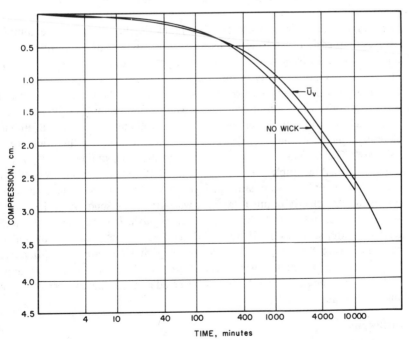

FIG. 6—*Compression versus time for consolidation test with no wick; $\bar{U}_v$ is the theoretical curve.*

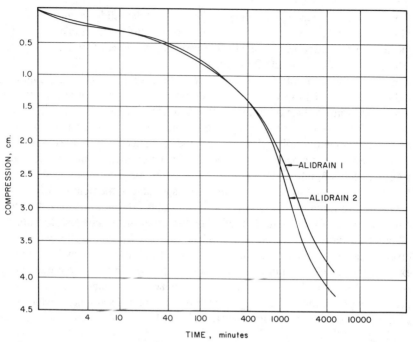

FIG. 7—*Compression versus time for two separate consolidation tests using Alidrain wicks.*

tlements greater than the no-wick test. The Franki-Kjellman wick test exhibited the least compression of all the wicks tested. This is due to the design of the drain itself; wide grooves allow the filter to be pressed into the core of the drain, reducing the drainage capacity of the wick.

Figure 9 plots the cumulative water release data from only the prefabricated drains for the seven tests discussed herein. The two Alidrain tests can easily be distinguished from the other four wicks after $t = 100$ min. The Franki-Kjellman wick drained the least water from the clay, the settlement rate for this wick being the slowest. The amount of water drained from the wick drainage port for the no wick test was negligible, as expected. Figure 10 plots the cumulative water release data from only the sand drainage layer for the seven tests discussed herein. The no-wick test clearly drained the most water from the sand layer. The other tests do not enable the wick drains to be distinguished on the basis of water release from the sand layer. There is little correlation between the amount of water drained from the wick and from the sand. Figure 9 indicates that a difference exists between the amount of water drained from each of the wicks, while Fig. 10 indicates that the sand layer drainage is affected by the respective performance of the wick drains.

Conclusions

The following generalized conclusions can be drawn from the results of the wick drain testing program:

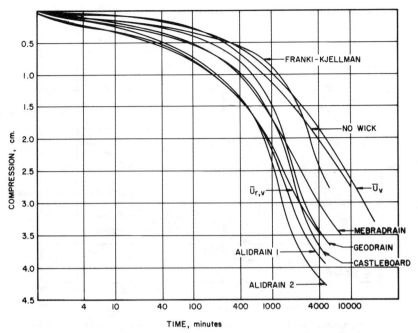

FIG. 8—*Compression versus time for wick drain testing program;* $\bar{U}_v$ *and* $\bar{U}_{r,v}$ *are theoretical curves.*

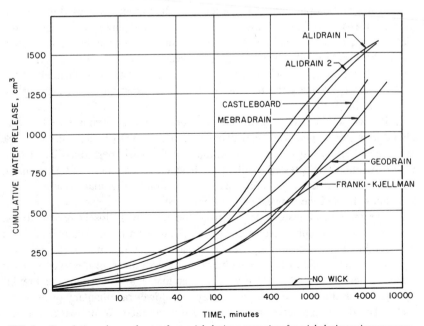

FIG. 9—*Cumulative release of water from wick drains versus time for wick drain testing program.*

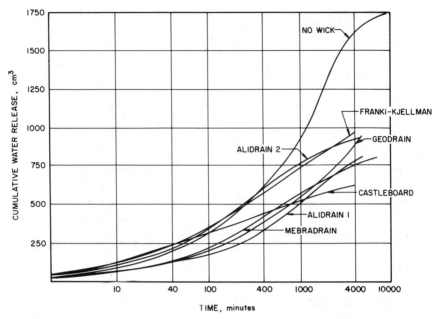

FIG. 10—*Cumulative release of water from sand drainage layer versus time for wick drain testing program.*

1. Consolidation proceeded at a faster rate when a wick was used than when no wick was used. This statement was always true at any time, t, for all the wicks except for the Franki-Kjellman drain. For a short time during the experiment, the Franki-Kjellman wick was observed to have only a small effect on the overall consolidation rate of the clay. The wide and shallow channels in the wick can be blocked by the pressure of the soil against the wick, causing the wick to be ineffective.

2. It was possible to compare the prefabricated wick drains based on the laboratory experiments, utilizing the consolidometer, and to distinguish each wick on the basis of the test results. The Alidrain wick clearly performed better than any of the other wicks tested, while the Castleboard, Geodrain, and Mebradrain performed in essentially a similar manner. The Franki-Kjellman wick drain experiment exhibited the slowest rate of compression.

3. The amount of water released from the wick was closely related to the observed rate of compression for each wick drain experiment. The two Alidrain tests exhibited the fastest compression rate as well as the most water released over time. The Castleboard, Mebradrain, and Geodrain were observed to have similar water release rates. At a time greater than $t = 1000$ min, the Franki-Kjellman wick drained water from the clay at a slower rate than any of the other wicks. The difference between the water discharge rates for the wicks was not

as substantial in the early part of the experiment (i.e., $t < 1000$ min) as it was in the latter part of the experiment.

4. An examination of water release rates from the sand drainage layer revealed that the no-wick test discharged the most water. The sand layer drained approximately one half the amount of the no-wick test when a wick was placed in the consolidometer. This reduction in the amount of water released is due to the fact that horizontal drainage toward the wick reduces the amount of consolidation that occurs due to vertical drainage. As stated in Conclusion 1 above, combined horizontal and vertical consolidation was always observed to proceed at a faster rate than vertical consolidation alone. Therefore Conclusion 4 is consistent with Conclusion 1.

5. The wicks could not be distinguished on the basis of water released from the sand layer. There are differences in the amount of water released, but all the wicks performed in a similar manner.

6. Repeatability was possible. It was next to impossible to distinguish the two Alidrain consolidation tests from a compression versus time plot. An examination of the rate at which water was expelled from the consolidometer verified this repeatability.

7. All wick drains with a loosely fitted filter jacket performed better than the Franki-Kjellman wick. The Castleboard's one-piece filter and core construction performed as well as the Geodrain and Mebradrain wicks but not as well as the studded core design of the Alidrain. The Alidrain core is one reason for superior drain performance. The grooved-core designs of the Mebradrain and Geodrain are more susceptible to lateral pressures, thus reducing the channel volume available for water conduction.

Table 1 lists various prefabricated drain technical specifications. The free surface of the Alidrain is approximately the same as that of the Mebradrain and Geodrain, but the free volume of the Alidrain is much greater than that of any other drain. Based on test results, it is concluded that the free-volume concept and core configuration of the Alidrain are reasons for its superior performance. The Alidrain is also thicker than any of the other drains, which is one reason why its free volume is substantially higher than that of the other drains.

References

[1] Kjellman, W. in *Proceedings,* 2nd International Conference on Soil Mechanics, Vol. 2, 1948, pp. 302–305.
[2] Terzaghi, K., *Erdbaumechanik auf bodenphysikalischer Grundlage,* Deuticke, Vienna, 1925.
[3] Barron, R. A., *Transactions of the American Society of Civil Engineers,* Vol. 113, 1948, pp. 718–742.
[4] Richart, F. E., *Journal of the Soil Mechanics and Foundations Division,* American Society of Civil Engineers, Vol. 83, No. SM3, July 1957, pp. 1301-1–1301-38.
[5] Carrillo, M., *Journal of Mathematics and Physics,* Vol. 21, No. 1, March 1942, pp. 1–9.
[6] Atkinson, M. S. and Eldred, P. J. L., "Symposium in Print on Vertical Drains," *Geotechnique,* Vol. 31, No. 1, March 1981, pp. 33–43.
[7] Hansbo, S., *Ground Engineering,* Vol. 12, No. 5, July 1979, pp. 16–25.

[8] Hansbo, S. in *Proceedings*, 10th International Conference on Soil Mechanics and Foundation Engineering, Vol. 3, 1981, pp. 677–682.

[9] Ludewig, N. M., "A Comparatory Laboratory Evaluation of Band Shaped Prefabricated Drains," thesis presented to The Cooper Union for the Advancement of Science and Art, in partial fulfillment for the degree of Master of Engineering, New York, 1984.

[10] Taylor, D. W., *Fundamentals of Soil Mechanics*, Wiley, New York, 1948, p. 237.

DISCUSSION

Anonymous reviewer (*written discussion*)—(1) Conclusions 2 and 3 indicate that the Alidrain wick performed better than the other wick drains tested based on the information shown in Figs. 8 and 9. These figures indicate the total settlement and water release to be the greatest for the Alidrain. However, this is not the true test for concluding the Alidrain to be superior. The true test should be to compare the values of times for a certain percent of consolidation to take place such as the standard curve fitting method of determining the time for 90% of consolidation (T_{90}) to take place. As the New York State Department of Transportation (NYS DOT) Soil Mechanics Bureau has found, variations in total settlement and water release such as shown in Figs. 8 and 9 can be attributed to the soil itself even when the soil used is a relatively uniform manufactured clay soil. This variation has been largely attributed to the method of placement of the soil in the consolidation chamber and the time that the mixed soil is allowed to cure prior to placement in the chamber.

(2) Conclusion 6 concerning repeatability is not properly supported based on the results of only two tests and the findings of NYS DOT indicated above.

(3) Conclusion 7 is also not properly supported based on the information stated above and testing done by NYS DOT Soil Mechanics Bureau. NYS DOT has found that the core of rigid cored wick drains possess several orders of magnitude more core flow capacity than flow of water reaching the wick drains due to soil consolidation.

V. A. Guido and N. M. Ludewig (*authors' closure*)—(1) In general, different prefabricated drains have different equivalent diameters, d_w. The authors feel it is for this reason the discusser indicates that the true test (to determine which prefabricated drain performed best) should be to compare the values of time for a certain percent consolidation to take place. It is shown in Table 1 that the Mebradrain, Geodrain, and Franki-Kjellman prefabricated drains all have the same d_w equal to 66 mm, while the Castleboard and Alidrain drains have d_w equal to 62 mm and 68 mm, respectively. The average d_w for the five drains considered is 66 mm, which is the value quoted by Atkinson and Eldred [6]. For the range of d_w's encountered for different prefabricated drains, the effect on the theoretical combined radial and vertical percent consolidation, $\bar{U}_{r,v}$ is negligible (the $\bar{U}_{r,v}$ curve in Fig. 8 is for $d_w = 66$ mm). Therefore, since the theoretical values of

$\bar{U}_{r,v}$ are virtually unaffected by the value of d_w for prefabricated drains, the observed laboratory curves of Fig. 8 (even if all of them do not have the same value of d_w) can be compared on a one-to-one basis.

Furthermore, the discusser indicates that to determine the time for a certain percent consolidation to take place the standard curve fitting method of determining the time for 90% of consolidation (T_{90}) could be used. This is inappropriate since the standard curve-fitting method mentioned above (square root of time method) is applicable only for one-dimensional (vertical) drainage and not for three-dimensional (radial and vertical) drainage, which exists with prefabricated drains. In addition, the standard time fitting method will yield a percent consolidation which is in the primary compression region. For a large number of soils secondary compression is as important as primary compression, if not more so. Therefore it is the total compression (both primary and secondary) which is of the greatest importance; one should want to know which prefabricated drain yields the greatest compression in the shortest period of time. The curves in Fig. 8 yield this information, since the data used to plot the curves were obtained until no further compression or water release was detected for a given pressure increment (at the end of primary and secondary compression).

In addition, the discusser indicates that the variations in total settlement and water release such as shown in Figs. 8 and 9 can be attributed to the soil itself. The authors disagree. If the soil is manufactured in a relatively uniform manner and allowed to cure for a specified time period from batch to batch, and the saturation of the clay is 100% when initially placed in the wick drain consolidometer, then the variations in total settlement and water release are attributable only to the differences in the prefabricated drains themselves. It is imperative that the water release from the prefabricated drains and from the sand drainage layer be collected separately. This is substantiated by Conclusion 5.

Therefore the authors feel Conclusions 2 and 3 are correct as originally stated.

(2) For two separate tests run with the Alidrain prefabricated drain, Fig. 7 indicates that is was possible to repeat the test results. More than two such tests are warranted to make a universal statement regarding repeatability. The authors did not intend for Conclusion 6 to be such a statement. In addition, the good agreement between the two curves in Fig. 7 is an indication of the homogeneity of the soil from two different batches of clay. (Mixing, curing, and placement of the clay in the wick drain consolidometer was consistent from batch to batch.) Therefore Conclusion 6 indicates that for the two Alidrain tests performed, repeatability was possible and dependent upon the homogeneous nature of the manufactured clay. If care is taken that the clay from batch to batch is homogeneous, there is no reason why consolidation tests performed with the same type of prefabricated drain cannot be repeated.

(3) The authors do not see the relevance to Conclusion 7 of the discusser's statement that the core of rigid cored wick drains possess several orders of magnitude more core flow capacity than flow of water reaching the wick drains due to soil consolidation. If one were using sand drains instead of prefabricated

drains, the sand in the drains would invariably have a much greater flow capacity than the surrounding clay soil. However, different gradations of sand would allow the consolidation process to occur at different rates. This would also be true for different prefabricated drains with different properties such as free volume. Therefore it was observed that the core configuration and prefabricated drain design itself affected the drains' performance (Figs. 8 and 9). Hence the statement in Conclusion 7 was made.

In conclusion, the authors feel that the three comments offered by the discusser are not justified. In an area such as the testing of prefabricated drains in the laboratory, where little if any previous work has been done, the use of such terms as "true test" is not warranted. We feel that our testing was innovative, well-conceived, and carefully performed. The discusser consistently refers to similar work done by the NYS DOT Soil Mechanics Bureau; at this early stage of research work in this field, who is to say that their results are superior to ours? We are sure that some of the procedures we developed to test the prefabricated drains are very different from those of the NYS DOT and vice versa. Any further research should incorporate the best ideas of both research endeavors.

L. David Suits,[1] Raymond L. Gemme,[1] and Joseph J. Masi[1]

Effectiveness of Prefabricated Drains on Laboratory Consolidation of Remolded Soils

REFERENCE: Suits, L. D., Gemme, R. L., and Masi, J. J., "**Effectiveness of Prefab-ricated Drains on Laboratory Consolidation of Remolded Soils,**" *Consolidation of Soils: Testing and Evaluation, ASTM STP 892,* R. N. Yong and F. C. Townsend, Eds., American Society for Testing and Materials, Philadelphia, 1986, pp. 663–683.

ABSTRACT: The Soil Mechanics Bureau of the New York State Department of Trans-portation has carried out an extensive testing program over the past two years to determine the relative effectiveness of several types of prefabricated "wick" drains in the laboratory and field environments. The test program and results, along with discussions, and the relationship of the test results to theoretical field performance are presented in this paper.

KEY WORDS: prefabricated drains, wick drains, consolidation, remolded soil, equivalent diameter, soil improvement, lateral drainage, coefficients of consolidation, time rate of consolidation, stabilization, geotextiles

One of the major concerns in geotechnical engineering is the settlement which occurs when a highway embankment or structure such as a bridge or building is built on compressible soils. The magnitude and the time rate of settlement are of concern for both engineering and economic reasons. For instance, if 305 mm (1 ft) of embankment settlement takes place over a long period of time (say 20 years) at a pile supported bridge abutment, it is not considered a major problem because early and frequent asphaltic patching would not be required at the bridge approach after construction. However, the reverse would be true if this settlement were to take place within a few years after construction. To eliminate the need for patching of approach embankments to pile supported abutments, waiting periods and surcharges are often employed, and the time rate of foundation settlement is increased.

In the past, an increase in the rate of foundation settlement was accomplished by installing sand drains in a problem area. With the advent of the use of geotextiles in geotechnical engineering, a new method or substitute for sand

[1] Soils Engineering Laboratory Supervisor, Associate Soils Engineer, and Assistant Soils Engineer, respectively, Soil Mechanics Bureau, New York State Department of Transportation, Albany, NY 12232.

drains has been developed. Prefabricated drains, or wick drains as they are called, are porous geotextiles wrapped around a drainage core of material, usually plastic. The geotextile allows water to enter the core and flow vertically through the core to a drainage blanket. The benefits of using these wick drains over sand drains include ease and speed of installation, lower material cost, and lower overall cost, even though more wick drains are needed in comparison to sand drains.

As more of these wick drains were introduced on the market, it became necessary for the New York State Department of Transportation (NYSDOT) to develop a test program for evaluating the effectiveness of each drain and to compare the effectiveness of one drain to another. This paper deals primarily with the laboratory determination of the effectiveness of wick drains.

Laboratory Tests

The total laboratory evaluation program undertaken by the Department's Soil Mechanics Bureau consisted of four tests:

1. *Large diameter consolidation (LDC) test*—This was the main test to determine the effectiveness of prefabricated drains as compared to regular sand drains. This test comprises most of the contents of this paper.
2. *Standard consolidation tests*—These were run to determine the vertical coefficient of consolidation.
3. *Crimp test*—This test was developed to determine if crimping the wick drains would decrease their effectiveness.
4. *Lateral pressure test*—This test was developed to determine if increases in lateral pressures on the wide sides of the wicks would decrease their effectiveness.

The LDC, crimp, and lateral pressure test procedures are too lengthy to be dealt with in this paper, but information can be obtained from the NYSDOT Soil Mechanics Bureau.

The program for the LDC tests consisted of using three types of remolded soil, so that the effect of soil type on wick drain performance could be determined. At the same time that the LDC tests were being performed, standard consolidation tests were also being performed on the soil types being tested. The purpose of the standard tests was to determine the vertical coefficients of consolidation (C_V) for the soils used.

In a remolded soil, the vertical coefficient of consolidation (C_V) should be equal to the horizontal coefficient of consolidation (C_H). The latter is one of the soil properties being determined from the LDC test.

Refer to Table 1 for descriptions of the soils used in this program.

Large Diameter Consolidation (LDC) Tests

Test Apparatus

The apparatus used for the LDC test program consisted of a 254-mm (10 in.) inside diameter by 559-mm (22-in.)-high PVC cylinder (Fig. 1). A piston-type

TABLE 1—*Average soil parameters derived from tests.*

Soil Type[a]	LDC Test[b]			Standard Consolidation Tests[b]		
	Moisture Content, %	C_h, m²/day	Settlement, mm	Moisture Content, %	C_v, m²/day
1. Organic silty clay (Westway soil)	50	0.003	44	52	0.004
2. Manufactured clay	42	0.004	51	39	0.006
3. Peat	338	0.004	122	354	0.001

[a] Soil types will be referred to as 1, 2, or 3 in remaining tables.
[b] Conversion factors: 25.4 mm = 1 in.; 0.092 m² = 1 ft².

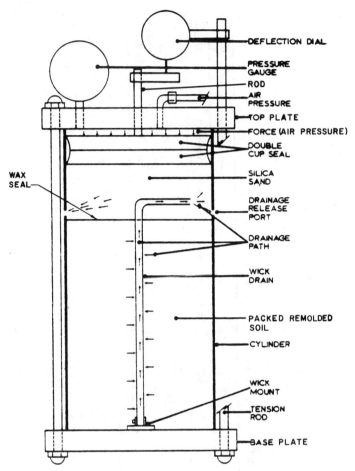

FIG. 1—*Wick drain consolidometer.*

device consisting of two 254-mm (10-in.)-diameter rubber cup seals placed back-to-back with a rod extending from their center up through the top acrylic plate was placed inside the cylinder as the load transfer unit.

The load was applied by introducing air pressure to the top of the piston assembly. The assembly served to distribute the load equally over the surface of the silica sand drainage blanket (described in the Test Procedure subsection). Drainage ports were drilled and spaced equally around the cylinder and located 152 mm (6 in.) from the top of the cylinder to allow drainage of the water from the soil due to consolidation. Additional ports had to be added for the peat soil to accommodate the large settlements which occurred (Table 1). Attached to the upper acrylic plate was a deflection dial which was used to measure the consolidation by recording the movement of the center rod attached to the cup seals.

Test Procedure

Since three of the above-described chambers had been fabricated, a "test series" was considered a series of tests performed on three sand drain specimens or three wick drain specimens of the same brand name. Enough of each soil type was blended together to fill six chambers (considered six batches of soil), ensuring a continuous testing program. After a sand drain or wick drain test series, each batch of soil was mixed back to its initial moisture content, allowed to cure, and reused on subsequent tests.

The setting up of the test consisted of placing the soil at a predetermined wet density around a 508-mm (20-in.)-long wick drain specimen. The soil was placed to a height of 381 mm (15 in.) in the chamber. Once the soil was at the desired wet density, a thin layer of molten wax was poured over the surface of the soil and allowed to cure. This acted as a moisture barrier so that the only water allowed to leave the soil was that which reached the wick drain horizontally and traveled vertically through the wick core to a 102-mm (4-in.) silica sand drainage blanket placed over the wax seal (Fig. 1). In order to reduce the effect of side friction between the silica sand and the inside chamber wall, the wall was coated with silicon spray prior to soil placement.

Once the system was assembled, a load of approximately 95.9 kPa (2000 psf) was placed on the soil, and time deflection readings were taken. The deflection readings were plotted on two curves. One curve consisted of dial reading versus the square root of time, and the other, dial reading versus the log of time. The plots were made as the test progressed to determine when the soil reached 90% of primary consolidation. Figures 2 and 3 are examples of these plots. The

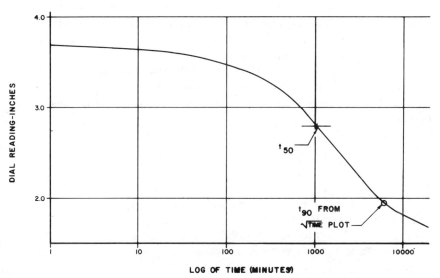

FIG. 2—*Typical LDC t_{50} results.*

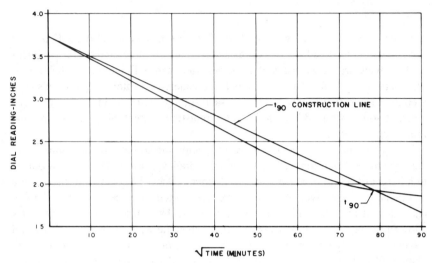

FIG. 3—*Typical LDC* t_{90} *results.*

deflection versus square root of time curve was used to determine 90% consolidation by the standard curve-fitting method. All tests were terminated at 90% consolidation.

Test Results

Analysis of the data consisted of two phases. The first phase was a determination of the C_H value of the soil. The second phase was to determine the range of equivalent sand drain diameters (d_w range) for the wick drains. A wick drain is said to be equivalent to a sand drain when it drains a soil at the same rate as a specified diameter sand drain with the same spacing.

Determination of C_H *of Soil*—The results of the test series ran on actual $51\pm$-mm-diameter ($2\pm$ in.) sand drains were used to determine the coefficient of horizontal consolidation (C_H) of the remolded soils. The standard sand drain equation to compute C_H is

$$C_H = T_{Hs}d_e^2/t_{90s} \qquad (1)$$

where

C_H = coefficient of horizontal consolidation (m²/day),
T_{Hs} = time factor for sand drains taken from Fig. 4,
t_{90s} = time in days for 90% consolidation of sand drains, and
d_e = effective drain spacing in metres.

For this computation, d_e is equal to the diameter of the test cylinder or 0.26 m

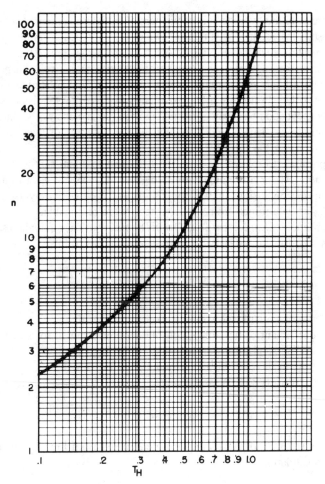

FIG. 4—*Relationship between* n *and* T_H *at 90% consolidation with radial drainage to sand drain or prefabricated drain.*

(0.833 ft). To determine the value of T_H at 90% consolidation, the following standard sand drain formula was used with n input into Fig. 4:

$$n = d_e/d_s \tag{2}$$

where

d_e = 0.26 m (0.833 ft),
d_s = $\sqrt{4W_s/\pi H \gamma_s}$ = actual computed sand drain diameter,
W_s = weight of sand used,
γ_s = loose unit weight of sand, and
H_s = height of sand column.

Table 1 presents the average soil parameters of moisture content, C_H and C_v for the three soils tested. It is seen that for two of the soils, C_H is about equal to C_v. With the peat soil, however, the C_v result is somewhat lower than C_H. After completion of the LDC test, the interior of the wick drains were inspected and were found to contain some organic fines with slight caking noticed on the surface of the geotextiles. From this it was postulated that the porous disks used in the standard consolidation test, being less porous than the geotextiles, became clogged with organic fines. This would have reduced the flow rate of water leaving the soil, thus reducing the effective C_v.

It is optimistic to expect C_H from the LDC test to equal C_v for the standard consolidation test for a remolded soil. However, C_v was not used in any of the calculations or conclusions drawn from the paper. They were only used to verify that the C_H as determined from the LDC tests were reasonable.

Determination of Equivalent Sand Drain Diameters—The equivalent sand drain diameters (d_w) for the wick drains were determined as follows:

$$T_{Hw} = C_H \, t_{90w} / d_e^2 \qquad (3)$$

where

T_{Hw} = time factor for wick drains taken from Fig. 4,
C_H = coefficient of horizontal consolidation determined for soil from LDC test on sand drains during Phase 1 Testing,
t_{90w} = time in days for 90% consolidation with wick drains,
d_e = 0.26 m (0.833 ft), and

$$d_w = d_e / n \qquad (4)$$

where d_e = 0.26 m (0.833 ft) and n is determined from Fig. 4 with T_{Hw} input at 90% consolidation.

Discussion

Table 2 summarizes the average and range of times for 90% consolidation for the sand drains and the various wick drains tested. Table 3 summarizes the wick drain equivalent sand drain diameters for each of the three soils used in the LDC test.

The average equivalent sand drain diameter results for all the wick drains tested generally varied between 38 mm (1.5 in.) and 64 mm (2.5 in.). For the same spacing, a 64 mm (2.5 in.) equivalent diameter wick will drain a soil approximately 20% faster than a 38 mm (1.5 in.) equivalent-diameter wick. This significance is minor when applied to normal applied field spacings. For instance, a 64 mm (2.5 in.) equivalent sand drain diameter with a 1.72 m (5.5 ft) drain spacing will drain the same soil equally as well as a 38 mm (1.5 in.) equivalent sand drain diameter with a 1.56 m (5 ft) drain spacing.

TABLE 2—t_{90} test results (days).

Prefabricated Drain Type	Soil 1		Soil 2		Soil 3[a]	
	Avg t_{90}	Range	Avg t_{90}	Range	Avg t_{90}	Range
2-in. sand drain	6.09	6.70–5.75	5.03	5.33–4.45	4.97	5.44–4.39
Ali drain	6.87	9.00–5.10	4.17	4.39–3.96	5.91	6.36–5.25
Vinylex	...	...	4.73	4.84–4.61	...	...
Castleboard	7.90	9.00–7.00	5.15	5.63–4.39	5.55	5.88–5.12
Bando	6.50	6.90–5.70	5.20	5.32–5.08	5.45	5.81–5.16
Desol (2 specimen test)	12.87	13.40–12.60	6.47	6.67–6.27	...	...
Mebra	5.58	7.20–4.80	4.06	4.44–3.85	...	...
Auger Flo-Drain (Clay)	7.17	8.10–5.50	6.00	6.33–5.81	...	...
Auger Flo-Drain (Silt)	8.10	8.30–8.00	4.98	5.63–4.23	...	...
Hitek (Original)[b]	6.93	8.50–5.00	...	...	...	...
Hitek (Gray)[c]	...	...	3.85	4.07–3.42	...	...
Hitek (Reemay)[c]	...	...	3.67	3.82–3.45	...	...
Amer-Drain	7.97	8.50–7.40	3.67	4.04–3.07	...	...
Geo-Drain (Fabric)[b]	6.87	8.00–5.90	...	...	...	...
Colbond CX-1000	6.07	6.30–5.70	4.11	4.61–3.45	...	...
Sol-Compact (Sleeve)	6.23	6.60–5.90	3.85	4.38–3.39	...	...

[a] The high, mid, and low equivalent sand drain diameter wick drains from Soil 1 were the only wick drains tested for Soil 3.

[b] Not available for testing in Soil 2.

[c] Not available for testing in Soil 1.

Implications

The results of the LDC tests, the first phase of the testing program, brought to light several areas which needed further investigation:

1. When the LDC testing system was dismantled, it was found that the stiffer wick drains had one or more crimps or bends in them. The questions raised were just what effect did these crimps have on the laboratory test results, and what effect would they have under field applications? The program which was developed to investigate these effects indicates that *crimping alone will not affect the performance of the wick drains.*

2. Some of the wick drains have thick geotextile core coverings and nonrigid cores. Therefore it became necessary to determine what effect lateral pressure has on the performance of these types of wick drains. We found that *lateral pressure alone seriously affected the performance of the soft* core wick drains.

3. We needed to know what effect a different soil type had on the performance of the wick drains, and so used a manufactured clay soil and a peat soil in performing a series of LDC tests. *The types of soil tested did not appear to significantly affect the average equivalent sand drain diameters determined from the LDC tests.*

TABLE 3—Equivalent sand drain diameters (millimetres).

Prefabricated Drain Type	Soil 1		Soil 2		Soil 3[a]	
	Avg Diameter	Range	Avg Diameter	Range	Avg Diameter	Range
Ali Drain	48.8	63.5–34.8	55.4	57.7–52.1	42.9	48.0–39.6
Vinylex	...	...	48.3	49.8–47.5	...	...
Castleboard	41.7	46.7–35.8	45.0	51.8–40.4	45.2	47.8–43.7
Bando	51.3	57.7–48.0	44.7	46.2–43.4	46.2	48.8–43.4
Desol[b]	20.1	20.6–18.8	35.1	36.8–33.3	...	...
Mebra	59.2	66.8–43.7	57.7	60.5–51.8	...	...
Auger Flo-Drain (Clay)	47.0	59.7–40.1	37.6	39.1–34.8	...	...
Auger Flo-Drain (Silt)	40.1	40.6–39.1	49.5	60.7–41.4	...	...
Hitek (Original)[c]	49.8	63.5–42.4	...	...	...	...
Hitek (Gray)[d]	...	...	60.4	66.8–57.2	...	...
Hitek (Reemay)[d]	...	...	64.0	67.1–62.0	...	...
Amer-Drain[d]	40.1	43.7–36.3	...	...	...	...
Geo-Drain (Fabric)[c]	49.0	56.4–40.6	...	...	...	...
Colbond CX 1000	54.6	57.7–52.8	57.4	66.8–50.8	...	...
Sol-Compact (Sleeve)	53.3	56.4–50.3	60.7	67.0–53.3	...	...

[a] The high, mid, and low equivalent sand drain diameters were the only wick drains tested for Soil 3.
[b] Desol Soil 1 test results based on 2 specimens. Test 3 malfunctioned due to equipment breakdown.
[c] Not available for testing in Soil 2.
[d] Not available for testing in Soil 1.

Crimp Tests

In the LDC test program it was found that under applied loading, severe crimps formed in several of the rigid core drains because the wick tends to get shorter as the surrounding soil settles. Two questions arose as a result:

1. What effect does a crimp have on the core flow capacities of the wick drains?
2. Was the crimp caused by the test system and would it actually occur under field applications?

The first question was answered with the development of equipment and procedures to study the effects of crimping on the core flow capacity of the wick drains. The test program used is described in this discussion, along with the test results. The second question can not be directly answered without examination of the wick drains in place. However, a theoretical approach was made to study the effects of crimping in the field. This approach is also described in this discussion.

Test Program

The test program consisted of placing a 0.62 m (2 ft) length of wick drain in the apparatus shown in Fig. 5. The vertical flow rate of water was then measured through the wick in both an uncrimped and crimped condition.

The wick drain was sealed against loss of water through the geotextile by wrapping the entire length of the wick with heat shrink plastic. A turn screw device was located at about midheight of the apparatus as shown in Fig. 5. Attached to the screw is a wedge-shaped device that forces (crimps) the wick drain against a 90° wedge-shaped receiver.

Testing consisted of determining the average flow rate of water through three specimens in both the uncrimped and crimped conditions. The flow rate was determined under a 622-mm (24.5 in.) head of de-aired water.

Discussion

Table 4 shows the results of the uncrimped and crimped flow rate for the various wick drains tested. Results of the test program indicate that there is a range in reduction of volumetric flow rate from 15 to 67% for the wick drains tested. Maximum reduction in volumetric flow occurred in the soft core (Auger) wick drain and the wick drain with no geotextile (DESOL). Also, a wick drain with the geotextile glued to the core (Bando) had a low value. To understand or determine what effect this change in flow rate would have on the LDC test results, it was necessary to compare the volumetric flow rate results in the crimp test to the volumetric flow rate of water reaching the wick drains due to consolidation of the soil in the LDC test. For the sake of expediency, only the organic silty

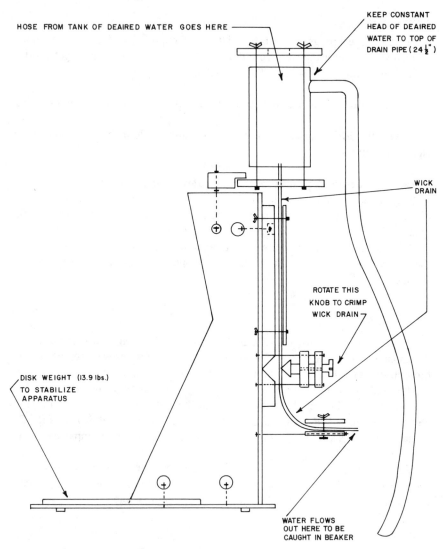

FIG. 5—*Wick drain crimp test apparatus.*

clay soil results will be shown here. However, similar results could be calculated for both the manufactured clay and peat soils tested.

Two volumetric flow rates were computed, one for an equivalent sand drain diameter of 76 mm (3 in.) for the wick drains and one for an equivalent sand drain diameter of 25.4 mm (1 in.) for the wick drains. These equivalent sand drain diameters represent the total range of results obtained from LDC wick drain tests.

The theoretical volume of water which the wick drains would have to transmit

TABLE 4—*Crimp test results (laboratory condition).*

	Flow Rate, cm³/s	
Drain Type	Avg Uncrimped	Avg Crimped
Auger Flo-Drain II (Silt)	45	17
Auger Flo-Drain I (Clay)	49	16
Mebra	256	170
Colbond CX 1000	112	75
Bando	77	27
Hitek (Original)	279	210
Hitek (Du Pont Reemay 2033)	248	219
Hitek (Gray)	232	208
Ali Drain	195	165
Castleboard	98	77
Desol	103	47
Amer-Drain	153	135
Vinylex	184	161

due to consolidation of the soil was determined based on a maximum settlement of 51 mm (2 in.) taking place in the organic silty clay soil in the LDC test. The volume of water anticipated was taken as the cross sectional area of the chamber times the anticipated settlement. This value was calculated to be 0.013 m³ (0.48 ft³).

The next step was to determine theoretical times for various percents of primary consolidation to occur. The percents selected were 10, 20, 50, and 90%. The volume of water previously determined was reduced by the appropriate percentage (10, 20, 50, and 90) in the calculation of the average volumetric flow rates. The times for the selected percents of consolidation were determined using Eqs 3 and 4. These values of time were divided into the volume of water previously determined to compute the average volumetric flow rate of water through the soil. Results of these computations are shown in Table 5.

As expected from consolidation theory, the indicated flow rates during the initial stages of consolidation were greater than for the later stages.

When comparing the flow rate results from Table 4 with the results from Table 5, it is evident that *even in the crimped state the wick drains possess a minimum of two orders of magnitude greater core flow rate capacity than the flow rate of water given up by the soil due to consolidation in the LDC test.*

TABLE 5—*Volumetric flow rate through soil (laboratory condition).*

	Time, days		Flow Rate, cm³/s	
% Consolidation	$dw = 76.2$ mm	$dw = 25.4$ mm	76.2 mm	25.4 mm
10	0.19	0.60	0.08	0.03
20	0.42	1.03	0.08	0.03
50	1.22	3.35	0.06	0.03
90	3.95	11.00	0.03	0.02

Lateral Pressure Tests

A maximum of 143.8 kPa (3000 psf) lateral soil pressure would be applied to the wick drains under typical field conditions. This is equivalent to the lateral soil pressure which could be expected on a wick drain installed beneath a 9.4-m (30-ft)-high embankment. It was therefore deemed necessary to devise a test system which would investigate the effect of lateral pressure on the flow rate capacity of the various wick drains tested.

Test Program

The test program consisted of placing a specimen of wick drain, sealed in heat-shrink plastic, in the device shown in Fig. 6.

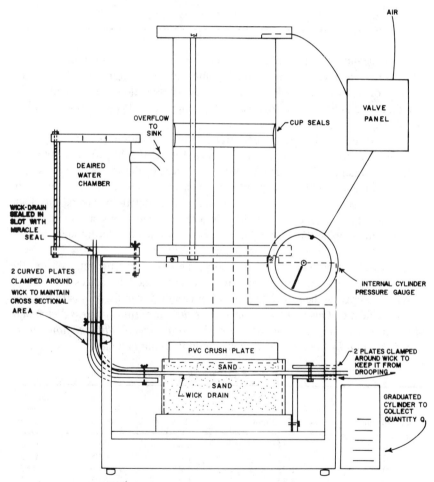

FIG. 6—*Wick drain lateral pressure test apparatus.*

A 453-mm (18.5-in.) head of de-aired water was then applied to the test specimen. The volumetric flow rate was determined with no lateral pressure applied to the specimen. The next step in the test was to apply incremental lateral loads through an air pressure system and measure the change in flow rate from the zero pressure condition.

The increments of pressure were 7.2, 17.7, 28.8, 40.3, and 55.6 kPa (150, 370, 600, 840, and 1160 psf). Owing to laboratory restrictions, 55.6 kPa (1160 psf) was the maximum applied lateral pressure. Table 6 shows the results of this testing program for each of the drains tested and each of the lateral pressures applied.

Discussion

The largest drop in flow rate occurred between 0 and 7.2 kPa (see Fig. 7 for a typical plot of results). This is believed to be caused by a seating action of sealing the heat-shrink plastic around the fabric, which probably allowed water to initially flow between the heat-shrink plastic and the fabric. At higher pressures this flow was cut off.

The typical linear relationship shown in Fig. 7 between 7.2 and 55.6 kPa (150 and 1160 psf) could be expected to continue to the anticipated maximum field pressure of 143.8 kPa (3000 psf). The results of making this assumption are also shown in Table 6.

Note that the soft core wick drains (Auger) reduced to a zero flow condition with the estimated application of the 143.8 kPa (3000 psf) lateral pressure.

Field Application

The assumptions made to determine what effect, if any, crimping and lateral pressure would have in a typical field application were as follows:

TABLE 6—*Lateral pressure test results.*

Drain Type	Flow Rate (cm³/s) at Loading (kPa) of						
	0	7.2	17.7	28.8	40.3	55.6	143.8 (Estimated)
Hitek (Gray)	184	174	172	169	167	165	140
Hitek (Du Pont Reemay)	184	174	170	169	167	164	146
Mebra	155	140	137	136	135	134	123
Vinylex	146	128	126	125	125	124	117
Amer-Drain	122	114	110	109	107	105	89
Bando	109	106	105	105	105	104	100
Ali Drain	109	99	95	92	90	88	68
Sol Compact	93	88	88	88	87	86	82
Colbond CX 1000	70	64	62	61	60	60	53
Desol	66	58	58	57	58	57	55
Castleboard	63	58	53	53	52	51	38
Auger Flo Drain I (Clay)	31	18	15	13	11	9.6	0
Auger Flo Drain II (Silt)	40	15	11	9.4	7.8	6.2	0

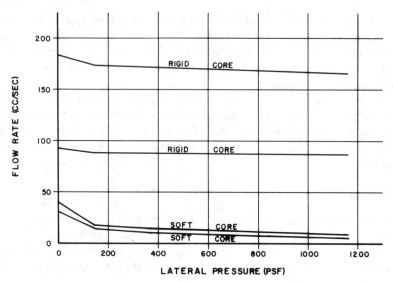

FIG. 7—*Typical lateral pressure test results (four wick drains).*

1. Soil C_H = 0.04 m²/day (0.4 ft²/day).
2. d_w = 51 mm (2 in.), d_e = 1.25 m (4 ft).
3. Settlement = 2.8 m (9 ft).
4. Length of wick drain = 18.75 m (60 ft).
5. Volume of water leaving soil due to initial 10% consolidation = 0.32 m (11.6 ft³).

The assumption of C_H was based on the observation that an undisturbed soil in New York State possesses a C_H value of approximately one order of magnitude greater than a remolded specimen of the same soil. The resultant maximum volumetric flow rate of water reaching the wick drains due to consolidation is shown in Table 7.

The length of the specimen tested in the laboratory program was 0.62 m (2 ft). For this short length of wick drain, the effect of friction was ignored, but for the 18.75-m (60 ft) length, it should not be ignored.

Using the Hazen and Williams formula for the effect of friction on the flow of liquids through pipes, the following relationship can be derived:

$$Q_2 = Q_1 \, (L_1/L_2)^{0.54} \tag{5}$$

where

Q_2 = volumetric field flow rate (cm³/s),
Q_1 = volumetric laboratory flow rate (cm³/s). Varies from 16 to 210 cm³/s (crimped condition, Table 4),

TABLE 7—Summary of crimp and lateral pressure test program.

Drain Type	Uncrimped Core Flow (Laboratory), cm³/s	% Loss in Core Flow Capacity			Net Core Flow Field, cm³/s	Max Field Flow to Wick Drain, cm³/s
		Lab Crimped	Lab Lateral Pressure[a]	Field Friction		
Auger Flo Drain I	49	67	100	81	0	3.2
Auger Flo Drain II	45	62	100	82	0	3.2
Hitek (Gray)	232	12	15	84	28	3.2
Hitek (Reemay)	248	12	16	84	29	3.2
Mebra	256	34	12	84	24	3.2
Vinylex	184	12	9	84	24	3.2
Amer-Drain	153	12	22	84	17	3.2
Bando	77	65	6	85	4	3.2
Ali Drain	195	15	31	84	18	3.2
Sol-Compact	93	b	7	b	b	3.2
Colbond CX 1000	112	24	17	84	14	3.2
Desol	103	54	5	83	8	3.2
Castleboard	98	21	34	84	8	3.2

[a] Estimated due to 143.8 kPa (3000 psf) lateral field pressure.
[b] Not tested.

L_2 = length of field wick drain = 18.75 m (60 ft), and
L_1 = length of laboratory wick drain = 0.62 m (2 ft).

The resultant range of field flow rate (Q_2) varied between 3 and 34 cm³/s. This represents an 80% reduction in flow rate from the laboratory crimped condition taking friction into account.

Table 7 includes a summary of wick drain core flow capacities. The net core flow rate estimated in the field takes into account the reductions in flow rates due to crimping, lateral pressure, and friction for each of the wick drains tested assuming a cumulative effect on core flow due to flow rate reductions.

When comparing these values of minimum core flow capacities of the wick drains to the amount of water reaching the cores due to consolidation of the soil, it can be seen that the soft core wick drains do not possess any core flow capacity and therefore will not handle any of the water flow to the wick drains due to soil consolidation.

Summary

Based on the results of the three test programs described, the following conclusions can be drawn:

1. Wick drains were effective in increasing the consolidation rate of compressible soils in the LDC tests.
2. For the soils tested, wick drains have equivalent sand drain diameters of from 25 to 76 mm (1 to 3 in.) as determined in the LDC test.
3. For the soils tested, the equivalent sand drain diameter range does not significantly affect the typical field wick drain spacing.
4. Crimping alone would not reduce the core flow capacity of any of the wick drains enough to hamper flow of water to the drains due to soil consolidation.
5. Taking into account large lateral pressure (± 144 kPa) applied in the field due to in-place and normally applied embankment loads, the core flow capacity of the soft core (Auger) wick drains were reduced to zero. This effect did not completely show up in the laboratory LDC tests on these wick drains since the estimated lateral pressure in the LDC test was only approximately 47.9 kPa (1000 psf) (see comparison of lateral pressure test results at 55.6 and 143.8 kPa loads in Table 6). The core flow capacity values at 55.6 kPa lateral pressure were still adequate to handle the water reaching the drains due to field consolidation (compare values of flow capacity in Table 6 to values of estimated field flow to wick drains in Table 7). However, the core flow capacities of these drains at 143.8 kPa lateral pressure was estimated to be zero.
6. A combination of crimping and lateral pressure will not seriously affect the performance of rigid core wick drains installed in a typical field situation (compare net flow capacity to maximum field flow to wick drains in Table 7).

Further Research Needs

The last two columns in Table 7 indicate that the core of rigid core wick drains will handle an estimated conservative field soil consolidation flow based on the following:

$$\text{Soil } C_H = 0.04 \text{ m}^2/\text{day } (0.4 \text{ ft}^2/\text{day})$$
$$d_w = 51 \text{ mm } (2 \text{ in.})$$
$$d_e = 1.25 \text{ m } (4 \text{ ft})$$
$$\text{Total Soil Settlement} = 2.8 \text{ m } (9 \text{ ft})$$

The resulting maximum soil consolidation flow equals 3.2 cm³/s (Table 7).

Conversely, a much lower soil consolidation flow was obtained in the laboratory LDC tests based on the following:

$$\text{Soil } C_H = 0.004 \text{ m}^2/\text{day } (0.04 \text{ ft}^2/\text{day})$$
$$d_w = 51 \text{ mm } (2 \text{ in.})$$
$$d_e = 0.26 \text{ m } (0.833 \text{ ft})$$
$$\text{Total Soil Settlement} = 51 \text{ mm } (2 \text{ in.})$$

The resulting maximum soil consolidation flow lies between 0.03 and 0.08 cm³/s (Table 5).

We therefore plan to run additional laboratory LDC tests on a faster draining soil to simulate field soil consolidation flow conditions. This will enable us to determine if the geotextiles surrounding the rigid wick drain cores will influence the larger expected field consolidation flow rates. We also plan to run lateral pressure tests at 143.8 kPa (3000 psf) to verify straight-line extrapolation between flow rate and pressure to this pressure level.

Bibliography

[1] Suits, L. D., *Consolidation Testing with Wick Drains*, 11th INDA Technical Symposium, International Nonwoven and Disposables Association, Sept. 1983.

DISCUSSION

M. P. Gambin[1] and P. P. Schmitt[1] (written discussion)—Soletanche Entreprise has been placing drains for several decades. Since 1977 we have been successfully using Desol wick drains, mostly in France but also in Germany and Mexico (through our subsidiary Cimesa). Although credit must be given to the

[1] Soletanche Entreprise, Nanterre, France.

authors for showing that large variations in drain parameters measured in the laboratory do not lead to big changes in drain design parameters (mostly spacing and time of pore water pressure dissipation), some unusual results must be discussed.

1. The ratio of equivalent diameters measured in Soil 1 (organic silty clay) and Soil 2 (manufactured clay) is between 0.8 and 1.2 for all 15 tested drains except Desol, for which it reaches 1.75. What is the explanation? Did the equipment breakdown mentioned for one of the Desol tests have any effect on this result? Did the drain buckling change the result?

As a matter of fact, the Desol drain is more rigid than other drains. Assuming a strain of 10% under a force of 1 kN (as shown by test results), the rigidity of a 38-cm-long sample constrained at both ends is

$$\frac{1.0}{0.10 \times 0.38} = 26 \text{ kN/m}$$

Rigidity of the surrounding soil sample is calculated with a settlement of 50 mm under a pressure of 96 kN/m² applied on a circular pad 254 mm in diameter:

$$\frac{96 \times \pi \times 0.254^2}{4 \times 0.05} = 97 \text{ kN/m}$$

Load transfer before buckling could then be

$$\frac{96 \times \pi \times 0.254^2}{4} \times \frac{26}{26 + 97} = 1 \text{ kN}$$

which is very large. The consolidation test can turn into a buckling test for a rigid drain and consequently one can draw the following conclusions: (1) the soil sample settlement is reduced for more rigid drains and this is taken as a delay in settlement, and (2) buckling occurs and prevents any comparison with *in situ* behavior.

2. The assumption of a more rigid behavior of the Desol drain compared with most other drains is confirmed by the bending tests. When subjected to a complete folding, the discharge capacity of Desol drains decreases by a factor of 0.46, a lower value than observed for the other drains. This is caused by the Desol drain's larger rigidity.

Is this folding actually occurring *in situ*? For instance, near the toe of the preloading embankment, horizontal movements are significant over a large depth. Consequently drain curvature is kept to a small value.

Even in the middle of the embankment, total settlement is not sufficient to induce a buckling effect as in the laboratory. If we use the example of the authors, where a settlement of 2.8 m is reported within a soft layer 18.75 m thick, strain is 15%. The French Centre d'Etudes Techniques de l'Equipement in Bordeaux has shown that for such a strain, the discharge capacity of the Desol drain only decreases by a factor of 3%; that is a value far from the 46% mentioned above with a complete folding. A paper discussing these laboratory test results has been submitted to the 3rd International Geotextile Conference. A paper by W. Morin, T. Shafer, and K. Gangopadhyay comparing *in situ* behavior of Desol and Mebra drains for the Seagirt Marine Terminal Project concluded that there was a close similarity between both drains.

To conclude: It can be stated that parameter variations observed at laboratory scale may have no effect at full scale. Unfortunately these variations, mostly caused by inadequate testing procedures, may mislead designers.

L. D. Suits et al (authors' closure)—In response to the first point made by Messrs. Gambin and Schmitt, the equipment breakdown for Soil 1 only occurred in one test apparatus. The results of that test were not included in the Summary. The only result of having the third test, if it had performed properly, would have been to increase the high to low value ratio of equivalent sand drain diameters.

In referring to drain buckling, Item 4 of the Summary states that crimping did not have an effect on the results of the LDC tests. Test settlements were very close to theoretically estimated settlements for the test. We feel that the cause for the higher ratio of equivalent sand drain diameters between Soil 1 and Soil 2 for the Desol drain can be attributed to the greater potential for clogging of the surface openings in the plastic Desol drain due to migration of organic fines from the organic soil (Soil 1). We feel this clogging potential is greater because of the lesser open surface of the Desol drain area as compared to a geotextile covered drain.

In relation to the second point, the rigid nature of the Desol drain is not a factor in its performance. For the soils tested the crimped core flow capacity of the Desol drain was still adequate to handle flow from the soil. We feel that the problem of the Desol drain in the laboratory is as explained above (i.e., the lesser surface open area which will accept flow from consolidation of the surrounding soil rather than the internal core flow capacity of the drain).

We are aware of the successful use of Desol drains in field applications where relatively large settlements have taken place. Our laboratory testing program was more severe in the evaluation of wick drains than what can be expected in the field. This is because of the instantaneous load application in the laboratory as opposed to gradual application of a load due to embankment construction in the field.

Toshihisa Adachi,[1] *Yoshinori Iwasaki,*[2] *Minoru Sakamoto,*[3] *and Seiji Suwa*[4]

A Case History: Settlement of Fill over Soft Ground

REFERENCE: Adachi, T., Iwasaki, Y., Sakamoto, M., and Suwa, S., "**A Case History: Settlement of Fill over Soft Ground,**" *Consolidation of Soils: Testing and Evaluation, ASTM STP 892,* R. N. Yong and F. C. Townsend, Eds., American Society for Testing and Materials, Philadelphia, 1986, pp. 684–693.

ABSTRACT: This paper presents a case history on unexpectedly large settlement of a weak alluvial foundation clayey soil improved by both preloading and by use of vertical drains. An important cause of the settlement is probably secondary compression. When preloading is used to improve a clayey foundation soil, the importance of taking into account the secondary compression is emphasized.

KEY WORDS: preloading, vertical drains, sand drains, rope drains, clay, secondary compression, differential settlement

A residential land construction project of about 50 ha was planned by a permanent fill 2 m thick on poor clayey ground which was reclaimed by drainage. From the soil exploration, the upper alluvial clay stratum, 10 to 15 m thick, was found to be underlain by sandy soil. A typical soil profile and soil properties of the clay stratum are given in Fig. 1. Based on the soil exploration and test results, preloading without vertical drains was selected for the strengthening and preconsolidation of the clay stratum. Surcharge loads (i.e., loads in excess of those to be applied by a permanent fill 2 m thick) were designed as 1.5 m high. To ensure against the slip failure of the ground, a grade of the side slopes was determined by 1:30 at the edges of the land construction area as shown in Fig. 2.

Seven months after the first filling to fill up the creeks ended, the second filling of 1 m thick started and was finished successfully. One month after the

[1] Professor, Department of Transportation Engineering, Kyoto University, Yoshida-Honmachi, Sakyo-ku, Kyoto, Japan.

[2] Head, Osaka Soil Test Laboratory, 8-4, Utsubo-Honmachi 1-chome, Nishi- ku, Osaka, Japan.

[3] Vice-Head, Osaka Soil Test Laboratory, 8-4, Utsubo-Honmachi 1-chome, Nishi-ku, Osaka, Japan.

[4] Senior Consulting Engineer, Osaka Soil Test Laboratory, 8-4, Utsubo-Honmachi 1-chome, Nishi-ku, Osaka, Japan.

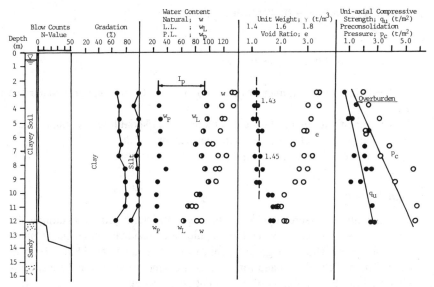

FIG. 1—*Typical soil profile and soil properties.*

second filling was completed, the third filling with the side slopes at the grade of 1:30 was started. Immediately after the finish of the third filling, a slip failure took place as shown in Fig. 2. To discover causes of the failure and devise a remedy, *in situ* vane shear tests and uniaxial compression tests were performed. A safety factor for the stability of 1.18 was obtained from the uniaxial compression test results, and of 1.43 from the *in situ* vane shear test results. The failure might be concluded to be due to the rate sensitive properties of clay. As the remedy, the installation of sand drains in the edges of the land and of rope drains in the remaining part were applied to accelerate the rate of settlement and to strengthen the ground. However, unexpectedly large residual settlement have been found to continue, exceeding the estimated consolidation time sharply.

We will explain the design of the preloading with vertical drains and the present situation of the settlement. Then, the cause of the large residual settlement will be discussed.

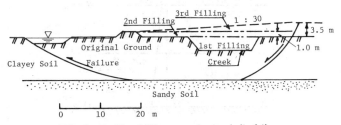

FIG. 2—*Profile of filling procedure and slip failure.*

Design of Preloading with Vertical Drains

Design Conditions

The design conditions are as follows:

1. The height of the permanent fill is 2.0 m above the original ground surface level.
2. The magnitude of the residual settlement (S_r) should be less than 10 cm after surcharge removal (i.e., under the permanent fill alone).
3. The length of time of the surcharge loads is six months.

Design Method

The magnitude of surcharge loads is properly determined by assuming the degree of consolidation to reduce the magnitude of residual settlement as much as possible. The stability of ground against the preloading should also be considered. The design of sand and rope drains (i.e., their diameter and spacing) is made based on the coefficient of consolidation of the clay.

Soil Conditions and Estimation of Settlement

As previously mentioned, the original ground was reclaimed by drainage; the upper weak alluvial clay stratum is 10 to 15 m thick and is underlain by a relatively firm sandy layer. A model of the ground shown in Fig. 3 was used in the settlement and stability analyses. The compression index (C_c) of 0.92 was used as the average value of the whole clay layer. Initially, the clay deposit was assumed to be in a normally consolidated state. The magnitude of settlement in the general area was calculated for each height of fill (i.e., H = 2.0, 2.5, 3.5, and 4.5 m, including the height of permanent fill) and is given in Table 1. Similarly, the magnitude of settlement of ground above creeks is given in Table 2.

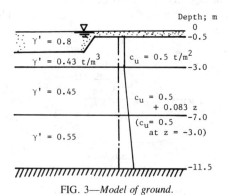

FIG. 3—*Model of ground.*

TABLE 1—*Surcharge versus consolidation settlement (general area).*

Surcharge (H), m	Consolidation Settlement (S), cm
2.0	73.1
2.5	91.9
3.5	129.0
4.5	158.8

TABLE 2—*Surcharge versus consolidation settlement (area above creeks).*

Surcharge (H), m	Consolidation Settlement (S), cm
2.0	53.3
2.5	66.7
3.5	90.4
4.5	108.3

Residual Settlement

The magnitude of surcharge is determined so that the difference between the settlement at 50% of degree of consolidation under a surcharge load and the total settlement under the permanent fill of 2 m is less than the allowable residual settlement, 10 cm. The relations between the magnitude of settlement at 50% of consolidation and the surcharge height are given in Fig. 4. One can see in the figure that the residual settlement becomes less than 10 cm if the surcharge is applied larger than 1.5 m (3.5 m = 2.0 m + 1.5 m). Thus 1.5 m of surcharge was determined.

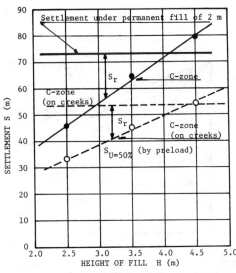

FIG. 4—*Settlement versus fill height relations (residual settlement).*

Time of Surcharge and Design of Vertical Drains

The length of time required to achieve a given amount of settlement under 1.5 m of surcharge load was decided to be six months. Barron's theory [1] for radial drainage was used to analyze the degree of consolidation of a clay layer. The coefficient of consolidation (c_v) is 4.0×10^{-4} cm²/s, but $c_h = 2\,c_v = 8.0 \times 10^{-4}$ cm²/s was used in the analysis because the relation $c_h = 2\,c_v$ is usually valid for a Japanese alluvial clay layer.

The diameter of sand drains was 40 cm, and the effective diameter of rope drains was 1.6 cm. The relations of time versus degree of consolidation are shown in Figs. 5a and 5b by using the spacing as a parameter. On the basis of these results, 2.2 m of spacing was determined for sand drains and 1.7 m for rope drains.

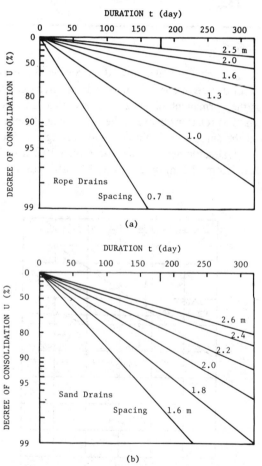

FIG. 5—*Relations of time versus degree of consolidation.* (a) *For area improved by rope drains.* (b) *For area improved by sand drains.*

Construction Procedure and Present Situation of Settlement

Construction based on the design started in 1975 and was finished in May 1980. One year later the lots were placed on sale, and some houses had been built by May 1981. It was found, however, that differential settlement between the areas improved by sand drains and rope drains caused damage to some houses, and unexpectedly large settlement magnitudes were also realized. Multilayer settlement gages were installed in August 1981. The settlement versus time curves after the installation of gages are shown in Fig. 6. Twice as much settlement took place in the area improved by sand drains than that by rope drains.

The construction of land improved by rope drains ended in 1977, while that by sand drains in 1980. This means that 1500 days had already passed before starting the measurements for the area improved by rope drains, and 450 days for that by sand drains. Taking account of these time passages before the start of measurement, we re-arranged the time-settlement curves (Fig. 7). The backward and forward estimations were made by Hoshino's method [2]. The backward estimation was that in the rope drains improved area 50 cm of settlement had occurred, while in the sand drains improved area 40 cm of settlement had occurred. From the forward estimation, 25 cm of further settlement is predicted. Figure 7 shows the differential settlement between the two differently improved areas. A profile of the uneven settlement is given in Fig. 8.

The settlement strain distribution curves versus depth are shown in Figs. 9a and 9b. Relatively uniform settlement strains have taken place, especially in the whole clay layer of the area improved by rope drains. The pore water pressure distribution curves of the area improved by rope drains are given in Fig. 10. The excess pore water pressure had dissipated.

Summary

Thus we have seen that the actual settlement exceeded the designed residual settlement by a large margin and that excess pore water pressure had dissipated. The major settlement might be caused by secondary compression.

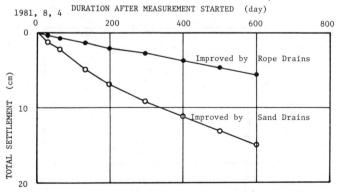

FIG. 6—*Settlement versus time curves.*

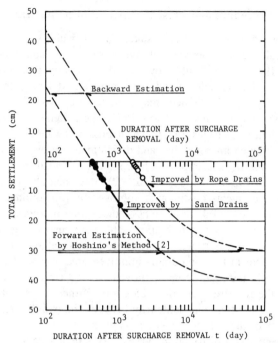

FIG. 7—*Settlement versus time curves (backward and forward estimations).*

The coefficient of secondary compression ($C_{\alpha e}$) of the clay is $3.6 \times 10^{-2}/$ log t from laboratory tests and $1.0 \times 10^{-2}/$log t from the field data (Fig. 7). These values are compared with those of Mexico City clay and Leda clay [3] in Fig. 11. This clay exhibits large secondary compression, although its organic content is less than 5%.

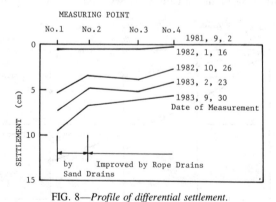

FIG. 8—*Profile of differential settlement.*

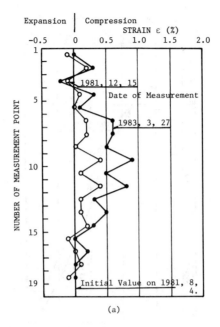

(a)

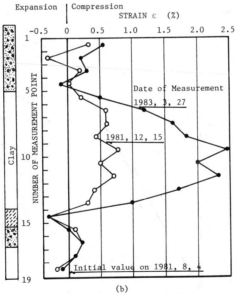

(b)

FIG. 9—*Settlement strain distribution.* (a) *For area improved by rope drains.* (b) *For area improved by sand drains.*

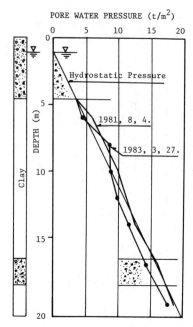

FIG. 10—*Pore water pressure distribution* (*in area improved by rope drains*).

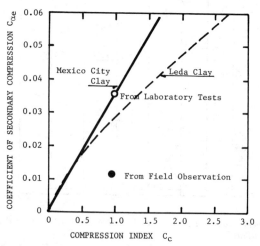

FIG. 11—*Relation between coefficient of secondary compression* ($C_{\alpha e}$) *and compression index* (C_c).

As mentioned by Mitchell [4], vertical drains are ineffective in a clay exhibiting large secondary compression. The differential settlement due to the difference in construction time and type of vertical drains should be also avoided.

References

[1] Barron, R. A., *Transactions of American Society of Civil Engineers*, No. 113, 1948, pp. 718–754.

[2] Hoshino, K., *Transactions of Japan Society of Civil Engineers*, Vol. 47, No. 7, 1962, pp. 63–67 (in Japanese).

[3] Ladd, C. C., Foott, R., Ishihara, K., Schllosser, F., and Poulos, H. G., "State-of-the-Art Report" in *Proceedings*, 9th International Conference on Soil Mechanics and Foundation Engineering, Vol. 2, 1977, pp. 421–429.

[4] Mitchell, J. K. and Katti, R. K., "State-of-the-Art Report" in *Proceedings*, 10th International Conference on Soil Mechanics and Foundation Engineering, Volume for General Reports, State-of-the-Art Reports and Lectures, 1981, pp. 261–317.

G. E. Bauer[1] and A. Z. El-Hakim[2]

Consolidation Testing— A Comparative Study

REFERENCE: Bauer, G. E. and El-Hakim, A. Z., "**Consolidation Testing—A Comparative Study,**" *Consolidation of Soils: Testing and Evaluation, ASTM STP 892,* R. N. Yong and F. C. Townsend, Eds., American Society for Testing and Materials, Philadelphia, 1986, pp. 694–710.

ABSTRACT: The knowledge of the preconsolidation pressure, p'_c, of an overconsolidated clay is of vital importance in the settlement analysis of structures. It has been found that the semigraphical methods to determine p'_c proposed by Casagrande, Burmister, and Schmertmann are not completely satisfactory for sensitive clays. Also, the standard oedometer test is not adequate to fully describe the strain-effective stress behavior of such clays. Renewed interest in this topic by several investigators has shown that the magnitude of the preconsolidation pressure is greatly affected by the rate of loading, type of loading, specimen size, and type of testing apparatus used.

This paper reports on the results of a comparative study of consolidation behavior using two types of clays and four different testing techniques. The first series of tests was carried out on undisturbed specimens of soft Leda (Champlain Sea) clay cut from block samples and the second series was performed on prepared and consolidated kaolin specimens. For each series, four different consolidation apparatuses were used: a standard oedometer, a modified Rowe cell, an Anteus apparatus, and triaxial K_0 tests. Both Leda clay and kaolin specimens were tested in parallel. Several sizes (diameters) of specimens and different loading techniques were used. Based on these test results, it was shown that the controlled gradient consolidation test using the Rowe cell with a softer membrane was the most suitable technique to establish a well-defined effective stress-strain curve. The controlled gradient test using the Anteus apparatus also gave a well-defined stress-strain relationship, but the operational procedures were found quite cumbersome. The K_0-triaxial test was quite time-consuming and an elaborate recording of data was required. The results from standard oedometer tests, even using a modified loading, did not yield continuous effective stress-strain plots and p'_c was difficult to determine. Also, a comparison of the coefficients of consolidation was made for the two clays and the various testing techniques employed.

KEY WORDS: consolidation, preconsolidation pressure, oedometer test, soft clay, effective stress, controlled gradient, comparative study, test apparatuses

Determination of the preconsolidation pressure of a clay is an important factor in any settlement analysis, and many investigators have focussed their attention on this problem [1–10]. These studies have shown that the slope of the load-

[1] Department of Civil Engineering, Carleton University, Ottawa, Ontario, Canada K1S 5B6.
[2] Department of Civil Engineering, Alexandria University, Alexandria, Egypt.

compression curve and the magnitude of the preconsolidation pressure are greatly influenced by the rate of loading, specimen size, type of test apparatus, initial preload, and to a large extent on the handling and preparation of the test specimen (stress relief and sample disturbance). A thorough review and discussion of these factors was given by El-Hakim [11] and will be cited here again as the need arises.

The scope of this study was to investigate the effect of different testing methods and procedures on the compression behavior of a sensitive soft clay. Four different types of consolidation machines were used: (a) standard oedometer, (b) Rowe cell, (c) Anteus compression machine, and (d) triaxial K_0 compression.

Different loading routines were also investigated, such as (a) conventional loading, (b) controlled gradient consolidation, and (c) constant-rate-of-strain loading. A parallel series of tests on remolded and consolidated kaolin soil was performed in order to study some factors affecting the magnitude of the preconsolidation pressure, such as sensitivity and stress history of the soil. The tests on the kaolin specimens were used mainly as a reference for the behavior of corresponding Leda clay specimens.

Soil Properties

Champlain Sea Clay (Leda Clay)

The undisturbed clay specimens used in this study were carefully trimmed from block samples of Leda or Champlain Sea clay. This sensitive marine clay was formed approximately 10 000 years ago and is characterized by its high water content and sensitive nature. This marine sediment is frequently encountered throughout the Ottawa and St. Lawrence River valleys and in the St. John drainage basin of the province of Quebec, Canada. The history of deposition and the mineralogical composition of this clay have been well documented in the literature [12–14]. The fabric of the Leda clay consists of an assembly of silt, clay size rock fragments, and clay minerals in which the arrangement of the clay platelets ranges from a completely random orientation (cardhouse structure) to a preferred orientation where the flat platelets are parallel to the bedding plane. The clay fabric develops considerable structural strength due to cementing, but once the structure is disturbed and the contact zones broken, the clay turns into a liquid when in a saturated condition. Sensitivities on the order of 20 to 100 are quite common, which classifies the clay as extremely quick.

It is therefore of extreme practical importance to determine, as realistically as possible, the preconsolidation pressure and the compressibility parameters of this clay. The preconsolidation pressure can be defined as the pressure at which a breakdown of structural bonds occurs associated with large deformations and generation of excess pore water pressures. The occurrence of landslides (flow slides), excessive settlement of structures, and bearing capacity failures are common in areas underlain by this clay.

The undisturbed block samples used in this study were cut from a depth of

4 m and were quite soft and completely saturated. A summary of the basic properties of the clay is given in Table 1.

A summary of properties of the kaolin clay is also given in Table 1. The chief constituent of the kaolin was the mineral kaolinite (62.4%). Other minerals were feldspar (8.3%), quartz (28.0%), and traces of limonite (1.3%). The kaolin powder was mixed with distilled water and consolidated in a top and bottom draining mold under a specified preconsolidation pressure. After completion of this consolidation, the clay sample was extruded and cut and trimmed to yield several specimens for the various testing apparatuses. The moisture contents were obtained before and after completion of the respective tests.

Apparatus and Test Procedure

As discussed earlier, the loading sequence, the magnitude of load increments, and the load duration can greatly affect the stress-strain relationship of clays, in particular of highly sensitive clays.

Four different testing apparatuses were used in this study.

Standard Oedometer

A conventional oedometer testing machine was employed. In order to investigate size effects of the sample on the deformation of the clays, two diameters of specimens were tested, 50.8 and 76.2 mm. For both series of tests the thickness of the specimens was 19.0 mm. The samples were confined in floating rings and were drained both at the top and bottom.

For sensitive clays [6,10] a modified loading procedure was recommended whereby each new load increment is applied at the end of primary consolidation and the increments should be of equal magnitude until the preconsolidation pressure is reached. Beyond that point each increment was left on for 24 h and was equal to one half of the previous load instead of doubling the load as recommended for the standard oedometer test. This procedure was adhered to for all the tests carried out in the conventional consolidation machine.

TABLE 1—*Geotechnical properties of the clays.*

Item	Leda Clay	Kaolin
Bulk unit weight (γ), kN/m^3	14.4 to 14.6	16.7
Natural moisture content (w), %	95 to 105	48 to 64
Plastic limit (w_p), %	28 to 30	28 to 32
Liquid limit (w_L), %	62 to 66	48
Plasticity index (I_p), %	34 to 36	18 to 20
Specific gravity (G_s)	2.77	2.65
Degree of saturation (S_r), %	98 to 100	96
Clay size particles, %	82	80
Activity	32	45

Rowe Cell

The main features of this cell were given by Shields [*15*] and by Rowe and Barden [*16*]. In the study on hand a 76.2-mm-diameter cell was used for a series of controlled gradient tests. The theory and procedures of the controlled gradient test were described in detail by Lowe et al [*8*] and Sällfors [*17*] as well as by El-Hakim [*11*].

In the controlled gradient test, top drainage was permitted, and a pressure transducer was connected to the bottom drainage outlet to measure the base pore water pressure. An automatic control unit was designed to maintain the pore water pressure at the base of the clay specimen within ± 0.07 kN/m^2 of the specified value. If the pore water pressure dropped due to consolidation below that predetermined value, an electric motor automatically raised the mercury pots to increase the pressure on the specimen until the pore water pressure at the base reached the specified value.

It was also found that the membrane supplied with the Rowe cell was rather stiff for the soft Leda clay. Therefore new membranes from a soft latex rubber compound were made in-house and the test results obtained were compared with the corresponding results from those using the standard membranes.

Anteus Apparatus

The design and various details of the apparatus were given by Lowe et al [*8*]. The special feature of this test consists in applying a back pressure to the soil specimen to ensure full saturation. Back pressures up to 1000 kN/m^2 can be specified. The load is applied hydrostatically and no drainage is permitted through the bottom porous stone. When the excess pore water pressure at the base of the specimen reaches some specified value in the range of 5 to 30 kN/m^2 it is maintained by continuous loading. A pacer regulates the load application to keep the set base pressure within ± 0.07 kN/m^2. The differential pore water pressure acting between the bottom and top of the clay causes seepage to the upper porous stone and the specimen consolidates. The total pressure in the load chamber is read generally by a gage. In this study the load was monitored by a pressure transducer. Knowing this pressure and the back pressure, the effective vertical stress on the specimen can be calculated. Compression of the soil specimen is observed by an extensometer.

Triaxial K$_0$ Compression

The design and the various details of the triaxial cell for K_0 testing were given by Bishop and Henkel [*18*]. The test yields continuous values for m_v, the coefficient of compressibility in one-dimensional consolidation. The K_0 test represents the special case of anisotropic consolidation with no lateral deformation of the soil. The confining pressure on the clay specimen is continuously adjusted to ensure that the specimen does not yield laterally. Before consolidation was started, a back pressure on the order of 300 kN/m² was applied to the specimen to saturate the soil. All other procedures were those specified for a conventional triaxial compression test.

Analysis and Comparison of Test Results

For the Leda clay a total of 34 consolidation tests were performed using the four test apparatuses outlined in the previous section. A summary of the tests is given in Table 2. A minimum of two and in most cases three identical tests were performed in order to check the repeatability of results. The repeatability was very good to excellent and the graphs given in this paper are therefore the average results of two or three identical tests.

Leda Clay

Standard Oedometer—Figure 1 shows a typical strain versus log effective pressure relationships. This curve consists basically of three parts: (1) An almost linear part, AB. Point B is close to the preconsolidation pressure. (2) A curved part, BC, indicating that the preconsolidation has been exceeded. And (3) again a linear portion, CD, representing the virgin compression of the clay. An unloading and reloading cycle is generally carried out in order that a field compression curve can be constructed as proposed by Schmertmann [3]. It should be kept in mind that if the standard incremental loading procedures were adhered to, one would obtain six points to define the load-deformation response of this clay. Generally only two to three points would be available to represent the curve below the preconsolidation pressure. This, of course, is in many cases not adequate to obtain a satisfactory value of p'_c by the graphical method.

Figure 2 shows a comparison between a compression curve from a standard load application and a relationship obtained by the loading procedure suggested by Bjerrum [10] from the Norwegian Geotechnical Institute (NGI). In the standard test there are only three points available below the preconsolidation to define this critical part of the curve, whereas in the NGI loading method one will generally obtain six to eight points. This is most important if the Casagrande graphical procedure is used to determine p'_c. Also, the time required to complete the test is generally three to four days compared to one week for the standard incremental loading.

Figure 3 compares the load-deformation response for two specimens of 50.8 and 76.2 mm diameter. Both specimens were of the same thickness, 19 mm. In both cases the preconsolidation pressure is approximately the same (52 kN/m²) using the NGI loading method. Bozozuk [17] found a small decrease in preconsolidation pressure from 62 to 60 kPa by increasing the area of the specimens from 40 to 60 cm². He attributed this difference to a greater sample disturbance due to handling and trimming. The authors feel that disturbance and trimming of soil from an undisturbed block sample should be the same if the size and weight difference is reasonable.

TABLE 2—Summary of testing program for Leda clay.

Apparatus	Test Type	Sample No.	No. of Tests	Base Water Pressure, kPa	Dimensions Diameter, mm	Dimensions Thickness, mm	Comment
Standard oedometer	STD NGI LIR = 0.5 LD = 24 h	A-1	3	0	76.2	19.0	...
		A-2	3	0	76.2	19.0	...
		A-3	2	0	76.2	19.0	...
		A-4	2	0	50.8	19.0	loading
		A-5	2	0	50.8	19.0	reloading
Rowe cell	CGT	B-1	3	20	76.2	19.0	new membrane
		B-2	3	20	76.2	19.0	old membrane
		B-3	3	10	76.2	19.0	new membrane
		B-4	3	10	76.2	19.0	old membrane
	Incremental	B-5	2	0	76.2	19.0	new membrane
Anteus	CGT	C-1	3	20	63.5	19.0	...
		C-2	3	10	63.5	19.0	...
K_0 triaxial	CRS	D	2	0	60	100	...

STD = standard.
LIR = load increment ratio.
CRS = constant rate of strain.
NGI = Norwegian Geotechnical Institute.
LD = load duration.
CGT = controlled gradient test.

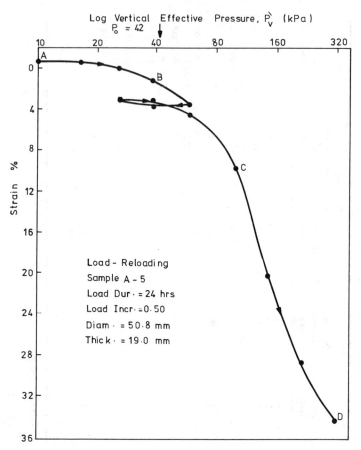

FIG. 1—*Standard oedometer test (Leda clay).*

In summary, the NGI incremental loading technique has several distinct advantages over the standard method insofar as that a well-defined curve is obtained, especially close to the preconsolidation pressure. In addition, the time required to complete the test is considerably less. The preconsolidation pressures obtained using different specimen sizes do not differ appreciably for specimens trimmed from undisturbed block samples.

Rowe Cell—The controlled gradient test method has a great advantage over the standard method in that a continuous effective stress-strain curve can be obtained, since the pore water pressure is known at any time. Figure 4 shows the results from controlled gradient tests using a 76.3-mm-diameter Rowe cell. One should note the well-defined curves. Therefore, the point of maximum curvature can be determined quite accurately and the Casagrande construction needs little interpretation. In most tests the pore water pressure was kept at a constant value of 20 kPa. The magnitude of the pore water pressure at the base has an effect on the preconsolidation pressure as demonstrated in Fig. 4. A higher base pore water pressure, which means a greater hydraulic gradient and a shorter consolidation time, yielded consistently higher preconsolidation pressures. This is in agreement with the observations made by others [5,7,9,17]. The difference was actually small; for example, increasing the base pore water pressure from 10 to 20 kPa increased the preconsolidation pressure from 56 to 58 kPa or by approximately 4%. The maximum increase in p'_c was about 7% due to an increase of pore water pressure at the base of 100%. A more marked influence on the load-deformation behavior and on the magnitude of the preconsolidation pressure was due to the membrane stiffness. This is shown in Fig. 5. The Rowe cell was outfitted by the manufacturer with a rather stiff rubber membrane through which

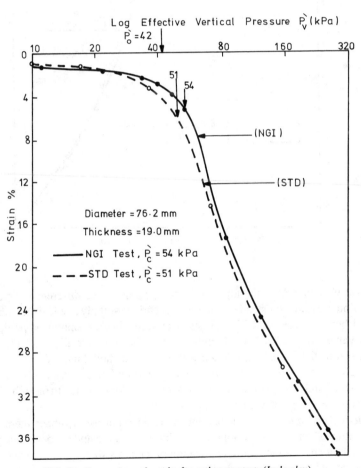

FIG. 2—*Comparison of results for oedometer tests (Leda clay).*

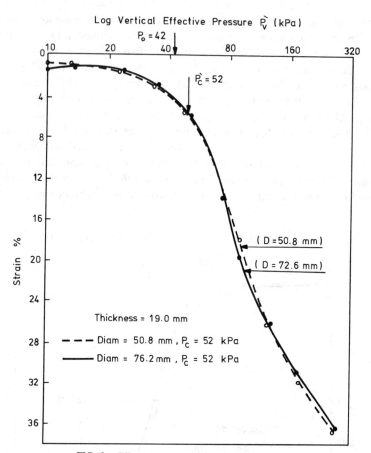

FIG. 3—*Effect of specimen size on* p'$_c$ (*Leda clay*).

the load was applied to the top of the soil specimen. The difference of preconsolidation pressures for standard membranes and the soft plyable latex membranes ranged between 16% and 23%. Tests with softer latex membranes yielded the lower values for p'_c. This clearly indicates that the commercially available membrane is too stiff for testing soft sensitive clays and should therefore be replaced by a softer type. The manufacture of the latex membrane is relatively easy and economical and was detailed by El-Hakim [11]. Membrane stiffness had little effect on the values of volume compressibility.

Anteus Apparatus—The Anteus apparatus does not use a rubber membrane, but a rigid platen, for load application similar to the standard oedometer test. The loading and back-pressure application procedures were quite involved in the somewhat old equipment used. The authors are quite aware that in the newer equipment the test procedure is simplified even though the method is the same. It is expected, therefore, that the new apparatus will not yield any better results even though the procedures are simpler.

The results from two controlled gradient tests (CGT) are given in Fig. 6 together with the results from a conventional oedometer test. Two base pore water pressures of 10 and 20 kPa were used. Both these tests gave higher preconsolidation pressures than the standard test. Again, one should be cautioned against placing too much reliance on the p'_c-value obtained from the standard test due to the very few test points which can be obtained in the vicinity of this critical value. A continuous record of the c_v-value could also be obtained in the controlled gradient test. The value for c_v above the preconsolidation was on the order of 1×10^{-7} m^2/s.

A comparison of consolidation curves from the Anteus apparatus and Rowe

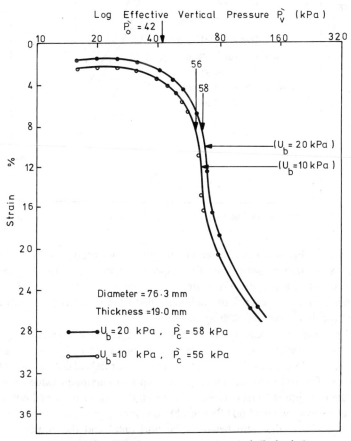

FIG. 4—*Effect of base pore water pressure on* p'$_c$ *(Leda clay).*

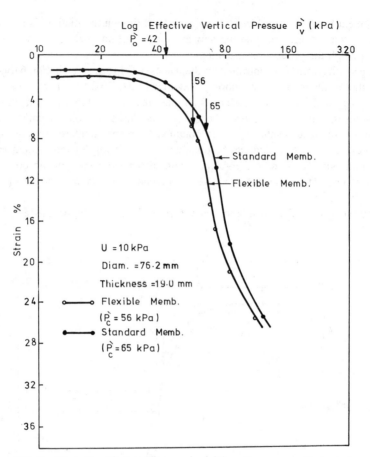

FIG. 5—*Effect of membrane stiffness on load-deformation response (Leda clay).*

cell is given in Fig. 7 for two different base pore water pressures. The difference in preconsolidation pressure from the respective curves is minimal. Therefore, the choice of test apparatus is a matter of preference, availability, and financial investment. Since both types of equipment had been modified, it is difficult to estimate the actual cost of each apparatus.

K_0-*Compression*—The clay specimens were tested in a triaxial compression cell in order to obtain their K_0 and preconsolidation values by the method recommended by Poulos and Davis [20]. The tests were made under strain-controlled conditions. The soil specimens were loaded up to a maximum value of 300 kPa during the one-dimensional compression. From the two tests carried out to completion an average value of 60 kPa was obtained for the preconsolidation pressure.

Figure 8 shows the ratio between the horizontal and the vertical effective stresses during the K_0 consolidation. This ratio is termed the coefficient of lateral earth pressure at rest and has little significance in settlement analysis, but is of importance in calculating earth pressures against unyielding structures. The mean value obtained was 0.65. The time required to complete one test was on the order of 10 to 12 days, and continuous adjustment of the confining pressure was required. From the time aspect alone this test is not well suited to commercial testing.

Kaolin Clay

Two types of specimens were made from the kaolin soil. The first series of tests was performed on normally consolidated soil and, in the second series, the soil was preconsolidated under a static vertical effective stress of 75 kPa. The number of tests carried out is summarized in Table 3. A total of 24 tests were performed on the kaolin using the various testing techniques detailed earlier.

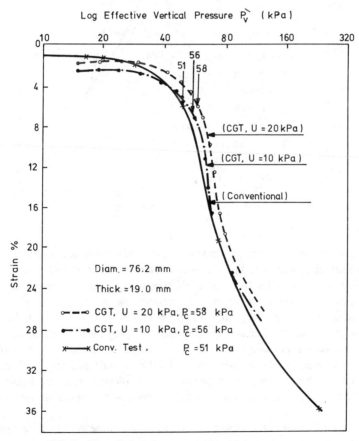

FIG. 6—*Controlled gradient test results with Anteus apparatus.*

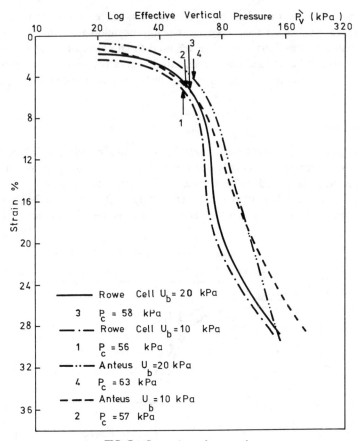

FIG. 7—*Comparison of test results.*

The initial water content of the soil had the most pronounced effect on the load deformation response. The minimum water content was 48% and the maximum value was 62% for fully saturated specimens. Figure 9 shows the strain versus effective stress behavior for these extreme cases.

The sample size had very little effect on the deformation characteristics of corresponding soil specimens. Loading and reloading cycles showed no effect whatsoever on the position and slope of the virgin compression portion of the curve. The preconsolidated specimens exhibited distinct and abrupt change of curvature at the preconsolidation pressure when plotted on a semilogarithmic scale. The curves obtained for different base pore water pressures were almost congruent.

With regard to the other aspects of the tests, it can be said that the tests on the kaolin specimens reconfirmed the observation made for the Leda soil, that the controlled gradient test results gave a completely defined effective stress-strain behavior of the soil.

Conclusions

From the comparative study carried out and from the analysis of the results, the following major conclusions are warranted:

1. The controlled gradient consolidation test using the Rowe cell with softer diaphragm is the most suitable laboratory technique to estimate the compressibility characteristics of the soft sensitive Leda clay.

2. The controlled gradient test using the Anteus apparatus also yields a well-defined stress-strain curve; however, the operating procedure of the apparatus used in this investigation was quite cumbersome.

3. The conventional oedometer consolidation test is the most used technique for determining the load-consolidation response of clay. Even using modified incremental loading will not give a well-defined stress-strain curve in the vicinity of the preconsolidation pressure. Smaller load increments would alleviate this problem somewhat, but in turn would prolong the testing time.

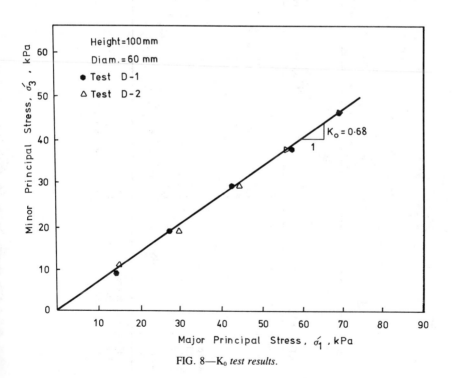

FIG. 8—K_0 test results.

TABLE 3—Summary of testing program for kaolin.

Apparatus	Test Type		Sample No.	No. of Tests	Water Contents, %	Base Water Pressure, kPa	Dimensions		Comments
							Diameter, mm	Thickness, mm	
Standard oedometer	STD		E-1	2	48	0	76.2	19.0	⋮
			E-2	2	48	0	50.8	19.0	⋮
			E-3	2	62	0	50.8	19.0	⋮
		LIR = 0.5	E-4	2	48	0	50.8	19.0	⋮
		LD = 24 h	E-5	2	48	20	76.2	19.0	preconsolidated
Rowe cell	CGT		F-1	3	48	20	76.2	19.0	new membrane
			F-2	3	48	10	76.2	19.0	preconsolidated
Anteus	CGT		G-1	3	48	20	63.5	19.0	preconsolidated kaolin
			G-2	3	48	10	63.5	19.0	preconsolidated kaolin
K_0 triaxial	CRS		H	2	48	0	60.0	100.00	preconsolidated kaolin

STD = standard test.
LD = load duration.
CRS = controlled rate of strain.
LID = load increment ratio.
CGT = controlled gradient test.

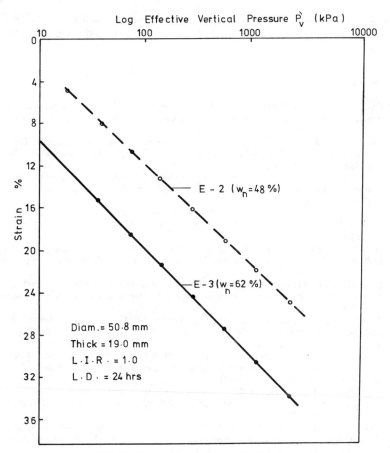

FIG. 9—*Compression behavior of kaolin clay.*

4. The K_0 triaxial compression test can be used to determine the consolidation parameters of sensitive clays. The drawbacks are that sophisticated and time-consuming testing and monitoring techniques are required. The advantages are that untrimmed cylindrical specimens from sample tubes can be used and that the application of back pressure would ensure complete saturation of the soil.

5. Very careful sampling, handling, and preparation procedures are required in order to minimize sample disturbance, which affects the consolidation behavior of this clay.

6. The preconsolidation pressure of the Leda clay is time-dependent. A shorter duration of dissipation of excess pore water pressure will yield higher values.

7. The Leda clay tested was found to be slightly overconsolidated with an overconsolidation ratio ranging from 1.3 to 1.5.

8. The coefficient of consolidation for both the Leda and kaolin clays was found to be independent of strain rates at stresses higher than the preconsolidation pressure.

References

[1] Casagrande, A., "The Determination of the Preconsolidation Load and Its Practical Significance," in *Proceedings, 1st International Conference on Soil Mechanics and Foundation Engineering,* Cambridge, MA, Vol. 3, 1936, p. 60.
[2] Burmister, D. M., "The Application of Controlled Test Methods in Consolidation Testing," in *Symposium on Consolidation Testing of Soils, ASTM STP 126,* American Society for Testing and Materials, Philadelphia, 1951, p. 83.
[3] Schmertmann, J. H., "Estimating the True Consolidation Behaviour from Laboratory Tests," *Journal of the Soil Mechanics and Foundation Division, Proceedings of the American Society of Civil Engineers,* Vol. 79, 1953.
[4] Leonards, G. A. and Ramiah, B. K., "Time Effects in Consolidation of Clay," in *Papers on Soils, ASTM STP 254,* American Society for Testing and Materials, Philadelphia, 1959, pp. 116–130.
[5] Hamilton, J. and Crawford, C. B., "Improved Determination of Preconsolidation Pressure," in *Papers on Soils, ASTM STP 254,* American Society for Testing and Materials, Philadelphia, 1959, pp. 254–271.
[6] Leonards, G. A. and Girault, P., "A Study of the One-Dimensional Consolidation Test," in *Proceedings,* 5th International Conference on Soil Mechanics and Foundation Engineering, Paris, Vol. 1, 1961, p. 213.
[7] Crawford, C. B., "Interpretation of the Consolidation Test," *Journal of the Soil Mechanics and Foundation Engineering Division, Proceedings of the American Society of Civil Engineers,* Vol. 90, 1964, pp. 87–109.
[8] Lowe, J., Joans, E., and Obrician, V., "Controlled Gradient Consolidation Test," *Journal of the Soil Mechanics and Foundation Division, Proceedings of the American Society of Civil Engineers,* Vol. 95, 1969, pp. 77–97.
[9] Eden, W. J., "Field Studies on the Consolidation Properties of Leda Clay," in *Proceedings,* 14th Canadian Soil Mechanics Conference and National Research Council Technical Memo No. 69, 1961, pp. 107–118.
[10] Bjerrum, L., "Problems of Soil Mechanics and Construction on Soft Clays and Structurally Unstable Soils," in *Proceedings,* 8th International Conference of Soil Mechanics and Foundation Engineering, Moscow, Vol. 3, 1973, pp. 111–159.
[11] El-Hakim, A. Z., "Consolidation Testing—A Comparative Study," M.E. thesis, Carleton University, Ottawa, ON, Canada, 1980.
[12] Gadd, N. R., "Surficial Geology of the Ottawa Map Area," Geological Survey of Canada, Publication No. 62-16, 1963, p. 4.
[13] Gillott, J. E., "Fabric of Leda Clay Investigated by Optical, Electron-Optical and X-Ray Diffraction Methods," *Engineering Geology,* Vol. 4, No. 2, 1970, pp. 133–153.
[14] Sangray, D. A., "Naturally Cemented Sensitive Soils," *Geotechnique,* Vol. 22, 1972, pp. 139–152.
[15] Shields, D. H., "Consolidation Tests," Technical Note, *Geotechnique,* Vol. 26, 1976, pp. 209–212.
[16] Rowe, P. W. and Barden, L. A., "A New Consolidation Cell," *Geotechnique,* Vol. 16, No. 2, 1966, pp. 116–124.
[17] Sällfors, G., "Preconsolidation Pressure of Soft, High Plastic Clays," Ph.D. thesis, Chalmers University, Göteborg, Sweden, 1975.
[18] Bishop, A. W. and Henkel, D. J., *The Measurement of Soil Properties in the Triaxial Test,* Edward Arnold, London, 1957.
[19] Bozozuk, M., "Effect of Sampling, Size and Storage on Test Results for Marine Clay," in *Sampling of Soil and Rock, ASTM STP 483,* American Society for Testing and Materials, Philadelphia, 1971, pp. 121–131.
[20] Poulos, H. G. and Davis, E. H., "Laboratory Determination of *In Situ* K_0," *Geotechnique,* Vol. 22, 1972, pp. 177–182.

Summary and Evaluation

Raymond N. Yong[1] and Frank C. Townsend[2]

Consolidation Testing and Evaluation: Problems and Issues

REFERENCE: Yong, R. N. and Townsend, F. C., "**Consolidation Testing and Evaluation: Problems and Issues,**" *Consolidation of Soils: Testing and Evaluation, ASTM STP 892*, R. N. Yong and F. C. Townsend, Eds., American Society for Testing and Materials, Philadelphia, 1986, pp. 713–718.

The techniques and procedures used in the determination of the one-dimensional consolidation properties of fine-grained saturated soils are presently described in two ASTM standards: (1) D 2435, "Test for One-Dimensional Consolidation Properties of Soils," and (2) D 4186, "Test for One-Dimensional Consolidation Properties of Soils Using Controlled-Strain Loading." The objectives of these standards are to determine engineering parameters to estimate the magnitude (e–p' relationship) and rate (C) of one-dimensional deformation of soils subjected to an effective stress change. The major distinction between the two standards is that the former permits full dissipation of pore pressure prior to application of the subsequent load, whilst the latter procedure measures the pore pressures during loading.

The summary presentation contained herein addresses some main concerns and issues. Full delineation of the various problems and laboratory case studies can be found in the state-of-the-art reports and the other technical papers contained in this volume.

General Areas of Concern

There are at least five distinct areas of concern:

1. *Standard technique:* The general sets of concern relate to such items as disturbance effects, sample storage, nonlinear stress-strain behavior, prior knowledge of initial stress conditions in the soil, and equipment constraints and influence.

2. *New techniques and procedures using the standard oedometer:* The main

[1] Geotechnical Research Centre, McGill University, Montreal, Quebec, Canada.
[2] Department of Civil Engineering, University of Florida, Gainesville, FL 32611.

items in this category concern methods of load application (e.g., rapid loading, constant gradient, controlled-strain or controlled-stress (and variations thereof), load increment ratios and durations).

3. *New equipment and associated techniques or procedures:* This category considers the use of new types of "consolidation devices" and associated techniques/procedures (e.g., biaxial and triaxial devices, centrifuge consolidation testing, slurry consolidometers, testing of wick drains, radial and axial drainage).

4. *Problem soils:* Using standard, modified, or new equipment/procedures, the testing of such materials as organic soils, slimes/sludges, gaseous soils, and unsaturated soils requires special attention.

5. *Novel or different methods of data scrutiny and reduction* for determination of "consolidation" properties and characteristics.

Standard Technique

The standard techniques described in ASTM D 2435 and D 4186, which cover testing methodologies and prescribed methods for calculation of C_v, C_c, and C_r from test measurements, are generally considered conventional tests. The record shows that the coefficient of consolidation C_v obtained varies with the level of effective stress established. Test results also show that C_v calculated from measured excess pore pressures differ significantly from C_v values determined from laboratory settlement-time measurements. The "bottom line" to this set of discussions is the realization that determination of consolidation performance parameters or properties is indeed sensitive to the effective stress-deformation relationship established. Not only is the method of laboratory determination of this relationship important, so is the methodology used in calculating this relationship.

The above problem bears testimony to the highly complex nature of clay soils and the need to obtain a clearer understanding of the mechanisms involved in the demonstration of load-deformation-time relationships. The influence of sample preparation and initial state of the soil, including stress and strain histories, on subsequent performance under load cannot be ignored or overlooked. The discussions provided by Olson in his state-of-the-art report and by some of the other papers in this volume show clearly this concern.

"Standard" Oedometer: New Techniques/Procedures

In recognition of many of the problems identified above, and in keeping with the desire to maintain a "standard" test cell as a base reference condition, new techniques and procedures have focussed on the scrutiny of load-deformation-time relationships. Several methods and procedures for load application have been described (e.g., controlled stresses, strains, time and pore pressure responses). The effect of any of these new techniques/procedures is generally demonstrated in terms of relationships for load and deformation, with time as a factor implicitly considered in the relationships or evaluated through data reduction methods. The evidence shows that the demonstrated load-deformation-time relationships between each of these new techniques/procedures, and also between these new techniques and the standard procedures, are not likely to be similar.

Recognizing that the end point of any of these techniques is the determination of the C_v, C_c, C_r, and p'_c values—via laboratory testing of representative samples—and recalling that clay soils are indeed complex materials, it is pertinent to ask: "Is there such a thing as a true value of C_v, C_c, C_r, or p'_c?" Given the realization of the highly variable state of clay soils, due in part to the compositional characteristics and in part to the response of these soils to various stress path loadings and boundary conditions, uniqueness as a condition in clay soils is not a realistic fact. Thus one now needs to ask whether research and development efforts should continue to seek "true" values, or concentrate on seeking a better appreciation of the many factors and conditions that participate in the production of a demonstrated load-deformation-time relationship. This attitude is motivated by the knowledge that the field environmental, initial, and boundary conditions are not generally modelled in laboratory consolidation testing and therefore, by extension, the C_v, C_c, C_r, and p'_c values existent in the field may not be similar to those determined from laboratory tests. It would appear that the new techniques and procedures which focus on methods of sample stressing are indeed directed towards this set of concerns. It now remains to expend further effort in evaluation and assessment of the significance of the laboratory-determined load-deformation-time relationships vis-à-vis field problems.

New Equipment, Associated Techniques, and Problem Soils

The development of new devices and associated testing techniques for the laboratory study of consolidation performance is generally prompted by one or more of the following motivations:

1. The desire and need to make laboratory test conditions conform with field loading and performance expectations (e.g., biaxial and triaxial consolidation devices, radial flow consolidation tests, restricted pore pressure dissipation tests).

2. The availability of new test techniques (e.g., centrifuge) or more sophisticated instrumentation and controls that permit measurements of various performance characteristics (and properties) not previously considered as measurable.

3. The need to address problem soils not heretofore considered as standard material for consolidation testing (e.g., organic soils, gaseous soils, slimes/sludges, unsaturated soils).

4. The development of new or extended concepts concerning load-deformation-time performance modelling and the availability of more sophisticated methods of analyses and computational tools.

Item (1) by and large speaks for itself. As one continues to further one's appreciation of the differences between laboratory and expected field performance characteristics of the soil, either because of site specificity or because of load boundary conditions, the normal response of the researcher or tester is to ensure that the test system closely matches the demands of the field situation. Thus, for example, biaxial and triaxial consolidation test systems with specific load application procedures that have been developed to model drainage conditions expected in the field can be readily seen to be a first-order adaptation of conventional procedures for assessment of the laboratory consolidation behavior of the soil. Not only are the questions of anisotropic properties handled with this system, so are the problems concerned with specification of vertical and horizontal stresses. Quite obviously, the load–deformation–time relationships obtained will be directly related to the test system itself and the procedures used; hence the C_v, C_c, C_r, and p'_c values calculated will be specific to the test. It is not likely that these values will match those obtained if the same soil samples were tested in the conventional manner using standard procedures. These points have been recognized by researchers and practitioners. It now remains to determine whether there is a need to establish (1) standard methods which will handle a representative range of conditions now being examined by these types of devices, and (2) recommended procedures for data reduction and computation of consolidation properties.

Another driving force for the development of new test systems and techniques is the need to determine consolidation properties of problem soils. Two examples are the consolidation behaviors of organic soils and soft sediments. Because of the anticipated large strain performance of each of the materials under load, and because of the high initial void ratios and water contents, conventional test systems and theories cannot properly handle the consolidation testing of these materials. Consolidation testing of organic soils using larger oedometers with specialized control on low load application has been attempted with varying degrees of success. The problems that arise are the interpretation of the load-deformation-time relationship obtained and the evaluation of the phenomenon of secondary compression.

Greater awareness of the capabilities of centrifuge testing has prompted researchers with centrifuge facilities to investigate the use of such a facility in conducting consolidation testing of very soft sediments. The combination of the various demands in centrifuge testing and soft sediment performance requires considerable effort in seeking a proper understanding of the types of results obtained and the kinds of analyses required to obtain representative consolidation properties.

A closely similar problem exists in the study of the settling performance of slimes and sludges. The slurry-state of these materials requires test samples sufficiently high (long) that the near self-weight characteristics of sedimentation/consolidation be properly assessed. The development and use of slurry consolidometers are seen as one direction to follow in the laboratory determination of pertinent sedimentation/consolidation properties. The coupling of the slurry consolidometer with centrifuge testing techniques is a further extension of this test procedure. Again, the problems arising with respect to data scrutiny and analyses need to be fully addressed.

Gaseous and unsaturated soils pose problems more directly associated with theory and experimental/testing. Because of the presence of the third phase (gas or air), one is left with two choices: (1) to introduce back–pressuring procedures to produce a fully saturated test sample (presuming this can indeed be successful) and to conduct the conventional consolidation test thereafter, or (2) to test the sample "as-is" and to seek methods or analytical models that would permit one to work with the test data obtained. It is not immediately clear if either of these procedures would provide one with a true picture of the problem behavior at hand. It is clear, however, that much work remains to be done in this particular area.

The introduction of man-made materials (i.e., wick drains) to enhance consolidation of clayey soils calls attention to new test techniques and equipment to evaluate performance, and a resurrection of radial flow theories for predictions. Tests measuring hydraulic conductivity of drains subjected to "kinks" and bends and *in situ* lateral stresses address this concern. Eventually, it is expected that standards will evolve for this particular area.

Different Methods of Data Scrutiny

With greater capability in measurement of conventional and "other" performance parameters, and with attention given to the testing of other kinds of soil material not previously considered in conventional consolidation tests, it is clear that different methods of data scrutiny and analysis are required. A case in point is the phenomenon of secondary compression. Not only is one interested in the characteristics of performance in the so-called secondary compression phase of "consolidation", but the specification of when such a phenomenon occurs is also of interest. The evidence at hand indicates that secondary compression characteristics are indeed influenced by the response performance characteristics of the material in the regular consolidation phase. This observation is not easily quantified, since the point at which secondary compression occurs cannot be readily identified. The problem lies as much in one's concept or definition of secondary compression as in one's belief that this phenomenon is an active partner in the description of the total e-log p curve derived from consolidation testing. Since classical consolidation theory deals with the phenomenon of pore pressure dissipation rates, the facile argument that everything that is not covered by classical consolidation theory is secondary compression provides little comfort to the researcher concerned with measurements that show "nonclassical" results.

At the opposite end of the spectrum is the sedimentation/consolidation problem. What is needed is to distinguish the void ratio at which sedimentation ceases and pore pressures (effective stresses) due to self-weight loading are generated. Obviously, classical consolidation theory is inappropriate for describing sedimentation phenomena.

Concluding Remarks: Problems and Issues

Various problems and issues can be deduced from the discussion dealing with the areas of concern. There are undoubtedly many other areas of concern that have not been addressed in the above discussion. A brief summary of pertinent issues can be given as follows:

1. *Load application procedures*—conventional and nonconventional test systems, criteria, data reduction and analyses, standardization.

2. *Pore pressure measurements*—routine measurements, criteria for use in control of loading, application in analysis of C_v, and C_c.

3. *Biaxial and triaxial test systems*—test procedures, standardization, test data scrutiny and analyses.

4. *Special tests for problem soils and man-made materials*—partly saturated, stiff residual, large particle soils, organic, gaseous, slimes, wick drains.

5. *Sedimentation and secondary compression*—specification and definition, influence on e-log p, evaluation, and prediction.

S. Leroueil[1] and M. Kabbaj[1]

General Discussion on Consolidation Theory and Testing

REFERENCE: Leroueil, S. and Kabbaj, M., **"General Discussion on Consolidation Theory and Testing,"** *Consolidation of Soils: Testing and Evaluation, ASTM STP 892,* R. N. Yong and F. C. Townsend, Eds., American Society for Testing and Materials, Philadelphia, 1986, pp. 719–723.

In the past 15 years, continuous oedometer testing techniques have been proposed to determine the compressibility characteristics of clays. These new tests are usually faster than the conventional oedometer test and, moreover, give continuous stress–strain curves. It is thus logical to wonder in 1985 whether or not we should abandon the conventional test in favor of continuous tests.

Essentially for the above reasons the authors would answer yes. Owing to strain rate effects on the behavior of natural clays, however, the results obtained from a "new" test will generally be different from those obtained from the conventional test (Fig. 1). Since our past experience is based entirely on the conventional test, it is necessary to calibrate the "new" test before using it in practice.

Various studies show that the effect of strain rate in natural clays is a very general phenomenon. As indicated in Table 1, strain rate effects were evidenced on a variety of clays with plasticity indices (I_p) between 8 and 105, liquidity indices (I_L) between 0.5 and 2.7, and conventional preconsolidation pressures ($\sigma'_{p\text{conv}}$) between 47 and 940 kPa. The strain rate effect is particularly evident in the normally consolidated range and at the preconsolidation pressure.

Considering Champlain clays, Leroueil et al [1] observed that the preconsolidation pressure–strain rate (σ'_p–$\dot{\epsilon}_v$) relationships had similar shapes. The σ'_p–$\dot{\epsilon}_v$ curves were thus normalized with respect to a reference strain rate taken arbitrarily equal to 4×10^{-6} s^{-1} (Fig. 2). As a first approximation, the resulting general relationship can be written:

$$\frac{\sigma'_p}{\sigma'_p(\dot{\epsilon}_v = 4 \times 10^{-6} \text{ s}^{-1})} = f(\dot{\epsilon}_v)$$

[1] Department of Civil Engineering, Laval University, Pavillon Pouliot, Quebec, Canada.

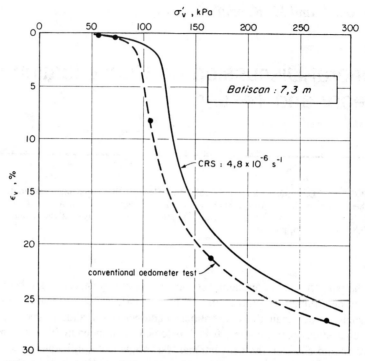

FIG. 1—*Typical compressibility curves obtained from conventional and CRS tests.*

TABLE 1—*Characteristics of clays presenting strain rate effects.*

Clays	I_p	I_L	$\sigma'_{p \ conv}$ (kPa)	S_t	Reference
Champlain clays:					
Ottawa clay	8	4	480	>50	6
14 clays	19 to 43	0.9 to 2.7	47 to 270	15 to 108	1
Louiseville	33	1.7	88	20	4
Other Canadian clays:					
Saint Jean–Vianney	16	1.4	940	≈100	7
Winnipeg	35 to 55	≈0.5	?	3	8
Broadback	11	2.1	175	≈200	authors' files
Other clays:					
Bäckebol (Sweden)	65	1.04	≈70	25	9
Belfast (N. Ireland)	20 to 40	≈0.75	?	8	8
Vallda (Sweden)	≈105	≈0.76	?	?	3

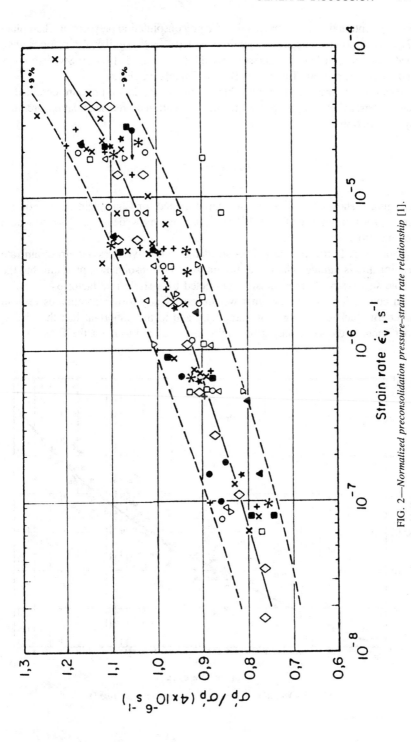

FIG. 2—*Normalized preconsolidation pressure–strain rate relationship* [1].

For conventional tests, it appears that the preconsolidation pressure of Champlain clays corresponds typically to a strain rate of 10^{-7} s^{-1}. Similar observations were made by Larsson [2] and Larsson and Sällfors (Fig. 8, [3]) on Swedish clays and by Silvestri et al (Table 2, [4]) on a Champlain clay.

From these remarks, it follows that the preconsolidation pressure obtained at a given strain rate can be related to the conventional preconsolidation pressure by the relation

$$\sigma'_{p\ conv} = \frac{\sigma'_p(\dot{\epsilon}_v)}{\alpha_2}$$

Figure 3 shows the α_2–$\dot{\epsilon}_v$ relationship obtained for Champlain clays. For example, for a strain rate of 2×10^{-6} s^{-1}, as used in Sweden, the α_2 factor would be equal to 1.18.

Moreover, Leroueil et al [5] found that the effective stress–strain–strain rate relationship is unique. Therefore, not only the preconsolidation pressure, but the whole stress–strain curve, must be corrected by a strain rate factor α_2.

In conclusion, we believe that we are now ready to use continuous tests in Southern Quebec. Owing to its simplicity and to the important fact that the α_2 coefficient is the same during the entire test, we tend to favor the CRS test.

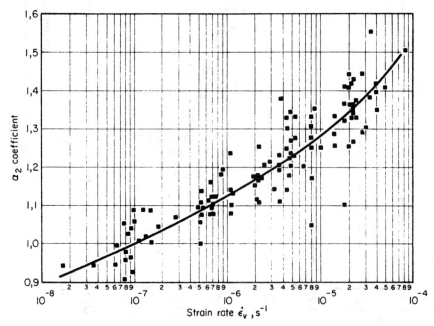

FIG. 3—*Variation of the* α_2 *coefficient with strain rate* [1].

References

[*1*] Leroueil, S., Tavenas, F., Samson, L., and Morin, P., "Preconsolidation Pressure of Champlain clays: Part II—Laboratory Determination," *Canadian Geotechnical Journal,* Vol. 20, No. 4, 1983, pp. 803–816.

[2] Larsson, R., "Drained Behaviour of Swedish Clays," Report 12, Swedish Geotechnical Institute, 1981.

[*3*] Larsson, R. and Sällfors, G., "Automatic Continuous Consolidation Testing in Sweden," this publication, pp. 299–328.

[*4*] Silvestri, V., Yong, R. N., Soulié, M., and Gabriel, F., "Controlled-Strain, Controlled-Gradient, and Standard Consolidation Testing of Sensitive Clays," this publication, pp. 433–450.

[5] Leroueil, S., Kabbaj, M., Tavenas, F., and Bouchard, R., "Stress–Strain–Strain Rate Relation for the Compressibility of Sensitive Natural Clays," *Geotechnique,* Vol. 35, No. 2, 1985, pp. 159–180.

[6] Crawford, C. B., "The Resistance of Soil Structure to Consolidation," *Canadian Geotechnical Journal,* Vol. 2, No. 2, 1965, pp. 90–97.

[7] Vaid, Y. P., Robertson, P. K., and Campanella, R. G., "Strain Rate Behaviour of Saint-Jean-Vianney Clay," *Canadian Geotechnical Journal,* Vol. 16, No. 1, 1979, pp. 34–42.

[8] Graham, J., Crooks, J. H. A., and Bell, A. L., "Time Effects on the Stress–Strain Behaviour of Natural Soft Clays," *Geotechnique,* Vol. 33, No. 3, 1983, pp. 327–340.

[9] Sällfors, G., "Preconsolidation Pressure of Soft High Plastic Clays," Ph.D. thesis, Chalmers University of Technology, Gothenburg, Sweden, 1975.

Author Index

Subject Index

CON2D, 125–126, 567, 568, 583, 585, 586–588, 591
  overprediction of immediate deformation, 587–588
CONMULT, 297
CONMULT model, 118, 390, 398
Consolidation
  average, 472
  behavior, 119, 610
  cell, 116
  characteristics variation, 77
  clay behavior, 379–383
  coefficient, radial, 29
  curves, 148–150
  finite difference equation, 407
  governing equations, 469–470
  law, 354
  seepage-induced, 503
  self-weight consolidation testing, 501–503
  Terzaghi's coefficient, 140–141
    constant, 145–146, 151
  time rate, 663
    effects of disturbance, 57
  See also Primary consolidation; Secondary consolidation
Consolidation line, virgin, 455–456, 458
  isotropic, 569
Consolidation parameters, 279, 548
  constant rate of strain test, 312
  continuous loading tests, 325
Consolidation pressure
  effect of nonclay content, 630
  void ratio relations, 631, 633–636
Consolidation ratio
  at bottom of specimen, 415
  at top of specimen, 414
  versus secondary consolidation rate, 90
Consolidation strain, clay layer, 85
Consolidation tests, 7–63
  antisotropically prepared specimens, 453
  associated techniques, 715–716
  boundary impedance, 19–21

case histories, 58–59
coefficient of consolidation (see Coefficient of consolidation)
comparative study, 694–710
  Anteus apparatus, 697
  comparison of test results, 706
  kaolin clay, 705–709
  Leda clay, test results, 698–705
  oedometer, 696
  Rowe cell, 697
  triaxial cell, 697–698
comparison of field observations and laboratory results, 58–61, 91–98
  coefficient of consolidation, 61
  compressions, 59–60
  long-term observations, 95
  Nottaway, Broadback, and Rupert River region, 93
  pressure/void curve, Olga clay, 93–94
  secondary consolidation effects, 92
  settlement, 59–60
    embankments, 92
    rate, 94–95
  stress-strain, 59–60
consolidometer, 453–454
constant rate of deformation tests, 47–49, 117
constant rate of strain, 79
defined, 7
different methods of data scrutiny, 717
disturbance, 54–57
  cemented soils, 55
  consolidation time rates, 57
  effect on coefficient of consolidation, 56
  effect on coefficient of secondary compression, 57
  effect on reduction in void ratio, 54–55
  thin-walled, fixed piston sampler, 55
early history, 8

profile, 685
properties, 8, 71, 177, 203, 243, 595, 685
remolded, 100
stratification and properties, 52
structure, 71, 100, 610
  bentonite-sand mixtures, 628–629
  resistance to compression, 99
tests, 71, 516
unsaturated (*see* Unsaturated soil)
unsaturated (*see* Unsaturated soil)
very soft, 405
volume change, 133
volumetric flow rate, 675
Soil index property dependent coefficient, 178–179
  versus liquidity index, 182
Solids content
  determination to dilute suspension, 122–123
  measurements, 500
Specimen preparation
  anisotropic, 453
  centrifuge model, 594–596
  isotropic, 452–453
  soft clay, 410, 412
  trimming, 331–334
Stabilization, 663
Standard incremental test (*see* Conventional-incremental-loading test)
Stiffness constants, 470, 479
Strain
  definition, 8–9
  distribution
    fill settlement, 691
    within specimen, 409
  equal, 645
  finite (*see* Finite strain)
  free, 645
  initial compressive, 44
  lateral (*see* Lateral strain measurement, ultrasonic method)
  versus log effective pressure, Leda clay, 698, 700
  natural, 532

nominal, 532
ratio of bottom to top strain, 408, 410, 413, 417–418
time relationship for peat, 523
total, increment, 385
vertical (*see* Vertical strain)
volumetric, 41, 264–267
Strain rate, 117–118, 170–174, 378
  comparison of test methods, 442–443
  comparisons of rates, 183
  constant gradient test, 326
  constant rate of strain test, 326
  effect on preconsolidation pressure, 83–85, 399
  elastic components, 382
  equation, 171–173, 179
    governing, 379
  natural clay, 719
  parameter determination, 173–174
  plastic strains, 382
  preconsolidation and, 313
  pressure relations, 443–446, 721, 722
  previous work, 171
  relationship to controlling parameters, 171–172
  shear stress relationship, 523–524
  strain relations, Batiscan clay, 390–391
  variation of α_2 coefficient, 722
Stress
  effective, 173
    average, 47, 301, 308
    distribution in centrifuge, 598
    mean, 48
    parabolic isochrone, 388
    in terms of volumetric strains, 470
    vertical, 123
  undrained paths under triaxial conditions, 600–601
  vertical (*see* Vertical stress)
  volume change relationship, 129
Stress-compression, 97–98
Stress-strain relationship, 8–9, 118, 302, 399, 402
  Acthafalaya clay, 19